Continued on back end papers

*Now available in a lower priced paperback edition in the Wiley Classics Library.

Univariate Discrete Distributions

Univariate Discrete Distributions

Second Edition

NORMAN L. JOHNSON

University of North Carolina
Chapel Hill, North Carolina

SAMUEL KOTZ

University of Maryland
College Park, Maryland

ADRIENNE W. KEMP

University of St Andrews
St Andrews, Scotland

A Wiley-Interscience Publication
JOHN WILEY & SONS, INC.
New York • Chichester • Brisbane • Toronto • Singapore

1803737

Library of Congress Cataloging in Publication Data:
Johnson, Norman Lloyd.
   Univariate discrete distributions / by Norman L. Johnson, Samuel
Kotz, Adrienne W. Kemp. -- 2nd ed.
      p. cm. -- (Wiley series in probability and mathematical
statistics. Probability and mathematical statistics)
   Rev. ed. of: Discrete distributions / Norman L. Johnson, Samuel
Kotz. 1969.
   "A Wiley-Interscience publication."
   Includes bibliographical references and index.
   ISBN 0-471-54897-9 (cloth)
   1. Distribution (Probability theory) I. Kotz, Samuel. II. Kemp,
Adrienne W. III. Johnson, Norman Lloyd. Discrete distributions.
IV. Title. V. Series.
QA273.6.J64   1992   92-11685

Printed in the United States of America

10 9 8 7 6 5 4 3 2 1

To
Regina Elandt–Johnson
Rosalie Kotz
David Kemp

Contents

2   Families of Discrete Distributions   69

3   Binomial Distribution   105

Preface

Since the publication of the first edition of this book in 1969, there have been very substantial advances in the field of discrete distributions; the amount of published research has at least doubled. The *Dictionary and Bibliography of Discrete Distributions*, compiled by G. P. Patil and S. W. Joshi (Edinburgh: Oliver and Boyd, 1968) appeared almost simultaneously with the publication of the first edition of this book. Their approach was largely complementary to our own, giving a useful listing of basic properties of discrete distributions, together with some three thousand references. In 1984 this was superceded by the much more extensive *Dictionary and Classified Bibliography of Statistical Distributions in Scientific Work*, Volume 1: *Discrete Distributions*, compiled by G. P. Patil, M. T. Boswell, S. W. Joshi and M. V. Ratnaparkhi (International Cooperative Publishing House: Fairland, Maryland, 1984). No comprehensive overview has since appeared.

Readers of our new edition will notice that there has been duplication of research effort in some areas of discrete distribution theory, including the naming of distributions. This somewhat uncoordinated development is at least partly due to fragmentation within the research literature; unfortunately there is no journal specifically devoted to the theory and methodology of statistical distributions. Over the past twenty years or so many of the leading statistical journals have devoted much space to other areas of research, and consequently many worthwhile contributions to distribution theory have appeared in a number of somewhat less well-known journals with smaller readerships.

An important feature of the last twenty years has been the impact of computer technology. For instance, there is now an increased emphasis on the generation of random variates and less attention to approximations. Above all, the past two decades have seen far greater emphasis on modeling; this has lead to a much greater understanding of relationships between distributions. The increasingly prevalent influence of Bayesian inferential methods on distribution theory should not be overlooked.

For these reasons the original authors (Norman L. Johnson and Samuel Kotz) felt, at the end of the 1980's, that the time was ripe for a new edition.

xv

Dr. Adrienne W. Kemp, with her wide experience and expertise in the field of discrete distributions, was invited to serve as coauthor of a completely revised second edition.

The main changes are

1. This volume deals only with *univariate* discrete distributions — material in the previous Chapter 11 will be transfered to a projected volume on multivariate distributions (discrete and continuous).

2. Chapter 1 has been extended and subdivided into three parts. Part A gives preliminary mathematical information; this takes into account the greater sophistication regarding mathematical notations and functions that is taken for granted by many present-day authors. Part B is no more than a reminder, with references, concerning basic probability and statistical theory. Part C is new; it gives a short introduction to the rapidly growing (and rapidly changing) field of computer generation of discrete pseudo-random variates.

3. Chapter 2 has been completely rewritten and is now devoted entirely to descriptions of the major families of discrete distributions, some of which have developed very significantly since 1969.

4. Each of Chapters 3 to 7 (on the binomial, Poisson, negative binomial, hypergeometric, and logarithmic distributions) has been given a similar structure in order to make access to information more user-friendly. Their final sections now contain a great deal of recent material on related distributions.

5. Chapters 8 and 9 have been reorganized — Chapter 8 now deals with mixtures of discrete distributions, while Chapter 9 concentrates on stopped-sum distributions. This rearrangement and consolidation takes into account the many advances in these areas, and consequently gives greater insight into interrelationships.

6. Chapter 10 includes material that it formerly contained regarding matching, occupancy, and runs distributions; its scope has been enlarged by including material from the fertile field of distributions of order k.

7. The new Chapter 11 describes a variety of distributions that do not fit very easily elsewhere in the book.

8. There is greater emphasis on the increasing relevance of Bayesian inference to discrete distribution theory, in particular with regard to the binomial and Poisson distributions (Chapters 3 and 4).

9. The references are very greatly increased in number and are now consolidated into a single bibliography with relevant chapter and section numbers.

10. A detailed index is provided.

In spite of the omission of the material in the original Chapter 11, it has been found necessary to increase the size of the volume considerably in order to retain the comprehensive and quasi-encyclopedic coverage characteristic of the first edition. Even so, keeping the size within reasonable limits has forced us to omit some interesting recent articles, and we apologize to their authors for this. We hope that the new edition will, in addition to its continued function as an up-to-date compendium, renew interest in this important area of statistical theory, and that it will contribute to further theoretical and practical advances.

Our special thanks go to Professor David Kemp, not only for his constant interest and invaluable advice (not always taken), but also for drawing our attention to a number of errors. He has constructed the extensive bibliography and has handled the electronic transfer of the draft material.

It is a pleasure to thank colleagues in the Mathematical and Computer Sciences Department of the University of St Andrews for their support, in particular Professor Richard Cormack and Mr. John Newton who carefully read and commented on draft versions of chapters of the book. The revision of the book has been assisted greatly by the very large number of reprints that numerous researchers have so generously supplied to the authors over the years; we would like to extend to them our sincere appreciation. We are also very grateful to Ms. June Maxwell, Ms. Peggy Ravitch and Dr. Halak Unal for their unfailing helpfulness in handling electronic communication in the United States.

A considerable amount of effort has been devoted to trying to ensure accuracy both in the text and in the bibliography. The authors would be glad if readers would bring to their attention any remaining errors.

N. L. JOHNSON
S. KOTZ
A. W. KEMP

List of Tables

Univariate Discrete Distributions

CHAPTER 1

Preliminary Information

INTRODUCTION

This work contains descriptions of many different distributions used in statistical theory and applications, each with its own pecularities distinguishing it from others. The book is intended primarily for reference. We have included a large number of formulas and results. Also we have tried to give adequate bibliographical notes and references to enable interested readers to pursue topics in greater depth. Although detailed proofs are not provided, we give outlines for proofs in many places to facilitate the use of the text for seminars or for self-study.

The same general ideas will be used repeatedly, so it is convenient to collect the appropriate definitions and methods in one place. This chapter does just that. The collection serves the additional purpose of allowing us to explain the sense in which we use various terms throughout the work. Only those properties likely to be useful in the discussion of statistical distributions are described. Definitions of exponential, logarithmic, trigonometric and hyperbolic functions are not given. Except where stated otherwise, we are using real (not complex) variables, and "log," like "ln," means natural logarithm (i.e., to base e).

A further feature of this chapter is material relating to formulas that will be used only occasionally; where appropriate, comparisons are made with other notations used elsewhere in the literature. In subsequent chapters the reader should refer back to this chapter when an unfamiliar and apparently undefined symbol is encountered.

A MATHEMATICAL PRELIMINARIES

A1 Factorial and Combinatorial Conventions

The number of different orderings of n elements is the product of n with all the positive integers less than n; it is denoted by the familiar symbol $n!$

(*factorial n*),

$$n! = n(n-1)(n-2)\cdots 1 = \prod_{j=0}^{n-1}(n-j). \tag{1.1}$$

The less familiar symbol $k!!$ means

$$(2n)!! = 2n(2n-2)\ldots 2,$$

where $k = 2n$.

The product of a positive integer with the next $k-1$ smaller positive integers is called a *descending (falling) factorial*; it will be denoted by

$$n^{(k)} = n(n-1)\cdots(n-k+1) \tag{1.2}$$

$$= \prod_{j=0}^{k-1}(n-j)$$

$$= \frac{n!}{(n-k)!}. \tag{1.3}$$

Note that there are k terms in the product and that $n^{(k)} = 0$ for $k > n$, where n is a positive integer. Readers are WARNED that there is no universal notation for descending factorials in the statistical literature. For example, Mood, Graybill and Boes (1974) use the symbol $(n)_k$ in the sense $(n)_k = n(n-1)\cdots(n-k+1)$, while Stuart and Ord (1987) write $n^{[k]} = n(n-1)\cdots(n-k+1)$.

Similarly there is no standard notation in the statistical literature for *ascending (rising) factorials*. Here in some contexts we will continue to use the notation of the earlier edition, namely

$$n^{[k]} = n(n+1)\cdots(n+k-1) \tag{1.4}$$

$$= \prod_{j=0}^{k-1}(n+j)$$

$$= \frac{(n+k-1)!}{(n-1)!}. \tag{1.5}$$

There is, however, a standard notation in the mathematical literature, where the symbol $(n)_k$ is known as *Pochhammer's symbol* after the German mathematician L. A. Pochhammer (1841–1920); it is used to denote

$$(n)_k = n(n + 1) \cdots (n + k - 1) \tag{1.6}$$

(this definition of $(n)_k$ differs from that of Mood et al.). We will use Pochhammer's symbol, meaning (1.6), as well as (1.4). Our aim will be to use whichever of the two symbols is nearer to that used in previous work on a particular topic.

The *binomial coefficient* $\binom{n}{r}$ denotes the number of different possible combinations of r items from n different items. We have

$$\binom{n}{r} = \frac{n!}{r!(n - r)!} = \binom{n}{n - r}; \tag{1.7}$$

also

$$\binom{n}{0} = \binom{n}{n} = 1, \tag{1.8}$$

and

$$\binom{n + 1}{r} = \binom{n}{r} + \binom{n}{r - 1}. \tag{1.9}$$

It is usual to define $\binom{n}{r} = 0$ if $r < 0$ or $r > n$. However,

$$\binom{-n}{r} = \frac{(-n)(-n - 1) \cdots (-n - r + 1)}{r!}$$

$$= (-1)^r \binom{n + r - 1}{r}. \tag{1.10}$$

The *binomial theorem* for an integer power n is

$$(a + b)^n = \sum_{j=0}^{n} \binom{n}{j} a^{n-j} b^j \tag{1.11}$$

(in the next section we shall find that it holds more generally for any real power n).

By equating coefficients of x in $(1 + x)^{a+b} = (1 + x)^a (1 + x)^b$, we obtain the well-known and useful identity known as *Vandermonde's theorem* (A. T. Vandermonde, 1735–1796):

$$\binom{a + b}{n} = \sum_{j=0}^{n} \binom{a}{j} \binom{b}{n - j}. \tag{1.12}$$

The *multinomial coefficient* is

$$\binom{n}{r_1, r_2, \ldots, r_k} = \frac{n!}{r_1! r_2! \cdots r_k!},$$ (1.13)

where $r_1 + r_2 + \cdots + r_k = n$.

The *multinomial theorem* is a generalization of the binomial theorem:

$$\left(\sum_{j=1}^{k} a_j \right)^n = \sum \left[n! \prod_{i=1}^{k} a_i^{n_i} \Big/ \prod_{i=1}^{k} n_i! \right],$$ (1.14)

where summation is over all sets of nonnegative integers $n_1, n_2, \ldots, n_k$ that sum to n.

A2 Gamma and Beta Functions

When n is <u>not</u> a positive integer, meaning can be given to $n!$, and hence to (1.2), (1.4), (1.7), (1.10) and (1.13), by defining

$$(n - 1)! = \Gamma(n),$$ (1.15)

where $\Gamma(n)$ is the *gamma function*.

The binomial theorem can thereby be shown to hold for any real power.

There are three equivalent definitions of the gamma function, due to L. Euler (1707–1783), C. F. Gauss (1777–1855), and K. Weierstrass (1815–1897), respectively.

Definition 1 (Euler)

$$\Gamma(x) = \int_0^\infty t^{x-1} e^{-t} dt, \qquad x > 0.$$ (1.16)

Definition 2 (Gauss)

$$\Gamma(x) = \lim_{n \to \infty} \left\{ \frac{n! n^x}{x(x+1) \cdots (x+n)} \right\}, \qquad x \neq 0, -1, -2, \ldots.$$ (1.17)

Definition 3 (Weierstrass)

$$\frac{1}{\Gamma(x)} = x e^{\gamma x} \prod_{n=1}^{\infty} \left\{ \left(1 + \frac{x}{n} \right) \exp \left(-\frac{x}{n} \right) \right\},$$ (1.18)

where γ is *Euler's constant*

$$\gamma = \lim_{n \to \infty} \left\{ 1 + \frac{1}{2} + \frac{1}{3} + \cdots + \frac{1}{n} - \ln n \right\} \cong 0.5772156649\ldots. \tag{1.19}$$

Using integration by parts, Definition 1 gives the recurrence relation for $\Gamma(x)$:

$$\Gamma(x + 1) = x\Gamma(x). \tag{1.20}$$

This enables us to define $\Gamma(x)$ over the entire real line, except where x is zero or a negative integer, as

$$\Gamma(x) = \begin{cases} \int\limits_0^{\infty} t^{x-1} e^{-t} dt, & x > 0 \\ x^{-1}\Gamma(x + 1), & x < 0, x \neq -1, -2, \ldots. \end{cases} \tag{1.21}$$

From Definition 1, $\Gamma(1) = 0! = 1$.
From Definition 3 it can be shown that $\Gamma(1/2) = \pi^{1/2}$; this implies that

$$\int_0^{\infty} \frac{e^{-t}}{t^{1/2}} dt = \sqrt{\pi};$$

hence, by taking $t = u^2$, we obtain

$$\int_0^{\infty} \exp\left(\frac{-u^2}{2}\right) du = \sqrt{\frac{\pi}{2}}. \tag{1.22}$$

Also, from $\Gamma(1/2) = \pi^{1/2}$, we have

$$\Gamma(n + \tfrac{1}{2}) = \frac{(2n)!\pi^{1/2}}{n!2^{2n}}, \tag{1.23}$$

Definition 3 and the product formula

$$\sin(\pi x) = \pi x \prod_{n=1}^{\infty} \left(1 - \frac{x^2}{n^2}\right) \tag{1.24}$$

together imply that

$$\Gamma(x)\Gamma(1 - x) = \frac{\pi}{\sin(\pi x)}, \qquad x \neq 0, -1, -2, \ldots. \tag{1.25}$$

Legendre's duplication formula (A.-M. Legendre, 1752–1833) is

$$\sqrt{\pi}\, \Gamma(2x) = 2^{2x-1}\Gamma(x)\Gamma(x + \tfrac{1}{2}), \qquad x \neq 0, -\tfrac{1}{2}, -1, -\tfrac{3}{2}, \ldots. \tag{1.26}$$

Gauss's multiplication theorem is

$$\Gamma(mx) = (2\pi)^{(1-m)/2} m^{mx-1/2} \prod_{j=1}^{m} \Gamma\left(x + \frac{j-1}{m}\right), \quad x \neq 0, -\frac{1}{m}, -\frac{2}{m}, -\frac{3}{m}, \ldots,$$

(1.27)

where $m = 1, 2, 3, \ldots$. This clearly reduces to Legendre's duplication formula when $m = 2$.

Many approximations for probabilities and cumulative probabilities have been obtained using various forms of *Stirling's expansion* (J. Stirling, 1692–1770) for the gamma function:

$$\Gamma(x+1) \approx (2\pi)^{1/2} (x+1)^{x+1/2} e^{-x-1}$$

$$\times \exp\left(\frac{1}{12(x+1)} - \frac{1}{360(x+1)^3} + \frac{1}{1260(x+1)^5} - \cdots\right),$$

(1.28)

$$\Gamma(x+1) \approx (2\pi)^{1/2} x^{x+1/2} e^{-x}$$

$$\times \exp\left(\frac{1}{12x} - \frac{1}{360x^3} + \frac{1}{1260x^5} - \frac{1}{1680x^7} + \cdots\right),$$

(1.29)

$$\Gamma(x+1) \approx (2\pi)^{1/2} (x+1)^{x+1/2} e^{-x-1}$$

$$\times \left(1 + \frac{1}{12(x+1)} + \frac{1}{288(x+1)^2} - \cdots\right),$$

(1.30)

$$\Gamma(x+1) \approx (2\pi)^{1/2} x^{x+1/2} e^{-x}$$

$$\times \left(1 + \frac{1}{12x} + \frac{1}{288x^2} - \frac{139}{51840x^3} - \frac{571}{2488320x^4} + \cdots\right).$$

(1.31)

These are divergent asymptotic expansions, yielding extremely good approximations. The remainder terms for (1.28) and (1.29) are each less in absolute value than the first term that is neglected, and they have the same sign.

Barnes' expansion (E. W. Barnes, 1874–1953) is less well-known, but it is useful for half-integers:

$$\Gamma(x + \tfrac{1}{2}) \approx (2\pi)^{1/2} x^x e^{-x} \exp\left(-\frac{1}{24x} + \frac{7}{2880x^3} - \frac{31}{40320x^5} + \cdots\right).$$

(1.32)

Also

$$\frac{\Gamma(x+a)}{\Gamma(x+b)} \approx x^{a-b}\left(1 + \frac{(a-b)(a+b-1)}{2x} + \cdots\right).$$

(1.33)

These also are divergent asymptotic expansions. Series (1.32) has accuracy comparable to (1.28) and (1.29).

The *beta function* $B(a,b)$ is defined by the *Eulerian integral of the first kind*:

$$B(a,b) = \int_0^1 t^{a-1}(1-t)^{b-1}dt, \qquad a > 0, b > 0. \tag{1.34}$$

Clearly $B(a,b) = B(b,a)$. Putting $t = u/(1+u)$ gives

$$B(a,b) = \int_0^\infty \frac{u^{a-1}du}{(1+u)^{a+b}}du, \qquad a > 0, b > 0. \tag{1.35}$$

The relationship between the beta and gamma functions is

$$B(a,b) = \frac{\Gamma(a)\Gamma(b)}{\Gamma(a+b)}, \qquad a, b \neq 0, -1, -2\ldots. \tag{1.36}$$

The derivatives of the logarithm of $\Gamma(a)$ are also useful, though they are not needed as often as the gamma function itself. The function

$$\psi(x) = \frac{d}{dx}\{\ln\,\Gamma(x)\} = \frac{\Gamma'(x)}{\Gamma(x)} \tag{1.37}$$

is called the *digamma function* (with argument x) or the *psi function*. Similarly

$$\psi'(x) = \frac{d}{dx}\{\psi(x)\} = \frac{d^2}{dx^2}\{\ln\,\Gamma(x)\} \tag{1.38}$$

is called the *trigamma function*, and generally

$$\psi^{(s)}(x) = \frac{d^s}{dx^s}\{\psi(x)\} = \frac{d^{s+1}}{dx^{s+1}}\{\ln\Gamma(x)\} \tag{1.39}$$

is called the $(s+2)$-*gamma function*. Extensive tables of the digamma, trigamma, tetragamma, pentagamma and hexagamma functions are contained in Davis (1933, 1935). Shorter tables are in Abramowitz and Stegun (1965).

The recurrence formula (1.20) for the gamma function yields the following recurrence formulas for the psi function:

$$\psi(x+1) = \psi(x) + x^{-1}$$

and
$$\psi(x+n) = \psi(x) + \sum_{j=1}^{n}(x+j-1)^{-1}, \qquad n = 1,2,3,\ldots. \tag{1.40}$$

Also
$$\psi(x) = \lim_{n\to\infty}\left[\ln(n) - \sum_{j=0}^{n}(x+j)^{-1}\right] \tag{1.41}$$

$$= -\gamma - \frac{1}{x} + \sum_{j=1}^{\infty} \frac{x}{j(x+j)} \qquad (1.42)$$

$$= -\gamma + (x-1) \sum_{j=0}^{\infty} [(j+1)(j+x)]^{-1} \qquad (1.43)$$

and

$$\psi(mx) = \ln(m) + \frac{1}{m} \sum_{j=0}^{m-1} \psi\left(x + \frac{j}{m}\right), \quad m = 1, 2, 3, \ldots, \qquad (1.44)$$

where γ is Euler's constant ($\cong 0.5772156649\ldots$)

An asymptotic expansion for $\psi(x)$ is

$$\psi(x) \approx \ln x - \frac{1}{2x} - \frac{1}{12x^2} + \frac{1}{120x^4} - \frac{1}{252x^6} + \cdots, \qquad (1.45)$$

and hence a very good approximation for $\psi(x)$ is

$$\psi(x) \doteq \ln(x - 0.5), \qquad (1.46)$$

provided that $x \geq 2$.

Particular values of $\psi(x)$ are

$$\psi(1) = -\gamma, \qquad \psi(\tfrac{1}{2}) = -\gamma - 2\ln(2) \doteq -1.963510\ldots.$$

A3   Finite Difference Calculus

The *displacement operator* E increases the argument of a function by unity:

$$E[f(x)] = f(x+1)$$
$$E[E[f(x)]] = E[f(x+1)] = f(x+2).$$

More generally $\qquad\qquad E^n[f(x)] = f(x+n) \qquad\qquad (1.47)$

for any positive integer n, and we interpret $E^h[f(x)]$ as $f(x+h)$ for any real h.

The *forward difference operator* Δ is defined by

$$\Delta f(x) = f(x+1) - f(x). \qquad (1.48)$$

Noting that

$$f(x+1) - f(x) = E[f(x)] - f(x) = (E-1)f(x), \qquad (1.49)$$

we have the *symbolic* (or *operational*) relation

$$\Delta \equiv E - 1. \tag{1.50}$$

If n is an integer, then the nth *forward difference* of $f(x)$ is

$$\Delta^n f(x) = (E - 1)^n f(x)$$

$$= \sum_{j=0}^{n} \binom{n}{j} (-1)^j E^{n-j} f(x)$$

$$= \sum_{j=0}^{n} \binom{n}{j} (-1)^j f(x + n - j). \tag{1.51}$$

Also, rewriting (1.50) as $E = 1 + \Delta$, we have

$$f(x + n) = (1 + \Delta)^n f(x)$$

$$= \sum_{j=0}^{n} \binom{n}{j} \Delta^j f(x). \tag{1.52}$$

Newton's forward difference (interpolation) formula (I. Newton, 1642–1727) is obtained by replacing n by h, where h may be any real number, and using the interpretation of $E^h[f(x)]$ as $f(x + h)$:

$$f(x + h) = (1 + \Delta)^h$$

$$= f(x) + h\Delta f(x) + \frac{h(h - 1)}{2!} \Delta^2 f(x) + \cdots. \tag{1.53}$$

The series on the right-hand side need not terminate. However, if h is small and $\Delta^n f(x)$ decreases rapidly enough as n increases, then a good approximation to $f(x + h)$ may be obtained with but few terms of the expansion. This expansion may then be used to interpolate values of $f(x + h)$, given values $f(x), f(x + 1), \ldots$, at unit intervals.

The *backward difference operator* ∇ is defined similarly, by the equation

$$\nabla f(x) = f(x) - f(x - 1) = (1 - E^{-1}) f(x). \tag{1.54}$$

Note that $\nabla \equiv \Delta E^{-1} \equiv E^{-1}\Delta$. There is a backward difference interpolation formula analogous to Newton's forward difference formula.

The *central difference operator* δ is defined by

$$\delta f(x) = f(x + \tfrac{1}{2}) - f(x - \tfrac{1}{2}) \tag{1.55}$$

$$= (E^{1/2} - E^{-1/2}) f(x).$$

Note that $\delta \equiv \Delta E^{-1/2} \equiv E^{-1/2}\Delta$. *Everett's central difference interpolation formula* (W. N. Everrett, 1924–)

$$f(x+h) = (1-h)f(x) + hf(x+1) - \tfrac{1}{6}(1-h)\{1 - (1-h)^2\}\delta^2 f(x)$$
$$- \tfrac{1}{6}h(1-h^2)\delta^2 f(x+1) + \cdots$$

is especially useful for computation.

Newton's forward difference formula (1.53) can be rewritten as

$$f(x+h) = \sum_{j=0}^{\infty} \binom{h}{j} \Delta^j f(x). \tag{1.56}$$

If $f(x)$ is a polynomial of degree N, this expansion ends with the term containing $\Delta^N f(x)$. In particular, putting $x = 0$, $h = x$, and $f(x) = x^n$,

$$x^n = \sum_{k=0}^{n} \binom{x}{k} \Delta^k 0^n = \sum_{k=0}^{n} \frac{S(n,k)x!}{(x-k)!}, \tag{1.57}$$

where $\Delta^k 0^n / k!$ in (1.57) means $\Delta^k x^n / k!$ evaluated at $x = 0$ and is called a *difference of zero*. The multiplier $S(n,k) = \Delta^k 0^n / k!$ of the descending factorials in (1.57) is called a *Stirling number of the second kind*.

Equation (1.57) can be inverted to give the descending factorials as polynomials in x with coefficients called *Stirling numbers of the first kind*:

$$\frac{x!}{(x-n)!} = \sum_{j=0}^{n} s(n,j)x^j. \tag{1.58}$$

These notations for the Stirling numbers of the first and second kind have won wide acceptance in the statistical literature. However, there are no standard symbols in the mathematical literature. Other notations are as follows:

Notations for the Stirling numbers:

1ST KIND	2ND KIND	REFERENCE
$s(n,j)$	$S(n,k)$	Riordan (1958)
$\binom{n-1}{j-1}B_{n-j}^{(n)}$	$\binom{n}{k}B_{n-k}^{(-k)}$	Milne-Thompson (1933)
	$\Delta^k 0^n / k!$	David and Barton (1962)
$S_n^{(j)}$	$\mathscr{S}_n^{(k)}$	Abramowitz and Stegun (1965)
S_n^j	$\mathfrak{S}_k^n$	Jordan (1950)
S_n^j	σ_n^k	Patil et al. (1984)

Both sets of numbers are nonzero only for $j = 0, 1, 2, \ldots, n$, $k = 0, 1, 2, \ldots, n$, $n > 0$. For given n or given k, the Stirling numbers of the first kind alternate in sign. The Stirling numbers of the second kind are always positive. An extensive tabulation of the numbers and details of their properties appear in Abramowitz and Stegun (1965) and in Goldberg et al. (1976). The numbers increase very rapidly as their parameters increase.

Useful properties are

$$\{\ln(1 + x)\}^j = j! \sum_{n=j}^{\infty} \frac{s(n, j) x^n}{n!}, \tag{1.59}$$

$$(e^x - 1)^k = k! \sum_{n=k}^{\infty} \frac{S(n, k) x^n}{n!}. \tag{1.60}$$

Also

$$s(n + 1, j) = s(n, j - 1) - n\, s(n, j), \tag{1.61}$$

$$S(n + 1, k) = kS(n, k) + S(n, k - 1), \tag{1.62}$$

and

$$\sum_{j=m}^{n} S(n, j) s(j, m) = \sum_{j=m}^{n} s(n, j) S(j, m) = \delta_{m,n}, \tag{1.63}$$

where $\delta_{m,n}$ is *Kronecker delta* (L. Kronecker, 1823–1891); that is, $\delta_{m,n} = 1$ for $m = n$ and zero otherwise.

Charalambides and Singh (1988) have written a very useful review and bibliography concerning the Stirling numbers and their generalizations.

Applying the difference operator Δ to the descending factorial $x^{(N)}$, we have

$$\begin{aligned}
\Delta x^{(N)} &= (x + 1)^{(N)} - x^{(N)} \\
&= (x + 1)x(x - 1) \cdots (x - N + 2) - x(x - 1)(x - 2) \cdots (x - N + 1) \\
&= \{(x + 1) - (x - N + 1)\}x(x - 1) \cdots (x - N + 2) \\
&= Nx^{(N-1)}.
\end{aligned} \tag{1.64}$$

Repeating the operation gives

$$\Delta^j x^{(N)} = N^{(j)} x^{(N-j)}, \qquad j \le N. \tag{1.65}$$

For $j > N$ we have $\Delta^j x^{(N)} = 0$.

A4   Differential Calculus

Next we introduce from the differential calculus the *differentiation operator* D, defined by

$$Df(x) = f'(x) = \frac{df(x)}{dx}. \tag{1.66}$$

More generally

$$D^j x^N = N^{(j)} x^{N-j}, \qquad j \leq N. \tag{1.67}$$

Note the analogy between (1.65) and (1.67). If the function $f(x)$ can be expressed in terms of a Taylor series, then the Taylor series is

$$f(x+h) = \sum_{j=0}^{\infty} \left(\frac{h^j}{j!} \right) D^j f(x). \tag{1.68}$$

The operator D acting on $f(x)$ formally satisfies

$$\sum_{j=0}^{\infty} \frac{(hD)^j}{j!} \equiv e^{hD}. \tag{1.69}$$

Comparing (1.53) with (1.68), we have (again formally)

$$e^{hD} \equiv (1 + \Delta)^h \qquad \text{and} \qquad e^D \equiv 1 + \Delta. \tag{1.70}$$

Although this is only a *formal* relation between operators, it gives exact results when $f(x)$ is a polynomial of finite order; it gives useful approximations in many other cases, especially when $D^j f(x)$ and $\Delta^j f(x)$ decrease rapidly as j increases.

Rewriting $e^D \equiv 1 + \Delta$ as $D \equiv \ln(1 + \Delta)$ we obtain a *numerical differentiation* formula

$$f'(x) = Df(x) = \Delta f(x) - \tfrac{1}{2}\Delta^2 f(x) + \tfrac{1}{3}\Delta^3 f(x) - \cdots. \tag{1.71}$$

(This is not the only numerical differentiation formula. There are others that are sometimes more accurate. This one is quoted as an example.)

Given a change of variable, $x = (1 + t)$, we have

$$[D^k f(x)]_{x=1+t} = D^k f(1 + t). \tag{1.72}$$

Consider now the *differentiation operator* θ, defined by

$$\theta f(x) = x D f(x) = x f'(x) = x \frac{df(x)}{dx}. \tag{1.73}$$

This satisfies

$$\theta^k f(x) = \sum_{j=1}^{k} S(k, j) x^j D^j f(x) \tag{1.74}$$

and

$$x^k D^k f(x) = \theta(\theta - 1) \ldots (\theta - k + 1) f(x). \tag{1.75}$$

Also

$$[\theta^k f(x)]_{x=e^t} = D^k f(e^t), \tag{1.76}$$

$$e^{-ct}[\theta^k f(x)]_{x=e^t} = (D + c)^k \{e^{-ct} f(e^t)\}, \tag{1.77}$$

and

$$x^c \theta^k \{x^{-c} f(x)\} = [e^{ct} D^k \{e^{-ct} f(e^t)\}]_{e^t=x} \tag{1.78}$$

$$= [(D - c)^k f(e^t)]_{e^t=x} \tag{1.79}$$

$$= (\theta - c)^k f(x). \tag{1.80}$$

The D and θ operators are useful for handling moment properties of distributions.

Lagrange's expansion (J. L. Lagrange, 1736–1813)) for the reversal of a power series assumes that if (1) $y = f(x)$, where $f(x)$ is regular in the neighborhood of x_0, (2) $y_0 = f(x_0)$, and (3) $f'(x_0) \neq 0$, then

$$x = x_0 + \sum_{k=1}^{\infty} \frac{(y - y_0)^k}{k!} \left[\frac{d^{k-1}}{dx^{k-1}} \left\{ \frac{x - x_0}{f(x) - y_0} \right\}^k \right]_{x=x_0}. \tag{1.81}$$

More generally

$$h(x) = h(x_0) + \sum_{k=1}^{\infty} \frac{(y - y_0)^k}{k!} \left[\frac{d^{k-1}}{dx^{k-1}} \left(h'(x) \left\{ \frac{x - x_0}{f(x) - y_0} \right\}^k \right) \right]_{x=x_0}, \tag{1.82}$$

where $h(x)$ is infinitely differentiable. (This expansion plays an important role in the theory of Lagrangian distributions, see Chapter 2, Section 5.2.)

L'Hôpital's rule (G. F. A. de L'Hôpital, 1661–1704) is useful for finding the limit of an indeterminate form. If $f(x)$ and $g(x)$ are functions of x for which $\lim_{x \to b} f(x) = \lim_{x \to b} g(x) = 0$, and if $\lim_{x \to b} \{f'(x)/g'(x)\}$ exists, then

$$\lim_{x \to b} \frac{f(x)}{g(x)} = \lim_{x \to b} \frac{f'(x)}{g'(x)}. \tag{1.83}$$

The use of the O, o notation (*Landau's notation*) (E. Landau, 1877–1938) is standard. We say that

$$f(x) = o(g(x)) \qquad \text{as } x \to \infty \quad \text{if} \quad \lim_{x \to \infty} \{f(x)/g(x)\} = 0$$

and $\qquad f(x) = O(g(x)) \qquad$ as $x \to \infty$ if $|f(x)/g(x)| < C$ $\qquad\qquad$ (1.84)

for some constant C and large x.

A5 Incomplete Gamma and Beta Functions, and Other Gamma-Related Functions

In statistical work we often encounter the *incomplete gamma function* $\gamma(a, x)$ and its complement $\Gamma(a, x)$; see, e.g., Khamis (1960) for a discussion of incomplete gamma function expansions of statistical distribution functions. These functions are defined by

$$\gamma(a, x) = \int_0^x t^{a-1} e^{-t} dt,$$

$$\Gamma(a, x) = \int_x^\infty t^{a-1} e^{-t} dt, \qquad x > 0; \qquad\qquad (1.85)$$

that is,

$$\gamma(a, x) + \Gamma(a, x) = \Gamma(a).$$

The notation $\Gamma_x(a) = \gamma(a, x)$ is also in use.

Infinite series formulas are

$$\gamma(a, x) = \sum_{n=0}^\infty \frac{(-1)^n}{n!} \frac{x^{a+n}}{a+n}$$

$$= a^{-1} x^a \, {}_1F_1[a; a+1; -x]$$

$$= a^{-1} x^a e^{-x} \, {}_1F_1[1; a+1; x], \qquad a \neq 0, -1, -2, \ldots, \qquad (1.86)$$

where ${}_1F_1[\cdot]$ is a confluent hypergeometric function; see Section A7.

The following recursion formulas are useful:

$$\gamma(a+1, x) = a\gamma(a, x) - x^a e^{-x},$$

$$\Gamma(a+1, x) = a\Gamma(a, x) + x^a e^{-x}. \qquad\qquad (1.87)$$

The *incomplete gamma function ratio*

$$\frac{\Gamma_x(a)}{\Gamma(a)} = \frac{\gamma(a, x)}{\Gamma(a)}$$

is used in the statistical literature more than $\Gamma_x(a) = \gamma(a, x)$ itself. (The word "ratio" is, alas, sometimes omitted.)

The function tabulated in Pearson's (1922) tables is

$$I(u, p) = \frac{\Gamma_{u\sqrt{p+1}}(p+1)}{\Gamma(p+1)}; \qquad\qquad (1.88)$$

it is given to seven decimal places for $p = -1(0.05)0(0.1)5(0.2)50$, with u at intervals of 0.1. Harter (1964) gave $I(u, p)$ to nine decimal places for $p = -0.5(0.5)74(1)164$ and u at intervals of 0.1. We note also the extensive tables of Khamis and Rudert (1965).

Pearson and Hartley (1976) [see also Abramowitz and Stegun (1965)] tabulated the function

$$Q(\chi^2|\nu) = \frac{\Gamma(\nu/2, \chi^2/2)}{\Gamma(\nu/2)} \tag{1.89}$$

(the upper tail of a χ^2 distribution), for

$$\chi^2 = 0.001(0.001)0.01(0.01)0.1(0.1)2(0.2)10(0.5)20(1)40(2)76$$
$$\nu = 1(1)30.$$

to five decimal places.

Just as we often need the incomplete gamma function, so we need also the *incomplete beta function*

$$B_p(a, b) = \int_0^p t^{a-1}(1-t)^{b-1} dt, \qquad 0 < p < 1, \tag{1.90}$$

and the *incomplete beta function ratio*

$$I_p(a, b) = \frac{B_p(a, b)}{B(a, b)}. \tag{1.91}$$

Again the word "ratio" is often omitted.

In terms of the hypergeometric function $_2F_1[\cdot]$ (cf. Section A6) we have

$$B_p(a, b) = \sum_{n=0}^{\infty} \frac{(1-b)_n p^{a+n}}{n!(a+n)}$$
$$= a^{-1} p^a {}_2F_1[a, 1-b; a+1; p]. \tag{1.92}$$

The incomplete beta function ratio $I_p(a, b)$ has the following properties:

$$I_p(a, b) = 1 - I_{1-p}(b, a), \tag{1.93}$$

$$I_p(k, n-k+1) = \sum_{j=k}^{n} \binom{n}{j} p^j (1-p)^{n-j}, \qquad 1 \le k \le n, \tag{1.94}$$

$$I_p(a, b) = p I_p(a-1, b) + (1-p) I_p(a, b-1), \tag{1.95}$$

$$(a+b-ap)I_p(a, b) = a(1-p)I_p(a+1, b-1) + bI_p(a, b+1), \tag{1.96}$$

$$(a+b)I_p(a, b) = a I_p(a+1, b) + b I_p(a, b+1). \tag{1.97}$$

Extensive tables of $I_p(a, b)$ to seven decimal places are contained in Pearson (1934), for $p = 0.01(0.01)1$; $a, b = 0.5(0.5)11(1)50$, $a \geq b$. These may be supplemented for small values of a by the tables of Vogler (1964). Both Pearson and Vogler give values for the complete beta function $B(a, b)$.

Pearson and Hartley (1976) have tabulated the percentage points of the F-distribution with upper tail

$$Q(F|\nu_1, \nu_2) = I_p(\nu_2/2, \nu_1/2), \tag{1.98}$$

where $p = \nu_2/(\nu_2 + \nu_1 F)$, for

$$Q(F|\nu_1, \nu_2) = 0.001, 0.005, 0.01, 0.025, 0.05, 0.1, 0.25, 0.5,$$

$$\nu_1 = 1(1)6, 8, 12, 15, 20, 30, 60, \infty,$$

$$\nu_2 = 1(1)30, 40, 60, 120, \infty,$$

to at least three significant digits; this table is quoted in Abramowitz and Stegun (1965).

The *Laplace transform*, (P. S. Laplace, 1749–1827), of a function $f(t)$ is defined as

$$F(p) = \int_0^\infty f(t)e^{-pt} dt. \tag{1.99}$$

The *error function* $\operatorname{erf}(x)$ is defined by

$$\operatorname{erf}(x) = \frac{2}{\sqrt{\pi}} \int_0^x \exp(-t^2) dt. \tag{1.100}$$

It is closely related to the *normal distribution function*,

$$\Phi(x) = \frac{1}{\sqrt{2\pi}} \int_{-\infty}^x \exp(-t^2/2) dt = 0.5\{1 + \operatorname{erf}(x/\sqrt{2})\}. \tag{1.101}$$

Its complement is $\operatorname{erfc}(x) = 1 - \operatorname{erf}(x)$. Sometimes one sees

$$\operatorname{Erf}(x) = 0.5\sqrt{\pi} \operatorname{erf}(x), \qquad \operatorname{Erfc}(x) = 0.5\sqrt{\pi} \operatorname{erfc}(x).$$

The *Bessel function of the first kind* $J_\nu(x)$ is

$$J_\nu(x) = \left(\frac{x}{2}\right)^\nu \sum_{j=0}^\infty \frac{(-x^2/4)^j}{j!\,\Gamma(\nu + j + 1)}, \tag{1.102}$$

where ν is the order of the function. The *modified Bessel function of the first kind* is

$$I_\nu(x) = (-i)^\nu J_\nu(ix) = \left(\frac{x}{2}\right)^\nu \sum_{j=0}^\infty \frac{(x^2/4)^j}{j!\,\Gamma(\nu + j + 1)}, \tag{1.103}$$

where $i = \sqrt{-1}$. Properties of these functions are given in Abramowitz and Stegun (1965).

A6 Gaussian Hypergeometric Functions

The *hypergeometric function*, or more precisely the *Gaussian hypergeometric function*, has the form

$$
\begin{aligned}
{}_2F_1[a,b;c;x] &= 1 + \frac{ab}{c\,1!}x + \frac{a(a+1)b(b+1)}{c(c+1)2!}x^2 + \cdots \\
&= \sum_{j=0}^{\infty} \frac{(a)_j(b)_j x^j}{(c)_j\, j!}, \qquad c \neq 0, -1, -2, \ldots, \qquad (1.104)
\end{aligned}
$$

where $(a)_j$ is Pochhammer's symbol (1.6). The suffices refer to the numbers of numerator and denominator parameters—there are two numerator parameters and one denominator parameter. Clearly ${}_2F_1[b,a;c;x] = {}_2F_1[a,b;c;x]$.

We will only be interested in the case where a,b,c and x are real. If a is a nonpositive integer, then $(a)_j$ is zero for $j > -a$, and the series terminates. When the series is infinite, it is absolutely convergent for $|x| < 1$ and divergent for $|x| > 1$. For $|x| = 1$, it is

1. absolutely convergent if $c - a - b > 0$,
2. conditionally convergent if $-1 < c - a - b \le 0$, $x = -1$,
3. divergent if $c - a - b \le -1$.

When $a = 1$ and $b = c$ (or $b = 1$ and $a = c$), the series becomes $1 + x + x^2 + \cdots$; hence the name "hypergeometric."

Gauss's summation theorem states that when $x = 1$,

$$
{}_2F_1[a,b;c;x] = \frac{\Gamma(c)\Gamma(c-a-b)}{\Gamma(c-a)\Gamma(c-b)} = \frac{B(c,c-a-b)}{B(c-a,c-b)}, \qquad (1.105)
$$

where $c - a - b > 0$, $c \neq 0, -1, -2, \ldots$.

When a is a nonpositive integer, $a = -n$ say, and $b = -u$, $c = v - n + 1$, this becomes *Vandermonde's theorem* (see Section A2):

$$
\sum_{j=0}^{n} \binom{u}{j}\binom{v}{n-j} = \binom{u+v}{n}. \qquad (1.106)
$$

The Gaussian hypergeometric function satisfies the second-order linear differential equation

$$
x(1-x)\frac{d^2y}{dx^2} + [c - (a+b+1)x]\frac{dy}{dx} - aby = 0, \qquad (1.107)
$$

or, equivalently, $[\theta(\theta + c - 1) - x(\theta + a)(\theta + b)]y = 0,$ (1.108)

where θ is the differentiation operator $x(d/dx)$; see Section A4.

The Gaussian hypergeometric function has been described as "the wooden plough of the nineteenth century"; it occurs frequently in mathematical applications because every linear differential equation of the second order, whose singularities are regular and at most three in number, can be transformed into the hypergeometric equation.

The derivatives are

$$\frac{d}{dx}{}_2F_1[a, b; c; x] = \frac{ab}{c}\,{}_2F_1[a + 1, b + 1; c + 1; x],$$

$$D^n{}_2F_1[a, b; c; x] = \frac{(a)_n(b)_n}{(c)_n}\,{}_2F_1[a + n, b + n; c + n; x].   (1.109)$$

Euler's integral for the function is

$${}_2F_1[a, b; c; x] = \frac{\Gamma(c)}{\Gamma(a)\Gamma(c - a)}\int_0^1 u^{a-1}(1 - u)^{c-a-1}(1 - xu)^{-b}du,   (1.110)$$

where $c > a > 0$. The function is also a Laplace transform:

$${}_2F_1[a, b; c; k/s] = \frac{s^b}{\Gamma(b)}\int_0^\infty e^{-su}u^{b-1}{}_1F_1[a; c; ku]du.   (1.111)$$

The *Euler transformations* are

$${}_2F_1[a, b; c; x] = (1 - x)^{-a}{}_2F_1[a, c - b; c; x/(x - 1)]   (1.112)$$

$$= (1 - x)^{-b}{}_2F_1[c - a, b; c; x/(x - 1)]   (1.113)$$

$$= (1 - x)^{c-a-b}{}_2F_1[c - a, c - b; c; x].   (1.114)$$

Hypergeometric representations of elementary functions are

$$(1 - x)^{-a} = {}_2F_1[a, b; b; x],$$

$$\ln(1 + x) = x\,{}_2F_1[1, 1; 2; -x],$$

$$\ln\left(\frac{1 + x}{1 - x}\right) = 2x\,{}_2F_1[1/2, 1; 3/2; x^2],$$

$$\arcsin(x) = x\,{}_2F_1[1/2, 1/2; 3/2; x^2],$$

$$\arctan(x) = x\,{}_2F_1[1/2, 1; 3/2; -x^2],$$

$$\left(1 + \frac{\sqrt{1 - x}}{2}\right)^{1-2a} = {}_2F_1[a, a - 1/2; 2a; x],$$

$$= \sqrt{1 - x}\,{}_2F_1[a, a + 1/2; 2a; x],$$

$$(1 + x)^{-2a} + (1 - x)^{-2a} = 2\,_2F_1[a, a + 1/2; 1/2; x^2],$$
$$(1 - x)^{-2a-1}(1 + x) = \,_2F_1[a + 1, 2a; a; x]. \tag{1.115}$$

A large number of special functions can also be represented as Gaussian hypergeometric functions. The incomplete beta function is

$$B_p(a, b) = a^{-1}p^a\,_2F_1[a, 1 - b; a + 1; p], \tag{1.116}$$

the *Legendre polynomials* are

$$P_n(x) = \,_2F_1[-n, n + 1; 1; (1 - x)/2], \tag{1.117}$$

the *Chebyshev polynomials* (P. L. Chebyshev, 1821–1894) are

$$T_n(x) = \,_2F_1[-n, n; 1/2; (1 - x)/2], \tag{1.118}$$
$$U_n(x) = (n + 1)_2F_1[-n, n + 2; 3/2; (1 - x)/2], \tag{1.119}$$

and the *Jacobi polynomials* (C. G. J. Jacobi, 1804–1851) are

$$P_n^{(a,b)}(x) = \binom{a + n}{n}\,_2F_1[-n, a + b + n + 1; a + 1; (1 - x)/2]. \tag{1.120}$$

For detailed studies of the Gaussian hypergeometric function, including recurrence relationships between contiguous functions, see Bailey (1935), Erdélyi et al. (1953, Vol.1), Slater (1966), and Luke (1975).

A7   Confluent Hypergeometric Functions (Kummer's Functions)

Notations vary for the *confluent hypergeometric series* (also known as *Kummer's series*; E. E. Kummer, 1810–1893). We have

$$\begin{aligned}
_1F_1[a; c; x] &= 1 + \frac{a}{c\,1!}x + \frac{a(a + 1)}{c(c + 1)2!}x^2 + \cdots \\
&= \sum_{j=0}^{\infty} \frac{(a)_j x^j}{(c)_j\, j!} \\
&= \lim_{|b|\to\infty} {}_2F_1[a, b; c; x/b], \qquad c \neq 0, -1, -2, \ldots, \tag{1.121}
\end{aligned}$$

where $(a)_j$ is Pochhammer's symbol.   Other notations for $_1F_1[a; c; x]$ are $M(a; c; x)$ and $\phi(a; c; x)$. The suffices in $_1F_1[a; c; x]$ emphasize that there is one numerator parameter and one denominator parameter. If a is a nonpositive integer, the series terminates. When c is not a nonpositive integer, the series converges for all real values of a, c and x. When $a = c$, $c > 0$, the series becomes the exponential series $1 + x + x^2/2! + x^3/3! + \cdots$.

The confluent hypergeometric function satisfies Kummer's differential equation

$$x\frac{d^2y}{dx^2} + (c - x)\frac{dy}{dx} - ay = 0, \qquad (1.122)$$

that is,

$$[\theta(\theta + c - 1) - x(\theta + a)]y = 0, \qquad (1.123)$$

where $\theta \equiv x(d/dx)$.

The derivatives of the confluent hypergeometric function are

$$\frac{d}{dx}{}_1F_1[a; c; x] = \frac{a}{c}{}_1F_1[a + 1; c + 1; x],$$

$$D^n{}_1F_1[a; c; x] = \frac{(a)_n}{(c)_n}{}_1F_1[a + n; c + n; x]. \qquad (1.124)$$

The following integral representation is useful:

$${}_1F_1[a; c; x] = \frac{\Gamma(c)}{\Gamma(a)\Gamma(c - a)} \int_0^1 u^{a-1}(1 - u)^{c-a-1}e^{xu}\,du, \qquad (1.125)$$

where $c > a > 0$.

Kummer's first theorem yields the transformation

$${}_1F_1[a; c; x] = e^x{}_1F_1[c - a; c; -x]. \qquad (1.126)$$

Kummer's second theorem is

$$e^{-x}{}_1F_1[a; 2a; 2x] = {}_0F_1[\ ; a + 1/2; x^2/4], \qquad (1.127)$$

where $a + 1/2$ is not a negative integer and

$${}_0F_1[\ ; c; x] = \lim_{|a| \to \infty}{}_1F_1[a; c; x/a] = \sum_{j=0}^{\infty}\frac{x^j}{(c)_j\,j!}, \qquad (1.128)$$

where $c \neq 0, 1, 2, \ldots$ and x is finite.

Kummer's differential equation is also satisfied by

$$\Psi(a, c; x) = x^{-a}{}_2F_0[a, a - c + 1;\ ; -1/x] \qquad (1.129)$$

$$= \frac{1}{\Gamma(a)} \int_0^\infty e^{-xu}u^{a-1}(1 + u)^{c-a-1}\,du, \qquad (1.130)$$

where $a > 0$, $x > 0$.

The following relationship holds:

$$\Psi(a,c;x) = \frac{\Gamma(1-c)}{\Gamma(a-c+1)} {}_1F_1[a;c;x]$$

$$+ \frac{\Gamma(c-1)x^{1-c}}{\Gamma(a)} {}_1F_1[a-c+1;2-c;x], \qquad (1.131)$$

provided that $c \neq 0, \pm 1, \pm 2, \ldots$. Also

$$\Psi(a,c;x) = x^{1-c}\Psi(a-c+1,2-c;x). \qquad (1.132)$$

Many functions that are important in distribution theory can be expressed in terms of the confluent hypergeometric function; for example, the incomplete gamma functions are

$$\gamma(a,x) = a^{-1}x^a {}_1F_1[a;a+1;-x],$$
$$\Gamma(a,x) = \Gamma(a) - a^{-1}x^a {}_1F_1[a;a+1;-x], \qquad (1.133)$$

and the *error functions* are

$$\mathrm{Erf}(x) = \frac{\sqrt{\pi}}{2}\mathrm{erf}(x) = 0.5\gamma(1/2,x^2)$$

$$= x\,{}_1F_1[1/2;3/2;-x^2] \qquad (1.134)$$

$$\mathrm{Erfc}(x) = \frac{\sqrt{\pi}}{2}\mathrm{erfc}(x) = \frac{\sqrt{\pi}}{2} - x\,{}_1F_1[1/2;3/2;-x^2]. \qquad (1.135)$$

The *Hermite polynomials* (Ch. Hermite, 1822–1901) as used in statistics are defined as

$$H_n(x) = \sum_{j=0}^{[n/2]} \frac{(-1)^j n! x^{n-2j}}{(n-2j)!j!2^j}, \qquad (1.136)$$

where $[\cdot]$ denotes the integer part. Hence

$$H_{2n}(x) = \frac{(-1)^n(2n)!}{n!2^n} {}_1F_1[-n;1/2;x^2/2],$$

$$H_{2n+1}(x) = \frac{(-1)^n(2n+1)!x}{n!2^n} {}_1F_1[-n;3/2;x^2/2]; \qquad (1.137)$$

see Stuart and Ord (1987, Secs. 6.14–6.15). Fisher (1951, p. xxxi) used the "modified" Hermite polynomials

$$H_n^*(x) = i^{-n} H_n(ix), \qquad \text{where } i = \sqrt{-1}. \qquad (1.138)$$

The Bessel functions $J_\nu(x)$, $I_\nu(x)$, and $K_\nu(x)$ (F. W. Bessell, 1784–1846), Whittaker functions (E. T. Whittaker, 1873–1956), Laguerre functions and polynomials (E. N. Laguerre, 1834–1886), and Poisson–Charlier polynomials (S. D. Poisson, 1781–1840, and C. L. Charlier, 1862–1939) can also all be represented as confluent hypergeometric functions.

Further details concerning some of these functions can be found in Section A11. Thorough coverage is in Erdélyi et al. (1953, Vols. 1 and 2) and in the book devoted to confluent hypergeometric functions by Slater (1960). Readers are WARNED, however, that most mathematical texts, including those by Erdélyi and by Abramowitz and Stegun, use slightly different notations for the Hermite polynomials (differing by powers of 2). Slater (1960), Rushton and Lang (1954), and Abramowitz and Stegun (1965) give useful tables.

A8 Generalized Hypergeometric Functions

The *generalized hypergeometric function* is a natural generalization of the Gaussian hypergeometric function. The series is defined as

$$
{}_pF_q[a_1,\dots,a_p;b_1,\dots,b_q;x] = {}_pF_q \left[\begin{array}{ccc} a_1,\dots,a_p; & x \\ b_1,\dots,b_q \end{array} \right] \qquad (1.139)
$$

$$
= \sum_{j=0}^{\infty} \frac{(a_1)_j \dots (a_p)_j x^j}{(b_1)_j \dots (b_q)_j \; j!}, \qquad (1.140)
$$

where $b_i \neq 0, -1, -2, \dots,$ $i = 1, \dots, q$.

There are p numerator parameters and q denominator parameters. Clearly the orderings of the numerator parameters and of the denominator parameters are immaterial. The simplest generalized hypergeometric series is

$$
{}_0F_0[-;-;x] = {}_0F_0[\; ; \; ;x] = 1 + x + \frac{x^2}{2!} + \frac{x^3}{3!} + \dots = e^x \qquad (1.141)
$$

(a blank indicates the absence of a parameter).

If one of the numerator parameters a_i, $i = 1, \dots, p$ is a negative integer, $a_1 = -n$ say, the series terminates and

$$
{}_pF_q \left[\begin{array}{ccc} -n, a_2, \dots, a_p; & x \\ b_1, \dots, b_q \end{array} \right]
$$

$$
= \sum_{j=0}^{n} \frac{(-n)_j (a_2)_j \dots (a_p)_j x^j}{(b_1)_j \dots (b_q)_j \; j!}, \qquad (1.142)
$$

$$= \frac{(a_2)_n \dots (a_p)_n (-x)^n}{(b_1)_n \dots (b_q)_n}$$

$$\times \ _{q+1}F_{p-1} \left[\begin{array}{c} -n, 1 - b_1 - n, \dots, 1 - b_q - n; \quad (-1)^{p+q-1}x^{-1} \\ 1 - a_2 - n, \dots, 1 - a_p - n; \end{array} \right].$$

$$(1.143)$$

When the series is infinite, it converges for $|x| < \infty$ if $p \le q$, it converges for $|x| < 1$ if $p = q + 1$, and it diverges for all x, $x \ne 0$ if $p > q + 1$. Furthermore, if

$$s = \sum_{i=1}^{q} b_i - \sum_{i=1}^{p} a_i,$$

then the series with $p = q + 1$ is absolutely convergent for $|x| = 1$ if $s > 0$, is conditionally convergent for $|x| = 1, x \ne 1$ if $-1 < s \le 0$, and is divergent for $|x| = 1$ if $s \le -1$.

The function is characterized as a power series $\sum_{j=0}^{\infty} A_j x^j$ by the property that A_{j+1}/A_j is a rational function of j.

The function satisfies the differential equation

$$\theta(\theta + b_1 - 1) \dots (\theta + b_q - 1)y = x(\theta + a_1) \dots (\theta + a_p)y, \qquad (1.144)$$

where θ is the differential operator $x(d/dx)$.

The derivatives are

$$\frac{d}{dx} {}_pF_q[a_1, \dots, a_p; b_1, \dots, b_q; x]$$

$$= \frac{a_1 \dots a_p}{b_1 \dots b_q} {}_pF_q[a_1 + 1, \dots, a_p + 1; b_1 + 1, \dots, b_q + 1; x], \qquad (1.145)$$

$$D^n {}_pF_q[a_1, \dots, a_p; b_1, \dots, b_q; x]$$

$$= \frac{(a_1)_n \dots (a_p)_n}{(b_1)_n \dots (b_q)_n} {}_pF_q[a_1 + n, \dots, a_p + n; b_1 + n, \dots, b_q + n; x]. \qquad (1.146)$$

The *Eulerian integral* generalizes to

$$_{p+1}F_{q+1}[a_1, \dots, a_p, c; b_1, \dots, b_q, d; x] =$$

$$\frac{\Gamma(d)}{\Gamma(c)\Gamma(d-c)} \int_0^1 u^{c-1}(1-u)^{d-c-1} {}_pF_q[a_1, \dots, a_p; b_1, \dots, b_q; xu]du. \qquad (1.147)$$

Also

$$_{p+1}F_q[a_1, \ldots, a_p, c; b_1, \ldots, b_q; x]$$

$$= \frac{1}{\Gamma(c)} \int_0^\infty e^{-u} u^{c-1} {}_pF_q[a_1, \ldots, a_p; b_1, \ldots, b_q; xu] du. \quad (1.148)$$

The product of two generalized hypergeometric functions can be expressed as a series in other generalized hypergeometric functions. So can generalized hypergeometric functions with arguments of the form $x = y + z$.

A generalized hypergeometric series tail-truncated after $m + 1$ terms can be represented as

$$_{p+1}F_{q+1}[a_1, \ldots, a_p, -m; b_1, \ldots, b_q, -m; x].$$

Head-truncation of the first k terms gives

$$\frac{(a_1)_k \ldots (a_p)_k x^k}{(b_1)_k \ldots (b_q)_k \, k!} \, {}_{p+1}F_{q+1}[a_1 + k, \ldots, a_p + k, 1; b_1 + k, \ldots, b_q + k, 1 + k; x].$$

Generalized hypergeometric representations of elementary functions include

$$e^x = {}_0F_0[-;-;;x] = {}_0F_0[\ ;\ ;x],$$
$$(1-x)^{-a} = {}_1F_0[a;-;x] = {}_1F_0[a;\ ;x],$$
$$\cos(x) = {}_0F_1[-;1/2;-x^2/4] = {}_0F_1[\ ;1/2;-x^2/4],$$
$$\sin(x) = x \, {}_0F_1[-;3/2;-x^2/4] = x \, {}_0F_1[\ ;3/2;-x^2/4],$$
$$\arctan(x) = x \, {}_2F_1[1/2, 1; 3/2; -x^2]. \quad (1.149)$$

Bessel functions can also be stated this way; for example,

$$J_\nu(x) = \frac{(x/2)^\nu}{\Gamma(\nu+1)} {}_0F_1\left[\ ;\nu+1; -x^2/4\right],$$

$$I_\nu(x) = \frac{(x/2)^\nu}{\Gamma(\nu+1)} {}_0F_1\left[\ ;\nu+1; x^2/4\right] \quad (1.150)$$

(see also Sections A5 and A7).

Extensive treatments of generalized hypergeometric functions (including further references) are provided in the books by Erdélyi (1953, Vol. 1), Rainville (1960), and Slater (1966). Certain useful integrals are in Erdélyi (1954, Vols. 1 and 2) and Exton (1978). More advanced special functions and their statistical applications have been studied by Mathai and Saxena (1973, 1978).

A9 Bernoulli and Euler Numbers and Polynomials

The *Bernoulli numbers* (J. Bernoulli, 1654–1705), $B_0, B_1, \ldots, B_r, \ldots$ are the coefficients of $1, t, \ldots, t^r/r!, \ldots$, in the expansion of $t(e^t - 1)^{-1}$. Numerical values are:

$$B_0 = 1, \; B_1 = -\frac{1}{2}, \; B_2 = \frac{1}{6}, \; B_4 = -\frac{1}{30}, \; B_6 = \frac{1}{42}, \; B_8 = -\frac{1}{30},$$

$$(1.151)$$

with $B_{2r+1} = 0$ for $r > 0$.

The *Bernoulli polynomials* $B_0(x), B_1(x), \ldots, B_r(x), \ldots$, are defined as the coefficients of $1, t, \ldots, t^r/r!, \ldots$, in the expansion of $te^{tx}(e^t - 1)^{-1}$, and so $B_r(0) = B_r$. A useful formula is

$$\sum_{j=1}^{n} j^r = (r+1)^{-1}\{B_{r+1}(n+1) - B_{r+1}\}.$$

$$(1.152)$$

These polynomials have the properties that

$$\frac{dB_r(x)}{dx} = rB_{r-1}(x),$$

$$(1.153)$$

and

$$B_r(x+h) = \sum_{j=0}^{r} \binom{r}{j} B_j(x) h^{r-j},$$

$$(1.154)$$

(symbolically $B_r(x+h) = (E+h)^r B_0(x)$ with the displacement operator E applying to the subscript.)

The first seven Bernoulli polynomials are

$$B_0(x) = 1,$$
$$B_1(x) = x - \tfrac{1}{2},$$
$$B_2(x) = x^2 - x + \tfrac{1}{6},$$
$$B_3(x) = x^3 - \tfrac{3}{2}x^2 + \tfrac{1}{2}x,$$
$$B_4(x) = x^4 - 2x^3 + x^2 - \tfrac{1}{30},$$
$$B_5(x) = x^5 - \tfrac{5}{2}x^4 + \tfrac{5}{3}x^3 - \tfrac{1}{6}x,$$
$$B_6(x) = x^6 - 3x^5 + \tfrac{5}{2}x^4 - \tfrac{1}{2}x^2 + \tfrac{1}{42}.$$

$$(1.155)$$

David et al. (1966) have tabulated the Bernoulli polynomials $B_n(x)$ for $n = 0(1)12$ and the Bernoulli numbers B_n for $n = 1(1)12$.

The *Euler polynomials* $E_r(x)$ are defined by the identity

$$2e^{tx}(e^t + 1)^{-1} \equiv \sum_{j=0}^{\infty} E_j(x)\frac{t^j}{j!}. \qquad (1.156)$$

Their properties include

$$E_n(x) + E_n(x + 1) = 2x^n, \qquad (1.157)$$

$$\frac{dE_n(x)}{dx} = nE_{n-1}(x). \qquad (1.158)$$

The *Euler numbers* E_r are defined as

$$E_r = 2^r E_r(1/2). \qquad (1.159)$$

They satisfy the symbolic formula

$$(E + 1)^n + (E - 1)^n = 0, \qquad (1.160)$$

with powers of E^m replaced by E_m. We find that $E_{2n+1} = 0$, and that the Euler numbers are all integers for r even:

$$E_0 = 1, \quad E_2 = -1, \quad E_4 = 5, \quad E_6 = -61, \quad E_8 = 1,385, \quad E_{10} = -50,521, \quad \text{etc.} \qquad (1.161)$$

Further values are given in Abramowitz and Stegun (1965).

The following symbolic relationships connect the Bernoulli and the Euler numbers:

$$E^{n-1} \equiv \frac{(4B - 1)^n - (4B - 3)^n}{2n},$$

$$E^{2n} \equiv \frac{4^{2n+1}(B - 1/4)^{2n+1}}{2n + 1}. \qquad (1.162)$$

If $m + n$ is odd, then

$$\int_0^1 B_m(x)B_n(x)dx = 0 = \int_0^1 E_m(x)E_n(x)dx. \qquad (1.163)$$

Both the polynomials $B_m(x)$, $B_n(x)$ and the polynomials $E_m(x)$, $E_n(x)$ are *orthogonal* over the interval $(0, 1)$ (see Section A11), with uniform weight function. For a full discussion of Bernoulli and Euler polynomials, we refer the reader to Nörlund (1923) and Milne-Thompson (1935). Abramowitz and Stegun give an excellent summary.

A10   Integral Transforms

The *exponential Fourier transform* (J. B. J. Fourier, 1768–1830),

$$\varphi(t) = \int_{-\infty}^{\infty} e^{itx} f(x) dx, \qquad (1.164)$$

gives the characteristic function of a distribution.

The *Laplace transform*

$$L(p) = \int_{0}^{\infty} e^{-px} f(x) dx \qquad (1.165)$$

(if it exists) yields the moment generating function $M(t)$ of a distribution with pdf $f(x)$ on the nonnegative real line by setting $t = -p$; that is, $M(t) = L(-t)$.

The *Mellin transform* (R. H. Mellin, 1854–1933), and its *inverse* are

$$H(s) = \int_{0}^{\infty} x^{s-1} f(x) dx \qquad (1.166)$$

$$f(x) = \frac{1}{2\pi i} \int_{c-i\infty}^{c+i\infty} x^{-s} H(s) ds. \qquad (1.167)$$

If $f(x)$ is a probability density function, then (1.166) gives the $(s-1)$th moment about the origin of a distribution on the nonnegative real line. Springer (1979) has demonstrated the key role of the Mellin transform and its inverse in the derivation of distributions of products, quotients, and other algebraic functions of independent random variables.

For a comprehensive coverage of these and other types of integral transforms, see Erdélyi et al. (1954, Vols. 1 and 2).

A11   Orthogonal Polynomials

If the polynomial $P_r(x)$ of degree r is a member of a family of polynomials $\{P_j(x)\}$, $j = 0, 1, \ldots$, and

$$\int_{-\infty}^{\infty} w(x) P_m(x) P_n(x) dx = 0 \qquad (1.168)$$

is satisfied whenever $m \neq n$, then the family of polynomials is said to be *orthogonal* with respect to the weight function $w(x)$. In particular cases, $w(x)$ may be zero outside certain intervals.

Two families of *orthogonal polynomials* have especial importance in distribution theory. These are the Hermite polynomials and the generalized Laguerre polynomials. The *Hermite polynomials* have the weight function

$$w(x) = e^{-x^2/2}. \qquad (1.169)$$

The rth Hermite polynomial is defined by

$$H_r(x) = (-1)^r e^{x^2/2} D^r (e^{-x^2/2}), \qquad r = 0, 1, \ldots, \tag{1.170}$$

It follows that

$$
\begin{aligned}
H_0(x) &= 1, \\
H_1(x) &= x, \\
H_2(x) &= x^2 - 1, \\
H_3(x) &= x^3 - 3x, \\
H_4(x) &= x^4 - 6x^2 + 3, \\
H_5(x) &= x^5 - 10x^3 + 15x,
\end{aligned}
$$

and generally

$$
\begin{aligned}
H_r(x) &= x^r - \frac{r(r-1)}{1! \cdot 2} x^{r-2} + \frac{r(r-1)(r-2)(r-3)}{2! \cdot 2^2} x^{r-4} - \cdots \\
&\quad + (-1)^j \frac{r!}{(r-2j)! j! \cdot 2^j} x^{r-2j} + \cdots \tag{1.171}
\end{aligned}
$$

(cf. Section A7). The series terminates after $j = [r/2]$, where $[r]$ denotes the largest integer less than or equal to r.

The *generalized Laguerre polynomials* have the weight function

$$
w(x) = \begin{cases} x^a e^{-x}, & x \geq 0, \ a > -1, \\ 0, & x < 0. \end{cases} \tag{1.172}
$$

The rth generalized Laguerre polynomial of order a is

$$
\begin{aligned}
L_r^{(a)}(x) &= \sum_{j=0}^{r} (-1)^j \binom{r+a}{r-j} \frac{x^j}{j!} \\
&= \binom{r+a}{r} {}_1F_1[-r; a+1; x]. \tag{1.173}
\end{aligned}
$$

The recurrence formula

$$x L_r^{(a+1)}(x) = (x - r) L_r^{(a)}(x) + (a + r) L_{r-1}^{(a)}(x) \tag{1.174}$$

is useful in computation.

The *Jacobi, Chebyshev, Krawtchouk, and Charlier polynomials* are other families of orthogonal polynomials that are occasionally used in statistical theory. The weight function for the Jacobi polynomial $P_n^{(a,b)}(x)$ is

$$
w(x) = \begin{cases} (1-x)^a(1+x)^b, & -1 \le x \le 1, \\ 0, & \text{otherwise.} \end{cases} \tag{1.175}
$$

The other three families have the following weight functions:

Chebyshev Polynomial $T_n(x)$

$$
w(x) = \begin{cases} (1-x^2)^{-1/2}, & -1 \le x \le 1, \\ 0, & \text{otherwise.} \end{cases} \tag{1.176}
$$

Chebyshev Polynomial $U_n(x)$

$$
w(x) = \begin{cases} (1-x^2)^{1/2}, & -1 \le x \le 1, \\ 0, & \text{otherwise.} \end{cases} \tag{1.177}
$$

Krawtchouk Polynomials

$$
w(x) = \begin{cases} \binom{n}{x} p^x (1-p)^{n-x}, & x = 0, 1, 2, \ldots, n, \\ 0, & \text{otherwise.} \end{cases} \tag{1.178}
$$

Charlier Polynomials

$$
w(x) = \begin{cases} e^{-\theta} \theta^x / x!, & x = 0, 1, 2, \ldots, \\ 0, & \text{otherwise.} \end{cases} \tag{1.179}
$$

Szegö (1939, 1959) is a standard work on orthogonal polynomials. Their properties are summarized in Abramowitz and Stegun (1965). Stuart and Ord (1987, Ch. 6) demonstrate some of their statistical uses.

A12 Miscellaneous Topics

The *Riemann zeta function* (G. F. B. Riemann, 1826–1866) is defined by the equation

$$\zeta(x) = \sum_{j=1}^{\infty} j^{-x}. \tag{1.180}$$

The series is convergent for $x > 1$, and it is only for these values of x that we shall use the function. A generalized form of the Riemann zeta function is defined by

$$\zeta(x, a) = \sum_{j=1}^{\infty} (j + a)^{-x}, \tag{1.181}$$

where $x > 1$ and $a > 0$.

An approximate formula for $\zeta(x)$ is

$$\zeta(x) \doteq 1 + \frac{2x^2 + 8.4x + 21.6}{(x - 1)(x + 7)2^{x+1}}. \tag{1.182}$$

Particular values are

$$\zeta(2) = \frac{\pi^2}{6} \quad \text{and} \quad \zeta(4) = \frac{\pi^4}{90}.$$

Values of $\zeta(n)$ for $n = 2(1)42$, to 20 decimal places, are given in Abramowitz and Stegun (1965).

A general formula, for even values of the argument, is

$$\zeta(2r) = \frac{(2\pi)^{2r}}{2\{(2r)!\}} |B_{2r}|, \tag{1.183}$$

where B_{2r} is a Bernoulli number.

The *modified Bessel function of the third kind*, $K_\nu(y)$, is defined as

$$K_\nu(y) = \frac{\pi}{2} \cdot \frac{I_{-\nu}(y) - I_\nu(y)}{\sin(\nu \pi)} \tag{1.184}$$

when ν is not an integer or zero. When ν is an integer or zero, the right hand side of this definition is replaced by its limiting value; see for example Abramowitz and Stegun (1965). $K_\nu(y)$ is sometimes called the modified Bessel function of the second kind in the statistical literature.

Useful properties are

$$K_{-\nu}(y) = K_\nu(y) \tag{1.185}$$

and the recurrence relation

$$K_{\nu+1}(y) = \frac{2\nu}{y} K_\nu(y) + K_{\nu-1}(y). \tag{1.186}$$

Heine's generalization of the hypergeometric series (H. E. Heine, 1821–1881) is known as a *basic hypergeometric series*, also as a *q-series* and as a *q-hypergeometric series*; it is defined as

$$_2\Phi_1(a,b;c;q,z)$$

$$= 1 + \frac{(1-a)(1-b)z}{(1-c)(1-q)} + \frac{(1-a)(1-aq)(1-b)(1-bq)z^2}{(1-c)(1-cq)(1-q)(1-q^2)} + \cdots$$

$$= \sum_{j=0}^{\infty} \frac{(a)_{q,j}(b)_{q,j}z^j}{(c)_{q,j}(q)_{q,j}}, \tag{1.187}$$

where $|q| < 1$, $|z| < 1$; there are two numerator parameters and one denominator parameter. By $(a)_{q,j}$ we mean

$$(a)_{q,0} = 1, \quad (a)_{q,j} = (1-a)(1-aq)\ldots(1-aq^{j-1}).$$

Readers are WARNED that there are several differing notations for this expression in the literature, for example

$(a)_{q,j}$ Bailey (1935)

$(a;q)_j$ Slater (1966), Andrews (1986)

$[a]_j$ Jackson

$[a;q,j]$ Exton (1983)

[for a complete list of F. H. Jackson's numerous publications, see Chaundy (1962)].

The general basic hypergeometric series (q-series) is

$$_A\Phi_B(a_1,\ldots,a_A;b_1,\ldots,b_B;q,z)$$

$$= {}_A\Phi_B \begin{bmatrix} a_1,\ldots, & a_A & ;q,z \\ b_1,\ldots, & b_B & \end{bmatrix}$$

$$= \sum_{j=0}^{\infty} \frac{(a_1)_{q,j}\ldots(a_A)_{q,j}z^j}{(b_1)_{q,j}\ldots(b_B)_{q,j}(q)_{q,j}}. \tag{1.188}$$

As $q \to 1$, we find that $(q^a)_{q,j}(1-q)^{-j} \to (a)_j$, where $(a)_j$ is Pochhammer's symbol. It follows that as $q \to 1$, a general basic hypergeometric series tends to a generalized hypergeometric series:

$$\lim_{q \to 1} {}_A\Phi_B(q^{a_1},\ldots,q^{a_A};q^{b_1},\ldots,q^{b_B};q,(1-q)^{B+1-A}z)$$

$$= {}_AF_B[a_1,\ldots,a_A;b_1,\ldots,b_B;z]. \tag{1.189}$$

Heine's theorem

$$_1\Phi_0(a; -; q, z) = {}_1\Phi_0(a; \ ; q, z) = \prod_{j=0}^{\infty} \frac{(1 - aq^j z)}{(1 - q^j z)} \tag{1.190}$$

follows from the relationship

$$(1 - z)_1\Phi_0(a; \ ; q, z) = (1 - az)_1\Phi_0(a; \ ; q, qz).$$

When $a = q^{-k}$ and k is a positive integer, Heine's theorem gives the following q-series analogue of the binomial theorem:

$$\prod_{j=0}^{k-1}(1 - q^{j-k}z) = {}_1\Phi_0(q^{-k}; \ ; q, z). \tag{1.191}$$

Another consequence of Heine's theorem is

$$_1\Phi_0(a; \ ; q, z)_1\Phi_0(b; \ ; q, z) = {}_1\Phi_0(ab; \ ; q, z). \tag{1.192}$$

Letting $a \to 0$ gives

$$_0\Phi_0(-; -; q, z) = {}_0\Phi_0(\ ; \ ; q, z) = \prod_{j=0}^{\infty}(1 - q^j z)^{-1}; \tag{1.193}$$

that is,

$$1 + \frac{z}{(1-q)} + \frac{z^2}{(1-q)(1-q^2)} + \cdots + \frac{z^j}{(1-q)\cdots(1-q^j)} + \cdots$$
$$= (1 - z)^{-1}(1 - qz)^{-1}(1 - q^2 z)^{-1}\ldots. \tag{1.194}$$

If z is replaced by $-z/a$ and $a \to \infty$, we obtain

$$1 + \frac{z}{(1-q)} + \frac{qz^2}{(1-q)(1-q^2)} + \cdots + \frac{q^{j(j-1)/2}z^j}{(1-q)\cdots(1-q^j)} + \cdots$$
$$= (1 + z)(1 + qz)(1 + q^2 z)\ldots. \tag{1.195}$$

B PROBABILITY AND STATISTICAL PRELIMINARIES

B1 Calculus of Probabilities

The probability of occurrence of an event E can be treated as a probability measure $\Pr(E)$, that is, as a set function defined on a measurable space $(\Omega, \mathcal{B})$, where Ω is a set of outcomes, and $\mathcal{B}$ is a σ-field of subsets of Ω. For $Pr(E)$ to be a probability measure, we require the following *probability axioms* to be satisfied:

1. $0 \le Pr(E) \le 1$.
2. $Pr(\Omega) = 1$.
3. If the events E_i are *mutually exclusive*, then $Pr(\bigcup_i E_i) = \sum_i Pr(E_i)$.

Probabilities defined in this way accord with the intuitive notion that the probability of an event E is the proportion of times that E might be expected to occur in repeated independent observations under specified conditions and that the probability of E therefore takes some value in the (closed) interval $[0, 1]$. The probability of an impossibility is taken to be zero, while the sum of the probabilities of all possibilities is deemed to be unity (the probability of a certainty). Given two events that cannot occur simultaneously, then the probability that one or other of them will occur is equal to the sum of their separate probabilities.

The compound event, "either E_1 or E_2 or both," is called the *logical sum* or *union* of E_1 and E_2 and is written symbolically as $E_1 + E_2$ or $E_1 \bigcup E_2$. (The two names and symbols refer to the same thing.)

The compound event, "both E_1 and E_2," is called the *logical product* or *intersection* of E_1 and E_2 and is written symbolically as $E_1 E_2$ or $E_1 \bigcap E_2$. (Again the two names and symbols refer to the same concept.)

These definitions can be extended to combinations of any number of events. Thus $E_1 + E_2 + \cdots + E_k$ or $\bigcup_{j=1}^{k} E_j$ means "at least one of $E_1, E_2, \ldots, E_k$," whilst $E_1 E_2 \cdots E_k$ or $\bigcap_{j=1}^{k} E_j$ means "every one of $E_1, E_2, \ldots, E_k$." By a natural extension we can form such compound events as $(E_1 \cup E_2) \cap E_3$ meaning "both E_3, and at least one of E_1 and E_2." By a further extension we can form compounds of enumerable infinities of events $\bigcup_{j=1}^{\infty} E_j$ and $\bigcap_{j=1}^{\infty} E_j$.

An important formula connecting probabilities of different, but related, events is

$$Pr(E_1 \cup E_2) = Pr(E_1) + Pr(E_2) - Pr(E_1 \cap E_2). \qquad (1.196)$$

The following extension of this formula is known in the literature on probability theory as *Boole's formula*:

$$Pr\left(\bigcup_{j=1}^{n} E_j\right) = \sum_{j=1}^{n} Pr(E_j) - \sum\sum Pr(E_{j_1} \cap E_{j_2})$$
$$+ \sum\sum\sum Pr(E_{j_1} \cap E_{j_2} \cap E_{j_3}) - \cdots + (-1)^{n-1} Pr\left(\bigcap_{j=1}^{n} E_j\right),$$

$$(1.197)$$

where a summation sign repeated m times means summation over all integers $j_1, j_2, \ldots, j_m$ subject to $1 \leq j_i \leq n$, $j_1 < j_2 < \cdots < j_m$. The *inclusion-exclusion principle* (see Chapter 10, Section 2) is important in the derivation of matching and occupancy distributions; it is closely related to Boole's formula.

The absolute values of the terms in (1.197) are nonincreasing. Boole's formula therefore enables bounds to be obtained for $\Pr(\bigcup_{j=1}^{n} E_j)$ by stopping at any two consecutive sets of terms. For example,

$$\sum_{j=1}^{n} \Pr(E_j) - \sum\sum \Pr(E_{j_1} \cap E_{j_2}) \leq \Pr\left(\bigcup_{j=1}^{n} E_j\right) \leq \sum_{j=1}^{n} \Pr(E_j). \qquad (1.198)$$

If $\Pr(E_1 \cap E_2) = 0$, the events E_1 and E_2 are said to be *mutually exclusive*. If every pair of the events $E_1, E_2, \ldots, E_n$ is mutually exclusive, then (1.197) becomes

$$\Pr\left(\bigcup_{j=1}^{n} E_j\right) = \sum_{j=1}^{n} \Pr(E_j). \qquad (1.199)$$

The mutually exclusive events $E_1, E_2, \ldots, E_n$ are said to be *exhaustive* if $\sum_{j=1}^{n} \Pr(E_j) = 1$.

The *conditional probability* of event E_1 given that E_2 has occurred, is denoted by $\Pr(E_1|E_2)$ and is given by

$$\Pr[E_1|E_2] = \frac{\Pr(E_1 \cap E_2)}{\Pr(E_2)}, \qquad (1.200)$$

where $Pr(E_2) > 0$; therefore

$$\Pr(E_1 \cap E_2) = \Pr(E_1)\Pr(E_2|E_1) = \Pr(E_2)\Pr(E_1|E_2). \qquad (1.201)$$

More generally,

$$\Pr\left(\bigcap_{j=1}^{n} E_j\right) = \Pr(E_1)\Pr(E_2|E_1)\Pr(E_3|E_1 \cap E_2) \cdots \Pr\left(E_n\Big|\bigcap_{j=1}^{n-1} E_j\right). \qquad (1.202)$$

The *Theorem of Total Probability* states that if $E_1, E_2, \ldots, E_n$ are mutually exclusive and exhaustive, then

$$\Pr(A) = \sum_{j=1}^{n} \Pr(A|E_j)\Pr(E_j). \qquad (1.203)$$

Bayes' theorem is an important consequence of (1.200) and (1.203) and is central to modern Bayesian methods of inference; see Section B3 later in this chapter.

If $\Pr(E_2|E_1) = \Pr(E_2)$, then E_2 is said to be *independent* of the event E_1 and (1.201) becomes

$$\Pr(E_1 \cap E_2) = \Pr(E_1)\Pr(E_2). \qquad (1.204)$$

We say that n events are *mutually independent* if, for every subset $\{E_{j_1}, E_{j_2}, \ldots, E_{j_k}\}$, $k \leq n$,

$$\Pr(E_{j_1} \cap E_{j_2} \cap \ldots E_{j_k}) = \prod_{i=1}^{k} \Pr(E_{j_i}). \qquad (1.205)$$

If E_j is independent of $\bigcap_{i=1}^{j-1} E_i$ for all $j \leq n$ (this is certainly true if the n events are mutually independent), then (1.202) simplifies to

$$\Pr(E_1 \cap E_2 \cap \ldots E_n) = \prod_{i=1}^{n} \Pr(E_j). \qquad (1.206)$$

The event "negation of E" is called the *complement* of E and is often denoted by $\overline{E}$. From *De Morgan's laws*,

$$\Pr(\overline{E}) = 1 - \Pr(E), \qquad (1.207)$$

$$\Pr(\overline{E_1 \cup E_2}) = \Pr(\overline{E_1} \cap \overline{E_2}), \qquad (1.208)$$

$$\Pr(\overline{E_1 \cap E_2}) = \Pr(\overline{E_1} \cup \overline{E_2}). \qquad (1.209)$$

B2 Real Variables

A *real random variable* X is a function from a sample space into the real numbers, with the property that for every outcome there is an associated probability $\Pr[X = x]$ which exists for all real values of x. Random variables (rv's) will be denoted throughout this work by uppercase letters. Realized values of a rv will be denoted by the corresponding lowercase letter.

The *cumulative distribution function* (cdf) of X, often just called the *distribution function* (DF), is defined as $\Pr[X \leq x]$ and regarded as a function of x; it is customarily denoted by $F_X(x)$.

Clearly $F_X(x)$ is a nondecreasing function of x, and $0 \leq F_X(x) \leq 1$. If $\lim_{x \to -\infty} F_X(x) = 0$ and $\lim_{x \to +\infty} F_X(x) = 1$, then the distribution is *proper*. We shall be concerned only with proper distributions.

The study of distributions is essentially a study of cumulative distribution functions. In all cases in these volumes the cumulative distribution function belongs to one of two classes, discrete or continuous, or it can be constructed by mixing elements from the two classes.

For *discrete distributions* $F_X(x)$ is a step function with only an enumerable number of steps. If the height of the step at x_j is p_j, then

$$\Pr[X = x_j] = p_j.$$

We call p_j a *probability mass function* (pmf), and we say that its *support* is the set $\{x_j\}$. If the distribution is proper, $\sum_j p_j = 1$. Random variables belonging to this class are called *discrete random variables*.

For *continuous distributions* $F_X(x)$ is absolutely continuous and can be expressed as an integral,

$$F_X(x) = \int_{-\infty}^{x} f_X(x)\,dx. \tag{1.210}$$

Any function $f_X(x)$ for which (1.210) holds for every x is a *probability density function* (pdf) of X. Random variables in this class are called *continuous random variables*.

The *survival function* is defined as $S_X(x) = \Pr[X > x] = 1 - F_X(x)$. The *hazard function (failure rate)* of a discrete distribution is $h_X(x) = \Pr[X = x]/\Pr[X \geq x]$; for a continuous distribution it is $h_X(x) = f_X(x)/S_X(x)$. In actuarial work $h_X(x)$ is known as the *force of mortality*.

A random variable is said to have an increasing hazard rate (IHR) if $h_X(x)$ is nondecreasing in x; it has a decreasing hazard rate (DHR) if $h_X(x)$ is nonincreasing in x. For these and other related terms such as NBU (new better than used), see for example Barlow and Proschan (1975).

When the subscripts for $f_X(x)$ and $F_X(x)$ are well-understood, they are often dropped, provided that this does not cause confusion.

The above concepts can be extended to the *joint distribution* of a finite number of rv's $X_1, X_2, \ldots, X_n$. The *joint cumulative distribution function* is

$$\Pr\left[\bigcap_{j=1}^{n}(X_j \leq x_j)\right] = F_{X_1, X_2, \ldots, X_n}(x_1, x_2, \ldots, x_n)$$

$$= F(x_1, x_2, \ldots, x_n). \tag{1.211}$$

If $\Pr\left[\bigcap_{j=1}^{n}(X_j = x_j)\right]$ is zero except for an enumerable number of sets of values $\{x_{1_i}, x_{2_i}, \ldots, x_{n_i}\}$, and

$$\sum_{i} \Pr\left[\bigcap_{j=1}^{n}(X_j = x_{j_i})\right] = 1,$$

then we have a *discrete joint distribution*. For such distributions

$$\sum_i \text{Pr}\left[\bigcap_{j=1}^{n-1}(X_j = x_j) \cap (X_n = x_{n_i})\right] = \text{Pr}\left[\bigcap_{j=1}^{n-1}(X_j = x_j)\right], \quad (1.212)$$

where the summation is over all values of x_{n_i} for which the probability is not zero.

If $F(x_1, x_2, \ldots, x_n)$ is absolutely continuous, then

$$F(x_1, x_2, \ldots, x_n) = \int_{-\infty}^{x_n} \int_{-\infty}^{x_{n-1}} \cdots \int_{-\infty}^{x_1} f(x_1, x_2, \ldots, x_n) dx_1 \ldots dx_n, \quad (1.213)$$

where $f(x_1, x_2, \ldots, x_n)$ (or strictly $f_{X_1, X_2, \ldots, X_n}(x_1, x_2, \ldots, x_n)$) is the *joint probability density function* of $X_1, X_2, \ldots, X_n$. For a continuous joint distribution

$$\int_{-\infty}^{\infty} f(x_1, x_2, \ldots, x_n) dx_n = f(x_1, x_2, \ldots, x_{n-1}). \quad (1.214)$$

By repeated summation or integration, it is possible in principle to obtain the joint distribution of any subset of $X_1, X_2, \ldots, X_n$; in particular the distributions of each separate X_j can be found. These are called *marginal distributions*.

The *conditional joint distribution* of $X_1, X_2, \ldots, X_r$, given $X_{r+1}, X_{r+2}, \ldots, X_n$ (i.e., the joint distribution of the subset of the first r rv's in the case where particular values have been given to the remaining $(n-r)$ variables), is defined as

$$\text{Pr}\left[\bigcap_{j=1}^{r}(X_j = x_j) \mid \bigcap_{i=r+1}^{n}(X_i = x_i)\right] = \frac{\text{Pr}\left[\bigcap_{j=1}^{n}(X_j = x_j)\right]}{\text{Pr}\left[\bigcap_{j=r+1}^{n}(X_j = x_j)\right]}, \quad (1.215)$$

(provided that $\text{Pr}[\bigcap_{j=r+1}^{n}(X_j = x_j)] > 0$) for discrete distributions, and by the pdf

$$f(x_1, x_2, \ldots, x_r \mid x_{r+1} \ldots, x_n) = \frac{f(x_1, x_2, \ldots, x_n)}{f(x_{r+1} \ldots, x_n)} \quad \text{when } f(x_{r+1}, \ldots, x_n) > 0,$$

$$= 0 \quad \text{when } f(x_{r+1}, \ldots, x_n) = 0,$$

$$(1.216)$$

for continuous distributions. (In (1.213), (1.214), and (1.216), subscripts for F and f have been omitted for convenience).

Usually a distribution depends on one (or more) *parameters*, say θ. When we want to emphasize the dependence of the distribution on the value of θ we write

$$F_X(x) = F_X(x; \theta) = F_X(x \mid \theta) \quad \text{and} \quad f_X(x) = f_X(x; \theta) = f_X(x \mid \theta). \quad (1.217)$$

B3 Bayes' Theorem

Bayesian methods of inference involve the systematic formulation and use of Bayes' theorem. These approaches are distinguished from other statistical approaches in that, prior to obtaining the data, the statistician formulates *degrees of belief* concerning the possible models that may give rise to the data. These degrees of belief are regarded as probabilities.

Suppose that $\{M_1, M_2, \ldots, M_k\}$ is a mutually exclusive and exhaustive set of possible probability models for the experimental situation of interest, and suppose that $\{D_1, D_2, \ldots, D_r\}$ is the set of possible outcomes when the experiment is carried out. Also let

1. $\Pr(M_i)$, $i = 1, \ldots, k$, be the probability that the correct model is M_i prior to learning the outcome of the experiment;
2. $\Pr(D_j)$, $j = 1, \ldots, r$, be the probability that the result of the experiment is the outcome D_j;
3. $\Pr(D_j|M_i)$ be the probability that model M_i will produce the outcome D_j;
4. $\Pr(M_i|D_j)$ be the probability that the model M_i is the correct model given that the experiment has had the outcome D_j.

Then by the definition of conditional probability

$$\Pr(M_i|D_j)\Pr(D_j) = \Pr(M_i \cap D_j) = \Pr(D_j|M_i)\Pr(M_i),$$

and by the Theorem of Total Probability

$$\Pr(D_j) = \sum_i \Pr(D_j|M_i)\Pr(M_i);$$

together these lead to the *discrete form of Bayes' theorem*

$$\Pr(M_i|D_j) = \frac{\Pr(D_j|M_i)\Pr(M_i)}{\sum_i \Pr(D_j|M_i)\Pr(M_i)}, \qquad (1.218)$$

where

$\Pr(M_i)$ is termed the *prior probability* of the model M_i,
$\Pr(D_j|M_i)$ is termed the *likelihood of the outcome D_j under the model M_i*, and
$\Pr(M_i|D_j)$ is termed the *posterior probability* of the model M_i given that the outcome D_j has occurred.

It follows that

$$\frac{\Pr(M_i|D_j)}{\Pr(\overline{M_i}|D_j)} = \frac{\Pr(D_j|M_i)\,\Pr(M_i)}{\Pr(D_j|\overline{M_i})\,\Pr(\overline{M_i})}. \qquad (1.219)$$

Because the ratio $\Pr(A)/\{1 - \Pr(A)\}$ is called the *odds on A*, the discrete form of Bayes' theorem is sometimes rephrased as "posterior odds are equal to the likelihood ratio times the prior odds."

Suppose now that the models do not form an enumerable set but instead are indexed by a parameter θ. Let $p(\theta)$ be the prior probability density of the parameter θ, let $p(x|\theta)$ be the likelihood that the experiment will yield an observed value x for the random variable X given the value of the parameter θ, and let $p(\theta|x)$ be the posterior probability density of θ given that the experiment has yielded the observation x. Then

$$p(\theta|x) = \frac{p(x|\theta)p(\theta)}{\int_\Theta p(x|\theta)p(\theta)d\theta}, \qquad (1.220)$$

where Θ is the set of all possible values of the parameter θ; that is, $\int_\Theta p(\theta)d\theta = 1$. This is the *continuous form of Bayes' theorem*; it is sometimes summarized as "posterior density is proportional to likelihood times prior density."

The resolution of the problem of assigning a prior distribution to the parameter θ by the use of Bayes' postulate caused controversy. *Bayes' postulate* is, in brief, the assumption that if there is no information to the contrary, then all prior probabilities are to be regarded as equal. This is known as the adoption of a *vague (diffuse, uninformative) prior* in contradistinction to an *informative prior* that takes into account positive empirical or theoretical information concerning the distribution of θ.

Helpful expository articles on Bayesian inference are the articles by Smith (1984) and Lindley (1990). Books that are written from a Bayesian standpoint give a fuller account; see, for example, Box and Tiao (1973). The assessment of prior densities is discussed in Hampton, Moore, and Thomas (1973), who also provide an extensive bibliography.

B4 Expected Values

The *expected value* of a mathematical function $g(X_1, X_2, \ldots, X_n)$ of $X_1, X_2, \ldots, X_n$ is defined as

$$E[g(X_1, X_2, \ldots, X_n)] = \sum_i g(x_{1_i}, x_{2_i}, \ldots, x_{n_i}) \Pr\left[\bigcap_{j=1}^n (X_j = x_{j_i})\right] \qquad (1.221)$$

for discrete distributions, and as

$$E[g(X_1, X_2, \ldots, X_n)]$$
$$= \int_{-\infty}^{\infty} \int_{-\infty}^{\infty} \cdots \int_{-\infty}^{\infty} g(x_1, x_2, \ldots, x_n) f(x_1, x_2, \ldots, x_n) dx_1 \ldots dx_n \qquad (1.222)$$

for continuous distributions.

In particular, when $n = 1$,

$$E[g(X)] = \sum_j g(x_j) \Pr[X = x_j] \quad \text{or} \quad \int_{-\infty}^{\infty} g(x)f(x)dx. \qquad (1.223)$$

If K is constant, then

$$E[K] = K,$$

$$E[Kg(X)] = KE[g(X)],$$

and
$$E[g_1(X_1) + g_2(X_2)] = E[g_1(X_1)] + E[g_2(X_2)]. \qquad (1.224)$$

More generally

$$E\left[\sum_{j=1}^{M} K_j g_j(X_1, X_2, \ldots, X_n)\right] = \sum_{j=1}^{M} K_j E[g_j(X_1, X_2, \ldots, X_n)]. \qquad (1.225)$$

These results apply to both discrete and continuous random variables. Conditional expected values are defined similarly, and formulas like (1.221) and (1.222) are valid for them.

The random variables X_1, X_2 are said to be *independent* if, for all real x_1, x_2, the events $(X_1 \le x_1), (X_2 \le x_2)$ are independent. The set $\{X_1, X_2, \ldots, X_n\}$ is a *mutually independent* set of random variables if, for any combination of values $x_1, x_2, \ldots, x_n$ assumed by the rv's $X_1, X_2, \ldots, X_n$, $\Pr[X_1 = x_1, \ldots, X_n = x_n] = \Pr[X_1 = x_1] \ldots \Pr[X_n = x_n]$. In this case

$$E\left[\prod_{j=1}^{k} g_j(X_j)\right] = \prod_{j=1}^{k} E[g_j(X_j)]. \qquad (1.226)$$

B5 Moments and Moment Generating Functions

The expected value of X^r, for r any real number, is termed the rth *uncorrected moment* (alternatively the rth *moment about zero*):

$$\mu_r'(X) = \mu_r' = E[X^r]. \qquad (1.227)$$

The *uncorrected moment generating function* (umgf), $M_X(t)$, if it exists (i.e., is finite), is the expected value $E[e^{tX}]$. If $\varphi(t)$ is the characteristic function of X (see Section B8 of this chapter), then $M(t) = \varphi(-it)$. When $M_X(t)$ exists for some interval $|t| < T$, where $T > 0$, then μ_r' is the coefficient of $t^r/r!$ in the Taylor expansion of $M_X(t)$:

$$M_X(t) = E[e^{tX}] = 1 + \sum_{r \ge 1} \frac{\mu_r' t^r}{r!}, \qquad (1.228)$$

The first uncorrected moment μ_1' (the *mean*) is often written as μ.

The uncorrected moments can also be obtained from the cumulative distribution function $F_X(x)$. If X is a continuous rv, then

$$E[X^r] = \int_0^\infty rx^{r-1}[1 - F_X(x) + (-1)^r F_X(-x)]dx. \qquad (1.229)$$

If X is discrete, taking values $0, 1, \ldots, b$ where b is finite or infinite, then

$$E[X^r] = \sum_{x=0}^{b-1}[(x+1)^r - x^r][1 - F(x)]. \qquad (1.230)$$

From the definition of the moment generating function,

$$M_{X+c}(t) = e^{ct}M_X(t). \qquad (1.231)$$

Moreover, if X_1 and X_2 are independent rv's, then

$$M_{X_1+X_2}(t) = M_{X_1}(t)M_{X_2}(t),$$

and

$$M_{X_1-X_2}(t) = M_{X_1}(t)M_{X_2}(-t). \qquad (1.232)$$

If $X_1, X_2, \ldots, X_n$ are mutually independent rv's, then

$$M_{X_1+X_2+\cdots+X_n}(t) = M_{X_1}(t)M_{X_2}(t)\ldots M_{X_n}(t). \qquad (1.233)$$

The rth moment about a constant a is $E[(X-a)^r]$. When $a = \mu$, we have the rth *moment about the mean* (also called the rth *central moment* or the rth *corrected moment*),

$$\mu_r(X) = \mu_r = E[(X-\mu)^r] = E[(X - E[X])^r]. \qquad (1.234)$$

The *central moment generating function* (cmgf), if it exists, is

$$E[e^{(X-\mu)t}] = e^{-\mu t}M_X(t) = 1 + \sum_{r\geq 1}\mu_r t^r/r! \qquad (1.235)$$

The first central moment μ_1 is always zero. The second central moment μ_2 is called the *variance* of X (written as Var(X)). The positive square root of this quantity is called the *standard deviation* and is often denoted by the symbols $\sigma(X)$ or σ_X. We have $\mu_2 \equiv \sigma_X^2$. The *coefficient of variation* (sometimes abbreviated to C. of V.) is σ_X/μ.

The *median* of a distribution is the value of the variate that divides the total frequency into equal halves. For a continuous distribution this is unique.

For a discrete distribution with $2N+1$ elements, the median is the value of the $(N+1)$th element; when there are $2N$ elements, there is ambiguity, and it is usual to define the median as the average of the Nth and $(N+1)$th elements.

If $f_X(x)$ is a continuous and twice differentiable pdf, then x is a *mode* if $df_X(x)/dx = 0$ and $d^2 f_X(x)/dx^2 < 0$. A discrete distribution has a mode at $X = x$ if $\Pr[X = x-c_1-1] < \Pr[X = x-c_1] \le \ldots \le \Pr[X = x] > \ldots \ge \Pr[X = x+c_2] > \Pr[X = x+c_2+1], 0 \le c_1, 0 \le c_2$. A distribution with only one mode is said to be *unimodal*; otherwise it is *multimodal*. A distribution with support $x \ge 0$ and a peak in frequency at $X = 0$ is sometimes said to have a *half-mode* at $X = 0$ and to be *sesquimodal*. Abouammoh and Mashhour (1981) have given necessary and sufficient conditions for a discrete distribution to be unimodal. Olshen and Savage (1970) have introduced the concept of α-unimodality for continuous distributions. For *discrete α-unimodality* (for discrete distributions), see Abouammoh (1987) and Steutel (1988).

Commonly used indices of the shape of a distribution are the *moment ratios*. The most important are

$$\alpha_3(X) = \sqrt{\beta_1(X)} = \mu_3(\mu_2)^{-3/2} \qquad \text{an index of } skewness, \quad (1.236)$$

$$\alpha_4(X) = \beta_2(X) = \mu_4(\mu_2)^{-2} \qquad \text{an index of } kurtosis, \quad (1.237)$$

and more generally

$$\alpha_r(X) = \mu_r(\mu_2)^{-r/2}. \qquad (1.238)$$

The α and β notations are both in use. Note that these moment ratios have the same value for any linear function $(A + BX)$ with $B > 0$. When $B < 0$, the absolute values are not altered, but ratios of odd order have their signs reversed.

It is often convenient to calculate the central moments μ_r from the uncorrected moments, and, less often, vice versa. Formulas for this involve the binomial coefficients:

$$\mu_r = E[(X - E[X])^r] = \sum_{j=0}^{r} (-1)^r \binom{r}{j} \mu'_{r-j} \mu^j; \qquad (1.239)$$

we refer the reader to Stuart and Ord (1987, p. 73) for further relevant formulas. In particular

$$\mu_2 = \mu'_2 - \mu^2,$$

$$\mu_3 = \mu'_3 - 3\mu'_2\mu + 2\mu^3,$$

$$\mu_4 = \mu'_4 - 4\mu'_3\mu + 6\mu'_2\mu^2 - 3\mu^4. \qquad (1.240)$$

For the inverse calculation

$$\mu_2' = \mu_2 + \mu^2,$$

$$\mu_3' = \mu_3 + 3\mu_2\mu + \mu^3,$$

$$\mu_4' = \mu_4 + 4\mu_3\mu + 6\mu_2\mu^2 + \mu^4. \tag{1.241}$$

The characterization of a distribution via its moment properties has been studied by several authors; see Johnson and Kotz (1990a, b) for a discussion of the methods that have been used and for new results.

Besides the uncorrected and the central moments there are *absolute moments*, defined as the expected values of the absolute values (moduli) of various functions of X. Thus the rth *absolute moment about zero* of X is

$$\nu_r'(X) = E[|X|^r], \tag{1.242}$$

while the rth *absolute central moment* is

$$\nu_r(X) = E[|X - E[X]|^r]. \tag{1.243}$$

If r is even, $\nu_r' = \mu_r'$ and $\nu_r = \mu_r$, but not if r is odd. Whereas $\mu_1 = 0$, in general $\nu_1 > 0$. We call ν_1 the *mean deviation* of X.

When studying discrete distributions, it is often advantageous to use the *factorial moments*. Those most commonly used are the descending factorial moments. The rth *descending factorial moment* of X is the expected value of $X!/(X - r)!$:

$$\mu_{[r]}' = E[X!/(X - r)!]. \tag{1.244}$$

Readers are WARNED that there are other notations in use for the descending factorial moments. In the first edition of this book $\mu_{(r)}$ was used. Patel, Kapadia and Owen (1976) use $\mu_{(r)}'$. The rth *ascending factorial moment* of X is $E[(X + r - 1)!/(X - 1)!]$.

Since $X!/(X - r)! = \sum_{j=0}^r s(r, j)X^j$, where $s(r, j)$ is the Stirling number of the first kind (see Section A3 near the start of this chapter), we find that

$$\mu_{[r]}' = \sum_{j=0}^{r} s(r, j)\mu_j'. \tag{1.245}$$

Thus

$$\mu_{[1]}' = \mu,$$

$$\mu_{[2]}' = \mu_2' - \mu,$$

$$\mu_{[3]}' = \mu_3' - 3\mu_2' + 2\mu,$$

$$\mu'_{[4]} = \mu'_4 - 6\mu'_3 + 11\mu'_2 - 6\mu,$$

$$\mu'_{[5]} = \mu'_5 - 10\mu'_4 + 35\mu'_3 - 50\mu'_2 + 24\mu,$$

$$\mu'_{[6]} = \mu'_6 - 15\mu'_5 + 85\mu'_4 - 225\mu'_3 + 274\mu'_2 - 120\mu. \tag{1.246}$$

Similarly

$$X^r = \sum_{j=0}^{r} \frac{S(r,j)X!}{(X-r)!},$$

where $S(r,j)$ are the Stirling numbers of the second kind (see Section A3), and so

$$\mu'_r = \sum_{j=0}^{r} S(r,j)\mu'_{[j]}. \tag{1.247}$$

Hence

$$\mu = \mu'_{[1]},$$

$$\mu'_2 = \mu'_{[2]} + \mu,$$

$$\mu'_3 = \mu'_{[3]} + 3\mu'_{[2]} + \mu,$$

$$\mu'_4 = \mu'_{[4]} + 6\mu'_{[3]} + 7\mu'_{[2]} + \mu,$$

$$\mu'_5 = \mu'_{[5]} + 10\mu'_{[4]} + 25\mu'_{[3]} + 15\mu'_{[2]} + \mu,$$

$$\mu'_6 = \mu'_{[6]} + 15\mu'_{[5]} + 65\mu'_{[4]} + 90\mu'_{[3]} + 31\mu'_{[2]} + \mu. \tag{1.248}$$

The (descending) *factorial moment generating function* (fmgf), if it exists, is $E[(1 + t)^X]$. The relationship between the factorial moment generating function and the probability generating function enables the probabilities of a discrete distribution to be expressed in terms of its factorial moments; see Section B9 of this chapter.

Finite difference methods for obtaining the moments of a discrete distribution are discussed in a series of papers and letters in *The American Statistician* in the early 1980's; see, in particular, Johnson and Kotz (1981), Chan (1982), and Khatri (1983).

B6   Cumulants and Cumulant Generating Functions

The logarithm of the uncorrected moment generating function of X is the *cumulant generating function* (cgf) of X. If the moment generating function exists, then so does the cumulant generating function. The coefficient of $t^r/r!$ in the Taylor expansion of the cgf is the *r*th *cumulant* of X and is denoted by the symbol $\kappa_r(X)$ or, when no confusion is likely to arise, by κ_r:

$$K_X(t) = \ln M_X(t) = \sum_{r \geq 1} \frac{\kappa_r t^r}{r!}. \tag{1.249}$$

Note that there is no term in t^0 in (1.249).

We have

$$K_{X+a}(t) = at + K_X(t). \tag{1.250}$$

Hence for $r \geq 2$ the coefficients of $t^r/r!$ in $K_{X+a}(t)$ and $K_X(t)$ are the same; that is, the cumulants for $r \geq 2$ are not affected by the addition of a constant to X. For this reason the cumulants have also been called *seminvariants* or *half-invariants*. Putting $a = -\mu$ shows that, for $r \geq 2$, the cumulants κ_r are functions of the central moments. In fact

$$\kappa_1 = \mu,$$
$$\kappa_2 = \mu_2,$$
$$\kappa_3 = \mu_3,$$
$$\kappa_4 = \mu_4 - 3\mu_2^2,$$
$$\kappa_5 = \mu_5 - 10\mu_3\mu_2. \tag{1.251}$$

Let $X_1, X_2, \ldots, X_n$ be independent rv's, and let $X = \sum_1^n X_j$; then, if the relevant functions exist,

$$K_X(t) = \sum_{j=1}^{n} K_{X_j}(t). \tag{1.252}$$

It follows from (1.252) that

$$\kappa_r\left(\sum_{j=1}^{n} X_j\right) = \sum_{j=1}^{n} \kappa_r(X_j) \qquad \text{for all } r; \tag{1.253}$$

that is, the cumulant of a sum equals the sum of the cumulants, which makes the name "cumulant" appropriate.

The logarithm of the (descending) factorial moment generating function is called the *factorial cumulant generating function* (fcgf). The coefficient of $t^r/r!$ in the Taylor expansion of this function is the rth *factorial cumulant*, $\kappa_{[r]}$:

$$\ln G(1+t) = \sum_{r \geq 1} \frac{\kappa_{[r]} t^r}{r!}. \tag{1.254}$$

Formulas connecting $\{\kappa_r\}$ and $\{\kappa_{[r]}\}$ parallel those connecting $\{\mu_r'\}$ and $\{\mu_{[r]}'\}$:

$$\kappa_1 = \kappa_{[1]} = \mu,$$

$$\kappa_2 = \kappa_{[2]} + \mu,$$

$$\kappa_3 = \kappa_{[3]} + 3\kappa_{[2]} + \mu, \tag{1.255}$$

$$\kappa_4 = \kappa_{[4]} + 6\kappa_{[3]} + 7\kappa_{[2]} + \mu, \quad \text{etc.} \tag{1.256}$$

Douglas (1980) has given a very full account of the relationships between the various types of moments and cumulants; see also Stuart and Ord (1987).

A sampling distribution arises as the distribution of some function of observations, taken over all possible samples from a particular distribution according to a specified sampling scheme. The moment properties of a sampling distribution can be expressed in terms of symmetric functions of the observations, known as k-statistics; these were introduced by Fisher (1929). The expected value of the univariate k-statistic of order r is the rth cumulant; see Stuart and Ord (1987, Ch. 12).

B7   Joint Moments and Cumulants

Moments of joint distributions, that is, quantities like $E\left[\prod_{j=1}^{n} X_j^{a_j}\right]$, are called *product moments (about zero)* and are denoted by $\mu'_{a_1 a_2 \ldots a_n}$. Quantities like

$$E[\prod_{j=1}^{n}(X_j - E[X_j])^{a_j}] = \mu_{a_1 a_2 \ldots a_n} \tag{1.257}$$

are called *central product moments* (sometimes *central mixed moments*).

The central product moment

$$\mu_{11} = E[(X_j - E[X_j])(X_{j'} - E[X_{j'}])] \tag{1.258}$$

is called the *covariance* of X_j and $X_{j'}$ and is denoted by $\text{Cov}(X_j, X_{j'})$. The *correlation* between X_j and $X_{j'}$ is defined as

$$\rho(X_j X_{j'}) = \rho_{jj'} = \frac{\text{Cov}(X_j, X_{j'})}{[\text{Var}(X_j)\text{Var}(X_{j'})]^{1/2}}. \tag{1.259}$$

(This is also sometimes written as $\text{Corr}(X_j X_{j'})$.) It can be shown that $-1 \le \rho_{jj'} \le 1$. If X_j and $X_{j'}$ are mutually independent, then $\text{Cov}(X_j X_{j'}) = 0 = \rho_{jj'}$; the converse is not necessarily true.

The *joint moment generating function* of $X_1, X_2, \ldots, X_n$ is defined as a function of n generating variables $t_1, t_2, \ldots, t_n$:

$$M_{X_1,\ldots,X_n}(t_1, t_2, \ldots, t_n) = M(t_1, t_2, \ldots, t_n) = E\left[\exp\sum_{j=1}^{n} t_j X_j\right]. \tag{1.260}$$

The *joint central moment generating function* is

$$E\left[\exp\left(\sum_{j=1}^{n} t_j(X_j - E[X_j])\right)\right] = \exp\left[-\sum_{j=1}^{n} t_j E[X_j]\right] M(t_1, t_2, \dots, t_n).$$

(1.261)

The *joint cumulant generating function* is $\ln M_{X_1 \dots X_n}(t_1, t_2, \dots, t_n)$. Use of these generating functions is similar to that for the single variable functions.

The *regression function* of a rv X on m other random variables, $X_1, X_2, \dots, X_m$, is defined as

$$E[X | X_1, X_2, \dots, X_m];$$

(1.262)

it is an important tool for the prediction of X from $X_1, X_2, \dots, X_m$. If (1.262) is a linear function of $X_1, X_2, \dots, X_m$, then the regression is called *linear* (or *multiple linear*). The variance of the conditional distribution of X given $X_1, X_2, \dots, X_m$ is called the *scedasticity*. If $\text{Var}(X | X_1, X_2, \dots, X_m)$ does not depend on $X_1, X_2, \dots, X_m$, then the conditional distribution is said to be *homoscedastic*.

Given that X_1 and X_2 are random variables, then their joint distribution is determined by the distribution of X_1 together with the conditional distribution of X_2, given X_1. There has been much research during the past two decades on characterizations based on regression properties; see, for instance, Korwar (1975) and Papageorgiou (1985). Kotz and Johnson (1990) have provided a good review concerning discrete distributions.

B8 Characteristic Functions

The *characteristic function* (cf) of a continuous distribution is defined as

$$\varphi(t) = E[e^{itX}] = \int_{-\infty}^{\infty} e^{itx} dF(x),$$

(1.263)

where $i = \sqrt{-1}$ and t is real. It is a complex-valued function. For a discrete distribution on the nonnegative integers, it is defined as

$$\varphi(t) = E[e^{itX}] = \sum_{j=0}^{\infty} e^{ijt} \Pr[X = j].$$

(1.264)

The characteristic function has great theoretical importance, particularly for continuous distributions. It is uniquely determined by the cumulative distribution function and exists for all distributions. It satisfies (1) $\varphi(0) = 1$, (2) $|\varphi(t)| \leq 1$, and (3) $\varphi(-t) = \overline{\varphi(t)}$, where the overline denotes the complex conjugate.

If the distribution with cdf $F(x)$ has finite moments μ'_j up to order n, then

$$\mu'_j = i^j \varphi^{(j)}(0), \qquad 1 \le j \le n, \qquad (1.265)$$

where $\varphi^{(j)}(0)$ is the jth derivative of $\varphi(t)$ evaluated at $t = 0$.

The cf uniquely determines the probability density function of a continuous distribution; we have

$$f(x) = \frac{1}{2\pi} \int_{-\infty}^{\infty} e^{-itx} \varphi(t) dt. \qquad (1.266)$$

Gauss (1900) called this *Ein Schönes Theorem der Wahrscheinlichkeitsrechnung*.

The corresponding inversion formula for discrete distributions on the nonnegative integers is

$$\Pr[X = x] = \frac{1}{2\pi} \int_{-\pi}^{\pi} e^{-itx} \varphi(t) dt. \qquad (1.267)$$

Lukacs (1970) gave further inversion formulas for continuous and for discrete distributions.

If X_1 and X_2 are independent rv's with cf's $\varphi_1(t)$ and $\varphi_2(t)$, respectively, then the cf of their sum $X_1 + X_2$ is the product of their cf's $\varphi_1(t)\varphi_2(t)$. Moreover the cf of their difference $X_1 - X_2$ is $\varphi_1(t)\varphi_2(-t)$.

Under very general conditions the cf for a limiting distribution is the limiting cf. If $\lim_{j \to \infty} \varphi_{X_j}(t) = \varphi_X(t)$, where $\varphi_X(t)$ is the cf of a random variable with cumulative distribution function F_X, then $\lim_{j \to \infty} F_{X_j}(x) = F_X(x)$.

A cf $\varphi(t)$ is said to be *infinitely divisible* if $\varphi(t) = \{\varphi_n(t)\}^n$ for all positive integer n, where $\varphi_n(t)$ is itself a cf. A cf is said to be *stable* if $\varphi(a_1 t)e^{itb_1}\varphi(a_2 t)e^{itb_2} = \varphi(a_3 t)e^{itb_3}$, where $a_i > 0$ for $i = 1, 2, 3$.

Key references concerning characteristic functions are Lukacs (1970, 1983) and Laha (1982).

B9 Probability Generating Functions

Suppose we have a nonnegative discrete random variable X which can, without loss of generality, be supposed to have nonzero probabilities only at nonnegative integer values. Let

$$P_j = \Pr[X = j], \qquad j = 0, 1, \ldots. \qquad (1.268)$$

If the distribution is proper, then $\sum_{j=0} P_j = 1$, and hence $\sum_{j=0} P_j z^j$ converges for $|z| \le 1$. (This is also true when the distribution is not proper since then $0 < \sum_{j=0} P_j < 1$. However, we will be concerned only with proper distributions.)

The *probability generating function* (pgf) of the distribution with probability mass function (1.268) (or equivalently of the rv X) is defined as

$$G(z) = \sum_{j=0}^{\infty} P_j z^j = E[Z^X].$$ (1.269)

Although it would be logical to use the notation $G_X(z)$ for the pgf of X, we will in general suppress the suffix when it is clearly understood.

The pgf is closely related to the cf; we have

$$\varphi_X(t) = E[e^{itX}] = G(e^{it}).$$ (1.270)

The pgf is defined by the probabilities; the uniqueness of a power series expansion implies that the pgf in turn defines the probabilities. We find that

$$P_j = \left[\frac{1}{j!}\frac{d^j G(z)}{dz^j}\right]_{z=0}, \qquad j = 0, 1, \ldots.$$ (1.271)

The rth moment, if it exists, is

$$\mu'_r = \sum_{j=0}^{\infty} j^r P_j$$ (1.272)

$$= \left[\frac{d^r G(e^t)}{dt^r}\right]_{t=0}, \qquad r = 1, 2, \ldots.$$ (1.273)

The factorial moment generating function (if it exists) is given by

$$E[(1+t)^X] = G(1+t).$$ (1.274)

When the pgf is known, therefore, successive differentiation of the pgf enables the (descending) factorial moments to be obtained in a straightforward manner as the coefficients of $t^r/r!$ in

$$G(1+t) = 1 + \sum_{r \geq 1} \frac{\mu'_{[r]} t^r}{r!};$$ (1.275)

that is, the rth (descending) factorial moment is

$$\mu'_{[r]} = \sum_{j=r}^{\infty} \frac{j!}{(j-r)!} P_j = \left[\frac{d^r G(z)}{dz^r}\right]_{z=1}$$ (1.276)

$$= \left[\frac{d^r G(1+t)}{dt^r}\right]_{t=0}.$$ (1.277)

Table 1.1   Relationships between the Probability Generating Function and Other Generating Functions

Probability generating function (pgf)	$G(z)$
Characteristic function (cf)	$\varphi(t) = G(e^{it})$
Uncorrected moment generating function	$M(t) = G(e^{t})$
Corrected moment generating function	$e^{-\mu t} M(t) = e^{-\mu t} G(e^{t})$
Factorial moment generating function	$G(1 + t)$
Cumulant generating function	$K(t) = \ln G(e^{t})$
Factorial cumulant generating function	$\ln G(1 + t)$

(The moments of a discrete distribution can be derived from its factorial moments as

$$\mu = \mu'_{[1]}, \qquad \mu_2 = \mu'_{[2]} + \mu - \mu^2, \quad \text{etc.;}$$

see Section B5.)

The following relationships hold between the probabilities and the factorial moments in the case of a discrete distribution:

$$\Pr[X = x] = \sum_{j \geq x} (-1)^{x+j} \binom{j}{x} \frac{\mu'_{[j]}}{j!} = \sum_{r \geq 0} (-1)^{r} \frac{\mu'_{[x+r]}}{x! r!} \tag{1.278}$$

[Fréchet (1939, 1943)];

$$\sum_{i \geq x} \Pr[X = i] = \sum_{j \geq x} (-1)^{x+j} \binom{j-1}{x-1} \frac{\mu'_{[j]}}{j!} \tag{1.279}$$

[Laurent (1965)].

The relationships between the pgf and various moment generating functions are summarized in Table 1.1.

If X_1 and X_2 are two independent rv's with pgf's $G_1(z)$ and $G_2(z)$, then the distribution of their sum $X = X_1 + X_2$ has the pgf $G(z) = E[z^X] = E[z^{X_1}]E[z^{X_2}] = G_1(z)G_2(z)$. This is called the *convolution* of the two distributions. Let A, B and C be the names of the distributions of X_1, X_2 and X; then we write C $\sim$ A*B.

More generally let $X_1, X_2, \ldots, X_n$ be mutually independent random variables with pgf's $G_1(z), G_2(z), \ldots, G_n(z)$, respectively. Then the pgf of $X = \sum_{i=1}^{n} X_j$ is

$$G_X(z) = \prod_{j=1}^{n} G_j(z). \tag{1.280}$$

The pgf for $(X_i - X_j)$ is $G_i(z)G_j(1/z)$.

The *joint probability generating function* of n discrete variables $X_1, X_2, \ldots, X_n$ is

$$G(z_1, z_2, \ldots, z_n) = E\left[\prod_{j=1}^{n} z_j^{X_j}\right], \qquad (1.281)$$

where

$$\Pr\left[\bigcap_{j=1}^{n}(X_j = a_j)\right] = P_{a_1, a_2, \ldots, a_n}, \qquad a_j = 0, 1, 2, \ldots. \qquad (1.282)$$

Let $r = \sum_{j=1}^{n} r_j$. Then the factorial moments of the distribution are given by

$$\mu'_{[r_1, r_2, \ldots, r_n]} = \left[\frac{\partial^r G(z_1, z_2, \ldots, z_n)}{\partial z_1^{r_1} \partial z_2^{r_2} \ldots \partial z_n^{r_n}}\right]_{z_1 = z_2 = \ldots = z_n = 1}. \qquad (1.283)$$

An historical account of the use of probability generating functions in discrete distribution theory has been given by Seal (1949b).

B10 Order Statistics

If $X_{(j)}$ is defined to be equal to the jth smallest of $X_1, \ldots, X_n$, then $X_{(1)}$, $X_{(2)}, \ldots, X_{(n)}$ are called the *order statistics* corresponding to $X_1, X_2, \ldots, X_n$. Evidently $X_{(1)} \leq X_{(2)} \leq \ldots \leq X_{(n)}$. For a continuous distribution the probability of a *tie* (two equal values) is zero, and therefore the definition is unambiguous. Ties may, however, occur given a discrete distribution; unambiguity can be achieved here also, provided that we interpret "jth smallest value" to mean "not more than $(j-1)$ smaller values <u>and</u> not more than $(n-j)$ larger values."

Among the n values of $X_1, X_2, \ldots, X_n$, the largest is $X_{(n)}$, and the smallest is $X_{(1)}$. Their difference, $w = X_{(n)} - X_{(1)}$, is called the *range*. If n is odd, the "middle" value $X_{((n+1)/2)}$ is called the *median*; see Section B5. When n is even, the median is not uniquely defined; often the observations are grouped, and reference is made to the *median class*. The closely related concepts of *hinges* and *fences* play a central role in exploratory data analysis (EDA); we refer the reader to Tukey (1977) and Emerson and Hoaglin (1983).

David's (1981) book provides an encyclopedic coverage of properties, statistical techniques, characterizations, and applications relating to order statistics from both continuous and discrete distributions. Harter (1988) has written an excellent introduction to the subject of order statistics; his paper gives definitions, examines the history and the importance of order statistics, and provides some key references. Balakrishnan (1986) extended recurrence relations for single and product moments of order statistics from the continuous to the discrete case.

There has been much work on characterizations of distributions using properties of their order statistics; see, for instance, Arnold and Meeden (1975), Shah and Kabe (1981), Hwang and Lin (1984), Khan and Ali (1987), and Lin (1987). Nagaraja (1990) has given a good account of work on order statistics for discrete distributions, particularly concerning characterizations of the geometric distribution, and extreme order statistics. Estimation methods based on order statistics are discussed in the book by Balakrishnan and Cohen (1991). Harter's *Chronological Annotated Bibliography of Order Statistics*, published by American Sciences Press, Columbus, Ohio, in eight volumes from 1983 onward, is the definitive bibliography.

B11 Truncation and Censoring

If values of the random variables $X_1, X_2, \ldots, X_n$ in a given region $\overline{R}$ are excluded, then the joint cdf of the variables is

$$
F(x_1, x_2, \ldots, x_n | R) = \Pr\left[\bigcap_{j=1}^{n} (X_j \le x_j) | (X_1, \ldots, X_n) \subset R\right]
$$

$$
= \frac{\Pr\left[\bigcap_{j=1}^{n} (X_j \le x_j) \cap ((X_1, \ldots, X_n) \subset R)\right]}{\Pr[(X_1, X_2, \ldots, X_n) \subset R]}. \quad (1.284)
$$

R is the complement of $\overline{R}$, and it comprises all the points that are not truncated. The distribution given by (1.284) is called a *truncated distribution*. All the quantities on the right-hand side of (1.284) can be calculated from the (unconditional) joint cdf $F(x_1, x_2, \ldots, x_n)$.

We shall usually be concerned with truncated distributions of single variables, for which R is a finite or infinite interval. If R is a finite interval with end points a and b inside the range of values taken by X, the distribution is *doubly truncated* (or *left-and-right truncated*, or *truncated below and above*); a and b are the *truncation points*.

If R consists of all values greater than a, then the distribution is said to be *truncated from below* or *left-truncated*; if R consists of all values less than b, then the distribution is said to be *truncated from above* or *right-truncated*. (The same terms are also used when R includes values equal to a or to b, as the case may be.)

If X' is a random variable having a distribution formed by doubly truncating the distribution of a continuous random variable X, then the pdf of X', in terms of the pdf and cdf of X, is

$$
f_{X'}(x') = \frac{f_X(x')}{F_X(b) - F_X(a)}, \qquad a \le x' \le b. \quad (1.285)
$$

A distinction needs to be made between truncation of a distribution and truncation of a sample. Truncation of a distribution occurs when a range of possible variate values either is ignored or is impossible to observe.

Truncation of a sample is commonly called *censoring*. Sometimes censoring is with respect to a fixed variate value; for instance, in a survival study it may be impossible within a limited time span to ascertain the length of survival of all the patients. When the existence of observations outside a certain range is known, but their exact value is unknown, the form of censoring is known as *type I censoring*.

When a predetermined number of order statistics are omitted from a sample the form of censoring is known as *type II*. If the ℓ smallest values $X_{(1)}$, $\ldots, X_{(\ell)}$ are omitted, it is *censoring from below* or *left censoring*; when the m largest values are omitted, it is *censoring from above* or *right censoring*. If both sets of order statistics are omitted, we have *double censoring*.

The term "truncation" is used in a different sense in sequential analysis, where it refers to the imposition of a cutoff point leading to cessation of the sequential sampling process before a decision has been reached.

B12   Mixture Distributions

A *mixture of distributions* is a superimposition of distributions with different functional forms or different parameters, in specified proportions.

Suppose that

$$\{F_j(x_1, x_2, \ldots, x_n)\}, \qquad j = 0, 1, 2, \ldots, m,$$

represents a set of different (proper) cumulative distribution functions, where m is finite or infinite. Suppose also that $a_j \geq 0$, $\sum_{j=0}^{m} a_j = 1$. Then

$$F(x_1, x_2, \ldots, x_n) = \sum_{j=0}^{m} a_j F_j(x_1, x_2, \ldots, x_n) \tag{1.286}$$

is a proper cumulative distribution function. This mixture of the distributions $\{F_j\}$ is *finite* or *infinite* according as m is finite or infinite.

For many of the mixture distributions in this book, the distributions to be mixed all have cumulative distribution functions of the same functional form, but are dependent on some parameter Θ. If Θ itself has a discrete distribution with pmf $\Pr[\Theta = \theta_j] = p_j$, $j = 0, 1, \ldots$, then the resultant mixture has the cdf

$$\sum_{j \geq 0} p_j F(x_1, x_2, \ldots, x_n | \theta_j).$$

If Θ has a continuous distribution with cdf $H_\Theta(\theta)$, then the resultant mixture has cdf

$$\int F(x_1,\ldots,x_n|\theta)\,dH_\Theta(\theta),$$

where integration is over all values of θ. In either case (discrete or continuous distribution of Θ) we have

$$F(x_1,x_2,\ldots,x_n) = E_\Theta[F_j(x_1,x_2,\ldots,x_n|\theta)], \qquad (1.287)$$

where the expectation is with respect to Θ. We call $F(x_1,x_2,\ldots,x_n)$ the *mixture distribution* and say that the distribution of Θ is the *mixing distribution*.

More generally the distribution of $X_1,X_2,\ldots,X_n$ may depend on several parameters $\Theta_1,\ldots,\Theta_k,\Theta_{k+1},\ldots,\Theta_m$, where $\Theta_1,\ldots,\Theta_k$ vary and $\Theta_{k+1},\ldots,\Theta_m$ are constant. The mixture distribution then has the cdf

$$F(x_1,x_2,\ldots,x_n|\theta_{k+1},\ldots,\theta_m) = E_{\Theta_1,\ldots,\Theta_k}[F(x_1,x_2,\ldots,x_n|\theta_1,\ldots,\theta_m)]. \quad (1.288)$$

Note that the parameters $\theta_1,\ldots,\theta_k$ do not appear in the mixture distribution because they have been summed out (for a discrete mixture) or integrated out (for a continuous distribution). The parameters $\Theta_{k+1},\ldots,\Theta_m$ have not been eliminated in this way.

B13   Variance of a Function

Given the moments of a rv X, suppose that we wish to obtain the moments of a mathematical function of X, that is, of $Y = h(X)$.

If exact expressions can be obtained and are convenient to use, this should of course be done. However, in some cases it may be necessary to use approximate methods. One approximate method is to expand $h(X)$ as a Taylor series about $E[X]$:

$$Y = h(E[X]) + (X - E[X])h'(E[X]) + (X - E[X])^2\frac{h''(E[X])}{2!} + \cdots. \quad (1.289)$$

Then, taking expected values of both sides of (1.289),

$$E[Y] = h(E[X]) + \text{Var}(X)\frac{h''(E[X])}{2} + R \qquad (1.290)$$

$$\doteq h(E[X]) + \text{Var}(X)\frac{h''(E[X])}{2}. \qquad (1.291)$$

Also

$$\{Y - h(E[X])\}^2 \doteq \{(X - E[X])h'(E[X])\}^2,$$

whence

$$\text{Var}(Y) \doteq \{h'(E[X])\}^2\text{Var}(X). \qquad (1.292)$$

This method of approximation has been used widely under a number of different names, for example, the *delta method*, the *method of statistical differentials*, and the *propagation of error*. The method assumes that the expected value of the remainder term R in (1.290) is small and that the higher-order central moments do not become large; otherwise the outcome may be very unreliable. The method will usually be more reliable for small values of $\text{Var}(X)$.

Equation (1.292) can be made the basis for an approximate *variance-stabilizing transformation*. If $\text{Var}(X)$ is a function $g(E[X])$ of $E[X]$, then $\text{Var}(Y)$ might be expected to be more nearly constant if $[h'(E[X])]^2 g(E[X])$ is a constant.

This will be so if

$$h(X) \propto \int^X \frac{dt}{[g(t)]^{1/2}}. \tag{1.293}$$

This suggests the use of $Y = h(X)$, with $h(X)$ satisfying (1.293), as a variance-stabilizing transformation. Often such transformations are also effective as *normalizing transformations* in that the distribution of Y is nearer to normality than that of X; see Chapter 12.

When $Y = X^a$, the method gives

$$E[X^a] \doteqdot \mu^a \left[1 + \frac{a(a-1)\sigma^2}{2\mu^2} \right],$$

$$\text{Var}(X^a) \doteqdot \mu^{2a-2} a^2 \sigma^2, \tag{1.294}$$

where $\mu = E[X]$ and $\sigma^2 = \text{Var}(X)$. The coefficient of variation of X^a is therefore very approximately $|a|(\sigma/\mu)$.

There are exact methods for obtaining the moments of a product of two rv's. For the quotient of two nonnegative rv's the delta method gives

$$E[X_1/X_2] \doteqdot \frac{\xi_1}{\xi_2} \left[1 + \frac{\text{Var}(X_2)}{\xi_2^2} - \frac{\text{Cov}(X_1, X_2)}{\xi_1 \xi_2} \right], \tag{1.295}$$

$$\text{Var}(X_1/X_2) \doteqdot \frac{\xi_1^2}{\xi_2^2} \left[\frac{\text{Var}(X_1)}{\xi_1^2} - \frac{2\text{Cov}(X_1, X_2)}{\xi_1 \xi_2} + \frac{\text{Var}(X_2)}{\xi_2^2} \right], \tag{1.296}$$

where ξ_1 and ξ_2 are the expected values of X_1 and X_2, respectively.

For further discussion and use of the delta method see Stuart and Ord (1987, Ch. 10).

B14 Geometrical Concepts

Occasionally multidimensional geometrical concepts will be used. These are obtained by extending familiar concepts from two and three dimensional

Cartesian coordinate geometry to higher dimensions. They are introduced merely to aid comprehension; they are not essential and can be replaced by analytical treatment.

When we say that the point with coordinates $(X_1, X_2, \ldots, X_n)$ is in a particular region, we mean that the coordinates $(X_1, X_2, \ldots, X_n)$ satisfy a certain set of inequalities. The two statements are equivalent.

We shall use the following concepts:

Hypersphere, meaning a region bounded by the *surface*

$$\sum_{j=1}^{n}(X_j - \xi_j)^2 = r^2,$$

where $\xi_1, \xi_2, \ldots, \xi_n$, and r are constants.

Inside the hypersphere, meaning

$$\sum_{j=1}^{n}(X_j - \xi_j)^2 \leq r^2.$$

Straight line, meaning

$$X_j - a_j = w\ell_j.$$

The direction cosine of the line relative to the axis of X_i, is denoted by ℓ_j, where $\sum \ell_j^2 = 1$.

Hyperellipsoid, meaning the region bounded by the surface

$$\sum_{j=1}^{n}\lambda_j(X_j - \xi_j)^2 = r^2,$$

where $\lambda_1, \lambda_2, \ldots, \lambda_n > 0, r > 0$.

Inside the ellipsoid, meaning

$$\sum_{j=1}^{n}\lambda_j(X_j - \xi_j)^2 < r^2.$$

Other concepts such as *hyperplane, simplex, and cylinder* will very seldom be used.

B15 Inference

In the past twenty years an immense amount of research has been devoted to statistical inference, both to theoretical developments and to practical aspects of inferential procedures. Computer-intensive methods are now feasible. The books by Barnett (1973) and Cox and Hinkley (1974) remain well worth reading for their lucid discussions of the many approaches to inference.

Readers of this book will occasionally meet references to results from goodness-of-fit, hypothesis testing, and decision theory. Relevant definitions and formulas are not included here. There are many well-written books on these topics, including those by D'Agostino and Stephens (1986), Rayner and Best (1989) (*goodness-of-fit*), DeGroot (1970), Cox and Hinkley (1974), Lehmann (1986) (*hypothesis testing*), and Berger (1985) (*Bayesian decision theory*). Kocherlakota and Kocherlakota (1986) concentrated on goodness-of-fit tests for discrete distributions.

Bayesian methods of inference continue to receive special attention and are widely applied. Many aspects of Bayesian statistics are discussed in the volume edited by Bernado, DeGroot, Lindley, and Smith (1988). Maritz and Lwin (1988) were concerned with empirical Bayes methods.

In this preliminary chapter we sketch only briefly some of the basic concepts and methods in estimation theory.

A cumulative distribution function that depends on the values of a finite number of quantities $\theta_1, \theta_2, \ldots, \theta_m$ (called *parameters*) is written $F(X_1, X_2, \ldots, X_n | \theta_1, \theta_2, \ldots, \theta_m)$. Often we want to estimate the values of these parameters. This is done using functions of the random variables $T_j \equiv T_j(X_1, X_2, \ldots, X_n)$ called *statistics*. When a statistic T_j is used to estimate a parameter θ_j, it is called an *estimator* of θ_j. An *estimate* is a realized value of an estimator for a particular sample of data.

A statistic T_j is said to be an *unbiased estimator* of the parameter θ_j if $E[T_j] = \theta_j$. If $E[T_j] \neq \theta_j$, the estimator is *biased*.

All distributions with finite means and variances possess unbiased estimators of their means and variances, namely

$$\bar{x} = \sum_{i=1}^{n} \frac{x_i}{n} \quad \text{and} \quad s^2 = \sum_{i=1}^{n} \frac{(x_i - \bar{x})^2}{n-1}, \tag{1.297}$$

respectively, where n is the sample size.

If T_j and T_j^* are both unbiased estimators of the same parameter θ_j, then any weighted average $wT_j + (1 - w)T_j^*$ is also an unbiased estimator of θ_j. An estimator is said to be *asymptotically unbiased* if $\lim_{n \to \infty} E[T_j] = \theta_j$.

The *relative efficiency* of two unbiased estimators is measured by the inverse ratio of their variances, that is to say, $\text{Var}(T_j^*)/\text{Var}(T_j)$ measures the efficiency of T_j relative to T_j^*. Comparisons of the efficiencies of biased estimators are often made on the basis of their mean squared errors. The *mean squared error* is defined to be

$$E[(T_j - \theta_j)^2] = \text{Var}(T_j) + \{E(T_j) - \theta_j\}^2. \tag{1.298}$$

If a measure of overall efficiency is required when several parameters $\theta_1, \theta_2, \ldots, \theta_m$ are being estimated by the unbiased estimators $T_1, T_2, \ldots, T_m$, respectively, then the *generalized variance* may be used. This is a determinant in which the element in the *j*th row and *j'*th column is $\text{Cov}(T_j, T_{j'}|\theta_1, \theta_2, \ldots, \theta_m)$. (Comparisons of the generalized variances of biased estimators are only meaningful if the biases are small enough to be neglected.)

A *consistent estimator* is one for which

$$\lim_{n \to \infty} \text{Pr}[|T_j - \theta_j| \geq c] = 0 \tag{1.299}$$

for all positive *c*. If T_j is unbiased, then it will also be consistent provided that

$$\lim_{n \to \infty} \text{Var}(T_j) = 0.$$

Consistency is an asymptotic property.

A *minimum variance unbiased estimator* (MVUE), T_j of θ_j, is an unbiased estimator of θ_j with a smaller variance than any other estimator of θ_j. If T_j is an unbiased estimator of θ_j, then the Cramér-Rao theorem states that the variance of T_j satisfies the *Cramér-Rao inequality*

$$\text{Var}(T_j) \geq \frac{1}{n\, E[(\partial \ln f(x)/\partial \theta_j)^2]}. \tag{1.300}$$

An MVUE may or may not not, however, attain this lower bound.

The *efficiency of an unbiased estimator* is the ratio of its variance to the Cramér-Rao lower bound. An estimator is called an *efficient estimator* if this ratio is unity; it is said to be an *asymptotically efficient estimator* if this ratio tends to unity as the sample size becomes large.

A *sufficient estimator* is one that summarizes from the sample of observations all possible information concerning the parameter; that is, no other statistic formed from the observations provides any more information. Such a statistic will exist if and only if the likelihood (see below) can be factorized into two parts, one depending only on the statistic and the parameters, and the other depending only on the sample observations. If an unbiased estimator has a variance equal to the Cramér-Rao lower bound, then it must be a sufficient estimator.

A family of distributions dependent on a vector of parameters Θ is said to be *complete* if $E_\Theta[h(T)] = 0$ for all values of the parameters implies that $\text{Pr}[h(T) = 0] = 1$ for all Θ, where $h(T)$ is a function of the observations and $E_\Theta[\cdot]$ denotes expectation with respect to the distribution with parameters Θ.

Stuart and Ord (1987) have given a careful and very full account of estimation principles, as well as details concerning the major types of estimation procedures.

The method of *maximum likelihood* is widely advocated. If observed values of $X_1, X_2, \ldots, X_n$ are $x_1, x_2, \ldots, x_n$, then their likelihood is

$$L(x_1, x_2, \ldots, x_n) = \Pr\left[\bigcap_{j=1}^{n}(X_j = x_j | \theta_1, \theta_2, \ldots, \theta_m)\right] \qquad (1.301)$$

for discrete distributions, and

$$L(x_1, x_2, \ldots, x_n) = f(x_1, x_2, \ldots, x_n | \theta_1, \theta_2, \ldots, \theta_m) \qquad (1.302)$$

for continuous distributions. In either case the values $\hat{\theta}_1 = T_1$, $\hat{\theta}_2 = T_2$, $\ldots$, $\hat{\theta}_m = T_m$ that maximize the likelihood are called *maximum likelihood estimators*. (Note that the $\hat{\theta}_j$'s are random variables.) If $X_1, X_2, \ldots, X_n$ are mutually independent and have identical distributions, then under rather general conditions

1. $\lim_{n \to \infty} E[\hat{\theta}_j | \theta_1, \theta_2, \ldots, \theta_m] = \theta_j, \qquad j = 1, 2, \ldots, m.$

2. Asymptotic estimates of the variances and covariances of the $\hat{\theta}_j$'s are given by the corresponding elements in the inverse of the information matrix evaluated at the maximum-likelihood values.

A *maximum likelihood estimator* (MLE) may or may not be unique, may or may not be unbiased, and need not be consistent. Nevertheless, MLE's possess certain attractive properties. Under certain mild regularity conditions they are asymptotically MVUE's and are also asymptotically normally distributed. The ML estimation method yields sufficient estimators whenever they exist. Also, if $\hat{\theta}$ is a MLE of θ and if $h(\cdot)$ is a function with a single-valued inverse, then the MLE of $h(\theta)$ is $h(\hat{\theta})$.

Maximizing the likelihood can usually be achieved by solving the equations

$$\frac{\partial L(x_1, x_2, \ldots, x_n | \theta_1, \theta_2, \ldots, \theta_m)}{\partial \theta_i} = 0, \qquad i = 1, 2, \ldots, m; \qquad (1.303)$$

these equations are called the *ML equations*. They are often intractable and require iteration (e.g., by the Newton-Raphson method) for their solution. The almost universal accessibility to cheap computing power has led to the development of a number of computer routines for ML estimation. Also the log-likelihood is usually negative, and so maximizing the likelihood is usually equivalent to minimizing the absolute value of the log-likelihood; this can be achieved by means of a computer optimization routine. The leading computer packages supply ML and suitable function optimization routines.

Reparameterization, where feasible, so that the new parameters are orthogonal, has been advocated by a number of authors; see for example, Cox and Reid (1987), Ross (1990), and Willmot (1988b).

The *method of moments* usually requires less onerous calculations than ML estimation, although the method cannot be guaranteed to give explicit estimators. The method is based on equating the first k *uncorrected sample moments* about zero, $m_r' = n^{-1} \sum_{j=1}^{n} x_j^r$, $r = 1, \ldots, k$, to the corresponding theoretical expressions for μ_r', where k is the number of unknown parameters. If preferred, the first k *central sample moments* $m_r = n^{-1} \sum_{j=1}^{n} (x_j - \bar{x})^r$, $r = 1, \ldots, k$, can be equated to the corresponding expressions for the central moments μ_r, or the first k factorial sample moments $m_{[r]}'$, $r = 1, \ldots, k$, can be set equal to $\mu_{[r]}'$; the three procedures give identical estimators. The equation obtained by equating the *sample mean* to the theoretical mean is called the *first-moment equation*.

The higher sample moments generally have large variances, however. This has led to the use of methods based on the quantiles (for continuous distributions) and on the mean and lowest observed frequencies (for discrete distributions). An alternative approach is to solve equations that are approximations to the ML equations. This approach has been discussed by Kemp (1986) for discrete distributions.

When the method of moments or a similar method leads to explicit estimators, they can be used to provide initial estimates for ML estimation.

The desirable properties of ML estimators have led to the development of a number of variants of ML estimation, especially for situations where a family of distributions is to be fitted to data, and hence there are a large number of parameters to be estimated.

> *Generalized ML estimators* exist and have near optimal properties in cases where ML estimators do not exist. In most other cases, though, the two methods give identical results; see, e.g., Weiss (1983).
>
> *Modified ML estimation* is used particularly for censored data. In many situations where ML estimation requires iteration, modified ML estimation gives explicit estimators; see, e.g., Tiku (1989).
>
> *Penalized ML estimation* is used in curve estimation. It involves a sacrifice of efficiency in order to achieve smooth fits; see, e.g., Silverman (1985).
>
> *Partial ML estimation* was introduced by Cox (1975) for analysing regression models involving explanatory variables, as a way to reduce the number of nuisance parameters; see, e.g., Kay (1985).
>
> *Conditional, marginal,* and *profile* likelihood procedures are methods somewhat similar to partial ML estimation; they are collectively described as *pseudo-likelihood* methods. They have probabilistic interpretations; see, e.g., Kalbfleisch (1986) and Barndorff-Neilsen (1991).
>
> *Quasi-likelihood* estimation is a nonlinear weighted least-squares method for generalized linear models. It is based on families of linear-exponential distributions. Only second-moment assumptions are used, and hence linear-exponentiality does not necessarily hold. The method yields equations similar to (1.303) and can be useful for situations with overdispersion; see, e.g., McCullagh (1991).

These methods lead to point estimates. Sometimes interval estimates are required. These are obtained by assigning a range of values, with a lower bound and an upper bound, within which it is asserted that the true value lies. In the *confidence interval* approach, the value of the parameter is regarded as fixed, and the limits for the interval are regarded as random variables. For the *fiducial limits* approach the parameter is treated as having a "fiducial probability" distribution which determines the limits. Robinson's (1982) introductory article on confidence intervals and regions gives useful references concerning the relationship between these two approaches, and also concerning links with Bayesian inference.

C COMPUTER GENERATION OF UNIVARIATE DISCRETE RANDOM VARIABLES

C1 General Comments

The methods of generating random variables that are discussed in the following sections all depend on an infinite sequence of random numbers $\{U_i\}$ uniformly distributed on $[0, 1]$. A suitable sequence is customarily generated by a computer, using a *pseudo-random number algorithm*, that is to say, an algorithm that generates a deterministic stream of numbers that appears to have the same relevant statistical properties as a sequence of truly random numbers.

Most of the earlier algorithms, such as the middle-square method, have long been abandoned as inadequate. In 1992 linear congruential generators (particularly multiplicative generators) are by far the most widely used. The use of shift register and generalized feedback shift register generators (which are based on linear recurrence modulo 2) is increasing. Interest is also growing in lagged-Fibonacci and in nonlinear congruential generators. The *period of a generator* is the length of the sequence of numbers that it produces before it starts to repeat itself. Much research is driven by the need for generators with longer periods than those currently in use. This need arises from the increasing speed of computers and the development of parallel processing. Currently a good introduction to different types of uniform random number generators is Lewis and Orav (1989); however, research is going ahead very fast indeed and information may rapidly become obsolete.

It cannot be stressed too highly that one should only use algorithms generating sequences that have passed a number of stringent tests, such as tests for independence between successive members of a sequence, tests for lengths of runs of increasing (or decreasing) values, permutation tests, tests for uniformity, and tests concerning pairs and k-tuples. To achieve effective randomness, some workers have recommended additional devices such as shuffling and the simultaneous use of two different generators.

The correct and efficient implementation of a generation algorithm on a specific computer requires considerable care and expertise. Readers are strongly advised to use sequences of random numbers from subroutine libraries such as IMSL and NAG that have had very thorough testing and extensive critical use. At the time of writing the undesirable nature of the built-in random number generators in many microcomputers is well recognized; see, e.g., Ripley (1987). This comment applies a fortiori to many pocket calculators.

In the remaining sections of this chapter we consider, first, the use of general methods for creating random variables from random numbers. Such methods include inversion of the cumulative distribution function and the alias method; in principle they can be applied to any univariate discrete distribution. Second, we document some generation algorithms that are specific to certain distributions; this material should be read in conjunction with the appropriate chapter later in this book. Special attention is given to the binomial and Poisson distributions because of their central role in discrete distribution theory.

Useful references concerning the computer generation of random variables are the books by Bratley, Fox and Schrage, (1987), Dagpunar (1988), Morgan (1984), and Ripley (1987). Devroye (1986) gives an encyclopedic coverage of the mathematical methodology of non-uniform random variate generation. A helpful general survey article is that by Boswell et al. (1993).

C2 Distributionally Non-Specific Generation Procedures

In recent years a number of very fast general methods for generating discrete random variables have been developed. Such methods are distribution non-specific, in the sense that they require tables relating to the actual values of the probability mass function, for example, a table of the cumulative distribution function, rather than a knowledge of the structural properties of the distribution. The set-up time for such tables usually takes very much longer than the generation time for a single output random variate, given that there are already suitable tables in computer memory. Fast general methods are therefore usually the method of choice when large numbers of random variables are required from a particular distribution with constant parameters. There follow brief descriptions of two such methods that are widely favoured and are suitable for discrete distributions.

The *inversion method* can be used for the generation of both continuous and discrete distributions. A uniform [0, 1] variate is generated and is transformed into a variate from the *target distribution* (the distribution of interest) by the use of a monotone transformation of the uniform cdf to the target cdf; this procedure is called inversion of the (target) cdf. The *table look-up method* is an adaptation of the inversion procedure that is particularly suitable for a discrete distribution and is very widely used. A set-up routine is required in which the cumulative probabilities for the target distribution are calculated correctly and are stored in computer memory. The cdf for a discrete distribution is of course a step-function, with step-jumps occurring at successive variate values and step-heights equal to successive probabilities. A uniform $[0,1]$ variate is generated, and the appropriate step-height interval within which it lies is sought using a search procedure. The variate value corresponding to this step-jump is then "returned" (i.e., made available) as a variate from the target distribution. The use of one of the many sophisticated search procedures that are now available can make this a very fast method.

Walker's (1974, 1977) *alias method* is based on the following theorem:

Every discrete distribution with probabilities $p_0, p_1, \cdots, p_{K-1}$ can be expressed as an equiprobable mixture of K two-point distributions.

First the probabilities for the target distribution must be calculated. Next a set-up procedure for constructing the K equiprobable mixtures is required; the information concerning these mixtures can be put into two arrays of size K by an ingenious method described in the books referenced in the preceding section. One uniform variate on $[0,1]$ is then used to choose a component in the equiprobable mixture, while a second uniform on $[0,1]$ decides which of the two points for that component should be returned as a target variable. Once the set-up algorithm has been implemented (this may take a non-trivial amount of computer time), the generation of large numbers of variates from the target distribution is very rapid. For implementation details for these two methods see, for example, Chen and Asau (1974) (for the indexed table look-up method) and Kronmal and Peterson (1979) (for the alias method).

If the order in which the variates are generated is immaterial, then Kemp and Kemp's (1987) *frequency-table method* provides an even faster approach in the fixed parameter situation. The method generates a sample of values in the form of a frequency table and is useful, for example, for studying the properties of estimators; the method does not attempt to provide a sequence of uncorrelated variate values.

Distributionally non-specific methods are not, however, suitable when the parameters of a distribution change from call to call to the computer generator. Consequently many different *distribution-specific generation methods* have been devised. Their underlying feature is the exploitation of certain structural properties of the distribution. They fall into two categories: methods that depend on a property specific to a particular distribution such as a characterization, and methods that can be adapted for a number of different distributions. Composition methods, acceptance-rejection methods, and acceptance-complement methods belong to the latter category.

In the *composition method* the target distribution is decomposed into two or more additive component distributions, with most of the probability mass assigned to distributions that are very easy to generate. At least two uniforms are needed for each output target variate—one uniform to select a component distribution, plus one or more uniforms to generate from the chosen component distribution.

The basic *acceptance-rejection method* also requires at least two uniforms for each output target variate. The choice of a suitable *envelope distribution* is critical—this should be very similar in shape to the target distribution and should also be very easy to generate. An excess number of variates is generated from the envelope distribution. For each generated envelope variate a second uniform variate is used in order to decide whether or not to retain it; an appropriate decision rule converts the envelope distribution into the target distribution. The length of computer time spent on making the decision can usually be markedly reduced by the use of a well-chosen simple initial screening rule that avoids the need for the complicated decision rule most of the time. The *acceptance-complement method* combines features of both of these methods.

Readers are referred to the texts mentioned at the end of the previous section for further details concerning such methods.

C3 Binomial Random Variables

If large numbers of random variables are required from a binomial distribution with constant parameters, then the ease of computation of its probabilities, coupled with the bounded support for the distribution, make non-specific methods very attractive.

However, when successive calls to the generator are for random binomial variates with changing parameters, then distribution-specific methods become important. A slow but very simple method is to simulate the flip of a biased coin n times and count the number of successes. When $p = 0.5$, it suffices to count the number of ones in a random uniformly distributed computer word of n bits. For $p \neq 0.5$, the method requires n uniforms per generated binomial variate, making it very slow (recycling uniform random numbers in order to reduce the number required is not generally recommended). An ingenious improvement (the beta, or median, method) was devised by Relles (1972); see also Ahrens and Dieter (1974).

Devroye (1986) gives two interesting waiting time methods based on the following features of the binomial distribution: First, let $G_1, G_2, \ldots$ be identically and independently distributed geometric random variables with parameter p, and let X be the smallest integer such that $\sum_{i=1}^{X+1} G_i > n$; then X is binomial with parameters n, p. Second, let $E_1, E_2, \ldots$ be identically and independently distributed exponential random variables , and let X be the smallest integer such that

$$\sum_{i=1}^{X+1} \frac{e_i}{n - i + 1} > -\ln(1 - p);$$

then X is binomial with parameters n, p. Both methods can be decidedly slow because of their requirement for very many uniform random numbers; however, as for the coin-flip method, their computer programs are very short.

Kemp's (1986) algorithm, based on inversion of the cdf by unstored search from the mode, competes favorably with the Ahrens and Dieter (1974) algorithm. Other unstored search programs are discussed in Kemp's paper.

Acceptance-rejection using a Poisson envelope was proposed by Fishman (1979). At the time of writing the fastest algorithm is Kachitvichyanukul and Schmeiser's (1988) algorithm BTPE. This is an intricate composition-acceptance-rejection algorithm. A recent simpler, but not quite so fast, algorithm is that of Stadlober (1991); this uses the ratio of two uniforms.

C4 Poisson Random Variables

The importance of the Poisson distribution in discrete distribution theory has led to major effort in the design of good and fast generators for Poisson variates. For a fixed value of the parameter θ a fast general method has much to commend it; see Section C2 of this chapter.

The generation of many small samples from different Poisson distributions requires the use of varying values of θ. Moreover Poisson variates with varying θ are useful for generating many other distributions, in particular those that can be interpreted as mixed Poisson distributions and those that are Poisson-stopped sum distributions. In such situations general methods involving the creation of tables for a specific value of θ are no longer suitable.

Generators for varying θ rapidly become outmoded. Here we mention only the two simplest methods and the current, 1992, state of the art. Devroye (1986) gives details of many other algorithms. Characterizations of the Poisson distribution form an important feature of certain methods.

The *exponential-gap* method was widely used in the past. It exploits the relationship between the Poisson distribution and exponential interarrival times in a homogeneous Poisson point process (see Chapter 4, Sections 2 and 8) by counting numbers of exponential variates. However, the use of this algorithm is no longer recommended, even for relatively low values of θ, because it is so dependent on the quality and speed of the requisite uniform generator.

Search-from-the-origin is a method based on inversion of the cdf. A single uniform random number U is compared first with $\Pr[X = 0]$, next with $\Pr[X = 0] + \Pr[X = 1]$, and so on, until U is exceeded by the cumulative probability $\sum_{j=0}^{k} \Pr[X = j]$. (The cumulative probabilities are calculated as required, using the current value of θ and the recursion formula $\Pr[X = j] = \Pr[X = j - 1]\theta/j$.) The value of k is then delivered as the generated Poisson variate. There are two versions of this algorithm, one using *build-up* of the probabilities, and another using *chop-down* of U; see, e.g., Kemp and Kemp

(1991). This method is useful for very low values of θ ($\theta < 5$ say), but it is very time-consuming for values of θ greater than, say, 20.

Kemp and Kemp's (1990) composition-search algorithm KPLOW is designed for low values of θ, ($\theta < 30$ say). The algorithm decomposes the Poisson distribution into three components,

$$e^{\theta(z-1)} = \sum_{j=0}^{2r-1} \pi_j z^j + \sum_{j=0}^{2r-1} (p_j - \pi_j) z^j + \sum_{j \geq 2r} p_j z^j, \qquad (1.304)$$

where r is the mode and π_j is a tight lower bound for p_j such that

$$\frac{\pi_j}{p_j} = \frac{\pi_r}{p_r} = \alpha < 1 \qquad \text{for } j = 0, 1, \ldots$$

(the determination of α is discussed in the paper). The first component is searched bilaterally from the mode r. In the rare situation where this fails to return a variate, it is necessary to search one or both of the other two components. Computer timings indicate that this algorithm is noticeably faster than search-from-the-origin.

Schmeiser and Kachitvichyanukul (1981) gave a very clear account of their composition-rejection algorithm PTPE and discussed various earlier methods. Ahrens and Dieter (1982) called their algorithm KPOISS; it is an acceptance-complement algorithm based for the most part on the normal distribution. Schmeiser and Kachitichyanukul (1981) gave timings for both algorithms, showing that they are both much faster than earlier algorithms.

Kemp and Kemp's (1991) algorithm KEMPOIS is based on Kemp's (1988b) modal approximations. It employs a unilateral search from the mode, with squeezing. For approximately $10 < \theta < 700$ it is faster than either PTPE or KPOISS.

C5 Negative Binomial Random Variables

Computer generation of random variables from a geometric distribution is very straightforward. One method is to exploit the waiting-time property. Consider a stream of uniform random variables. Then geometric random variables can be generated by counting the number of uniforms needed to obtain a uniform less than p (the number of failures needed to obtain the first success). Devroye (1986) considers that "for $p \geq 1/3$ the method is probably difficult to beat in any programming environment."

A second way to generate a geometric random variable G is by analytic inversion of the cdf. Let U be a uniform random variable. Then $G = [\ln(U)/\ln(1-p)]$ (where $[\cdot]$ denotes the integer part). If a stream of exponential random variables is available, then discretizing the exponential (E) gives $G = [-E/\ln(1-p)]$. Devroye notes (in an exercise) that there may be an accuracy problem for low values of p, and that one way that this may be overcome is via the expansion

$$\ln(1 - p) = \frac{2}{c} \left(1 + \frac{1}{3c^2} + \frac{1}{5c^4} + \cdots \right),$$

where $c = 1 - 2/p$ (c is negative).

The negative binomial with an integer parameter $k = N$ can be generated as the sum of N geometric random variables. Except for low values of N (say $N = 2, 3, 4$), this method cannot be advocated as it requires many uniforms for a single output negative binomial random variable. This argument applies a fortiori to the use of the sum of a Poisson number of logarithmic random variables.

The method generally recommended for generating negative binomial random variables with changing parameters is to generate Poisson random variables with random parameters drawn from a gamma distribution; see, e.g., Algorithm NB3 in Fishman (1978). For fixed parameters the use of a fast general method, such as indexed table look-up, alias, or frequency table, is recommended.

Three simple stochastic models that can be used to generate *correlated* negative binomial random variables have been described by Sim and Lee (1989). Two of their methods are based on the autoregressive scheme of the first-order Markovian process. The third uses the Poisson process from a first-order autoregressive gamma sequence.

C6 Hypergeometric Random Variables

Computer generation of classical hypergeometric random variables has been discussed in detail by Kachitvichyanukul and Schmeiser (1985). When the parameters remain constant, the alias method of Walker (1977) and Kronmal and Peterson (1979) is a good choice. Kachitvichyanukul and Schmeiser gave an appropriate program with safeguards to avoid underflow.

The simplest of all algorithms needs a very fast uniform generator. It is based on a sequence of trials in which the probability of success depends on the number of previous successes; that is, it uses the model of finite sampling without replacement. The number of successes in a fixed number of trials is counted. Fishman (1973) and McGrath and Irving (1973) have given details.

Fishman's (1978) algorithm requires a search of the cdf. Kachitvichyanukul and Schmeiser (1985) suggest ways in which the speed of the method can be improved.

Devroye (1986) indicated in an exercise how hypergeometric rv's can be generated by rejection from a binomial envelope distribution.

For large-scale simulations with changing parameters the current, 1992, state of the art is to use Kachitvichyanukul and Schmeiser's algorithm H2PE. Generation is by acceptance/rejection from an envelope consisting of a uniform with exponential tails. The execution time is bounded over the range of parameter values for which the algorithm is intended, that is, over the range $M - \max(0, n - N + Np) \geq 10$, where M is the mode of the distribution. Kachitvichyanukul and Schmeiser recommended inversion of the cumulative distribution function for other parameter values.

While the negative hypergeometric distribution could be generated by inverse sampling without replacement for a fixed number of successes, it would seem preferable to generate it using its beta-binomial model with the good extant beta and binomial generators.

Similarly a hypergeometric Type IV distribution could be generated as a beta mixture of negative binomials; see Chapter 6, Section 2.3.

C7 Logarithmic Random Variables

Efficient ways of generating random variables from a logarithmic distribution with a fixed value of θ, by searching a stored table of the cumulative probabilities, are discussed in Devroye (1986). Devroye also (in his Chapter 3) describes other very efficient general methods, suitable for fixed θ, such as the alias and acceptance-complement methods.

When θ varies, a distribution-specific method is required. Kemp et al. (1979) gave a build-up unstored-search procedure similar to Fishman's method for the Poisson distribution; Kemp (1981b) presented an algorithm for an unstored chop-down search procedure. Very high values of θ ($\theta > 0.99$) are commonplace in ecological applications of the logarithmic distribution. With this in mind, Kemp (1981b) also gave a generation algorithm based on the mixed shifted-geometric model for the distribution (see Chapter 7, Section 2). She showed that two variants of this mixed shifted-geometric method make enormous savings in computer time for very high values of θ.

Devroye (1986) gave details of all these algorithms and also presented two other, seemingly less attractive, algorithms. Shanthikumar's (1985) *discrete thinning method* (for distributions with hazard rate bounded below unity) and *dynamic thinning method* (for decreasing failure rate distributions) are interesting; Devroye showed how the discrete thinning method can be used for the logarithmic distribution. He also gave an algorithm based on rejection from an exponential distribution.

C H A P T E R 2

Families of Discrete Distributions

1 LATTICE DISTRIBUTIONS

In Chapter 1, Section B2, the class of discrete distributions was defined as having a cdf that is a step function with only a denumerable number of steps. According to this definition the class has considerable variety. For example the distribution defined by

$$\Pr[X = r/s] = (e - 1)^2 (e^{r+s} - 1)^{-1}, \qquad (2.1)$$

where r and s are relatively prime positive integers, is a discrete distribution. Its expected value is $e[1 - \ln(e - 1)]$ and all positive moments are finite. However, it is not possible to write down the values of r/s of X in ascending order of magnitude, though it is possible to enumerate them according to the values of r and s.

Most of the discrete distributions used in statistics belong to a much narrower class, the *lattice distributions*. In these distributions the intervals between the values of any one random variable for which there are nonzero probabilities are all integral multiples of one quantity (which depends on the random variable). Points with these coordinates thus form a lattice. By an appropriate linear transformation it can be arranged that all variables take values that are integers. For most of the discrete distributions that we will discuss, the values taken by the random variable cannot be negative.

There are a number of ways of classifying nonnegative lattice distributions. Classification into broad classes helps us to understand the multitude of available distributions and the relationships between them. The mere number of distributions in a class is not, in itself, a measure of importance. What is important is the inclusion of as wide a variety as possible within a specific class, with emphasis on the existence of special properties that apply to all members of the class. This is helpful when constructing models, and for the derivation of methods of analysis by analogy with known techniques for closely related distributions.

69

2 POWER SERIES DISTRIBUTIONS

2.1 Generalized Power Series Distributions

The very broad class of power series distributions includes many of the most common distributions. Membership of the class confers a number of special properties.

A distribution is said to be a *power series distribution* (PSD) if its probability mass function can be written in the form

$$\Pr[X = x] = \frac{a_x \theta^x}{\eta(\theta)}, \qquad x = 0, 1, \dots, \qquad \theta > 0, \tag{2.2}$$

where $a_j \geq 0$, and $\eta(\theta) = \sum_{x=0}^{\infty} a_x \theta^x$. In (2.2) θ is the *power parameter* of the distribution, and $\eta(\cdot)$ is the *series function*.

Continuous and discrete frequency functions of the form $f(x) = a_x \eta(x, \theta)/A(\theta)$, where the a_x depend only on x and not on θ, $\eta(x, \theta) = e^{-\alpha x} = \theta^x$, and $A(\theta)$ is the sum or integral of $a_x \eta(x, \theta)$ over the sample space, were studied by Tweedie (1947, 1965), who called such distributions "Laplacian"; a more modern term is "linear exponential". A power series distribution is a *discrete linear exponential distribution*.

The term "power series distribution" is generally credited to Noack (1950). Kosambi (1949) and Noack showed that many important discrete distributions belong to this class. They also investigated its moment and cumulant properties. The definition (2.2) was extended to multivariate distributions by Khatri (1959).

Patil (1961, 1962a) allowed the set of values that the variate can take to be any nonempty enumerable set S of nonnegative integers. Patil called this extended class *generalized power series distributions* (GPSD). Estimation and other properties of GPS distributions have been explored further in Patil (1962a, b, c, 1964a).

Among distributions of major importance belonging to this class are the binomial (Chapter 3), Poisson (Chapter 4), negative binomial (Chapter 5), and logarithmic (Chapter 7) distributions, and their related multivariate distributions. Furthermore, if a generalized power series distribution is truncated, then the truncated version is also a generalized power series distribution. Also the sum of n mutually independent rv's, each having the same generalized power series distribution, has a distribution of the same class, with series function $[\eta(\theta)]^n$. The probability generating function for (2.2) is

$$G(z) = \frac{\eta(\theta z)}{\eta(\theta)}. \tag{2.3}$$

Kosambi and Noack obtained the following results:

for the binomial distribution $\quad\quad\quad\quad\quad \eta(\theta) = (1 + \theta)^n$, n a positive integer,

for the Poisson distribution $\quad\quad\quad\quad\quad \eta(\theta) = e^\theta$,

for the negative binomial distribution $\quad \eta(\theta) = (1 - \theta)^{-k}$, $k > 0$,

and for the logarithmic distribution $\quad\quad \eta(\theta) = -\ln(1 - \theta)$.

The moment generating function for a power series distribution is

$$G(e^t) = \frac{\eta(\theta e^t)}{\eta(\theta)}, \tag{2.4}$$

and hence the mean and variance are

$$E[X] = \mu = \mu(\theta) = \theta \frac{d}{d\theta}[\ln \eta(\theta)] \tag{2.5}$$

$$\text{Var}(X) = \mu_2 = \theta \frac{d\mu}{d\theta} = \theta^2 \frac{d^2}{d\theta^2}[\ln \eta(\theta)] + \mu; \tag{2.6}$$

see Kosambi (1949). Kosambi showed that $\mu = \mu_2$ characterizes the Poisson distribution among power series distributions; see also Patil (1962a). The higher moments are

$$E[X^{r+1}] = \mu'_{r+1} = \theta \frac{d}{d\theta}[\mu'_r] + \mu\mu'_r, \tag{2.7}$$

$$E[(X - \mu)^{r+1}] = \mu_{r+1} = \theta \frac{d}{d\theta}[\mu_r] + r\mu_2\mu_{r-1} \tag{2.8}$$

[Craig (1934); Noack (1950)].

The factorial moment generating function is

$$G(1 + t) = \frac{\eta(\theta + \theta t)}{\eta(\theta)}, \tag{2.9}$$

and so the rth factorial moment is

$$\mu'_{[r]} = \frac{\theta^r}{\eta(\theta)} \frac{d^r}{d\theta^r}[\eta(\theta)] \tag{2.10}$$

and $\quad\quad\quad\quad \mu'_{[r+1]} = (\mu - r)\mu'_{[r]} + \theta \frac{d}{d\theta}[\mu'_{[r]}] \tag{2.11}$

[Patil (1961)].

From (2.4) and (2.9) the cumulant and factorial cumulant generating functions are

$$\ln G(e^t) = \ln \left[\frac{\eta(\theta e^t)}{\eta(\theta)} \right] \tag{2.12}$$

and
$$\ln G(1+t) = \ln\left[\frac{\eta(\theta+\theta t)}{\eta(\theta)}\right].$$
(2.13)

The cumulants satisfy the equation

$$\kappa_{r+1} = \theta\frac{d\kappa_r}{d\theta}$$
(2.14)

and the factorial cumulants satisfy

$$\kappa_{[r+1]} = \theta\frac{d\kappa_{[r]}}{d\theta} - r\kappa_{[r]}$$
(2.15)

[Khatri (1959)].

From these recurrence relations it can be seen that if $\kappa_1 = \kappa'_{[1]} = \mu$ is known *as a function of* θ, then all the cumulants (and so all the moments) are determined from this one function. Alternatively, the variance or a higher cumulant might be given as a function of θ [Tweedie and Veevers (1968)]. An even more remarkable result was obtained by Khatri (1959). According to this, knowledge of the first two moments (or, equivalently, the first two factorial moments, cumulants, or factorial cumulants) as functions of a parameter ω is sufficient to determine the whole distribution, given that it is a power series distribution.

For suppose that $\kappa_1 = y_1(\omega)$ and $\kappa_2 = y_2(\omega)$. Then, from (2.14),

$$y_2(\omega) = \theta\frac{dy_1}{d\theta} = \theta\frac{dy_1}{d\omega}\frac{d\omega}{d\theta}$$
(2.16)

and from (2.5),

$$y_1(\omega) = \theta\frac{d[\ln \eta(\theta)]}{d\theta};$$
(2.17)

that is,

$$\frac{d\ln \theta}{d\omega} = \frac{1}{y_2(\omega)}\frac{dy_1}{d\omega},$$
(2.18)

$$\frac{d[\ln \eta(\theta)]}{d\omega} = \frac{y_1(\omega)}{y_2(\omega)}\frac{dy_1}{d\omega}.$$
(2.19)

Apart from multiplicative constants this pair of equations determines θ and $\eta(\theta)$ as functions of ω. Khatri pointed out that no other pair of consecutive cumulants possesses this property.

Integrals for the tail probabilities have been obtained by Joshi (1974, 1975), who showed that given a family of power series distributions with support $\{0, 1, 2, \ldots, n\}$ or $\{0, 1, 2, \ldots\}$ and power parameter θ where $0 < \theta < \rho$, there exists a family of absolutely continuous distributions with support $(0, \rho)$ such that the lower tail probabilities of the power series distributions are equal to the upper tail probabilities of the family of continuous distributions. Three well-known examples are the relationships of

1. Poisson to gamma tail probabilities,
2. binomial to beta-of-the-second-kind tail probabilities,
3. negative binomial to beta tail probabilities.

Consider now the estimation properties of power series distributions.
Suppose that $x_1, x_2, \ldots, x_N$ are N independent observations from the same power series distribution, defined by (2.2). The likelihood function is then

$$\prod_{j=1}^{N} \left\{ \frac{a_{x_j} \theta^{x_j}}{\eta(\theta)} \right\} = \theta^T [\eta(\theta)]^{-N} \prod_{j=1}^{N} a_{x_j}, \tag{2.20}$$

where $T = \sum_{j=1}^{N} x_j$, and the maximum likelihood estimator, $\hat{\theta}$, of θ satisfies the first-moment equation

$$\hat{\theta}^{-1} \sum_{j=1}^{N} x_j - \frac{N \eta'(\hat{\theta})}{\eta(\hat{\theta})} = 0. \tag{2.21}$$

The ML estimator is thus a function of $T = \sum_{j=1}^{N} x_j$ and does not depend on the x_j's in any other way.

The conditional distribution of $X_1, \ldots, X_N$, given $\sum_{j=1}^{N} X_j$, does not depend on θ; that is, $T = \sum_{j=1}^{N} x_j$ is sufficient for θ. It is also complete, since it has a generalized power series distribution and the equation

$$\sum_{j=0}^{\infty} A_j [\eta(T)\theta]^j = 0 \qquad \text{for all } \theta \tag{2.22}$$

implies that $\eta(T) = 0$. Results for the asymptotic variance and for the bias in $\hat{\theta}$ were obtained by Patil (1962a, b).

A minimum variance unbiased estimator (MVUE) exists if and only if the power series distribution has support $\{a_1 + a_2 x\}$ where $x = 0, 1, \ldots$, and a_1 and a_2 are nonnegative integers. It follows that whilst the logarithmic and left-truncated Poisson distributions are MVU estimable, the binomial and right-truncated Poisson are not; see Roy and Mitra (1957) and Tate and Goen (1958). When it exists the MVUE of θ is

$$\theta^* = \frac{b \left(\sum_{j=1}^{N} x_j - 1 \right)}{b \left(\sum_{j=1}^{N} x_j \right)} \qquad \text{if } \sum_{j=1}^{N} x_j > 0,$$

$$\theta^* = 0 \qquad \text{if} \ \sum_{j=1}^{N} x_j = 0, \qquad (2.23)$$

where $b(k)$ is the coefficient of θ^k in the expansion of $[\eta(\theta)]^N$ [Roy and Mitra (1957)]. Properties associated with minimum variance unbiased estimators for power series distributions have been obtained by Patil (1963) and Patil and Joshi (1970). Estimators based on ratios and moments were studied in Patil (1962c). Abdul-Razak and Patil (1986) have investigated Bayesian inference for power series distributions. Eideh and Ahmed (1989) have investigated tests based on the Kullback-Leibler information measure, for a one-parameter power series distribution.

Generalized power series distributions (GPSD) *with two parameters*, that is, with

$$a_x = a_x(\lambda) \quad \text{and} \quad \eta(\theta) = \eta(\theta, \lambda) = \sum_x a_x(\lambda)\theta^x,$$

were studied in Patil (1964a) and were found to possess many of the properties of one-parameter power series distributions; see also Khatri (1959) and Douglas (1980).

Siromoney (1964) defined the *general Dirichlet series distribution* by the formula

$$\Pr[X = x] = \frac{a_x e^{-\lambda_x \psi}}{\xi(\psi)}, \qquad x = 0, 1, 2, \dots, \qquad (2.24)$$

where $\xi(\psi) = \sum_{j=0}^{\infty} a_j e^{-\lambda_j \psi}$ (supposing the series to converge). Putting $\lambda_x = x$ gives a power series distribution.

2.2 Modified Power Series Distributions

Modified power series distributions (MPSD) form an extension of the class of generalized power series distributions. They were created by R. C. Gupta (1974) by replacing θ^x in (2.2) by $\{u(\theta)\}^x$; that is, their probability mass functions have the form

$$\Pr[X = x] = \frac{a_x \{u(\theta)\}^x}{\eta(\theta)}, \qquad (2.25)$$

where the support of X is $0, 1, 2, \dots$, or a subset thereof, $a_x \geq 0$, and $u(\theta)$ and $\eta(\theta)$ are positive, finite, and differentiable. MPSD, like GPSD, are linear exponential.

The series function is now

$$\eta(\theta) = \sum_x a_x \{u(\theta)\}^x. \qquad (2.26)$$

Differentiating with respect to θ gives

$$\frac{d\,\eta(\theta)}{d\theta} = \sum_x x a_x \{u(\theta)\}^{x-1} \frac{d\,u(\theta)}{d\theta}, \tag{2.27}$$

and so

$$E[X] = \mu = \mu(\theta) = \sum_x \frac{x a_x \{u(\theta)\}^x}{\eta(\theta)}$$

$$= \frac{u(\theta)}{\eta(\theta)} \frac{\eta'(\theta)}{u'(\theta)}. \tag{2.28}$$

In a similar manner, differentiating

$$\mu'_r = \sum_x x^r \frac{a_x \{u(\theta)\}^x}{\eta(\theta)} \tag{2.29}$$

with respect to θ yields

$$\mu'_{r+1} = \frac{u(\theta)}{u'(\theta)} \frac{d\mu'_r}{d\theta} + \mu\mu'_r. \tag{2.30}$$

Also
$$\mu_{r+1} = \frac{u(\theta)}{u'(\theta)} \frac{d\mu_r}{d\theta} + r\mu_2 \mu_{r-1}, \tag{2.31}$$

and
$$\mu'_{[r+1]} = \frac{u(\theta)}{u'(\theta)} \frac{d\mu'_{[r]}}{d\theta} + (\mu - r)\mu'_{[r]}. \tag{2.32}$$

In particular

$$\mathrm{Var}(X) = \frac{u(\theta)}{u'(\theta)} \frac{d\mu}{d\theta}. \tag{2.33}$$

R. C. Gupta (1974) also obtained a relationship between the cumulants and the moments of a modified power series distribution. Further moment properties have been obtained by R. C. Gupta (1975b, 1984), P. L. Gupta (1982), Gupta and Singh (1981), and Kumar and Consul (1979). Tripathi, Gupta, and Gupta (1986) have studied the incomplete moments of MPSD.

The class of MPSD reduces to the class of GPSD when $u(\theta) = \theta$; the above formulas become the corresponding ones in Section 2.1.

Steyn's (1980, 1984) *two–parameter* and *multiparameter power series distributions* (see Chapter 11, Section 17) are important extensions of MPSD.

When $u(\theta)$ is invertible (e.g., by a Lagrangian expansion; see Section 5.2 below) and θ can be expressed as $\theta = \psi(u(\theta))$, the probability generating function can be written as

$$G(z) = \frac{\eta(\psi(u(\theta))z)}{\eta(\psi(u(\theta)))}. \tag{2.34}$$

Maximum–likelihood estimation for MPSD has been researched by Gupta (1975a), who showed that the ML estimate of θ is given by

$$\mu(\hat{\theta}) = \bar{x}. \tag{2.35}$$

He obtained general expressions for the bias and asymptotic variance of $\hat{\theta}$ and showed that it is unbiased only in the case of the Poisson distribution.

Minimum variance unbiased estimation has been studied when the support of the distribution is known by Jain and Gupta (1973), Gupta (1977), Jani (1978b), Jani and Shah (1979a, b), and Kumar and Consul (1980). It has been studied when the support is unknown by Jani (1977, 1978a, b), Patel and Jani (1977), and Kumar and Consul (1980). MVUE for the probability mass function has been investigated in Gupta and Singh (1982). Famoye and Consul (1989) have studied confidence intervals for modified power series distributions.

A number of characterizations have been obtained; for example, Jani (1978b) has shown that a discrete distribution is MPSD if and only if its cumulants satisfy the recurrence relation

$$\kappa_{r+1} = \frac{u(\theta)}{u'(\theta)} \frac{d\kappa_r}{d\theta}. \tag{2.36}$$

An integral expression for the tail probabilities of a MPSD in terms of absolutely continuous distributions has been derived by Jani and Shah (1979a). Gupta and Tripathi (1985) have given a number of other references to work on MPSD.

Consul (1990c) has studied the subclass of MPSD for which $u(\theta) = \mu$ (the mean of the distribution); the subclass contains several well–known distributions, such as the binomial, Poisson, negative binomial, and Lagrangian Poisson distributions. When $\mu = \theta$, (2.33) becomes

$$\text{Var}(X) = \mu_2 = \frac{u(\theta)}{u'(\theta)} = \frac{u(\mu)}{u'(\mu)}; \tag{2.37}$$

(2.30) to (2.32) and (2.36) simplify on setting $u(\mu)/u'(\mu) = \mu_2$.

Consul has given several characterizations for this subclass. From (2.28) and (2.37),

$$\eta(\mu) = \exp\left[\int \frac{\mu \, d\mu}{\mu_2}\right] \tag{2.38}$$

and

$$u(\mu) = \exp\left[\int \frac{d\mu}{\mu_2}\right]. \tag{2.39}$$

Hence, for specified support, the variance μ_2 as a function of the mean completely determines the distribution within this subclass (the constants induced by the integrations in (2.38) and (2.39) can be shown not to affect the resultant distribution).

Kosambi (1949) and Patil (1962a) have proved that equality of the mean and variance within the family of power series distributions characterizes the Poisson distribution; Consul (1990c), nevertheless, put forward a possible counterexample. Gokhale (1980) strengthened the mean–variance result by proving that within PSD, $\mu_2 = m(1 - m/c)$ if and only if X has a binomial, Poisson, or negative binomial distribution according to whether c is a positive integer, zero, or negative, respectively. Other results in Consul (1990c) concern characterizations where the variance is a particular cubic function of the mean. The question "Do there exist some new families of discrete probability distributions [in the subclass of MPSD for which the mean is the power parameter] such that the variance equals a fourth degree function of the mean m?" was posed by Consul (1990c); it remains as yet unanswered.

A number of interesting modified power series distributions that are not GPSD are, however, Lagrangian distributions. These include the Lagrangian Poisson, the Lagrangian negative binomial, and the Lagrangian logarithmic distributions; see Chapter 9, Section 11, Chapter 5, Section 12, and Chapter 7, Section 11. The two classes, MPSD and Lagrangian, clearly overlap. Consul (1981) has helped to clarify the situation by introducing a yet broader class of Lagrangian distributions; see Section 5.2 below.

3 DIFFERENCE EQUATION SYSTEMS

3.1 Katz and Extended Katz Families

Pearson (1895) noted that for the hypergeometric distribution (see Chapter 6),

$$\frac{p_x - p_{x-1}}{p_{x-1}} = \frac{a - x}{b_0 + b_1 x + b_2 x(x - 1)}, \tag{2.40}$$

where a, b_0, b_1 and b_2 are parameters, $p_x = \Pr[X = x]$, and x takes a range of positive integers. He used this as a starting point for obtaining (by a limiting process) the differential equation defining the *Pearson system of continuous distributions*; see Chapter 12. Pearson apparently did not pursue the development of a *discrete* analogue of his continuous system.

The difference equation (2.40) was used by Carver (1919) for smoothing actuarial data. Although Carver (1923) obtained expressions for the parameters in terms of the moments, he did not attempt a thorough examination of the discrete distributions arising from (2.40). A detailed study of the special case $b_0 = b_1$, $b_2 = 0$ was undertaken by Katz in his (1945) thesis, in abstracts [Katz (1946, 1948)], and in a longer form in Katz (1965). A comprehensive exploration of distributions satisfying (2.40) awaited the work of Ord (1967a, b, 1972).

Katz' restrictions $b_0 = b_1$, $b_2 = 0$ give the *Katz family* of distributions, with

$$\frac{p_{x+1}}{p_x} = \frac{\alpha + \beta x}{1 + x}, \qquad x = 0, 1, \ldots, \tag{2.41}$$

where $\alpha > 0$, $\beta < 1$ ($\beta \geq 1$ does not yield a valid distribution). If $\alpha + \beta n < 0$, then p_{n+j} is understood to be equal to zero for all $j > 0$. From (2.41),

$$(x + 1)^{r+1} p_{x+1} = (\alpha + \beta x)(x + 1)^r p_x ; \tag{2.42}$$

summing both sides with respect to x gives

$$\mu'_{r+1} = \sum_{j=0}^{r} \binom{r}{j} (\alpha \mu'_j + \beta \mu'_{j+1}), \tag{2.43}$$

whence

$$\mu = \frac{\alpha}{1 - \beta}, \tag{2.44}$$

$$\mu'_2 = \frac{\alpha + (\alpha + \beta)\mu}{1 - \beta};$$

that is

$$\mu_2 = \frac{\alpha}{(1 - \beta)^2}. \tag{2.45}$$

Furthermore

$$\mu_3 = \mu_2(2c - 1)$$

and

$$\mu_4 = 3\mu_2^2 + \mu_2(6c^2 - 6c + 1), \tag{2.46}$$

where $c = \mu_2/\mu = (1 - \beta)^{-1}$.

Katz showed that $\beta < 0$, $\beta = 0$, and $0 < \beta < 1$ (that is, $0 < c < 1$, $c = 1$, and $1 < c$) give rise to the binomial, Poisson, and negative binomial distributions, with parameters $n = -\alpha/\beta$, $p = \beta/(\beta - 1)$ for the binomial, $\theta = \alpha$ for the Poisson, and $k = \alpha/\beta$, $P = \beta$ for the negative binomial, respectively. Katz suggested the reparameterization $\xi = \alpha/(1 - \beta)$ $(= \mu)$ and $\eta = \beta/(1 - \beta)$ $(= \{\sigma^2 - \mu\}/\mu)$, and gave the ML equations for the distributions in terms of these new parameters.

A major motivation for the Katz (1965) paper was the problem of discriminating between these three distributions when a given set of data is known to come from one or other of them. Katz found that for these three distributions $Z = (s^2 - \bar{x})/\bar{x}$ is very approximately normally distributed with mean $(c - 1)$ and variance $2/N$ where N is the sample size. Tests for $H_0 : \mu_2 = \mu$ against $H_1 : \mu_2 < \mu$ (or alternatively $H_1 : \mu_2 > \mu$) are tests for a Poisson null hypothesis against a binomial alternative hypothesis (or a negative binomial alternative hypothesis).

The pgf for the Katz family is

$$G(z) = \left(\frac{1 - \beta z}{1 - \beta} \right)^{-\alpha/\beta}. \tag{2.47}$$

The factorial moment generating function is

$$G(1+t) = \left(\frac{1-\beta-\beta t}{1-\beta}\right)^{-\alpha/\beta} \tag{2.48}$$

and the r'th factorial moment is

$$\mu'_{[r]} = \left(\frac{\alpha}{\beta}\right) \cdots \left(\frac{\alpha}{\beta}+r-1\right)\left(\frac{\beta}{1-\beta}\right)^r$$

$$= \left(\frac{\alpha+r\beta-\beta}{1-\beta}\right)\mu'_{[r-1]} \tag{2.49}$$

for $r \geq 1$, with $\mu'_{[0]} = 1$.

Ottestad (1939) had previously used the ratio $\mu'_{[r+1]}/\mu'_{[r]}$ as the basis for a graphical discrimination test; the slope of the sample estimate of $\mu'_{[r+1]}/\mu'_{[r]}$ plotted against r can be expected to be positive, zero, or negative for the negative binomial, Poisson, and binomial distributions, respectively.

Guldberg's (1931) approach to deciding whether any of the three distributions is appropriate was to show that

$$\frac{(x+1)p_{x+1}}{p_x} + \frac{(\mu-\mu_2)x}{\mu_2} = \alpha = \frac{\mu^2}{\mu_2}; \tag{2.50}$$

that is,

$$T_x\mu_2 + (\mu-\mu_2)x = \mu^2, \tag{2.51}$$

where $T_x = (x+1)p_{x+1}/p_x$. The sample values

$$\frac{s^2(x+1)f_{x+1}/f_x + (\bar{x}-s^2)x}{\bar{x}^2} \tag{2.52}$$

(where f_x is the frequency of the observation x in the sample) should therefore be approximately equal to unity.

The mean deviation for the three distributions in the Katz family can be obtained by summing (2.42), with $r = 0$, over all values of x not less than m where $m = [\mu]$ (the largest integer not greater than μ). Since $\alpha + \beta\mu = \mu_2(1-\beta)$ from (2.44) and (2.45), this gives

$$E[|X-\mu|] = 2p_m\left(\frac{\alpha+\beta m}{1-\beta}\right) \doteq 2p_m\mu_2, \tag{2.53}$$

with an error of $2p_m(m-\mu)\beta/(1-\beta)$. The formula is exact if μ is an integer, and also when $\beta = 0$; see Bardwell (1960) and Kamat (1965).

Consider now the Katz family with the restriction $b_0 = b_1$ removed. This gives the *extended Katz family* of Gurland and Tripathi (1975) and Tripathi and Gurland (1977). The relationship between the probabilities now has the form

$$\frac{p_{x+1}}{p_x} = \frac{\alpha + \beta x}{\gamma + x},\tag{2.54}$$

yielding an extended hypergeometric distribution with pgf

$$G(z) = \frac{{}_2F_1[\alpha/\beta, 1; \gamma; \beta z]}{{}_2F_1[\alpha/\beta, 1; \gamma; \beta]};\tag{2.55}$$

see Chapter 6, Section 11.

When $\beta \to 0$ in (2.54), the pgf becomes

$$G(z) = \frac{{}_1F_1[1; \gamma; \alpha z]}{{}_1F_1[1; \gamma; \alpha]};\tag{2.56}$$

this is the hyper–Poisson distribution of Bardwell and Crow (1964) and Crow and Bardwell (1965); see Chapter 4, Section 12.4. An extended form of this distribution with

$$\frac{p_{x+1}}{p_x} = \frac{\alpha(\rho + x)}{(1 + x)(\gamma + x)}\tag{2.57}$$

and pgf

$$G(z) = \frac{{}_1F_1[\rho; \gamma; \alpha z]}{{}_1F_1[\rho; \gamma; \alpha]}\tag{2.58}$$

has been studied by Gurland and Tripathi (1975) and Tripathi and Gurland (1977, 1979); see Chapter 4, Section 11.4.

Tripathi and Gurland have designated these four families of distributions: K (for Katz), EK (for extended Katz), CB (for Crow and Bardwell), and ECB (for extended Crow and Bardwell). All four are special forms of Kemp's generalized hypergeometric probability family of distributions; see Section 4 in this chapter. Tripathi and Gurland have examined various types of estimators, especially minimum χ^2 estimators. They have been particularly concerned with the problem of selecting an appropriate member from a combined set of distributions (e.g., K, EK, ECB), when fitting data for which a specific model is unclear.

Sundt and Jewell (1981) and Willmot (1988a) have created the *Sundt and Jewell family* of distributions. Their papers are in actuarial journals as their interest lies in actuarial applications. The family is a modified form of the Katz family. The recurrence relationship for the probabilities is

$$\frac{p_{x+1}}{p_x} = \frac{a + b + ax}{1 + x}, \quad x = 1, 2, 3, \ldots;\tag{2.59}$$

this is the same as the Katz relationship (2.41) for all x except for the exclusion of $x = 0$.

The pgf for the Sundt and Jewell family has the form

$$G(z) = c + (1-c)\left\{\frac{1-az}{1-a}\right\}^{-(a+b)/a} \tag{2.60}$$

$$= c + (1-c)H(z),$$

where $H(z)$ is the pgf for a Katz distribution. A necessary restriction on c is $H(0)\{H(0) - 1\}^{-1} \leq c < 1$. The Sundt and Jewell family therefore contains the modified binomial, modified Poisson, and modified negative binomial distributions (Chapter 8, Section 2.2) as well as the binomial, Poisson, and negative binomial distributions themselves (Chapters 3, 4, and 5, respectively). It also includes the logarithmic distribution (Chapter 7), the zero–modified logarithmic distribution (Chapter 7, Section 10, and Chapter 8, Section 2.2), and Engen's extended negative binomial distribution (Chapter 5, Section 12.2).

3.2 Ord's Family

Ord's (1967a, b, 1972) difference equation family—*Ord's family*—comprises all the distributions that satisfy

$$\Delta p_{x-1} = p_x - p_{x-1} = \frac{(a-x)p_{x-1}}{b_0 + b_1 x + b_2 x(x-1)} \tag{2.61}$$

$$= \frac{(a-x)p_x}{(a+b_0) + (b_1 - 1)x + b_2 x(x-1)}, \tag{2.62}$$

where $p_x = \Pr[X = x]$ and x takes some range of integer values $\{T\}$.

The nature of the roots of $b_0 + b_1 x + b_2 x(x-1)$ is central to Ord's classification. We saw in Section 3.1 that $b_0 = b_1$, $b_2 = 0$ yields Katz' family; Ord labeled these IIIB, IIIP and IIIN for the binomial, Poisson and negative binomial, respectively. He noted furthermore that $b_2 = 0$, $b_0 \neq b_1$, leads to a system of "hyper" distributions, in particular to the hyper–Poisson distribution; see Chapter 4, Section 12.4.

For $b_2 \neq 0$, the denominator of (2.61) has two roots, with the possibilities:

Type I: one root zero, the other nonzero, and range finite,
Type VI: one root zero, the other negative, and infinite range,
Type IV: both roots imaginary.

Ord's Type II distributions are symmetrical forms of Ord's Type I. His Type V is a limiting form of Type IV. The numbering system for Ord's discrete family therefore differs somewhat (though not radically) from the numbering system for Pearson's continuous family.

Ord's more detailed level of classification depends on the roots of the denominator of (2.62), or rather on the value of

$$\kappa = \frac{(b_1 - b_2 - 1)^2}{4(a + b_0)b_2};$$
(2.63)

for $0 \leq \kappa \leq 1$ the roots are imaginary. Table 2.1 summarizes the types of distributions available in Ord's family.

Ord (1967a) published a (β_1, β_2) chart, where β_1 and β_2 are the usual indices of skewness and kurtosis, analogous to the (β_1, β_2) chart for Pearson's continuous system. However, he later considered this to be not very useful in the discrete case [see Ord (1985)] and turned his attention to the measures $I = \mu_2/\mu$ and $S = \mu_3/\mu_2$. A diagram of the (S, I) regions was published by Ord (1967b, 1972). Note that I is necessarily positive. For the Katz family $S = 2I - 1$, with $0 < I < 1$, $I = 1$, and $1 < I$ for the binomial, Poisson, and negative binomial, respectively. Also $S < 2I - 1$ gives the beta-binomial, and $S > 2I - 1$ yields the hypergeometric or the beta-Pascal according to whether, in addition, $S < 1$ or $S > 1$.

The parameters of the distributions in the family are determined by the first three moments; see Ord (1967b). The (S, I) diagram can therefore be used as an aid for the selection of an appropriate distribution for a given data set.

Ord's preferred selection method [Ord (1967a)] uses a plot of $u_r = rf_r/f_{r-1}$ against r, where f_r is a sample frequency; some degree of smoothing of the u_r may be helpful. The exact relationships that are satisfied by rp_r/p_{r-1} are summarized in Tables 2.2 and 2.3. Although sample ratios rf_r/f_{r-1} cannot be expected to satisfy these relationships exactly, Ord anticipated that sample plots will give a fair indication of an appropriate type of distribution.

Properties of members of the family were reviewed by Ord (1972). He included the following properties: modality, Janardan and Patil (1970); mean deviation, Kamat (1966) and Ord (1967c); mean difference, Katti (1960) and Ord (1967c); incomplete moments, Guldberg (1931) and Kamat (1965); moment estimation and MLE, Carver (1923) and Ord (1967b).

Bowman, Shenton, and Kastenbaum (1991) have studied an extension of Ord's family with

$$p_x = \left(1 + \frac{\alpha - x}{C_0 + C_1 y + C_2 y^2}\right) p_{x-1},$$
(2.64)

where $y = x - \mu = x - E[X]$; the ratio of successive probabilities is here the ratio of two quadratic expressions in x. There are two possibilities: (1) distributions defined on the non-negative integers (finite or infinite) and (2) distributions defined on the negative integers as well. There are four parameters, α, C_0, C_1, and C_2. The authors identified certain forms of these distributions and found it reasonably straightforward to estimate the four

Table 2.1 The Types of Distributions Derived by Ord (1967a)

Type	Name	P_x	Criteria	Range	Comments
I(a)	Hypergeometric	$\binom{Np}{x}\binom{Nq}{n-x}/\binom{N}{n}$	$I<1, \kappa>1$	$[0,m]$ $m=\min(n,Np)$	J- or bell-shaped
I(b)	Negative hypergeometric or beta-binomial	$\binom{k+x-1}{x}\binom{N-k-x}{Np-x}/\binom{N}{Np}$	$\kappa<0$	$[0,Np]$	J- or bell-shaped
I(e)	—	$\binom{A}{x}\binom{C}{B-x}/\binom{A+C}{B}$	$\kappa>1$	$[0,\infty)$	A, C noninteger but have the same integral part
I(u)	—	$a\left\{\binom{A}{C+x}\binom{B}{D-x}\right\}^{-1}$	$\kappa>1$	$[0,n], n<D$	U-shaped
VI	Beta-Pascal	$\frac{A}{(k+A)}\binom{k+x-1}{x}\binom{A+B-1}{A}/\binom{k+A+B+x-1}{k+A}$	$I>1, \kappa>1$	$[0,\infty)$	J- or bell-shaped
IV	—	$aQ(x,a,d)/Q(x,k+a,b), x>0$; similar expression for $x<0$	$0<\kappa<1$	$(-\infty,\infty)$	k a positive integer; bell-shaped
II(a) II(b) II(u)	As for Type I(·)		$I<1, \kappa=1$ $\kappa=0$ $\kappa=1$	As type I(·)	Symmetric forms of type I(·)
V	—	As Type IV, but $b=0$	$\kappa=0$	$[0,\infty)$ or $(-\infty,\infty)$	Limiting form of IV
III(B)	Binomial	$\binom{n}{x}p^x(1-p)^{n-x}$ $(0<p,q; p+q=1)$	$I<1, \kappa\to\infty$	$[0,n]$	Limiting form of I(a), I(b)
III(N)	Negative binomial or Pascal	$\binom{k+x-1}{x}p^k(1-p)^x$	$I>1, \kappa\to\infty$	$[0,\infty)$	Limiting form of I(b), VI
III(P)	Poisson	$e^{-m}m^x/x!$	$I=1, \kappa\to\infty$	$[0,\infty)$	Limiting form of III(B), III(N)
VII	Discrete Student's t	$\alpha\left[\prod_{j=1}^k\{(j+x+a)^2+b^2\}\right]^{-1}$	$0<\kappa<1$	$(-\infty,\infty)$	"Nearly" symmetric form of IV

Notes: 1. $Q(x,a,d)=(a^2+d^2)\{(a+1)^2+d^2\}\cdots\{(a+x)^2+d^2\}$.
2. α is a constant such that the total probability is 1.
3. $\kappa=(b_1-b_2-1)^2[4b_2(b_0+a)]^{-1}$.
4. $I =$(variance)/(mean)
Adapted from Ord (1967a).

83

Table 2.2   Intercept and Slope of the Plot of xf_x/f_{x-1} against x for Certain Discrete Distributions

Distribution	Intercept (c_0)	Slope (c_1)
Binomial	$(n+1)p/q$	$-p/q$
Negative binomial	$(k-1)q$	q
Poisson	λ	0
Logarithmic	$-q$	q
Rectangular	0	1

From Ord (1967a).

Table 2.3   Beginning and End Points of the Hypergeometric Distribution Curves Compared with the Binomial and Negative Binomial Lines

Distribution	Beginning	End
Hypergeometric	Above binomial	Below binomial
Beta-binomial	Below binomial Above negative binomial	Above binomial Below negative binomial
Beta-Pascal	Below negative binomial	Above negative binomial

From Ord (1967a).

parameters using the first four moments (assuming that they exist) in case (1). For case (2) slowly converging series may be encountered. Bowman, Shenton, and Kastenbaum investigated further extensions with recurrence relations such as

$$p_x = \left(1 + \frac{\alpha - x}{C_0 + C_1 y + C_2 y^2 + C_3 y^3 + C_4 y^4}\right) p_{x-1}, \qquad (2.65)$$

with $x = \pm 1, \pm 2, \ldots$, $y = x - \mu = x - E[X]$; these extensions require six or more parameters. Their *discrete normal distribution* received special attention. They did not consider applications to empirical data, or estimation problems.

4   KEMP FAMILIES

4.1   Generalized Hypergeometric Probability Distributions

The distributions in Kemp's (1968a, b) very broad family of *generalized hypergeometric probability distributions* (GHPD or HP-distributions for short) have many useful properties. Their pgf's have the form $_pF_q[\lambda z]/_pF_q[\lambda]$, where

$$_pF_q[a_1, \ldots, a_p; b_1, \ldots, b_q; \lambda z]$$

$$= 1 + \frac{a_1 \cdots a_p(\lambda z)}{b_1 \cdots b_q 1!} + \frac{a_1(a_1+1) \cdots a_p(a_p+1)(\lambda z)^2}{b_1(b_1+1) \cdots b_q(b_q+1)2!} + \cdots$$

$$= \sum_{j \geq 0} \frac{(a_1 + j - 1)! \cdots (a_p + j - 1)!(b_1 - 1)! \cdots (b_q - 1)!(\lambda z)^j}{(a_1 - 1)! \cdots (a_p - 1)!(b_1 + j - 1)! \cdots (b_q + j - 1)!j!} \quad (2.66)$$

$$= \sum_{j \geq 0} \left[\left\{ \prod_{i=1}^{p}(a_i)_j \right\} (\lambda z)^j \Big/ \left[j! \prod_{i=1}^{q}(b_i)_j \right] \right\};$$

this is known in the theory of special functions as a generalized hypergeometric function (see Chapter 1, Section A8). Various models for HP-distributions are discussed later in this section. These distributions are difference equation distributions; when $\lambda \neq 1$ they are power series distributions. They are also limiting forms as $q \to 1$ of *q-probability series distributions* (see for example, Chapter 4, Section 12.6); these have pgf's of the form

$$G(z) = \frac{{}_A\Phi_B[a_1, \ldots, a_A; b_1 \ldots, b_B; q, \lambda z]}{{}_A\Phi_B[a_1, \ldots, a_A; b_1 \ldots, b_B; q, \lambda]}, \quad (2.67)$$

where ${}_A\Phi_B[\cdot]$ is a basic hypergeometric function (see Chapter 1, Section A12). The symbol q is used conventionally in both (2.66) and (2.67), but with different meanings.

A second family of distributions, *generalized hypergeometric factorial moment distributions* (GHFD or HF-distributions for short), have pgf's of the form $_pF_q[\lambda(z-1)]$; see Kemp (1968a) and Kemp and Kemp (1974). This family includes not only some of the most common distributions but also a number of important matching and occupancy distributions; see Chapter 10, Sections 3 and 4.1. Their properties and modes of genesis are examined in Section 4.2 of this chapter.

The two families overlap. Their intersection includes some of the most common distributions, including the binomial, Poisson, negative binomial, and hypergeometric [Potts (1953)]; such distributions have duplicate sets of properties, including very simple relationships for both their probabilities and their factorial moments.

Both families are contained within an even more general class of distributions whose pgf's have the form $_pF_q[\lambda z + \xi]/_pF_q[\lambda + \xi]$. These are *generalized hypergeometric recast distributions*, (GHRD or HR-distributions for short); see the end of Section 4.2. They arise from certain ascertainment models.

In Section 2.1 of this chapter we saw that power series distributions have pgf's of the form

$$G(z) = \frac{\eta(\lambda z)}{\eta(\lambda)} = \sum_{x \geq 0} \alpha_x \lambda^x z^x \Big/ \sum_{x \geq 0} \alpha_x \lambda^x, \quad (2.68)$$

where λ is the power parameter and α_x is a function of x independent of λ. Suppose now that α_{x+1}/α_x is the ratio of two polynomials in x with all roots real:

$$\frac{\alpha_{x+1}}{\alpha_x} = \frac{(a_1 + x)\dots(a_p + x)}{(b_1 + x)\dots(b_q + x)(b_{q+1} + x)} \tag{2.69}$$

(the coefficients of the highest powers of x in these polynomials can be assumed to be unity without loss of generality concerning the form of the resultant distribution). Then

$$\frac{\Pr[X = x + 1]}{\Pr[X = x]} = \frac{(a_1 + x)\dots(a_p + x)\lambda}{(b_1 + x)\dots(b_q + x)(b_{q+1} + x)}. \tag{2.70}$$

Assuming that X is a counting variable with $\Pr[X = 0] \neq 0$, then when $\eta(\lambda)$ is a convergent series in λ the support of X is $0, 1, \dots$, and when $\eta(\lambda)$ is a terminating series in λ the support is $0, 1, \dots, n$. The corresponding pgf is

$$G(z) = K \left\{ 1 + \frac{a_1 \cdots a_p(\lambda z)}{b_1 \cdots b_{q+1}} + \frac{a_1(a_1 + 1) \cdots a_p(a_p + 1)(\lambda z)^2}{b_1(b_1 + 1) \cdots b_{q+1}(b_{q+1} + 1)} + \cdots \right\}$$

$$= \frac{{}_{p+1}F_{q+1}[1, a_1, \dots, a_p; b_1, \dots, b_{q+1}; \lambda z]}{{}_{p+1}F_{q+1}[1, a_1, \dots, a_p; b_1, \dots, b_{q+1}; \lambda]}. \tag{2.71}$$

This is a particular HP-distribution. Clearly the numerator parameters are commutable; so are the denominator ones. Suppose now that one of the denominator parameters in (2.69) is equal to unity, say $b_{q+1} = 1$; then (2.70) and (2.71) become

$$\frac{\Pr[X = x + 1]}{\Pr[X = x]} = \frac{(a_1 + x)\dots(a_p + x)\lambda}{(b_1 + x)\dots(b_q + x)(x + 1)} \tag{2.72}$$

and

$$G(z) = K \left\{ 1 + \frac{a_1 \cdots a_p(\lambda z)}{b_1 \cdots b_q 1!} + \frac{a_1(a_1 + 1) \cdots a_p(a_p + 1)(\lambda z)^2}{b_1(b_1 + 1) \cdots b_q(b_q + 1)2!} + \cdots \right\}$$

$$= \frac{{}_pF_q[a_1, \dots, a_p; b_1, \dots, b_q; \lambda z]}{{}_pF_q[a_1, \dots, a_p; b_1, \dots, b_q; \lambda]}, \tag{2.73}$$

this is a general HP-distribution [see Kemp (1968a, b)]. Note that (2.71) can be treated as a special case of (2.73), whereas it might be thought that the reverse holds.

     The relationship (2.70) for successive probabilities seems to have appeared for the first time in a paper by Guldberg (1931), who concentrated on recurrence relationships for the moments for the special cases corresponding to the binomial, Poisson, Pascal and hypergeometric distributions; see also Qvale (1932). Restating (2.70) in the form

$$\frac{\Pr[X=x+1] - \Pr[X=x]}{\Pr[X=x]} = \frac{\{(a_1 + x)\dots(a_p + x)\lambda\} - \{b_1 + x)\dots(b_{q+1} + x)\}}{(b_1 + x)\dots(b_{q+1} + x)}$$

(2.74)

emphasizes the relationship between HP-distributions and difference equation distributions. The Katz family corresponds to (2.72) with $p = 1$, $q = 0$, while the extended Katz family corresponds to (2.70) with $p = 1$, $q = 0$; see Section 3.1. For Tripathi and Gurland's extended Bardwell-Crow family $p = 1$, $q = 1$ in (2.72); again see Section 3.1. Ord's difference equation family (Section 3.2) has $p = 2$, $q = 1$, $\lambda = 1$ in (2.70), but note that Ord's family contains distributions for which the variable takes negative values; such distributions are no longer power series distributions. Kemp and Kemp's (1956a) earlier "generalized" hypergeometric distributions with $p = 2$, $q = 1$, $\lambda = 1$ in (2.72) are GHPD and are included within Ord's family; they utilise the Gaussian hypergeometric function of Chapter 1, Section A6 (see Chapter 6, Section 2.4). (The Gaussian hypergeometric function is of course a special form of the generalized hypergeometric function of Chapter 1, Section A8.)

Kemp (1968a, b) showed that terminating distributions can arise for all nonnegative integer values of p and q, provided that at least one of the a_i is a negative integer and that the other parameters are chosen suitably. A shifted HP-distribution (with support $m + 1, m + 2, \dots$) may arise if one of the b_j is a negative integer. The existence of nonterminating members of the family depends on the sign of $p - q - 1$; they exist

1. when $p - q - 1 < 0$, provided that $\lambda > 0$ and $a_i, b_j > 0 \ \forall i, j$ (except that a pair of a_i, b_j may be negative provided that they lie in the interval between two consecutive nonpositive integers),
2. when $p - q - 1 = 0$, provided that $0 < \lambda < 1$ and $a_i, b_j > 0 \ \forall i, j$ (except as above),
3. when $p - q - 1 = 0$ and $\lambda = 1$, provided that $\sum_{i=1}^{p} a_i < \sum_{j=1}^{q} b_j$ and $a_i, b_j > 0 \ \forall i, j$ (except as above).

Nonterminating members do not exist when $p > q + 1$.

Table 2.4 is taken from Dacey (1972). It gives a summary of named probability distributions in Patil and Joshi (1968) with pgf's expressible in terms of $_pF_q(\lambda z)$. As the table shows, a very large number of distributions belong to the family.

The reversed form of a GHPD is also a GHPD. Truncating the first m probabilities of a GHPD gives a shifted GHPD; truncating all except the first m probabilities gives a new (terminating) GHPD.

Table 2.4   Named Probability Distributions in Patil and Joshi (1968) with pgf's Expressible in Terms of $_pF_q[\lambda z]$

	Name and number of distribution	pgf
[1]	Binomial	$C\,_1F_0[-n;-\lambda z], \quad p=\lambda/(1+\lambda)$
[2]	Gen. Poisson binomial	$\prod_{j=1}^{n} C\,_1F_0[-1;-\lambda_j z]$
[6]	Poisson	$C\,_0F_0[\lambda z]$
[7]	Poisson type	$Cz^b\,_0F_0[\lambda z^a]$
[8]	Displaced Poisson	$C\,_1F_1[1;r+1;\lambda z]$
[11]	Hyper-Poisson	$C\,_1F_1[1;\lambda;\theta z]$
[13i]	Geometric	$C\,_1F_0[1;qz]$
[13ii]	Geometric	$Cz\,_1F_0[1;qz]$
[14i]	Pascal	$Cz^k\,_1F_0[k;qz]$
[14ii]	Pascal	$C\,_1F_0[k;qz]$
[15]	Negative binomial	$C\,_1F_0[k;qz]$
[16]	Pólya-Eggenberger	$C\,_1F_0[a;pz], \quad a=h/\theta,\ p=\theta/(1+\theta)$
[17]	Logarithmic	$Cz\,_2F_1[1,1;2;\theta z]$
[18]	Modified logarithmic	$C\,_1F_0[-1;-\lambda g(z)], \quad \delta=1/(1+\lambda),$
		$g(z)=C'z\,_2F_1[1,1;2;\theta z]$
[23]	Stirling, first kind	$\{Cz\,_2F_1[1,1;2;\theta z]\}^n$
[24]	Stirling, second kind	$\{Cz\,_1F_1[1;2;\theta z]\}^n$
[26]	Hypergeometric	$C\,_2F_1[-n,-M;N-M-n+1;z]$
[27i]	Inverse hypergeometric	$Cz^k\,_2F_1[M-N,k;k-N;z]$
[27ii]	Inverse hypergeometric	$C\,_2F_1[M-N,k;k-N;z]$
[28]	Negative hypergeometric	$C\,_2F_1[-n,M;M-N-n+1;z]$
[29]	Pólya	$C\,_2F_1[-n,p/\gamma;-n+1-q/\gamma;z]$
[30]	Inverse Pólya	$C\,_2F_1[k,q/\gamma;k+1;z]$
[31-38]	Other hypergeometric[†]	$C\,_2F_1[-n,-a;b-n+1;z]$
[43]	Poisson-rectangular	$\{C\,_0F_0[\beta z]-C\,_0F_0[\alpha z]\}/(\alpha-\beta)\,_1F_0[-1;z]$
[44]	Hermite	$C\,_0F_0[\lambda g(z)], g(z)=(\lambda_1 z+\lambda_2 z^2)/(\lambda_1+\lambda_2)$
[47]	Neyman Type A	$C\,_0F_0[\lambda_1 g(z)], g(z)=C'\,_0F_0[\lambda_2 z]$
[49]	Neyman Type B	$C\,_0F_0[\lambda_1 g(z)], g(z)=\,_1F_1[1;2;\lambda_2(z-1)]$
[50]	Neyman Type C	$C\,_0F_0[\lambda_1 g(z)], g(z)=\,_1F_1[1;3;\lambda_2(z-1)]$
[51]	Beall-Rescia	$C\,_0F_0[\lambda_1 g(z)], g(z)=\,_1F_1[1;\beta+1;\lambda_2(z-1)]$
[52]	Gurland's gen. Neyman	$C\,_0F_0[\lambda_1 g(z)], g(z)=\,_1F_1[\alpha;\alpha+\beta;\lambda_2(z-1)]$
[53]	Thomas	$C\,_0F_0[\lambda_1 g(z)], g(z)=C'z\,_0F_0[\lambda_2 z]$
[54]	Poisson-binomial	$C\,_0F_0[\lambda g(z)], g(z)=C'\,_1F_0[-n;-\mu z],\ p=\mu/(1+\mu)$
[55]	Pólya-Aeppli	$C\,_0F_0[\lambda_1 g(z)],\ g(z)=C'\,_1F_0[1;\lambda_2 z]$
[56]	Poisson-negative binomial	$C\,_0F_0[\lambda g(z)],\ g(z)=C'\,_1F_0[k;qz]$
[57]	Waring	$C\,_2F_1[1,k;k+\rho+1;z]$
[58]	Yule	$C\,_2F_1[1,1;\rho+2;z]$
[62]	Uniform	$Cz\,_2F_1[-n+1,1;-n+1;z]$
[68]	Distribution of exceedances	$C\,_2F_1[-n,M;M-N-n;z]$
[69]	Factorial	$C\,_2F_1[1,n-\lambda+1;\lambda+1;z]$
[72]	Ising-Stevens	$C\,_2F_1[-n_1+1,-n_2+1;2;z]$
[73]	Extended hypergeometric	$C\,_2F_1[-n_1,-r;n_2-r+1;\theta z],\ n_2\geq r$
		$Cz^{r-n_2}\,_2F_1[-n_2,r-n_1-n_2;r-n_2+1;\theta z],\ n_2<r$

Note: The notation of Patil et al. (1984) is used, and their dictionary should be consulted for constraints on parameter values.

[†] n is here not necessarily an integer.

From Dacey (1972).

The pgf satisfies the differential equation

$$\theta \prod_{j=1}^{q} (\theta + b_j - 1)\, G(z) = \lambda z \prod_{i=1}^{p} (\theta + a_i) G(z), \tag{2.75}$$

where θ is the differential operator $z\, d/dz \equiv zD$. Using

$$z^k D^k \equiv \theta(\theta - 1)\dots(\theta - k + 1) \quad \text{and} \quad \theta^k \equiv \sum_{j=1}^{k} S(k, j) z^j D^j,$$

where $S(k, j)$ are Stirling numbers of the second kind, (2.75) can be transformed into a differential equation of the form

$$f_1(D)G(z) = \lambda f_2(D)G(z). \tag{2.76}$$

The factorial moment generating function satisfies

$$f_1(D)G(1 + t) = \lambda f_2(D)G(1 + t), \tag{2.77}$$

and by identifying the coefficient of $t^i/i!$ in this equation we get a recurrence formula for the factorial moments that involves at most $\max(p + 1, q + 2)$ of them.

The generating function for the moments about c satisfies

$$(D+c) \prod_{j=1}^{q} (D+c+b_j - 1) \left\{ e^{-ct} G(e^t) \right\} = \lambda e^t \prod_{i=1}^{p} (D+c+a_i) \left\{ e^{-ct} G(e^t) \right\}; \tag{2.78}$$

identifying the coefficient of $t^i/i!$ in this equation yields a recurrence formula for the moments about c. The uncorrected and the corrected moments are given by $c = 0$ and $c = \mu_1' = \mu$, respectively. Kemp (1968a,b) investigated these and other moment properties, including ones for the cumulants, factorial cumulants, and incomplete moments. Over–, under–, and equi–dispersion aspects of GHPD have been studied by Tripathi and Gurland (1979).

Three results concerning limiting forms of GHPD are

$$\lim_{v \to \pm\infty} \frac{{}_{p+1}F_q[a_1, \dots, a_p, u + v; b_1, \dots, b_q; \lambda z/v]}{{}_{p+1}F_q[a_1, \dots, a_p, u + v; b_1, \dots, b_q; \lambda/v]}$$

$$= \frac{{}_pF_q[a_1, \dots, a_p; b_1, \dots, b_q; \lambda z]}{{}_pF_q[a_1, \dots, a_p; b_1, \dots, b_q; \lambda]}, \tag{2.79}$$

$$\lim_{v \to \pm\infty} \frac{pF_{q+1}[a_1, \ldots, a_p; b_1, \ldots, b_q, u + v; \lambda v z]}{pF_{q+1}[a_1, \ldots, a_p; b_1, \ldots, b_q, u + v; \lambda v]}$$

$$= \frac{pF_q[a_1, \ldots, a_p; b_1, \ldots, b_q; \lambda z]}{pF_q[a_1, \ldots, a_p; b_1, \ldots, b_q; \lambda]}, \quad (2.80)$$

$$\lim_{v \to \pm\infty} \frac{p+1F_{q+1}[a_1, \ldots, a_p, v; b_1, \ldots, b_q, u + kv; \lambda k z]}{p+1F_{q+1}[a_1, \ldots, a_p, v; b_1, \ldots, b_q, u + kv; \lambda k]}$$

$$= \frac{pF_q[a_1, \ldots, a_p; b_1, \ldots, b_q; \lambda z]}{pF_q[a_1, \ldots, a_p; b_1, \ldots, b_q; \lambda]}. \quad (2.81)$$

Many limit theorems in the literature can be proved using these results. The Poisson is an ultimate limiting form for distributions with infinite support; the binomial is an ultimate limiting form for terminating ones.

Results concerning gamma-type and beta-type mixtures of GHPD have been derived by Kemp (1968a) and Tripathi and Gurland (1979); they are described in Chapter 8, Section 3.4.

Weighted distributions are distributions that are modified by the method of ascertainment; see Rao (1965), Patil et al. (1986), and Chapter 3, Section 12.4. Let the sampling chance (i.e., weighting factor) for a value x be γ_x. Then if the original distribution has pgf $\sum_x p_x z^x$, the ascertained (weighted) distribution has pgf

$$G(z) = \sum_x \gamma_x p_x z^x \Big/ \sum_x \gamma_x p_x.$$

If $\gamma_x = \gamma^x$, then for all power series distributions the form of the ascertained distribution is the same as the original. For a HP-distribution with pgf (2.73), the ascertained pgf is

$$G(z) = \frac{pF_q[a_1, \ldots, a_p; b_1, \ldots, b_q; \gamma\lambda z]}{pF_q[a_1, \ldots, a_p; b_1, \ldots, b_q; \gamma\lambda]}. \quad (2.82)$$

Suppose now that $\gamma_x = x$; then (2.73) is ascertained as

$$G(z) = z \frac{pF_q[a_1 + 1, \ldots, a_p + 1; b_1 + 1, \ldots, b_q + 1; \lambda z]}{pF_q[a_1 + 1, \ldots, a_p + 1; b_1 + 1, \ldots, b_q + 1; \lambda]}. \quad (2.83)$$

More generally, when $\gamma_x = x!/(x - r)!$, (2.73) becomes

$$G(z) = z^r \frac{pF_q[a_1 + r, \ldots, a_p + r; b_1 + r, \ldots, b_q + r; \lambda z]}{pF_q[a_1 + r, \ldots, a_p + r; b_1 + r, \ldots, b_q + r; \lambda]}. \quad (2.84)$$

Particular HP-distributions can arise from nonequilibrium stochastic processes; for example, the Poisson process yields the Poisson distribution (Chapter 4, Section 2) and the non-homogeneous linear growth process yields the negative binomial distribution (Chapter 5, Section 3).

Consider now the time-homogeneous Markov process with Kolmogorov differential equations

$$\frac{dp_0(t)}{dt} = \beta_1 p_1(t) - \alpha_0 p_0(t),$$

$$\frac{dp_i(t)}{dt} = \beta_{i+1} p_{i+1}(t) - (\beta_i + \alpha_i)p_i(t) + \alpha_{i-1}p_{i-1}(t), \qquad (2.85)$$

where $\alpha_i = P(i)/Q(i)$, $\beta_i = R(i)/S(i)$, and $P(i)$, $Q(i)$, $R(i)$, and $S(i)$ are polynomials in i with all real roots such that α_i does not become infinite and β_i does not become zero. Then the equilibrium solution, if it exists, is given by

$$\frac{p_i}{p_{i-1}} = \frac{\alpha_{i-1}}{\beta_i} = \frac{P(i-1)S(i)}{Q(i-1)R(i)}. \qquad (2.86)$$

This can be rewritten as

$$\frac{p_i}{p_{i-1}} = \frac{(i+a_1-1)\cdots(i+a_p-1)\rho}{(i+b_1-1)\cdots(i+b_q-1)},$$

and hence the equilibrium pgf is

$$G(z) = \frac{{}_{p+1}F_q[1, a_1, \ldots, a_p; b_1, \ldots, b_q; \rho z]}{{}_{p+1}F_q[1, a_1, \ldots, a_p; b_1, \ldots, b_q; \rho]}. \qquad (2.87)$$

This can be interpreted both in terms of a birth-death process and as a busy-period distribution; see Kemp (1968a) and Kemp and Newton (1990). Kapur (1978a,b) has studied the special case $Q(x) = S(x) = 1$ [see Srivastava and Kashyap (1982)].

Applications of HP-distributions to topological features of drainage basins have been made by Dacey (1975).

4.2 Generalized Hypergeometric Factorial Moment Distributions

Generalized hypergeometric factorial moment distributions (GHFD or HF-distributions for short) have pgf's of the form

$$G(z) = {}_pF_q[a_1, \ldots, a_p; b_1, \ldots, b_q; \lambda(z-1)] \qquad (2.88)$$

[Kemp (1968a)], and hence their factorial moments are generated by

$$G(1+t) = {}_pF_q[a_1, \ldots, a_p; b_1, \ldots, b_q; \lambda t]; \qquad (2.89)$$

Table 2.5   Named GHF-Distributions with pgf's Expressible as $_pF_q[\lambda(z-1)]$

Name	Form	Reference
	Terminating	
Binomial	$_1F_0[-n;\ ;p(1-z)]$	
Hypergeometric	$_2F_1[-n,-Np;-N;1-z]$	
Negative		
hypergeometric	$_2F_1[-n,a;a+b;1-z]$	
Discrete rectangular	$_2F_1[-n,1;2;1-z]$	
Chung-Feller	$_2F_1[-n;1/2;1/2;1-z]$	
Pólya	$_2F_1[-n,M/c;(M+N)/c;1-z]$	
Matching distribution	$_1F_1[-n;-n;z-1]$	
Gumbel's matching		
distribution	$_1F_1[-n;-n;\lambda(z-1)]$	Fréchet (1943)
Laplace-Haag	$_1F_1[-n;-N;M(z-1)]$	Fréchet (1943)
Anderson's matching		
distribution	$_1F_{N-1}[-n;-n,\ldots,-n;(-1)^N(z-1)]$	Anderson (1943)
Stevens-Craig	$_nF_{n-1}[1-k,\ldots,1-k;-k,\ldots,-k;1-z]$	Patil et al. (1984)
(Coupon collecting)		
STERRED rectangular	$_3F_2[-n,1,1;2,2;1-z]$	Kemp and Kemp (1969b)
	Nonterminating	
Poisson	$_0F_0[\ ;\ ;\lambda(z-1)]$	
Negative binomial	$_1F_0[k;\ ;p(z-1)]$	
Poisson∧beta	$_1F_1[a;a+b;\lambda(z-1)]$	
Type H$_2$	$_2F_1[k,a;a+b;\lambda(z-1)]$	Gurland(1958), Katti(1966)
STERRED geometric	$_2F_1[1,1;2;q(z-1)/(1-q)]$	Kemp and Kemp (1969b)

Note: See Kemp and Kemp (1974) for constraints on the parameters.

$$\text{thus} \qquad \mu'_{[r]} = \frac{(a_1+r-1)!\cdots(a_p+r-1)!(b_1-1)!\cdots(b_q-1)!\lambda^r}{(a_1-1)!\cdots(a_p-1)!(b_1+r-1)!\cdots(b_q+r-1)}. \qquad (2.90)$$

Kemp and Kemp (1969b) commented that many common discrete distributions have pgf's of this form and gave a number of examples; see also Gurland (1958), Katti (1966), Kemp (1968a), Kemp and Kemp (1974), and Tripathi and Gurland (1979). In Section 4.1 it was possible to give exhaustive conditions under which (2.73) is a pgf. It is more difficult, however, to determine criteria for a set of values to be the factorial moments for a valid discrete distribution [see, e.g., Fréchet (1940)] and hence to state the conditions for (2.88) to be a pgf, though some progress has been made by Kemp (1968a) and Tripathi and Gurland (1979). Interest has centered on the identification of HF-distributions rather than their complete enumeration. Table 2.5 gives lists of terminating and non-terminating GHFD. Note the overlap with Table 2.4. Given that $G(1+t) = \sum_{r\geq 0}\mu'_{[r]}t^r/r!$, we have

$$G(z) = \sum_{j \geq 0} \sum_{x=0}^{j} \frac{\mu'_{[j]}(-1)^{j-x}z^x}{x!(j-x)!}$$

$$= \sum_{x \geq 0} \sum_{k \geq 0} \frac{\mu'_{[x+k]}(-1)^k z^x}{x!k!}$$

(cf. Chapter 10, Section 2). The probabilities for HF-distributions are therefore

$$Pr[X = x] = \sum_{k \geq 0} \left[\frac{\prod_{i=1}^{p}[(a_i + x + k - 1)!/(a_i - 1)!}{\prod_{j=1}^{p}[(b_j + x + k - 1)!/(b_j - 1)!} \cdot \frac{\lambda^{x+k}(-1)^k}{x!k!} \right]$$

$$= \frac{\lambda^x \prod_{i=1}^{p} \Gamma(a_i + x)/\Gamma(a_i)}{x! \prod_{j=1}^{q} \Gamma(b_j + x)/\Gamma(b_j)}$$

$$\times {}_pF_q[a_1 + x, \ldots, a_p + x; b_1 + x, \ldots, b_q + x; -\lambda]. \qquad (2.91)$$

A simpler method of handling the probabilities is via a recursion formula. The factorial moment generating function satisfies the differential equation

$$\theta \prod_{j=1}^{q}(\theta + b_j - 1) \, G(1+t) = \lambda t \prod_{i=1}^{p}(\theta + a_i)G(1+t), \qquad (2.92)$$

where θ is now the differential operator $td/dt \equiv tD$; θ satisfies the symbolic equations

$$t^k D^k \equiv \theta(\theta - 1) \ldots (\theta - k + 1) \quad \text{and} \quad \theta^k \equiv \sum_{j=1}^{k} S(k,j)t^j D^j,$$

where $S(k,j)$ is a Stirling number of the second kind (see Chapter 1, Section A3). The pgf therefore satisfies

$$\left[\theta \prod_{j=1}^{q}(\theta + b_j - 1) \right]_{t=z-1} G(z) = \lambda(z-1) \left[\prod_{i=1}^{p}(\theta + a_i) \right]_{t=z-1} G(z). \qquad (2.93)$$

Restated in terms of the operator D, a factor of $(z-1)$ can be removed, and identifying the coefficient of $z^i/i!$ in the resultant equation gives a recurrence formula involving at most $\max(p+1, q+2)$ of the probabilities.

Equation (2.93) has the form

$$h_1(z, \theta)G(z) = \lambda h_2(z, \theta)G(z). \qquad (2.94)$$

The generating function for moments about c can be obtained by substituting e^t for z and $(c + D)$ for θ in (2.94); identification of the coefficient of $t^i/i!$

then leads to a recurrence formula for the moments about c, where $c = 0$ and $c = \mu_1' = \mu$ give the uncorrected and central moments, respectively. Examples of the use of these recurrence relations and results for the cumulants and factorial cumulants are in Kemp and Kemp (1974). Tripathi and Gurland (1979) have investigated over–, under–, and equi–dispersion aspects of GHFD.

The distribution with pgf

$$G(z) = {}_pF_q[a_1, \ldots, a_p; b_1, \ldots, b_q; \lambda(z-1)], \qquad (2.95)$$

is the limiting form as $v \to \pm\infty$ for those with pgf's:

$$G(z) = {}_{p+1}F_q[a_1, \ldots, a_p, u+v; b_1, \ldots, b_q; \lambda v^{-1}(z-1)], \qquad (2.96)$$

$$G(z) = {}_pF_{q+1}[a_1, \ldots, a_p; b_1, \ldots, b_q, u+v; \lambda v(z-1)], \qquad (2.97)$$

and

$$G(z) = {}_{p+1}F_{q+1}[a_1, \ldots, a_p, v; b_1, \ldots, b_q, u+kv; \lambda k(z-1)]. \qquad (2.98)$$

As in the case of HP-distributions, the Poisson and binomial are ultimate limiting forms for nonterminating and terminating HF-distributions, respectively. For applications to matching and occupancy distributions, see Kemp (1978b).

Contagion models and time-dependent stochastic processes leading to GHFD are discussed in Kemp (1968a).

Beta and gamma mixtures of GHFD yield another GHFD, subject to existence conditions, see Chapter 8, Section 3.4.

The relationship in inventory decision theory between a demand distribution and its STER distribution was expressed by Patil and Joshi (1968) in the form of an integral; see Chapter 11, Section 12. Kemp and Kemp (1969b) showed that applying Patil and Joshi's result to a demand distribution with pgf (2.88) shifted to support $1, 2, \ldots$ yields another GHFD with support $0, 1, \ldots$ and altered parameters.

In the last section we saw that a weighted GHPD, with sampling chance (weighting factor) γ_x set equal to x, is a GHPD on $1, 2, \ldots$ with the original numerator and denominator parameters increased by unity. A parallel result holds for GHFD; see Kemp (1974). Suppose, however, that the sampling chance has the form $\gamma_x = \gamma^x$. Then the distribution with pgf (2.88) is ascertained not as a GHFD, but as a GHRD (a generalized hypergeometric recast distribution) with pgf

$$G(z) = \frac{{}_pF_q[a_1, \ldots, a_p; b_1, \ldots, b_q; \lambda(\gamma z - 1)]}{{}_pF_q[a_1, \ldots, a_p; b_1, \ldots, b_q; \lambda(\gamma - 1)]} \qquad (2.99)$$

[Kemp (1974)].

Pgf's for GHRD can be stated in the general form

$$G(z) = \frac{{}_pF_q[a_1, \ldots, a_p; b_1, \ldots, b_q; \lambda z + \xi]}{{}_pF_q[a_1, \ldots, a_p; b_1, \ldots, b_q; \lambda + \xi]} \tag{2.100}$$

(see Section 4.1 of this chapter); $\xi = 0$ gives a GHFD, and $\xi = -\lambda$ gives a GHPD.

Haight's accident distribution [Haight (1965)] with pgf

$$G(z) = \frac{{}_2F_1[N + b, b; b + 1; (Tz - T - a)/t]}{{}_2F_1[N + b, b; b + 1; -a/t]} \tag{2.101}$$

and Kemp's (1968c) *limited risk* ${}_cP_p$ *distribution* with pgf

$$G(z) = \frac{{}_1F_1[b; b + 1; \psi Tz - \psi T - \psi a]}{{}_1F_1[b; b + 1; -\psi a]} \tag{2.102}$$

both belong to the GHRD family. The *burnt fingers distribution* is closely related. This was derived by Greenwood and Yule (1920) on the assumption that the occurrence of the first accident reduces the probability of future accidents; see also McKendrick (1926) and Irwin (1953). Kemp (1968c) showed that its pgf is

$$G(z) = \frac{1 - p + pz\,{}_1F_1[b; b + 1; \psi Tz - \psi T - \psi a]}{1 - p + p\,{}_1F_1[b; b + 1; -\psi a]}, \qquad 0 < p < 1. \tag{2.103}$$

5 DISTRIBUTIONS BASED ON LAGRANGIAN EXPANSIONS

5.1 Otter's Multiplicative Process

The multiplicative process corresponding to the equation

$$\mathcal{G}(z, w) \equiv z f(w) - w = 0, \tag{2.104}$$

where $f(w)$ is the pgf of the number of segments from any vertex in a rooted tree with numbered vertices, and $w = \mathcal{P}(z)$ is the pgf for the number of vertices in the rooted tree, was first studied by Otter (1949). He showed that the number of vertices n segments removed from the root can be interpreted as the number of members in the nth generation of a branching process, and hence that his process has applications in the study of population growth, in the spread of rumors and epidemics, and in nuclear chain reactions.

Otter took as his first example

$$f(w) = p_0 + p_1 w + p_2 w^2. \tag{2.105}$$

This yields

$$w = \mathcal{P}(z) = \frac{1 - p_1 z - [(1 - p_1 z)^2 - 4p_0 p_2 z^2]^{1/2}}{2p_2 z}. \tag{2.106}$$

Kemp and Kemp (1969a) showed that when $f(w)$ is the pgf for a binomial distribution with pgf

$$f(w) = (p + qw)^2, \tag{2.107}$$

we have $p_1^2 = 4p_0 p_2$, and hence Otter's multiplicative process gives a lost-games distribution (Chapter 11, Section 10) with pgf

$$w = \mathcal{P}(z) = \frac{\{1 - (1 - 4pqz)^{1/2}\}^2}{4q^2 z}. \tag{2.108}$$

Other models that are based on Otter's process and lead to particular lost-games distributions were also investigated by Kemp and Kemp (1969a).

Neyman and Scott (1964) recognized the relevance of Otter's multiplicative process to their study of stochastic models for the total number of individuals infected during an epidemic started by a single infective individual; see also Neyman (1965). Their epidemiological model also uses (2.107).

For his second example Otter took

$$f(w) = e^{\lambda(z-1)} \tag{2.109}$$

and, using a Lagrangian reversion technique, found that for $\lambda < 1$, $\mathcal{P}(z) = \sum_x P_x z^x$, where

$$P_x = \frac{e^{-n\lambda}(n\lambda)^{n-1}}{n!}; \tag{2.110}$$

this is the Borel–Tanner distribution; see Chapter 9, Section 11.

Otter proved that in the general case, where $f(w) = \sum_{j=0}^{\infty} p_j w^j$, the probabilities for the pgf $\mathcal{P}(z) = \sum_x P_x z^x$ can be obtained as

$$P_x = \frac{1}{x!} \left[\frac{\partial^{x-1}}{\partial w^{x-1}} (f(w))^x \right]_{w=0} \tag{2.111}$$

by a Lagrangian expansion.

5.2 Lagrangian Distributions

Lagrangian expansions for the derivation of expressions for the probabilities of certain discrete distributions have been used for many years. Early examples include Otter's multiplicative process (see the previous section) and

Haight and Breuer's (1960) treatment of the Borel-Tanner (Tanner-Borel) distribution (for the latter see Chapter 9, Section 11).

The potential of this technique for deriving distributions and their properties has been systematically exploited by Consul and Shenton and their co-workers, beginning with a group of key papers in the early 1970's. Research concerning this class of distributions continues to be very prolific. We begin by mentioning the most important of these distributions.

The *Lagrangian binomial distribution* (see Chapter 3, Section 12) was obtained by Mohanty (1966) as the distribution of the number of failures x encountered in getting $\beta x + n$ successes given a sequence of independent Bernoulli trials. It was given a queueing process interpretation by Takacs (1962) and Mohanty (1966). Jain and Consul (1971) derived an analogous Lagrangian negative binomial distribution (see Chapter 5, Section 12.3).

The *Lagrangian Poisson distribution* was obtained by Consul and Jain (1973a) as a limiting form of the *Lagrangian negative binomial distribution*; its very close relationship to a shifted Borel-Tanner distribution has been increasingly appreciated (see Chapter 9, Section 11, for the Lagrangian Poisson and Borel-Tanner distributions). Consul's (1989) book on the Lagrangian Poisson distribution highlights how intensively it has been studied for its many properties, and also for its various modes of genesis.

The *Lagrangian logarithmic distribution* (see Chapter 7, Section 11) was introduced into the literature by Jain and Gupta (1973).

The nature of the generalization process for these distributions was clarified in two important papers by Consul and Shenton (1972, 1973), see also Consul (1983).

Consider, first, the most basic Lagrangian distributions. Let $g(z)$ be a pgf defined on the nonnegative integers with $g(0) \neq 0$. Then the numerically smallest root, $z = \ell(u)$, of the equation $z = ug(z)$, defines a pgf $z = \ell(u)$ with the following Lagrangian expansion in powers of u:

$$z = \ell(u) = \sum_{x=1}^{\infty} \left[\frac{u^x}{x!} \frac{\partial^{x-1}}{\partial z^{x-1}} (g(z))^x \right]_{z=0}. \qquad (2.112)$$

The corresponding probabilities are

$$\Pr[X = x] = \frac{1}{x!} \left[\frac{\partial^{x-1}}{\partial z^{x-1}} (g(z))^x \right]_{z=0}. \qquad (2.113)$$

Notice that there are changes in notation compared with the notation for Otter's multiplicative process. Otter's key transformation

$$z\, f(w) - w = 0$$

becomes $\qquad\qquad u\, g(z) - z = 0.$

Examples of such *basic Lagrangian-type distributions* are as follows:

1. the *geometric distribution* with support $1, 2, \ldots$, for which $g(z) = 1 - p + pz$, $0 < p < 1$ (that is, $g(z)$ is the pgf for a Bernoulli rv);
2. the *Borel-Tanner distribution*, for which $g(z)$ is Poissonian, with $g(z) = e^{\lambda(z-1)}$, $0 < \lambda$;
3. the *Haight distribution* [Haight (1961a)] with index parameter equal to unity, for which $g(z)$ is geometric with pgf $g(z) = (1 + P - Pz)^{-1}$, $0 < P$;
4. the *random walk distribution* with pgf

$$z = \ell(u) = (1 - \sqrt{1 - 4pqu^2})/(2qu),$$

for which $g(z)$ is a Bernoulli-doublet distribution with pgf $g(z) = 1 - p + pz^2$, $0 < p < 1$;
5. the *Consul distribution* with pmf

$$\Pr[X = x] = \frac{1}{x} \binom{mx}{x-1} \left(\frac{\theta}{1-\theta} \right)^{x-1} (1-\theta)^{mx},$$

for which $g(z)$ is binomial with pgf $g(z) = (1 - \theta + \theta z)^m$, where m is a positive integer [see Consul and Shenton (1975) and Consul (1983)].

Consul and Shenton (1975) showed that all Lagrangian distributions of this basic type are closed under convolution, and that their first four cumulants, $\kappa_i = D_i$, $i = 1, 2, 3, 4$, are

$$\kappa_1 = D_1 = \mu = \left[\frac{1}{1 - g'(z)} \right]_{z=1} = \frac{1}{1 - G_1},$$

$$\kappa_2 = D_2 = G_2 \mu^3,$$

$$\kappa_3 = D_3 = G_3 \mu^4 + 3G_2^2 \mu^5,$$

$$\kappa_4 = D_4 = G_4 \mu^5 + 10G_3 G_2 \mu^6 + 15G_2^3 \mu^7, \tag{2.114}$$

where G_i is the ith cumulant of the distribution with pgf $g(z)$.

Consul and Shenton also gave formulas for the higher cumulants and for the uncorrected moments, and they interpreted these distributions as busy-period distributions in queueing theory.

Second, we come to *general Lagrangian-type distributions*. Suppose that $f(z)$ is another pgf for which

$$\left[\frac{\partial^{x-1}}{\partial z^{x-1}} \left\{ (g(z))^x \frac{\partial f(z)}{\partial z} \right\} \right]_{z=0} \geq 0 \qquad \text{for } x \geq 1. \tag{2.115}$$

(Note that this is not the same as Otter's $f(\cdot)$.) Then the pgf for the general Lagrangian distribution formed from $f(z)$ and $g(z)$, where $z = ug(z)$, is

$$f(\ell(u)) = f(0) + \sum_{x>0} \frac{u^x}{x!} \left[\frac{\partial^{x-1}}{\partial z^{x-1}} \left\{ (g(z))^x \frac{\partial f(z)}{\partial z} \right\} \right]_{z=0}, \qquad (2.116)$$

and its probabilities are

$$\Pr[X = 0] = f(0),$$

$$\Pr[X = x] = \frac{1}{x!} \left[\frac{\partial^{x-1}}{\partial z^{x-1}} \left\{ (g(z))^x \frac{\partial f(z)}{\partial z} \right\} \right]_{z=0}, \qquad x > 0. \qquad (2.117)$$

Consul and Shenton (1972) presented in table form the outcomes from more than a dozen combinations of specific pgf's for $g(z)$ and $f(z)$.

The following theorems hold:

1. Distributions with $f(z) = z^n$ are n-fold convolutions of the basic Lagrangian-type distribution.
2. The general distribution with $f(z) = g(z)$ is the same as the corresponding basic Lagrangian-type distribution, except that it is shifted one step to the left.
3. The general Lagrangian distribution can be derived by randomizing the index parameter m in the distribution with pgf z^m according to another distribution with pgf $f(z)$.
4. All the general Lagrangian distributions corresponding to the same basic Lagrangian distribution are closed under convolution.

Limiting forms of the distributions were considered in Consul and Shenton (1973), where it was proved that under one set of limiting conditions all discrete Lagrangian distributions tend to normality, and that under another set of limiting conditions they tend to inverse Gaussian distributions.

Consider now the moment properties of general Lagrangian distributions. These can be obtained straightforwardly via their cumulants. Let F_r be the rth cumulant for the pgf $f(z)$ as a function of z, and let D_r be the rth cumulant for the basic Lagrangian distribution obtained from $g(z)$ (as in the equations (2.114)). Then

$$\kappa_1 = F_1 D_1,$$

$$\kappa_2 = F_1 D_2 + F_2 D_1^2,$$

$$\kappa_3 = F_1 D_3 + 3F_2 D_1 D_2 + F_3 D_1^3,$$

$$\kappa_4 = F_1 D_4 + 3F_2 D_2^2 + 4F_2 D_1 D_3 + 6F_3 D_1^2 D_2 + F_4 D_1^4 \qquad (2.118)$$

[Consul and Shenton (1975)].

Minimum variance unbiased estimation for Lagrangian distributions has been examined by Consul and Famoye (1989).

In Consul (1981) the author considerably widened the scope of Lagrangian distributions by removing the restriction that $g(z)$ and $f(z)$ be pgf's. Instead, $g(z)$ and $f(z)$ are assumed to be two functions that are successively differentiable, with $g(1) = f(1) = 1$, $g(0) \neq 0$, and $0 \leq f(0) < 1$. As an example Consul instanced the use of $g(z) = (1 - p + pz)^{3/2}$ and $f(z) = (1 - p + pz)^{1/2}$; this leads to a valid distribution.

The advantage of Consul's extended definition of a Lagrangian distribution is that it generates a wider class of distributions that encompasses not only Patil's generalized power series distributions but also Gupta's modified power series distributions. However, models that have been constructed for general Lagrangian distributions will not on the whole be valid for distributions in the extended class.

The application of Lagrangian-type distributions in the theory of *random mappings* has been researched by Berg and Mutafchiev (1990) and Berg and Nowicki (1991).

Devroye (1992) has studied the computer generation of Lagrangian-type variables.

5.3 Gould and Abel Distributions

Gould series distributions are generated by expressing a suitable function as a series of Gould polynomials. Suppose that $A(s;r)$ is a positive function of a parameter s that may also depend on a second parameter r; assume also that

$$A(s;r) = \sum_{x=0}^{\infty} a(x;r)s(s + rx - 1)_{x-1}, \qquad (2.119)$$

where $(s, r) \in S_0 \times R_0$, $S_0 = \{s; |s| < s_0\}$, $R_0 = \{r; |r| < r_0\}$ and the coefficients $a(x;r)$, $x = 0, 1, 2, \ldots$, are independent of s. Then

$$\Pr[X = x; s, r] = p(x; s, r) = [A(s;r)]^{-1}a(x;r)s(s + rx - 1)_{x-1}, \qquad (2.120)$$

$x = 0, 1, 2, \ldots$, satisfies the properties of a probability mass function, provided that the parameter space $S_0 \times R_0$ is restricted to $S \times R$ in such a way that the terms of the expansion (2.119) are nonnegative; both $\{a(x;r) \geq 0, 0 \leq s < s_0, 0 \leq r < r_0\}$ and $\{(-1)^x a(x;r) \geq 0, -s_0 < s \leq 0, -r_0 < r \leq 0\}$ are possibilities. The polynomials

$$G_x(s;r) = s(s + rx - 1)_{x-1}, \quad x = 1, 2, \ldots,$$

$$= s(s + rx - 1)(s + rx - 2)\ldots(s + rx - x + 1),$$

$$G_0(s;r) = 1,$$

were introduced by Gould (1962) and were called Gould polynomials by Roman and Rota (1978).

The family has been developed by Charalambides (1986a), who found that they occur in fluctuations of sums of interchangeable random variables. He showed that they have applications concerning the busy period in queueing processes and the time to emptiness in dam and storage processes. The family includes the generalized binomial and negative binomial distributions of Jain and Consul (1971) and Consul and Shenton (1972); these distributions have the pmf

$$\Pr[X = x] = \frac{s}{s + rx}\binom{s + rx}{x}p^x(1 - p)^{s+rx-x}, \quad x = 0, 1, 2, \ldots, \quad (2.121)$$

where the parameters satisfy (1) s, r positive integers and $0 < p < \min\{1, 1/r\}$, and (2) $s, r < 0$ and $1/r < p < 0$, respectively. Other members of the family are the *quasi–hypergeometric I* and *quasi–Pólya I* distributions of Consul (1974) [also Janardan (1975)] with pmf

$$\Pr[X = x] = \binom{n}{x}\frac{s(s + rx - 1)_{x-1}(m + r(n - x))_{n-x}}{(s + m + rn)_n}, \quad x = 0, 1, 2, \ldots, n, \quad (2.122)$$

where s, r, m are positive integers and s, r, m are negative real numbers, respectively, and the *quasi–hypergeometric II* and *quasi–Pólya II* distributions of and Consul and Mittal (1975), with pmf

$$\Pr[X = x] = \binom{n}{x}\frac{s(s + rx - 1)_{x-1}m(m + r(n-))_{n-x-1}}{(s + m)(s + m + rn - 1)_{n-1}}, \quad x = 0, 1, 2, \ldots, n, \quad (2.123)$$

where again s, r, m are positive integers and negative real numbers, respectively.

Charalambides has obtained closed–form expressions for the pgf and the factorial moments of such distributions. He has also shown that the generalized binomial distribution with pmf (2.121) is the only member of the Gould series family that is closed under convolution.

More recently Charalambides (1990) has explored the use of Abel series. An *Abel series distribution* arises when a positive function $A_b(\theta; \lambda)$ is expanded as

$$A_b(\theta; \lambda) = \sum_{x=0}^{\infty} a_b(x; \lambda)\theta(\theta + \lambda x)^{x-1}, \quad (2.124)$$

where $(\theta, \lambda) \in \Theta_0 \times \Lambda_0$, $\Theta_0 = \{\theta; |\theta| < \rho_1\}$, $\Lambda_0 = \{\lambda; |\lambda| < \rho_2\}$, and the coefficients $a_b(x; \lambda)$, $x = 0, 1, 2, \ldots$, are independent of the parameter θ. Suppose that the parameter space $\Theta_0 \times \Lambda_0$, is restricted to $\Theta \times \Lambda$ in such a way that the terms of (2.124) are all nonnegative; then the pmf

$$\Pr[X = x] = [A_b(\theta; \lambda)]^{-1}a_b(x; \lambda)\theta(\theta + \lambda x)^{x-1}, \quad x = 0, 1, 2, \ldots, \quad (2.125)$$

is a valid pmf. The "generalized Poisson" distribution of Consul and Jain (1973a, b) is a member of the family; see Chapter 9, Section 11. So are the *quasi–binomial I and II distributions*, with pmf's

$$\Pr[X = x] = (1 + \theta)^{-n} \binom{n}{x} (1 - \lambda x)^{n-x} \theta (\theta + \lambda x)^{x-1} \tag{2.126}$$

[Consul (1974)] and

$$\Pr[X = x] = \binom{n}{x} \frac{(1 + \lambda n - \lambda x)^{n-x-1} \theta (\theta + \lambda x)^{x-1}}{(1 + \theta)(1 + \theta + \lambda n)^{n-1}} \tag{2.127}$$

[Consul and Mittal (1975)], respectively, where in both cases $x = 0, 1, 2, \ldots, n$ and $0 < \theta < \infty$, $0 \le \lambda < 1/n$.

Charalambides has given new modes of genesis for these distributions and has investigated their occurrence in insurance risk, queueing theory, and dam and storage processes. In his study of their properties he has obtained formulas for their factorial moments and has shown that, within the family of Abel distributions, Consul and Jain's "generalized Poisson" distribution is the only one that is closed under convolution.

Nandi and Dutta (1988) have developed a family of *Bell distributions* based on a wide generalization of the Bell numbers.

Janardan (1988) and Charalambides (1991) have investigated *generalized Eulerian distributions*. These have pgf's involving generalized Eulerian polynomials; their factorial moments can be stated in terms of noncentral Stirling numbers.

6 FACTORIAL SERIES DISTRIBUTIONS

Whereas power series distributions are based on expansions of the pgf as a Taylor series and modified power series distributions often relate to Lagrangian expansions, the family of *factorial series distributions* (FSD) is based on expansions of the pgf as a factorial series, using forward differences of a defining function $A(N)$; see Berg (1974, 1975, 1983a). More specifically, let $A(N)$ be a real function of an integer parameter N, and suppose that $A(N)$ can be expressed as a factorial series in N with nonnegative coefficients,

$$A(N) \equiv \sum_{x=0}^{N} \binom{N}{x} [\Delta^x A(N)]_{N=0} = \sum_{x=0}^{N} \frac{N!}{(N-x)!} a_x, \tag{2.128}$$

with $a_x \geq 0$ for $x = 0, 1, \ldots, N$, a_x not involving N. Then the corresponding FSD has pgf

$$G(z) = \sum_{x=0}^{N} \frac{N! a_x z^x}{(N-x)!} \qquad (2.129)$$

and pmf

$$\Pr[X = x] = \binom{N}{x} \frac{\Delta^x A(0)}{A(N)} = \frac{N! a_x}{(N-x)! A(N)}, \qquad x = 0, 1, \ldots, N, \quad (2.130)$$

where $\Delta^x A(0)$ is understood to mean $[\Delta^x A(N)]_{N=0}$.

Berg (1983a) commented that the class of FSD is the discrete parameter analogue of the class of PSD; he pointed out that the two classes have certain properties in common.

Consideration of the rv $Z = N - X$ leads to the following expression for the factorial moments of Z:

$$E\left[\frac{Z!}{(Z-r)!}\right] = \frac{N!}{(N-r)!} \frac{A(N-r)}{A(N)}, \qquad (2.131)$$

and hence the factorial moments of X can be proved to be

$$E\left[\frac{X!}{(X-r)!}\right] = \mu'_{[r]} = \frac{N!}{(N-r)!} \frac{\nabla^r A(N)}{A(N)}, \qquad (2.132)$$

where ∇ denotes the backward difference operator. This gives

$$\mu = N \cdot \frac{A(N) - A(N-1)}{A(N)}$$

$$\mu_2 = N(N-1) \cdot \frac{A(N-2)}{A(N)} + \frac{NA(N-1)}{A(N)} - \left[\frac{NA(N-1)}{A(N)}\right]^2. \quad (2.133)$$

The genesis of FSD's [Berg (1983a)] can be understood by considering a finite collection of exchangeable events, $E_1, \ldots, E_N$ (where the probability of the occurrence of r events out of N depends only on N and r, and not on the order in which the events might occur). Let $S_{r,N}$ denote the probability that $N - r$ specific events fail to occur. Then by the inclusion–exclusion principle (see Chapter 10, Section 2), the probability that exactly Z events do not occur is

$$\Pr[Z = z] = \sum_{j=N-z}^{N} (-1)^{N-z-j} \binom{j}{N-z} \binom{N}{j} \overline{S_{N-r,N}}$$

$$= \binom{N}{z} \Delta^z \overline{S_{0,N}}. \qquad (2.134)$$

The probability that exactly x of the events occur is obtained by putting $z = N - x$ in (2.134). If $\overline{S_{0,N}}$ has the property of factorizing into two parts, one depending only on r and the other only on N, the outcome is a FSD.

Examples of FSD include the binomial and hypergeometric distributions [Berg (1974] and the Stevens–Craig and matching distributions [Berg (1978)]. Related distributions are the Waring, generalized Waring, and Yule distributions, and Marlow's factorial distribution [Marlow (1965); Berg (1983a)]; see the index for references to these distributions elsewhere in this volume.

Inference for FSD's has been studied in a series of papers by Berg; see Berg (1983a). Subject to a restriction on N, a unique MVUE for N can be obtained [Berg (1974)]; moment and ML estimation were also discussed and the problem raised as to when an ML solution exists. Estimation theory for FSD's was applied in this paper to certain types of capture–recapture sampling. The close relationship between estimation for a zero–truncated power series distribution and estimation for a FSD was investigated by Berg (1975). The concept of a moment distribution was applied to the family of FSD's by Berg (1978).

Generalized factorial series distributions are defined in Berg (1983b) as having pmf's of the form

$$\Pr[X = x] = \frac{(N - m)!}{(M - m - x)!} \frac{a_x}{A(N; n(x))}, \qquad x = 0, 1, \ldots, N - m, \quad (2.135)$$

where m is an integer such that $0 < m < N$. The paper discusses their relationship to *snowball sampling* and to the *Reed–Frost chain binomial model*. The extra parameter m has to do with the starting value of the process, while $n(x)$ specifies the stopping rule of the sampling process.

CHAPTER 3

Binomial Distribution

1 DEFINITION

The *binomial distribution* can be defined, using the binomial expansion

$$(q + p)^n = \sum_{x=0}^{n} \binom{n}{k} p^k q^{n-k} = \sum_{x=0}^{n} \frac{n!}{k!(n-k!)} p^k q^{n-k},$$

as the distribution of a random variable X for which

$$\Pr[X = x] = \binom{n}{x} p^x q^{n-x}, \qquad x = 0, 1, 2, \ldots, n, \qquad (3.1)$$

where $q + p = 1$, $p > 0$, $q > 0$ and n is a positive integer. Occasionally a more general form is used in which the variable X is transformed to $a + bX$, where a and b are real numbers with $b \neq 0$.

When $n = 1$, the distribution is known as the *Bernoulli distribution*.

The cf of the binomial distribution is $(1 - p + pe^{it})^n$, and the pgf is

$$G(z) = (1 - p + pz)^n = (q + pz)^n, \qquad 0 < p < 1,$$

$$= \frac{{}_1F_0[-n; \; ; -pz/q]}{{}_1F_0[-n; \; ; -p/q]} \qquad (3.2)$$

$$= {}_1F_0[-n; \; ; p(1 - z)]. \qquad (3.3)$$

The mean and variance are

$$\mu = np \quad \text{and} \quad \mu_2 = npq. \qquad (3.4)$$

The distribution is a power series distribution with finite support. From (3.2) it is a generalized hypergeometric probability distribution and from (3.3) it is a generalized hypergeometric factorial moment distribution. It is a

105

member of the exponential family of distributions (when n is known), and it is an Ord and also a Katz distribution; for more details see Section 4 of this chapter.

2 HISTORICAL REMARKS AND GENESIS

If n independent trials are made, and in each there is probability p that the outcome E will occur, then the number of trials in which E occurs can be represented by a random variable x having the binomial distribution with parameters n, p. This situation occurs when a sample of fixed size n is taken from an *infinite* population where each element in the population has an equal and independent probability p of possesion of a specified attribute.The situation also arises when a sample of fixed size n is taken from a *finite* population where each element in the population has an equal and indepen-dent probability p of having a specified attribute and elements are sampled independently and sequentially with replacement.

The binomial distribution is one of the oldest to have been the subject of study. The distribution was derived by James Bernoulli (in his treatise *Ars Conjectandi*, published in 1713), for the case $p = r/(r + s)$ where r and s are positive integers. Earlier Pascal had considered the case $p = 1/2$. In his *Essay*, published posthumously in 1764 , Bayes removed the rational restriction on p by considering the position relative to a randomly rolled ball of a second ball randomly rolled n times. The early history of the distribution is discussed, inter alia, by Boyer (1950), Stigler (1986), Edwards (1987), and Hald(1990).

A remarkable new derivation as the solution of the simple birth-and-emigration process was given by McKendrick (1914). The distribution may also be regarded as the stationary distribution for the Ehrenfest model [Feller (1957)]. Haight (1957) has shown that the $M/M/1$ queue with balking gives rise to the distribution, provided that the arrival rate of the customers when there are n customers in the queue is $\lambda = (N - n)N^{-1}(n + 1)^{-1}$ for $n < N$ and zero for $n \geq N$ (N is the maximum queue size).

3 MOMENTS

The moment generating function is $(q + pe^t)^n$, and the cumulant generat-ing function is $n \ln(q + pe^t)$. The factorial cumulant generating function is $n \ln(1+pt)$, whence $\kappa_{[r]} = n(r-1)!p^r$. The factorial moments can be obtained straightforwardly from the fmgf which is $(1 + pt)^n$. We have

$$\mu'_{[r]} = \frac{n!p^r}{(n - r)!};$$

i.e.
$$\mu'_{[1]} = \mu = np,$$

$$\mu'_{[2]} = n(n-1)p^2,$$

$$\mu'_{[3]} = n(n-1)(n-2)p^3, \qquad \text{etc.} \tag{3.5}$$

From $\mu'_r = \sum_{j=0}^{r} S(r,j)\mu'_{[j]}$ (see Chapter 1, Section B5), it follows that the rth moment about zero is

$$\mu'_r = E[X^r] = \sum_{j=0}^{r} \frac{S(r,j)n!p^r}{(n-r)!}. \tag{3.6}$$

In particular

$$\mu'_1 = np,$$

$$\mu'_2 = np + n(n-1)p^2,$$

$$\mu'_3 = np + 3n(n-1)p^2 + n(n-1)(n-2)p^3,$$

$$\mu'_4 = np + 7n(n-1)p^2 + 6n(n-1)(n-2)p^3 + n(n-1)(n-2)(n-3)p^4.$$

$$\tag{3.7}$$

Hence (or otherwise) the central moments can be obtained. The lower-order central moments are

$$\mu_2 = \sigma^2 = npq,$$

$$\mu_3 = npq(q-p),$$

$$\mu_4 = 3(npq)^2 + npq(1-6pq). \tag{3.8}$$

The moment ratios $\sqrt{\beta_1}$ and β_2 are

$$\sqrt{\beta_1} = (q-p)(npq)^{-1/2},$$

$$\beta_2 = 3 + (1-6pq)(npq)^{-1}. \tag{3.9}$$

For a fixed value of p (and so of q) the (β_1, β_2) points fall on the straight line

$$\frac{\beta_2 - 3}{\beta_1} = \frac{1 - 6pq}{(q-p)^2} = 1 - \frac{2pq}{(q-p)^2}. \tag{3.10}$$

As $n \to \infty$, the points approach the limit $(0, 3)$.

Note that the same straight line is obtained when p is replaced by q. The two distributions are mirror images of each other, so they have identical values of β_2 and the same absolute value of $\sqrt{\beta_1}$. The slope $(\beta_2 - 3)/\beta_1$ is always less than 1. The limit of the ratio as p approaches 0 or 1 is 1.

For $p = q = 0.5$ the binomial distribution is symmetrical and $\beta_1 = 0$. For $n = 1$, the point (β_1, β_2) lies on the line $\beta_2 - \beta_1 - 1 = 0$. (Note that for *any* distribution, $\beta_2 - \beta_1 - 1 \geq 0$.)

Romanovsky (1923) derived the following recursion formula for the central moments:

$$\mu_{r+1} = pq\left(nr\mu_{r-1} + \frac{d\mu_r}{dp}\right). \tag{3.11}$$

A somewhat similar relation holds for moments about zero

$$\mu'_{r+1} = pq\left\{\left(\frac{n}{q}\right)\mu'_r + \frac{d\mu'_r}{dp}\right\}. \tag{3.12}$$

Kendall (1943) used differentiation of the cf to derive the relationship

$$\mu_r = npq\sum_{j=0}^{r-2}\binom{r-1}{j}\mu_j - p\sum_{j=0}^{r-2}\binom{r-1}{j}\mu_{j+1}. \tag{3.13}$$

A simpler recursion formula holds for the cumulants

$$\kappa_{r+1} = pq\frac{\partial\kappa_r}{\partial p}, \qquad r \geq 1. \tag{3.14}$$

Formula (3.12) also holds for the incomplete moments, defined as

$$\mu'_{j,k} = \sum_{i=k}^{n}i^j\binom{n}{i}p^iq^{n-i}.$$

The mean deviation is

$$\nu_1 = E[|X - np|] = 2n\binom{n-1}{[np]}p^{[np]+1}q^{n-[np]}, \tag{3.15}$$

where $[\cdot]$ denotes the integer part [Bertrand (1889); Frisch (1924); Frame (1945)]. Diaconis and Zabell (1991) have discussed the provenance and import of this formula and other equivalent formulas. They found that ν_1 is an increasing function of n, but that ν_1/n is a decreasing function of n. Johnson's (1957) article led to a number of generalizations. Using Stirling's approximation for $n!$,

$$\nu_1 \doteq \left(\frac{2npq}{\pi}\right)^{1/2}\left[1 + \frac{(np - [np])(nq - [nq])}{2npq} - \frac{1 - 2pq}{12npq}\right];$$

this shows that the ratio of the mean deviation to the standard deviation approaches the limiting value $(2/\pi)^{1/2} \doteq 0.798$ as $n \to \infty$.

Katti's (1960) method for obtaining the absolute moments of general order about m is to write the mgf in the form

$$E\left[e^{t|X-m|}\right] = \psi(t) - \psi(-t) + E\left[e^{t(m-X)}\right],$$

whence $$E\left[|X-m|^{2r+1}\right] = 2\psi^{(2r+1)}(0) - E\left[(X-m)^{2r+1}\right]. \qquad (3.16)$$

All inverse moments of the binomial distribution (i.e., $E(X^{-r})$ with $r = 1, 2, \ldots$) are infinite because $\Pr[X = 0] > 0$. Inverse moments of the *positive binomial distribution* (formed by zero-truncation) are discussed in Section 11.

Direct manipulation of the definition of the *inverse factorial moment*, as in Stancu (1968), yields

$$E\left[\left\{(X+r)^{(r)}\right\}^{-1}\right] = \left\{(n+r)^{(r)}\right\}^{-1} p^{-r}\left[1 - \sum_{y=0}^{r-1}\binom{n+r}{y}p^y q^{n+r-y}\right]. \qquad (3.17)$$

Chao and Strawderman (1972) obtained a general result for $E\left[(X+a)^{-k}\right]$ in terms of a k-fold multiple integral of $t^{-1}E\left[t^{X+a-1}\right]$, and they applied this to the binomial distribution. A modification of their approach enabled Lepage (1978) to express the inverse (ascending) factorial moments $R_x(a, k) = E\left[\left\{(X+a)\cdots(X+a+k-1)\right\}^{-1}\right]$ as a k-fold multiple integral of $E\left[t^{X+a-1}\right]$. Cressie et al. (1981) obtained $E\left[(X+a)^{-k}\right]$ both as a k-fold multiple integral of $E\left[e^{-t(X+a)}\right]$, and also from a single integral of $t^{k-1}E\left[e^{-t(X+a)}\right]$. Jones (1987) developed an analogous single-integral result for $R_x(a, k)$, namely

$$R_x(a, k) = \{\Gamma(k)\}^{-1}\int_0^1 (1-t)^{k-1}E\left[t^{X+a-1}\right]dt,$$

and compared the different (though equivalent) expressions that are yielded by the different approaches for $R_x(a, k)$ for the binomial distribution.

4 PROPERTIES

The binomial distribution belongs to a number of families of distributions and hence possesses the properties of each of the families.

It is a distribution with finite support. As defined by (3.1), it consists of $(n+1)$ nonzero probabilities associated with the values $0, 1, 2, \ldots, n$ of the random variable X. The ratio

$$\frac{\Pr[X = x+1]}{\Pr[X = x]} = \frac{(n-x)p}{(x+1)q}, \qquad x = 0, 1, \ldots, n-1, \qquad (3.18)$$

shows that $\Pr[X = x]$ increases with x so long as $x < np - q = (n + 1)p - 1$, and decreases with x if $x > np - q$. Hence as x increases, $\Pr[X = x]$ increases to a maximum value at the integer x satisfying $(n + 1)p - 1 < x \leq (n + 1)p$ and thereafter decreases. When $(n + 1)p$ is an integer,

$$\Pr[X = (n + 1)p - 1] = \Pr[X = (n + 1)p].$$

The distribution is therefore unimodal. When $p < (n+1)^{-1}$, the mode occurs at the origin.

The distribution is a member of the exponential family of distributions with respect to $p/(1 - p)$, since

$$\Pr[X = x] = \exp\left[x \log\left\{\frac{p}{1 - p}\right\} + \log\binom{n}{x} + n \log(1 - p)\right].$$

Morris (1982, 1983) has shown that it is one of the six subclasses of the natural exponential family for which the variance is at most a quadratic function of the mean; he used this property to obtain unified results and to gain insight concerning limit laws. Unlike the other five subclasses, however, it is not infinitely divisible (no distribution with finite support can be infinitely divisible).

Because $\Pr[X = x]$ is of the form

$$\frac{b(x)\theta^x}{\eta(\theta)}, \qquad \theta > 0, \; x = 0, 1, \ldots, n,$$

where $\theta = p/(1 - p)$ [Kosambi (1949); Noack (1950)], the distribution belongs to the important family of power series distributions (see Chapter 2, Section 2). Patil has investigated these in depth; see, e.g., Patil (1986). He has shown that for the binomial distribution

$$\frac{\theta \eta^{(r+1)}(\theta)}{\eta^{(r)}(\theta)} = \mu - pr, \tag{3.19}$$

where $\eta(\theta) = \sum_x b(x)\theta^x$. Integral expressions for the tail probabilities of power series distributions were obtained by Joshi (1974, 1975), who thereby demonstrated the duality between the binomial distribution and the beta distribution of the second kind. Indeed

$$\sum_{x=r}^{n}\binom{n}{x}p^x q^{n-x} = I_p(r, n - r + 1) = \Pr\left[F \leq \frac{v_2 p}{v_1 q}\right], \tag{3.20}$$

where F is a random variable that has an F-distribution with parameters $v_1 = 2r$, $v_2 = 2(n - r + 1)$; see Raiffa and Schlaifer (1961).

Berg (1974, 1983a) has explored the properties of the closely related family of factorial series distributions with

$$\Pr[X = x] = \frac{n^{(x)}c(x)}{x!h(n)},$$

to which the binomial distribution can be seen to belong by taking $c(x) = (p/q)^x$.

Expression (3.18) shows that the binomial distribution belongs to the discrete Pearson system [Katz (1945, 1965); Ord (1967b)]. Tripathi and Gurland (1977, 1979) have examined methods for selecting from those distributions having

$$\frac{\Pr[X = x + 1]}{\Pr[X = x]} = \frac{A + Bx}{C + Dx + Ex^2}$$

a particular member such as the binomial.

Kemp (1968a, 1968b) has shown that the binomial is a generalized hypergeometric distribution with pgf

$$G(z) = \frac{{}_1F_0[-n; \; ;pz/(p-1)]}{{}_1F_0[-n; \; ;p/(p-1)]}.$$

Moreover, because the ratio of successive factorial moments is $(n - r)p$, the distribution is also a generalized hypergeometric factorial moment distribution with pgf ${}_1F_0[-n; \; ;p(1 - z)]$ [Kemp (1968a); Kemp and Kemp 1974)]. Its membership of those families enabled Kemp and Kemp to obtain differential equations and associated difference equations for the pgf and various moment generating functions, including the generating functions for the incomplete and the absolute moments.

The binomial is an increasing failure-rate distribution [Barlow and Proschan (1965)]. The Mill's ratio for a discrete distribution is defined as $\sum_{j \geq x} \Pr[X = j]/\Pr[X = x]$, and therefore it is the reciprocal of the failure rate. Diaconis and Zabell (1991) showed that the Mill's ratio for the binomial distribution satisfies

$$\frac{x}{n} \leq \frac{\sum_{j=x}^{n} \Pr[X = j]}{\Pr[X = x]} \leq \frac{x(1 - p)}{x - np},$$

provided that $x > np$. The binomial distribution is also a monotone likelihood-ratio distribution. The *skewness* of the distribution is positive if $p < \frac{1}{2}$ and is negative if $p > \frac{1}{2}$. The distribution is symmetrical if and only if $p = \frac{1}{2}$.

Denoting $\Pr[X \leq c]$ by $L_{n,c}(p)$, Uhlmann (1966) has shown that, for $n \geq 2$,

$$L_{n,c}\left(\frac{c}{n-1}\right) > \frac{1}{2} > L_{n,c}\left(\frac{c+1}{n+1}\right) \qquad \text{for} \quad 0 \leq c < \frac{n-1}{2},$$

$$L_{n,c}\left(\frac{c}{n-1}\right) = \frac{1}{2} = L_{n,c}\left(\frac{c+1}{n+1}\right) \qquad \text{for} \quad c = \frac{n-1}{2},$$

and $\qquad L_{n,c}\left(\frac{c+1}{n+1}\right) > \frac{1}{2} > L_{n,c}\left(\frac{c}{n-1}\right) \qquad \text{for} \quad \frac{n-1}{2} < c \le n. \quad (3.21)$

The distribution of the *standardized binomial variable*

$$X' = \frac{X - np}{\sqrt{npq}}$$

tends to the unit normal distribution as $n \to \infty$; that is, for any real numbers α, β (with $\alpha < \beta$)

$$\lim_{n \to \infty} \Pr[\alpha < X' < \beta] = \frac{1}{\sqrt{2\pi}} \int_\alpha^\beta e^{-u^2/2} du. \qquad (3.22)$$

This result is known as the *De Moivre-Laplace theorem*. It forms a starting point for a number of approximations in the calculation of binomial probabilities; these will be discussed in Section 6.

In Section 1 we noted that the pgf of X is $(q + pz)^n$. If X_1, X_2 are independent random variables having binomial distributions with parameters n_1, p and n_2, p, respectively, then the pgf of $X_1 + X_2$ is $(q + pz)^{n_1}(q + pz)^{n_2} = (q + pt)^{n_1+n_2}$. Hence $X_1 + X_2$ has a binomial distribution with parameters $n_1 + n_2, p$. This property is also apparent on interpreting $X_1 + X_2$ as the number of occurrences of an outcome E having constant probability p in each of $n_1 + n_2$ independent trials.

The distribution of X_1, conditional on $X_1 + X_2 = x$, is

$$\Pr[X_1 = x_1 | x] = \frac{\binom{n_1}{x} p^{x_1} q^{n_1-x} \binom{n_2}{x-x_1} p^{x-x_1} q^{n_2-x+x_1}}{\binom{n_1+n_2}{x} p^x q^{n_1+n_2-x}}$$

$$= \binom{n_1}{x_1}\binom{n_2}{x-x_1} \bigg/ \binom{n_1+n_2}{x}, \qquad (3.23)$$

where $\max(0, x-n_2) \le x_1 \le \min(n_1, x)$. This is a *hypergeometric distribution*; see Chapter 6.

The distribution of the difference $X_1 - X_2$ is

$$\Pr[X_1 - X_2 = x] = \sum_{x_1} \binom{n_1}{x_1}\binom{n_2}{x_1 - x} p^{2x_1-x} q^{n_1+n_2-2x_1+x}, \qquad (3.24)$$

where the summation is between the limits $\max(0, x) \le x_1 \le \min(n_1, n_2 + x)$.

When $p = q = 0.5$,

$$\Pr[X_1 - X_2 = x] = \binom{n_1 + n_2}{n_2 + x} 2^{-n_1 - n_2}, \qquad -n_2 \le x \le n_1, \qquad (3.25)$$

so that $X_1 - X_2$ has a binomial distribution of the more general form mentioned in Section 1.

From the De Moivre-Laplace theorem and the independence of X_1 and X_2, it follows that the distribution of the standardized difference

$$\{X_1 - X_2 - p(n_1 - n_2)\}\{pq(n_1 + n_2)\}^{-1/2}$$

tends to the unit normal distribution as $n_1 \to \infty$, $n_2 \to \infty$ (whatever the ratio n_1/n_2). A similar result also holds when X_1 and X_2 have binomial distributions with parameters n_1, p_1 and n_2, p_2 with $p_1 \ne p_2$; however, the conditional distribution of X_1, given $X_1 + X_2 = x$, is no longer hypergeometric. Its distribution has been studied by Stevens (1951), and also by Hannan and Harkness (1963) who developed asymptotic normal approximations.

Springer (1979) has examined the distribution of products of discrete independent random variables; he used as an illustration the product of two such binomial variables with parameters n_1, p_1 and n_2, p_2, where $n_1 = n_2 = 2$.

5   ORDER STATISTICS

As is the case for most discrete distributions, order statistics based on observed values of random variables with a common binomial distribution are not often used. Mention may be made, however, of discussions of binomial order statistics by Gupta (1965), Khatri (1962), and Siotani (1956); see also David (1981). Tables of the cumulative distribution of the smallest and largest order statistic and of the range (in random samples of sizes 1(1)20) are in Gupta (1960b), Siotani and Ozawa (1958), and Gupta and Panchapakesan (1974). These tables can be applied in selecting the largest binomial probability among a set of k, based on k independent series of trials. This problem has been considered by Somerville (1957) and by Sobel and Huyett (1957).

Gupta (1960b) and Gupta and Panchapakesan (1974) have tabulated the mean and variance of the smallest and largest order statistic. Balakrishnan (1986) has given general results for the moments of order statistics from discrete distributions, and he has discussed the use of his results in the case of the binomial distribution.

6 APPROXIMATIONS, BOUNDS, AND TRANSFORMATIONS

6.1 Approximations

The binomial distribution is of such importance in applied probability and statistics that it is frequently necessary to calculate probabilities based on this distribution. Although the calculation of sums of the form

$$\sum_{x} \binom{n}{x} p^x q^{n-x}$$

is straightforward, it can be tedious, especially when n and x are large, and when there are a large number of terms in the summation. It is not surprising that a great deal of attention and ingenuity have been applied to constructing useful approximations for sums of this kind.

The *normal approximation* to the binomial distribution (based on the De Moivre-Laplace theorem)

$$\Pr[\alpha < (X - np)(npq)^{-1/2} < \beta] \doteq \frac{1}{\sqrt{2\pi}} \int_{\alpha}^{\beta} e^{-u^2/2} du = \Phi(\beta) - \Phi(\alpha) \quad (3.26)$$

has been mentioned in Section 4 of this chapter. This is a relatively crude approximation, but it can be useful when n is large. Numerical comparisons have been published in a number of textbooks [e.g., Hald (1952)].

A marked improvement is obtained by the use of a *continuity correction*. The following normal approximation is used widely on account of its simplicity:

$$\Pr[X \le x] \doteq \Phi(\{x + 0.5 - np\}/\{npq\}^{1/2}); \quad (3.27)$$

its accuracy for various values of n and p was assessed by Raff (1956) and by Peizer and Pratt (1968), who used the absolute and the relative error, respectively. Various rules of thumb for its use have been recommended in various standard textbooks. Two such rules of thumb are:

(1) use when $np(1 - p) > 9$,

and (2) use when $np > 9$, for $0 < p \le 0.5 \le q$.

Schader and Schmid (1989) have carried out a numerical study of these two rules which showed that, judged by the absolute error, rule 1 guarantees increased accuracy at the cost of a larger minimum sample size. Their study also showed that for both rules the value of p strongly influences the error. For fixed n the maximum absolute error is minimized when $p = q = 1/2$; it is reasonable to expect this since the normal distribution is symmetrical whereas the binomial distribution is symmetrical only when $p = 1/2$. The maximum value that the absolute error can take (over all values of n and p) is $0.140(npq)^{-1/2}$; Schader and Schmid showed that under rule 1 it decreases from $0.0212(npq)^{-1/2}$ to $0.0007(npq)^{-1/2}$ as p increases from 0.01 to 0.5.

Decker and Fitzgibbon (1991) have given a table of inequalities of the form $n^c \geq k$, for different ranges of p and particular values of c and k, that yield specified degrees of error when (3.27) is employed.

For *individual* binomial probabilities, the normal approximation with continuity correction, (3.27), gives

$$\Pr[X = x] \doteq (2\pi)^{-1/2} \int_{(x-0.5-np)/\sqrt{(npq)}}^{(x+0.5-np)/\sqrt{(npq)}} e^{-u^2/2} du. \qquad (3.28)$$

A nearly equivalent approximation is

$$\Pr[X = x] \doteq \frac{1}{\sqrt{npq}} \frac{1}{\sqrt{2\pi}} \exp\left[-\frac{1}{2}\frac{(x - np)^2}{npq}\right]; \qquad (3.29)$$

see Prohorov (1953) concerning its accuracy.

Approximation (3.27) can be improved still further by replacing α, β on the right hand-side of (3.26) by

$$\{[\alpha\sqrt{npq} + np] - 0.5 - np\}/\sqrt{npq}, \qquad (3.30)$$

$$\{[\alpha\sqrt{npq} + np] + 0.5 - np\}/\sqrt{npq}, \qquad (3.31)$$

respectively, where $[\cdot]$ denotes the integer part. A very similar approximation was given by Laplace (1820).

Peizer and Pratt (1968) and Pratt (1968) have developed a normal approximation formula for $\sum_{j=0}^{x} \binom{n}{j} p^j q^{n-j}$ in which the argument of $\Phi(\cdot)$ is

$$\frac{x + 2/3 - (n + 1/3)p}{\{(n + 1/6)pq\}^{1/2}}$$

$$\times \frac{1}{\delta_x}\left[2\left\{\left(x+\frac{1}{2}\right)\ln\left(\frac{x+1/2}{np}\right) + \left(n-x-\frac{1}{2}\right)\ln\left(\frac{n-x-1/2}{nq}\right)\right\}\right]^{1/2} (3.32)$$

where $\delta_x = (x+1/2-np)/\sqrt{npq}$. This gives good results that are even better when the multiplier $x + 2/3 - (n + 1/3)p$ is increased by $\{(x+1)^{-1}q - (n-x)^{-1}p + (n+1)^{-1}(q-1/2)\}/50$. With this adjustment, the error is less than 0.1% for $\min(x + 1, n - x) \geq 2$.

Cressie (1978) has suggested a slightly simpler formula with the argument of $\Phi(\cdot)$ equal to

$$\frac{(q - p)}{6(npq)^{1/2}} + \left\{1 - \frac{(1 - pq)}{36(npq)}\right\} \delta_x - \frac{(q - p)}{6(npq)^{1/2}}\delta_x^2 + \frac{(5 - 14pq)}{72(npq)}\delta_x^3. \qquad (3.33)$$

This is not as accurate as Peizer and Pratt's improved formula, and the gain in simplicity is slight. Samiuddin and Mallick (1970) have used the argument

$$\frac{(n - x - 1/2)(x + 1/2)}{n} \left\{ \ln \left(\frac{x + 1/2}{np} \right) - \ln \left(\frac{n - x - 1/2}{nq} \right) \right\}, \qquad (3.34)$$

which has some points of similarity with Peizer and Pratt's formula. This approximation is considerably simpler but not as accurate.

Borges (1970) found that

$$Y = (pq)^{-1/6}(n + 1/3)^{1/2} \int_p^y [t(1 - t)]^{-1/3} dt, \qquad (3.35)$$

where $y = (x + 1/6)/(n + 1/3)$, is approximately unit normally distributed; tables for the necessary beta integral are in Gebhardt (1971). This was compared numerically with other approximations by Gebhardt (1969). Another normal approximation is that of Ghosh (1980).

Kemp (1986) has obtained an approximation for the *modal probability* based on Stirling's expansion (Chapter 1, Section A2) for the factorials in the pmf,

$Pr[X = m]$

$$\doteq \frac{e}{\sqrt{2\pi}} \left\{ \frac{a}{bc} \right\}^{1/2} \exp \left[n \ln \left\{ \frac{a(1 - p)}{c} \right\} + m \ln \left\{ \frac{cp}{b(1 - p)} \right\} \right.$$

$$\left. + \frac{1}{12} \left\{ \frac{1}{a} - \frac{1}{b} - \frac{1}{c} \right\} - \frac{1}{360} \left\{ \frac{1}{a^3} - \frac{1}{b^3} - \frac{1}{c^3} \right\} + \frac{1}{1260} \left\{ \frac{1}{a^5} - \frac{1}{b^5} - \frac{1}{c^5} \right\} \right],$$

$$(3.36)$$

where m is the mode, $a = n + 1$, $b = m + 1$ and $c = n - m + 1$. He reported that it gives at least eight-figure accuracy.

Littlewood (1969) has made an exhaustive analysis of binomial sums. He obtained complicated asymptotic formulas for $\ln\{\sum_{j=x}^n \binom{n}{j} p^j q^{n-j}\}$ with uniform bounds of order $O(n^{-3/2})$ for each of the ranges

$$n \left(p + \frac{q}{24} \right) \le x \le n(1 - n^{-1/5}p)$$

and $$n(1 - n^{-1/5}) \le x \le n,$$

and also for $$np \le x \le n \left(1 - \frac{q}{2} \right).$$

A number of approximations to binomial probabilities are based on the equation

$$\Pr[X \geq x] = \sum_{j=x}^{n} \binom{n}{j} p^j q^{n-j}$$

$$= B(x, n-x+1)^{-1} \int_0^p t^{x-1}(1-t)^{n-x} dt$$

$$= I_p(x, n-x+1) \tag{3.37}$$

(this formula can be established by integration by parts). Approximation methods can be applied either to the integral or to the incomplete beta function ratio $I_p(x, n-x+1)$.

Bizley (1951) and Jowett (1963) have pointed out that since there is an exact correspondence between sums of binomial probabilities and probability integrals for certain central F distributions (see Sections 4 and 8.3 of this chapter), approximations developed for the one distribution are applicable to the other, provided that the values of the parameters correspond appropriately.

The *Camp-Paulson* approximation (Chapter 26) was developed with reference to the F-distribution. When applied to the binomial distribution it gives

$$\Pr[X \leq x] \doteq (2\pi)^{-1/2} \int_{-\infty}^{Y(3\sqrt{Z})^{-1}} e^{-u^2/2} du, \tag{3.38}$$

where

$$Y = \left[\frac{(n-x)p}{(x+1)q}\right]^{1/3} \left(9 - \frac{1}{n-x}\right) - 9 + \frac{1}{(x+1)},$$

$$Z = \left[\frac{(n-x)p}{(x+1)q}\right]^{1/3} \left(\frac{1}{n-x}\right) + \frac{1}{(x+1)}.$$

The maximum absolute error in this approximation cannot exceed $0.007(npq)^{-1/2}$.

A natural modification of the normal approximation that takes into account asymmetry, is to use a Gram-Charlier expansion with one term in addition to the leading (normal) term. The maximum error is now $0.056(npq)^{-1/2}$. It varies with n and p in much the same way as for the normal approximations, but is usually substantially smaller (about 50%). The relative advantage, however, depends on x as well as on n and p. For details, see Raff (1956).

If $n \to \infty$ and $p \to 0$ in such a way that $np = \theta$ remains finite and constant, then $\Pr[X = x] \to e^{-\theta}\theta^x/x!$; that is, the limiting form is the Poisson distribution (see Chapter 4). This is the basis for the *Poisson approximation* to the binomial distribution

$$\Pr[X \le x] \doteq e^{-np} \sum_{j=0}^{x} \frac{(np)^j}{j!}, \tag{3.39}$$

which has been used widely in inspection sampling. The maximum error is practically independent of n and approaches zero as p approaches zero.

In more detail, Anderson and Samuels (1967) have shown that this gives an underestimate if $x \ge np$, and an overestimate if $x \le np/(1+n^{-1})$. Thus the Poisson approximation tends to overestimate tail probabilities at both tails of the distribution. The absolute error of approximation increases with x for $0 \le x \le (np+0.5) - \sqrt{(np+0.25)}$ and decreases with x for $(np+0.5) + \sqrt{(np+0.25)} \le x \le n$.

Rules of thumb for the use of (3.39) that have been recommended by various authors are summarized in Decker and Fitzgibbon (1991), together with their own findings regarding levels of accuracy. For practical work they advise the use of the normal approximation (3.27) when $n^{0.31}p \ge 0.47$; for $n^{0.31}p < 0.47$ they advise using the Poisson approximation (3.39).

Simon and Johnson (1971) were able to use a result due to Vervaat (1969) to show that if $n \to \infty$ and $p \to 0$ with $np = \theta$, then

$$\sum_{j=0}^{\infty} \left| \binom{n}{j} p^j q^{n-j} - \frac{e^{-\theta} \theta^j}{j!} \right| h(j) \to 0 \tag{3.40}$$

for any $h(j)$ for which $\sum_{j=0}^{\infty} \theta^j (j!)^{-1} h(j)$ converges.

Ivchenko (1974) has studied the ratio

$$\left[\sum_{j=0}^{x} \binom{n}{j} p^j q^{n-j} \right] \Bigg/ \left[\sum_{j=0}^{x} \frac{e^{-np}(np)^j}{j!} \right]. \tag{3.41}$$

Hald (1967) and Steck (1973) have constructed Poisson approximations to cumulative binomial probabilities by seeking solutions in θ to the equation

$$\sum_{j=0}^{x} \binom{n}{j} p^j q^{n-j} = \sum_{j=0}^{x} \frac{e^{-\theta} \theta^j}{j!}. \tag{3.42}$$

Steck gave bounds for θ, while Hald obtained the approximation

$$\theta \doteq \frac{(n - x/2)p}{(1 - p/2)}. \tag{3.43}$$

The accuracies of a number of Poisson approximations to the binomial distribution have been studied by Morice and Thionet (1969) and by Gebhardt (1969). Gebhardt used as an index of accuracy the maximum absolute difference between the approximate and the exact cdf's. Romanowska (1978) has made similar comparisons using the sum of the absolute differences between approximate and exact values.

The Poisson Gram-Charlier approximation for the cumulative distribution function is

$$\Pr[X \le x] \doteq \sum_{j=0}^{x} [P(j,np) + 0.5(j - np)\Delta P(j,np)], \qquad (3.44)$$

where $P(j,np) = e^{-np}(np)^j/j!$ and the forward difference operator Δ operates on j.

Kolmogorov's approximation is

$$\Pr[X \le x] \doteq \sum_{j=0}^{x} [P(j,np) - 0.5np^2 \nabla^2 P(j,np)]. \qquad (3.45)$$

It is a form of Gram-Charlier Type B expansion. The next term is $np^3 \nabla^3 P(j,np)/3$. A detailed comparison of these two approximations is given in Dunin-Barkovsky and Smirnov (1955).

Galambos (1973) has given an interesting *generalized* Poisson-approximation theorem. Let $S_x(n)$ denote the sum of the $\binom{n}{x}$ probabilities incurred by different sets of x among n events $(E_1, \ldots, E_n)$. Then the conditions $S_1(n) \to a$ and $S_2(n) \to a^2/2$ as $n \to \infty$ are sufficient to ensure that the limiting distribution of the number of events that have occurred is Poissonian with parameter a.

Molenaar (1970a) has provided a systematic review of the whole field of approximations among binomial, Poisson, and hypergeometric distributions. This is an important source of detailed information on relative accuracies of various kinds of approximation. For "quick work" his advice is to use

$$\Pr[X \le x]$$
$$\doteq \Phi(\{4x + 3\}^{1/2}q^{1/2} - \{4n - 4x - 1\}^{1/2}p^{1/2}) \qquad \text{for } 0.05 < p < 0.93,$$
$$\doteq \Phi(\{2x + 1\}^{1/2}q^{1/2} - 2\{n - x\}^{1/2}p^{1/2}) \qquad \text{for } p \le 0.5, \ p \ge 0.93,$$

and

$$\Pr[X \le x] \doteq \sum_{j=0}^{x} \frac{e^{-\lambda}\lambda^j}{j!} \qquad (3.46)$$

with $\lambda = (2n - x)p/(2 - p)$ for p "small" (Molenaar suggests $p \le 0.4$ for $n = 3$, $p \le 0.3$ for $n = 30$, and $p \le 0.2$ for $n = 300$).

Wetherill and Köllerstrom (1979) have derived further interesting and useful inequalities among binomial, Poisson, and hypergeometric probabilities, with special reference to their use in the construction of acceptance sampling schemes.

6.2 Bounds

In the previous section we gave approximations to various binomial probabilities; in this section we examine bounds. Generally approximations are closer to the true values than bounds. Nevertheless, bounds provide one-sided approximations, and they often give useful limits to the magnitude of an approximation error.

Feller (1945) showed that if $x \geq (n+1)p$, then

$$\Pr[X = x] \leq \Pr[X = m] \exp\left[-\frac{p\{x - (n+1)p + 1/2\}^2}{2(n+1)pq} \right.$$

$$\left. + \left\{ m - (n+1)p + \frac{1}{2} \right\}^2 \right], \tag{3.47}$$

where m is the integer defined by $(n+1)p - 1 < m \leq (n+1)p$ and

$$\binom{n}{m}\left(\frac{m+1}{n+1}\right)^m \left(1 - \frac{m+1}{n+1}\right)^{n-m} \leq \binom{n}{m} p^m q^{n-m} \leq \binom{n}{m}\left(\frac{m}{n}\right)^m \left(1 - \frac{m}{n}\right)^{n-m}. \tag{3.48}$$

There are a number of formulas giving bounds on the probability $\Pr[|X/n - p| \geq c]$ where c is some constant, that is, on the probability that the difference between the relative frequency X/n and its expected value p will have an absolute value greater than c. A discussion is given by Kambo and Kotz (1966) and Krafft (1969).

Uspensky (1937) showed that

$$\Pr[|X/n - p| \geq c] < 2\exp(-nc^2/2); \tag{3.49}$$

Lévy (1954) established that

$$\Pr[|X/n - p| \geq c] < 2c^{-1}\exp(-2nc^2) \tag{3.50}$$

and that

$$\Pr[|X/n - p| \geq c] < 2c^{-1}n^{-1/2}\exp(-2nc^2), \tag{3.51}$$

(provided that $p, q \geq \max(4n^{-1}, 2c)$).

Okamoto (1958) found that

$$\Pr[X/n - p \geq c] < \exp\left(-\frac{nc^2}{2pq}\right) \quad \text{for } p \geq 1/2$$

and $\qquad \Pr[X/n - p \le -c] < \exp\left(-\dfrac{nc^2}{2pq}\right) \qquad$ for $p \le 1/2,$ $\qquad$ (3.52)

and further that

$$\Pr[\sqrt{X/n} - \sqrt{p} \ge c] < \exp(-2nc^2)$$

and $\qquad \Pr[\sqrt{X/n} - \sqrt{p} \le -c] < \exp(-2nc^2).$ $\qquad$ (3.53)

Kambo and Kotz (1966) and Krafft (1969) have improved Okamoto's bounds, obtaining, for example,

$$\Pr[X/n - p \ge c] < \exp(-2nc^2 - 4nc^4/3),$$
$$\Pr[X/n - p \le -c] < \exp(-2nc^2 - 4nc^4/3) \qquad (3.54)$$

if $0 \le c < 1 - p,$

$$\Pr[X/n - p \ge c] < \exp\left(-\frac{nc^2}{2pq} - \frac{4}{9}nc^4\right) \qquad \text{if} \quad p \ge 1/2,$$

$$\Pr[X/n - p \le -c] < \exp\left(-\frac{nc^2}{2pq} - \frac{4}{9}nc^4\right) \qquad \text{if} \quad p \le 1/2 \qquad (3.55)$$

(compare (3.52)), and

$$\Pr[\sqrt{X/n} - \sqrt{p} \ge c]$$
$$< \exp[-2nc^2/q - (2/3)nc^3(1 + \sqrt{p}/q + 2\sqrt{p}/q^2)]$$
$$< \exp[-2nc^2/q - (2/3)nc^3]. \qquad (3.56)$$

Kambo and Kotz also obtained the following improvement on Lévy's bound (3.51):

$$\Pr[|X/n - p| \ge c] < \sqrt{2}(c\sqrt{n})^{-1} \exp(-2nc^2 - 4nc^4/3) \qquad (3.57)$$

if $p, q \ge \max(4/n, 2c)$ and $n > 2.$

Both upper and lower bounds for $\Pr[X \ge x]$ have been obtained by Bahadur (1960). Starting from the hypergeometric series representation

$$\Pr[X \ge x] = \binom{n}{x} p^x q^{n-x} \times q \ {}_2F_1[n + 1, 1; x + 1; p], \qquad (3.58)$$

he obtained

$$\frac{q(x + 1)}{x + 1 - (n + 1)p}\left(1 + \frac{npq}{(x - np)^2}\right)^{-1} \le \frac{\Pr[X \ge x]}{\binom{n}{x}p^x q^{n-x}} \le \frac{q(x + 1)}{x + 1 - (n + 1)p}. \qquad (3.59)$$

Slud (1977) has developed further inequalities starting with the inequalities

$$\sum_{j=x+1}^{\infty} \frac{e^{-np}(np)^j}{j!} \leq \Pr[X \geq x+1] \quad \text{for } x \leq \frac{n^2 p}{n+1}, \tag{3.60}$$

$$\sum_{j=x}^{\infty} \frac{e^{-np}(np)^j}{j!} \geq \max\left[\Pr[X \geq x], 1-\Phi\left(\frac{x-np}{\sqrt{(npq)}}\right)\right] \quad \text{for } x \geq (np+1), \tag{3.61}$$

and

$$\Pr[X \geq x] \geq \sum_{j=x}^{\infty} e^{-np}(np)^j/j! \geq 1 - \Phi\left(\frac{x-np}{\sqrt{(npq)}}\right) \quad \text{for } x \leq np. \tag{3.62}$$

The second inequality in (3.62) is valid for *all* x. Slud improved (3.62) by replacing $\sqrt{np}$ by $\sqrt{npq}$ when either $np \leq x \leq nq$, or *alternatively* $np \leq x \leq n$ and $0 \leq p \leq 1/4$. When $np \leq x \leq n$ and $x \geq 2$, we have

$$\Pr[X = x] \leq \Phi\left(\frac{x-np+1}{\sqrt{npq}}\right) - \Phi\left(\frac{x-np}{\sqrt{npq}}\right).$$

Extensive calculations supported Slud's conjectures.

Prohorov (1953) quoted the following upper bound on the total error for the Poisson approximation to the binomial:

$$\sum_{j=0}^{\infty}\left|\binom{n}{j}p^j q^{n-j} - \frac{e^{-np}(np)^j}{j!}\right| \leq \min\{2np^2, 3p\} \tag{3.63}$$

(see Sheu (1984) for a relatively simple proof). Guzman's (1985) numerical studies suggest that in practice this bound is rather conservative.

We note the following inequalities for the ratio of a binomial to a Poisson probability when the two distributions have the same expected value:

$$e^{np}\left(1 - \frac{x}{n}\right)^x (1-p)^n \leq \frac{\binom{n}{x}p^x q^{n-x}}{e^{-np}(np)^x/x!} \leq e^{np}(1-p)^{n-x}. \tag{3.64}$$

Neuman's (1966) inequality is

$$\Pr[X \leq np] > \frac{1}{2} + \frac{1+q}{3\sqrt{2\pi}}(npq)^{-1/2}$$

$$- \frac{3q^2 + 12q + 5}{48}(npq)^{-1} - \frac{1+q}{36\sqrt{2\pi}}(npq)^{-3/2}. \tag{3.65}$$

6.3 Transformations

Methods of transforming data to satisfy the requirements of the normal linear model generally seek either to stabilize the variance, or to normalize the errors, or to remove interactions in order to make effects additive. Transformations are often used in the hope that they will at least partially fulfill more than one objective.

A widely used variance stabilization transformation for the binomial distribution is

$$u(X/n) = \arcsin \sqrt{X/n}. \tag{3.66}$$

Anscombe (1948) showed that replacing X/n by $(X + 3/8)/(n + 3/4)$ gives better variance stabilization; moreover it produces a random variable that is approximately normally distributed with expected value $\arcsin(\sqrt{p})$ and variance $1/(4n)$. Freeman and Tukey (1950) suggested the transformation

$$u(X/n) = \arcsin \sqrt{X/(n+1)} + \arcsin \sqrt{(X+1)/(n+1)};$$

this leads to the same approximately normal distribution. Tables for applying this transformation were provided by Mosteller and Youtz (1961). For p close to 0.5, Bartlett (1947) suggested the transformation

$$u(X/n) = \ln \left(\frac{X}{n - X} \right).$$

7 COMPUTATION AND TABLES

Recursive computation of binomial probabilities is straightforward. Since

$$\Pr[X = n] \gtrless \Pr[X = 0] \qquad \text{according as} \qquad p \gtrless 0.5,$$

forward recursion from $\Pr[X = 0]$ using

$$\Pr[X = x + 1] = \frac{(n - x)p}{(x + 1)q} \Pr[X = x]$$

is generally recommended for $p \leq 0.5$, and backward recursion from $\Pr[X = n]$ using

$$\Pr[X = x - 1] = \frac{xq}{(n - x + 1)p} \Pr[X = x]$$

for $p > 0.5$. Partial summation of the probabilities then gives the tail probabilities. When all the individual probabilities are required, computation with low overall rounding errors will result when an assumed value is taken for $\Pr[X = x_0]$ and both forward and backward recursion from x_0 are used; the resultant values must then be divided by their sum in order to give the true probabilities. Either the integer part of $n/2$ or an integer close to the mode of the distribution would be a sensible choice for x_0.

If only some of the probabilities are required, then recursion from the mode can be achieved using Kemp's (1986) very accurate approximation for the modal probability that was given in the previous section. This is the basis of his method for the computer generation of binomial random variables; see Chapter 1, Section C3.

There are a number of tables giving values of individual probabilities and sums of these probabilities. We first note that tables of the incomplete beta function ratio [Pearson (1934)] contain values to eight decimal places of $\Pr[X \geq k] = I_p(k, n-k+1)$ for $p = 0.01(0.01)0.99$ and $\max(k, n-k+1) \leq 50$. Other tables are

Biometrika Tables for Statisticians
  [E. S. Pearson and H. O. Hartley (1976)]
  $p = 0.01, 0.02(0.02)0.1(0.01)0.5$ and $n = 5(5)30$.
Tables of the Binomial Probability Distribution
  [National Bureau of Standards (1950)]
  $p = 0.01(0.01)0.50$ and $n = 2(1)49$.
Binomial Tables
  [H. G. Romig (1953)]
  $p = 0.01(0.01)0.50$ and $n = 50(5)5100$
  (this supplements the tables of the National Bureau of Standards).

Details of some other tables of binomial probabilities are given in the first edition of this book.

Bowerman and Scheuer (1990) have devised a method for computing $\Pr[X \geq x]$ (the binomial survival function) that is designed to avoid underflow and overflow problems; it is especially suitable for large n.

Among auxiliary tables are tables of the standard deviation $\{pq(n_1^{-1} + n_2^{-1})\}^{1/2}$ of the difference between two independent binomial proportions with common parameter p. Stuart (1963) gave tables from which values of this function to four decimal places can be obtained, for $p = 0.01, 0.05, 0.075, 0.1, 0.15, 0.2, 0.25, 0.3, 0.4, 0.5$ and for 20 values of each of n_1 and n_2 ranging from 25 to 5,000.

Nomographs for calculating sums of binomial probabilities have been developed; see, e.g., Larson (1966). Such nomographs have also been constructed and labeled for the equivalent problem of calculating values of the incomplete beta function ratio [Hartley and Fitch (1951)].

8 ESTIMATION

8.1 Model Selection

The use of binomial probability paper in exploratory data analysis is described by Hoaglin and Tukey (1985). Binomiallydistributed data should produce a straight line with slope and intercept that can be interpreted in terms of estimates of the parameters.

Other graphical methods include

1. a plot of the ratio of sample factorial cumulants $\tilde{\kappa}_{[r+1]}/\tilde{\kappa}_{[r]}$ against successive low values of r [Hinz and Gurland (1967); see also Douglas (1980)];

2. a plot of xf_x/f_{x-1} against successive low values of x, where f_x is the observed frequency of x [Ord (1967a)] see Table 2.2;

3. marking the position of $(\tilde{\kappa}_3/\tilde{\kappa}_2, \tilde{\kappa}_2/\tilde{\kappa}_1)$ on Ord's (1970) diagram of distributions, where $\tilde{\kappa}_1$, $\tilde{\kappa}_2$ and $\tilde{\kappa}_3$ are the first three sample cumulants.

Further graphical methods have been developed by Gart (1970) and Grimm (1970).

8.2   Point Estimation

Usually n is known. The method of moments, maximum likelihood, and minimum χ^2 estimators of p are then all equal to $\bar{x}/n$. This estimator is unbiased. Its variance is $pq \left(\sum_{j=1}^{k} n_j \right)^{-1}$, which is the Cramer-Rao lower bound for unbiased estimators of p; the estimator is in fact the minimum variance unbiased estimator of p. Its expected absolute error has been investigated by Blyth (1980).

An approximately median-unbiased estimator of p is

$$\frac{\bar{x}+1/6}{n+1/3}$$

[Crow (1975); see also Birnbaum (1964)]. A helpful expository account of estimation for p (including Bayesian estimation) is in Chew (1971). A useful summary of results is in Patel, Kapadia and Owen (1976).

Estimation of certain functions of p (when n is known) has also been investigated. Sometimes an estimate of $\Pr[\alpha < X < \beta]$ is required. The minimum variance unbiased estimator of this polynomial function of p is

$$\sum_{\alpha < \xi < \beta} \binom{n}{\xi}\binom{n(N-1)}{T-\xi} \Big/ \binom{nN}{T} \tag{3.67}$$

with $T = \sum_{j=1}^{N} x_j$ and ξ taking integer values. From this expression it can be seen that the minimum variance unbiased estimator of a probability $\Pr[X \in \omega]$, where ω is *any* subset of the integers $\{0, 1, \ldots, n\}$, has the same form with the range of summation $\alpha < \xi < \beta$ replaced by $\xi \in \omega$. Rutemiller (1967) has studied the estimator of $\Pr[X = 0]$ in some detail, giving tables of its bias and variance. Pulskamp (1990) has shown that the MVUE of $\Pr[X = x]$ is admissible under quadratic loss when $x = 0$ or n but is inadmissible otherwise. He has conjectured that the ML estimator is always admissible.

Another function of p for which estimators have been constructed is $\min(p, 1 - p)$. A natural estimator to use, given a single observed value x is $\min(x/n, 1 - x/n)$. The moments of this statistic have been studied by Greenwood and Glasgow (1950), and the cumulative distribution by Sandelius (1952).

Cook, Kerridge and Pryce (1974) have shown that given a single observation x, then $\psi(x) - \psi(n)$ is a useful estimator of $\ln(p)$; they also showed that

$$|E[\psi(x) - \psi(n)] - \ln(p)| < \frac{q^{n+1}p}{n+1},\tag{3.68}$$

where $\psi(y)$ is the derivative of $\ln\Gamma(y)$ (i.e., the psi function of Chapter 1, Section A2). They also obtained an "almost unbiased" estimator of the entropy $p\ln(p)$ and used the estimator to construct an estimator of the entropy for a multinomial distribution.

Unbiased sequential estimation of $(1/p)$ has been studied by Gupta (1967), Sinha and Sinha (1975), and Sinha and Bose (1985). DeRouen and Mitchell (1974) have constructed minimax estimators for linear functions of $p_1, p_2, \ldots, p_r$ corresponding to r different (independent) binomial variables.

Suppose now that $X_1, X_2, \ldots, X_k$ are independent binomial rv's, and that X_j has parameters n_j, p, where $j = 1, 2, \ldots, k$. Then, given a sample of k observations $x_1, \ldots, x_k$, comprising one from each of the k distributions, the maximum likelihood estimator of p is the overall relative frequency

$$\hat{p} = \left(\sum_{j=1}^{k} x_j\right) \bigg/ \left(\sum_{j=1}^{k} n_j\right).\tag{3.69}$$

Moreover $\sum_{j=1}^{k} x_j$ is a sufficient statistic for p. Indeed, since $\sum_{j=1}^{k} X_j$ has a binomial distribution with parameters $\sum_{j=1}^{k} n_j, p$, the analysis is the same as for a single binomial distribution.

The above discussion assumes that $n_1, n_2, \ldots, n_k$ (or at least $\sum_{j=1}^{k} n_j$) are known. The problem of estimating the values of the n_j's has been studied by Student (1919), Fisher (1941), Hoel (1947), and Binet (1953); for more historical details see Olkin, Petkau, and Zidek (1981).

Given a single observation of a random variable X having a binomial distribution with parameters n, p, then, if p is known, a natural estimator for n is x/p. This is unbiased and has variance nq/p.

The equation for the ML estimator $\hat{n}$ of n, when p is known, is

$$\sum_{j=0}^{R-1} A_j(\hat{n} - j)^{-1} = -N \ln(1 - p),\qquad (3.70)$$

where A_j is the number of observations that exceed j and $R = \max(x_1, \ldots, x_N)$ [Haldane (1941)]. When N is large

$$\sqrt{N}\,\mathrm{Var}(\hat{n}) \doteq \left\{ \sum_{j=1}^{n} \left(\Pr[X = j] \sum_{i=0}^{j-1} (n - i)^{-2} \right) \right\}^{-1}.\qquad (3.71)$$

The consistency of this estimator was studied by Feldman and Fox (1968).

Dahiya (1981) has constructed a simple graphical method for obtaining the maximum likelihood estimate of n, incorporating the integer restriction on n. In Dahiya (1986) he examined the estimation of m (an integer) when $p = \theta^m$ and θ is a known constant.

Suppose now that $X_1, X_2, \ldots, X_k$ are independent random variables all having the same binomial distribution with parameters n, p. Then equating the observed and expected first and second moments gives the moment estimators $\tilde{n}$ and $\tilde{p}$ of n and p as the solutions of $\bar{x} = \tilde{n}\tilde{p}$ and $s^2 = \tilde{n}\tilde{p}\tilde{q}$. Hence

$$\tilde{p} = 1 - \frac{s^2}{\bar{x}},\qquad (3.72)$$

$$\tilde{n} = \frac{\bar{x}}{\tilde{p}}.\qquad (3.73)$$

Note that if $\bar{x} < s^2$, then $\tilde{n}$ is negative, suggesting that a binomial distribution is an inappropriate model.

Continuing to ignore the limitation that n must be an integer, the maximum likelihood estimators $\hat{n}$, $\hat{p}$ of n and p satisfy the equations

$$\hat{n}\hat{p} = \bar{x},\qquad (3.74)$$

$$\sum_{j=0}^{R-1} A_j(\hat{n} - j)^{-1} = -N \ln(1 - \bar{x}/\hat{n}),\qquad (3.75)$$

where A_j is the number of observations that exceed j and $R = \max(x_1, \ldots, x_N)$. The similarity between (3.73) and (3.74) arises because the binomial distribution is a power series distribution; see Chapter 2, Section 2.1. Unlike the method of moments equations, the ML equations require iteration for their solution. DeRiggi (1983) has proved that a maximum likelihood solution exists if and only if the sample variance is less than $\bar{x}$ and that, if a solution exists, it is unique.

If N is large,

$$\operatorname{Var}(\hat{n}) \doteq \frac{n}{N} \left\{ \sum_{j=2}^{n} \left(\frac{p}{q} \right)^{j} \frac{(j-1)!(N-j)!}{j(N-1)!} \right\}^{-1}, \tag{3.76}$$

and the asymptotic efficiency of $\tilde{n}$, relative to $\hat{n}$, is

$$\left\{ 1 + 2 \sum_{j=1}^{n-1} \left(\frac{p}{q} \right)^{j} \frac{j!(N-j-1)!}{(j+1)(N-2)!} \right\}^{-1} \tag{3.77}$$

[Fisher (1941)].

Olkin, Petkau and Zidek (1981) found theoretically and through a Monte Carlo study that when both n and p are to be estimated, the method of moments and ML estimation both give rise to estimators that can be highly unstable; they suggested more stable alternatives based on (1) ridge-stabilization and (2) jackknife stabilization. Blumenthal and Dahiya (1981) also recognized the instability of the ML estimator of n, both when p is unknown and when p is known; they too gave an alternative stabilized version of ML estimation.

In Carroll and Lombard's (1985) study of the estimation of population sizes for impala and waterbuck, these authors stabilized ML estimation of n by integrating out the nuisance parameter p, using a beta distribution with parameters a and b. The need for a stabilized estimator of n has been discussed by Casella (1986).

The idea of minimizing the likelihood as a function of n, with p integrated out, can be interpreted in a Bayesian context. For a helpful description of the principles underlying Bayesian estimation for certain discrete distributions, including the binomial, see Irony (1992). Geisser (1984) has discussed and contrasted ways of choosing a prior distribution for binomial trials.

An early Bayesian treatment of the problem of estimating n is that of Draper and Guttman (1971). For p known they chose as a suitable prior distribution for n the rectangular distribution with pmf $1/k$, $1 \leq n \leq k$, k some large preselected integer. For p unknown, they again used a rectangular prior for n and, like Carroll and Lombard, adopted a beta prior for p, thus obtaining a marginal distribution for n of the form

$$p(n \mid x_1, \ldots, x_N) \propto \frac{(Nn - T + b - 1)!}{(Nn + a + b - 1)!} \prod_{j=1}^{N} \frac{n!}{(n - x_j)!}, \tag{3.78}$$

where $\max(x_i) \le n \le k$, $T = \sum_{j=1}^{N} x_i$. Although this does not lead to tractable analytical results, numerical results are straightforward to obtain. Kahn (1987) has considered the tailweight of the marginal distribution for n after integrating out p, when n is large; he has shown that the tailweight is determined solely by the prior density on n and p. This led him to recommend caution when adopting specific prior distributions. Hamedani and Walter (1990) have reviewed both Bayesian and non-Bayesian approaches to the estimation of n.

Empirical Bayes methods have been created by Walter and Hamedani (1987) for unknown p, and by Hamedani and Walter (1990) for unknown n. In their 1990 paper they used an inversion formula and Poisson-Charlier polynomials to estimate the prior distribution of n; this can then be smoothed if it is thought necessary. Their methods are analogous to the Bayes-empirical Bayes approach of Deely and Lindley (1981). Barry (1990) has developed empirical Bayes methods, with smoothing, for the simultaneous estimation of the parameters p_i for many binomials in both one-way and two-way layouts.

8.3 Confidence Intervals

The binomial distribution is a discrete distribution, and so it is not generally possible to construct a confidence interval for p with an exactly specified confidence coefficient using only a set of observations.

Let $x_1, x_2, \ldots, x_N$ be values of independent random binomial variables with exponent parameters $n_1, n_2, \ldots, n_N$ and common second parameter p. Then approximate $100(1 - \alpha)\%$ limits may be obtained by solving the following equations for p_L and p_U:

$$\sum_{j=T}^{n} \binom{n}{j} (p_L)^j (1 - p_L)^{n-j} = \frac{\alpha}{2} \tag{3.79}$$

$$\sum_{j=0}^{T} \binom{n}{j} (p_U)^j (1 - p_U)^{n-j} = \frac{\alpha}{2}, \tag{3.80}$$

where

$$n = \sum_{j=1}^{N} n_j \quad \text{and} \quad T = \sum_{j=1}^{N} x_j.$$

This is the standard approach of Clopper and Pearson (1934) and Pearson and Hartley (1976).

The values of p_L and p_U depend on T, and the interval (p_L, p_U) is an *approximate* $100(1 - \alpha)\%$ *confidence interval* for p. Values of p_L and p_U can be found in the following publications: Mainland (1948), Clark (1953), Crow (1956, 1975), Pachares (1960), and Blyth and Hutchinson (1960). Also

Pearson's (1934) tables of the incomplete beta function ratio can be used to solve (3.79), provided that $\max(x, n - x + 1) \leq 50$, and to solve (3.80), provided that $\max(x + 1, n - x) \leq 50$. The identity

$$I_p(x, n - x + 1) = \sum_{j=x}^{n} \binom{n}{j} p^j q^{n-j}$$

is used.

Equation (3.79) can be rewritten in the form

$$I_{p_L}(x, n - x + 1) = \frac{\alpha}{2}.$$

Because the lower $100\beta\%$ point, $F_{\nu_1, \nu_2, \beta}$, of the F-distribution with ν_1, ν_2 degrees of freedom satisfies the equation

$$I_c(\nu_1/2, \nu_2/2) = \beta,$$

where $c = \nu_1 F / (\nu_1 + \nu_2 F)$ (Chapter 26), it follows that

$$p_L = \frac{\nu_1 F_{\nu_1, \nu_2, \alpha/2}}{\nu_2 + \nu_1 F_{\nu_1, \nu_2, \alpha/2}} \tag{3.81}$$

with $\nu_1 = 2x$, $\nu_2 = 2(n - x + 1)$. There is a similar formula for p_U, with $\alpha/2$ replaced by $1 - \alpha/2$ and $\nu_1 = 2(x + 1)$, $\nu_2 = 2(n - x)$. Tables of percentage points for the F distribution can therefore be used to obtain values of p_L and p_U [Satterthwaite (1957)]. Charts from which confidence limits for p can be read off are given in Pearson and Hartley's (1976) Biometrika Tables.

A confidence interval does not have to have probability $(1 - \alpha)$ of covering p and *equal* probabilities $\alpha/2$ and $\alpha/2$ of lying entirely above or entirely below p. Alternative methods for constructing confidence intervals for the binomial parameter p have been examined by Angus and Schafer (1984) and Blyth and Still (1983). The latter authors gave a one-page table of 95% and 99% confidence intervals with certain desirable properties such as monotonicity in x, monotonicity in n, and invariance with respect to the transformation $X \to (n - X)$ and the induced transformation $p \to 1 - p$.

Approximations to p_L and p_U can be obtained using a normal approximation to the binomial distribution (Section 6.1). The required values are the roots of

$$(x - np)^2 = \lambda_{\alpha/2}^2 np(1 - p), \tag{3.82}$$

where

$$(2\pi)^{-1/2} \int_{\lambda_{\alpha/2}}^{\infty} e^{-u^2/2} du = \frac{\alpha}{2}.$$

More accurate values can be obtained using continuity corrections. An in-depth comparison of the accuracy of various normal approximations to confidence limits for the binomial parameter p has been carried out by Blyth (1986).

Nayatani and Kurahara (1964) considered that the crude, though often used, approximation

$$\frac{x}{n} \pm \lambda_{\alpha/2} \left[\frac{x}{n}\left(1 - \frac{x}{n}\right)\right]^{1/2} \qquad (3.83)$$

can be used, provided that $np > 5$, $p \leq 0.5$.

Stevens' (1950) method for exact confidence statements for discrete distributions involves the use of a uniform random number u where $0 \leq u \leq 1$. Let $Y = X + U$. Then

$$\Pr[Y \geq y] = \Pr[X > x] + \Pr[X = x]\Pr[U \geq u], \qquad (3.84)$$

yielding an exact lower bound; similarly for an exact upper bound. Kendall and Stuart (1961) have applied the procedure to the binomial distribution.

Patel, Kapadia and Owen (1976) have quoted formulas for an asymptotic confidence interval for p when n is large and known.

When n trials are conducted and no success is observed, a $100(1 - \alpha)\%$ confidence interval for p is

$$\{0 \leq p \leq 1 - \alpha^{1/n}\}.$$

Louis (1981) has commented that this interval is exactly the $100(1 - \alpha)\%$ Bayesian prediction interval based on a uniform prior distribution for p. Also he reinterpreted $S_n = n(1 - \alpha^{1/n})$ as the "number of successes" in a future experiment of the same size.

The conservativeness of conventional Clopper-Pearson confidence intervals, especially for sample sizes less than 100, is demonstrated very clearly in Figure 1 of Brenner and Quan (1990). These authors have developed an alternative, Bayesian, technique for obtaining confidence limits, based on the special beta prior $f(p) = (1 + \beta)p^\beta$ (a power-function prior distribution) and on the restraint that the length of the interval should be as short as possible. Their exact confidence limits were obtained numerically and are presented graphically.

The sample mean is a complete and sufficient statistic for the nuisance parameter p, and hence approximations applied to the conditional likelihood

$$L(n) = \prod_{i=1}^{N} \binom{n}{x_i} \bigg/ \binom{Nn}{x_1 + \cdots + x_N}$$

can be used to find approximate confidence limits for n and to derive an approximate test for $H_0 : n = n_0$ against $H_1 : n > n_0$ [Hoel (1947)].

Bain, Engelhardt and Williams (1990) have applied a partition-generating algorithm to this conditional likelihood; by generating all possible ordered outcomes, they have obtained tables of the confidence limits for n without the use of either approximations or simulation. Comparison with Hoel's approximate bounds showed that the latter are appropriate more widely than had previously been thought.

Little attention has been given to the problem of deriving confidence limits for n when p is known. Approximate confidence limits (n_L, n_U) for n can be obtained by solving the equations

$$\sum_{j=0}^{x} \binom{n_U}{j} p^j (1-p)^{n_U-j} = \frac{\alpha}{2}$$

$$\sum_{j=0}^{x} \binom{n_L}{j} p^j (1-p)^{n_L-j} = 1 - \frac{\alpha}{2}. \tag{3.85}$$

Hald and Kousgaard (1967) have given appropriate tables.

8.4 Model Verification

Goodness-of-fit tests for discrete distributions have not been researched as extensively as those for continuous distributions. Two techniques for the binomial distribution that are very widely used are Pearson's χ^2 goodness-of-fit test, and Fisher's binomial index-of-dispersion test. These are described in detail in, for example, Lloyd (1984). Rayner and Best (1989) have discussed goodness-of-fit tests in general and have provided details of their own smooth goodness-of-fit procedure given the assumption of a binomial distribution.

Louv and Littell (1986) have studied how to combine a number of one-sided tests concerning the parameter p. The need to test $H_0 : p_i = p_{i0}$ for all i, against $H_1 : p_i < p_{i0}$ for at least one i, arose in a study concerning equal employment opportunity. Louv and Littell investigated median significance levels, asymptotic relative efficiencies, and accuracy of null distribution approximations for six different procedures. They made a number of specific recommendations; in particular they advocated the use of the Mantel-Haenszel statistic to detect a consistent pattern of departure from the null hypothesis.

D'Agostino, Chase, and Belanger (1988) have examined tests for the equality of p for two independent binomial distributions by generating a complete enumeration of all possible sample configurations for some 660 combinations of p, n_1 and n_2. This enabled them to construct the exact distributions of the test statistics for the following four tests: Pearson's χ^2 with Yates' continuity correction; Fisher's exact test; Pearson's χ^2 without Yates' correction; and the two-independent-samples t-test. The first two tests were found to be extremely conservative. They reached the conclusion that the use of the two-independent-samples t-test should be encouraged.

9   CHARACTERIZATIONS

Lukacs (1965) characterized the binomial distribution by the property of constancy of regression of a particular polynomial statistic on the sample mean.

The starting point for a number of other characterization results is the Rao-Rubin theorem [Rao and Rubin (1964)]. This supposes that X is a discrete random variable with values $0, 1, \ldots$, and that $d(r|n)$ is the probability that the value n of X is reduced by a damage (ruin) process to the value r; the resultant random variable, also taking the values $0, 1, \ldots$, is denoted by Y. Suppose also that

$$\Pr[Y = s | \text{ruined}] = \Pr[Y = s | \text{not ruined}]. \tag{3.86}$$

Then, (1) if

$$d(r|n) = \binom{n}{r} \pi^r (1 - \pi)^{n-r}$$

for all n, where π is fixed, it follows that X is a Poisson variable. Moreover, (2) if X is a Poisson random variable with parameter λ and (3.86) is satisfied for all λ, then the ruin process is binomial. This result has considerable practical importance.

Yet other characterizations have stemmed from a general theorem of Patil and Seshadri (1964). A corollary of their theorem is that if the conditional distribution for X given $X + Y$ is hypergeometric with parameters m and n, then X and Y have binomial distributions with parameters (m, θ) and (n, θ), respectively.

Kagan, Linnik and Rao (1973) have developed a number of characterization theorems by considering binomial random walks on a lattice of integer points in the positive quadrant, corresponding to sequential estimation plans with Markovian stopping rules.

Let X be a nonnegative rv with pmf p_x and hazard (failure) rate $h_x = p_x / \sum_{j=x}^{n} p_j$. Then Ahmed (1991) has shown that X has a binomial distribution with parameters n and p iff

$$E[X|X \geq x] = np + qxh_x, \tag{3.87}$$

and a Poisson distribution with parameter λ iff

$$E[X|X \geq x] = \lambda + xh_x. \tag{3.88}$$

Haight (1972) has tried to unify various isolated characterizations for discrete distributions by the use of Svensson's (1969) theorem; this states that if X is a random variable with support on the nonnegative integers, then there exists a two-dimensional random variable (X, Y) with the property that the pgf of $\Pr[X = x | X + Y]$ is of the form $G(z) = (1 - p + pz)^n$ iff

$$\pi_{X+Y}(z) = \pi_X((z - 1 + p)/p),$$

where $\pi_Y(z)$ is the generating function of Y.

10 APPLICATIONS

The binomial distribution arises whenever underlying events have two possible outcomes, the chances of which remain constant. The importance of the distribution has extended from its original application in gaming to many other areas.

Its use in genetics arises because the inheritance of biological characteristics depends on genes that occur in pairs; see, e.g., Fisher and Mather's (1936) analysis of data on straight versus wavy hair in mice. Another, more recent, application in genetics is the study of the number of nucleotides that are in the same state in two DNA sequences [Kaplan and Risko (1982)].

The number of defectives found in random samples of size n from a stable production process is a binomial variable; acceptance sampling is a very important application of the test for the mean of a binomial sample against a hypothetical value.

Seber (1982b) has given a number of instances of the use of the binomial distribution in animal ecology, for example, in mark-release estimation of the size of an animal population. Boswell, Ord, and Patil (1979) gave applications in plant ecology. Cox and Snell (1981) provided a range of data analysis examples; the binomial distribution underlies a number of them.

The importance of the binomial distribution in model building is evidenced by Khatri and Patel (1961), Katti (1966), and Douglas (1980). Johnson and Kotz (1977) have given many instances of urn models with Bernoulli trials.

The binomial distribution is the sampling distribution for the test statistic in both the sign test and McNemar's test.

It is also a limiting form for a number of other discrete distributions [see Kemp (1968 a,b)], and hence, for suitable values of their parameters, it may be used as an approximation. For instance, it is often taken as an approximation to the hypergeometric distribution. The latter distribution (see Chapter 6) is appropriate for sampling without replacement from a finite population of size M consisting of two types of entities. Table 3.1 indicates the way in which the binomial limit is approached as $M \rightarrow \infty$ when the sample size n is 10 and $p = 0.5$.

Table 3.1 Comparison of Hypergeometric and Binomial Probabilities

	Hypergeometric			Binomial
x	M=50	M=100	M=300	∞
0	0.0003	0.0006	0.0008	0.0010
1	0.0050	0.0072	0.0085	0.0098
2	0.0316	0.0380	0.0410	0.0439
3	0.1076	0.1131	0.1153	0.1172
4	0.2181	0.2114	0.2082	0.2051
5	0.2748	0.2539	0.2525	0.2461

Although appealing in their simplicity, the assumptions of independence and constant probability for the binomial distribution are not often precisely satisfied. Published critical appraisals of the extent of departure from these assumptions in actual situations are rather rare, though an interesting discussion concerning the applicability of the distribution to mortality data has been provided by Seal (1949a). Nevertheless, the model often gives a sufficiently accurate representation to enable useful inferences to be made. Even when the assumptions are known to be invalid, the binomial model provides a reference mark from which departures can be measured. Distributions related to the binomial by various relaxations of the assumptions are described in Section 12.

11 TRUNCATED BINOMIAL DISTRIBUTIONS

A *doubly truncated binomial distribution* is formed by omitting both the values of x such that $0 \leq x < r_1$ and also the values of x such that $n - r_2 < x \leq n$ (with $0 < r_1 < n - r_2 < n$). For the resulting distribution

$$\Pr[X = x] = \binom{n}{x} p^x q^{n-x} \bigg/ \sum_{j=r_1}^{n-r_2} \binom{n}{x} p^x q^{n-x} , \qquad x = r_1, \ldots, n - r_2. \quad (3.89)$$

A *singly truncated binomial distribution* is formed if *only* the values $0, 1, \ldots, r_1 - 1$, where $r_1 \geq 1$, *or* the values $n - r_2 + 1, \ldots, n$, where $r_2 \geq 1$, are omitted.

The distribution formed by omission of only the zero class, giving

$$\Pr[X = x] = \binom{n}{x} \frac{p^x q^{n-x}}{(1 - q^n)}, \qquad x = 1, 2, \ldots, n, \quad (3.90)$$

is called the *positive binomial* distribution. (Sometimes, however, the untruncated binomial distribution has been referred to as the positive binomial in order to distinguish it from the negative binomial distribution, for which see Chapter 5.)

The rth moment about zero of a rv having the positive binomial distribution (3.90) is equal to $\mu'_r/(1 - q^n)$, where μ'_r is the rth moment about zero for the untruncated binomial distribution. In particular, the expected value is $np/(1 - q^n)$ and the variance is $npq/(1 - q^n) - n^2p^2/(1 - q^n)^2$. It is not possible to obtain a simple expression for the negative moments [Stephan (1945)], though Grab and Savage (1954) noted the approximation

$$E[X^{-1}] \doteq (np - q)^{-1} \tag{3.91}$$

which has two figure accuracy for $np > 10$. Mendenhall and Lehman (1960) obtained approximate formulas by first approximating to the positive binomial with a (continuous) beta distribution (see Chapter 24), making the first and second moments agree. They found that

$$E[X^{-1}] \doteq \frac{1 - 2/n}{np - q}, \tag{3.92}$$

$$\operatorname{Var}(X^{-1}) \doteq \frac{(1 - 1/n)(1 - 2/n)q}{(np - q)^2(np - q - 1)}. \tag{3.93}$$

Formula (3.92) gives two figure accuracy for $np > 5$.

Situations in which n is known and it is required to estimate p from data from a doubly truncated binomial distribution are uncommon. An interesting example of a practical application has, nevertheless, been described by Newell (1965).

Finney (1949) showed how to calculate the maximum likelihood estimator of p; this was the method used by Newell for the data of his example.

Shah (1966) gave a method of estimating p using the sample moments calculated from a sample of N observed values, $x_1, x_2, \ldots, x_N$, each having the distribution (3.89). The first three moments about zero of (3.89) are

$$\mu'_1 = A - B + np,$$
$$\mu'_2 = r_1 A - (n - r_2 + 1)B + \mu'_1(n - 1)p + np,$$
$$\mu'_3 = r_1^2 A - (n - r_2 + 1)^2 B + \mu'_2(n - 2)p + \mu'_1(2n - 1)p + np, \tag{3.94}$$

where

$$A = r_1 q \binom{n}{r_1} p^{r_1} q^{n - r_1} \bigg/ \sum_{j=r_1}^{n-r_2} \binom{n}{j} p^j q^{n-j} \tag{3.95}$$

and

$$B = (n - r_2 + 1)q \binom{n}{n - r_2 + 1} p^{n - r_2 + 1} q^{r_2 - 1} \bigg/ \sum_{j=r_1}^{n-r_2} \binom{n}{j} p^j q^{n-j}. \tag{3.96}$$

The moment estimator of p is obtained by eliminating A and B from these equations and replacing μ'_s, $s = 1, 2, 3$, by $\sum_{j=1}^{N} x_j^s / N$. Shah calculated the asymptotic efficiency of this moment estimator relative to the ML estimator to be over 90% for the case $r_1 = r_2 = 1$. This value seems remarkably high, especially since his method uses the third sample moment.

The singly truncated (positive) binomial (3.90) is of frequent occurrence in demographic enquiries wherein families are chosen for investigation on the basis of an observed "affected" individual, so that there is at least one such individual in each family in the study. The maximum likelihood estimator $\hat{p}$ of p, based on N independent observations from the distribution (3.90), satisfies the first-moment equation

$$\bar{x} = \frac{n\hat{p}}{1 - \hat{q}^n}, \tag{3.97}$$

where $\hat{q} = 1 - \hat{p}$. An alternative estimator proposed by Mantel (1951) is

$$\tilde{p} = \frac{\bar{x} - f_1/N}{n - f_1/N},$$

where f_1/N is the observed relative frequency of unity.

For large samples

$$\text{Var}(\hat{p}) = \frac{pq(1 - q^n)^2}{Nn(1 - q^n - npq^{n-1})}, \tag{3.98}$$

$$\text{Var}(\tilde{p}) = \frac{pq(1 - q^n)(1 - 2q^{n-1} + npq^{n-1} + q^n)}{Nn(1 - q^{n-1})^2}. \tag{3.99}$$

The asymptotic efficiency of $\tilde{p}$ relative to $\hat{p}$ would seem to be at least 95%; see Gart (1968).

When p is large the effect of truncation is small and $\tilde{p}$ differs little from $\hat{p}$. Thomas and Gart (1971) carried out a thorough study of $\hat{p}$ and $\tilde{p}$, and they concluded that $\tilde{p}$ has comparable accuracy and is less biased than $\hat{p}$. They suggested certain refinements.

If the sample observations come from positive binomial distributions with a common value of p, but differing parameters $n_1, \ldots, n_N$, an estimator analogous to $\tilde{p}$ is $(\bar{x} - f_1/N)/(\bar{n} - f_1/N)$, where $\bar{n} = \sum_{j=1}^{N} n_j / N$.

12 OTHER RELATED DISTRIBUTIONS

12.1 Limiting Forms

The central importance of the binomial distribution in statistics is shown by the fact that it is related to a wide variety of standard distributions. We now summarize some of the relationships concerning limiting forms that appear elsewhere in the book. Mixtures of binomial distributions are discussed in the context of mixture distributions in Chapter 8, Sections 2.4 and 3.3. Rao's damage model receives attention in Chapter 9, Section 2.

The limiting form of the *standardized binomial distribution* as $n \to \infty$ is the *normal distribution* (this is the De Moivre-Laplace theorem; see Sections 4 and 6.1 of this chapter). If $n \to \infty$ and $p \to 0$ with $np = \theta$ (fixed), the *Poisson distribution* is obtained (Chapter 4, Section 2). The binomial distribution is itself a *limiting form of the hypergeometric distribution* (Chapter 6, Section 4).

Numerical relationships between the binomial, Poisson and hypergeometric distributions have been studied by a number of authors; see Section 6 of this chapter, also Chapter 6, Section 5.

12.2 Poissonian Binomial, Lexian, and Coolidge Schemes

The model for *Poissonian binomial sampling* is sometimes called a *Poisson trials model*. It gives rise to a form of distribution known in the earlier literature [e.g. in Aitken's (1945) useful and concise account] as the *binomial distribution of Poisson*. The distribution has also sometimes been called the *Poisson binomial distribution*, an appellation that leads to confusion with the Poisson-binomial distribution of Chapter 9, Section 5, with pgf $\exp\{\lambda(1 - p + pz)^n - \lambda\}$.

Poisson (1837) was interested in the problem of n trials with p varying from trial to trial. Consider k independent throws of n dice where p_{ij} is the probability of success for die number i on throw number j; each toss of each die is then a Bernoulli trial with probability p_{ij}. A binomial model holds when $p_{ij} = p$ (constant) for all i, j. For Poissonian binomial sampling we require that $p_{ij} = p_i$ for all k throws of die number i; we also require that the dice always behave independently and that not all of the p_i are equal. The pgf for the number of successes per throw is no longer $(1 - p + pz)^n$ but instead is

$$G_P(z) = \prod_{i=1}^{n}(q_i + p_i z), \quad 0 \le p_i \le 1 \ \forall i, \ q_i = 1 - p_i. \qquad (3.100)$$

Because the pgf can be regarded as a convolution of Bernoulli pgf's, the cumulants are the sums of the individual Bernoulli cumulants:

$$\mu = \sum_{i=1}^{n} p_i = n\bar{p}, \quad \text{say,}$$

$$\mu_2 = \sum_{i=1}^{n} p_i(1 - p_i) = n\bar{p}(1 - \bar{p}) - \sum_{i=1}^{n}(p_i - \bar{p})^2 = n\bar{p}(1 - \bar{p}) - n\sigma_w^2, \quad \text{say,}$$

$$\mu_3 = \sum_{i=1}^{n} p_i(1 - p_i)(1 - 2p_i), \qquad \text{etc.,} \qquad\qquad (3.101)$$

where σ_w^2 is the within-throw variance of the p_i. In an interesting and lucid article Nedelman and Wallenius (1986) have discussed the somewhat surprising result that the variance is less than that for a binomial distribution with parameters $n, \bar{p}$, that is, with the same mean. The variance is greatest when $p_1 = p_2 = \ldots = p_n$, and is least when $\sum_{i=1}^{n} p_i = n\bar{p}$ with some of the p_i equal to 0 and the remainder equal to 1.

As n becomes large in such a way that the largest p_i tends to zero but the sum $\sum_{i=1}^{n} p_i = \theta$ remains constant, the limiting form of the binomial distribution of Poisson is a Poisson distribution with parameter θ [Feller (1968, p. 282)].

Brainerd (1972), in a study of textual type and token counts, showed that for his model I the Markov chain with transition probabilities

$$\Pr[X_1 = j + 1 | X_0 = j] = 1,$$

$$\Pr[X_n = j + 1 | X_{n-1} = j] = g(n - 1) > 0 \quad \text{for } n > 1,$$

$$\Pr[X_n = j | X_{n-1} = j] = 1 - g(n - 1), \qquad\qquad (3.102)$$

leads to Poisson's binomial scheme with pgf

$$G_P(z) = \prod_{i=0}^{n-1} [1 + (z - 1)g(i)]; \qquad\qquad (3.103)$$

when $g(i) = e^{-\alpha i}$,

$$G_P(z) = \prod_{i=0}^{n-1} (1 - e^{-\alpha i} + e^{-\alpha i} z) \qquad\qquad (3.104)$$

and the p_i are in geometric progression [Gani (1975)].

Thomas and Taub (1975) drew attention to a different problem — the problem of computing rapidly the probability of obtaining x hits from a sequence of n shots, when each shot has a different probability of success due to changes in the distance from the target. Danish and Hundley's (1979) technical note gives a computer algorithm for this. Thomas and Taub's (1982) algorithm is recursive and considerably faster; they applied it to the case where the probability of success is assumed to increase linearly (the nearer the target becomes, the greater the probability that a shot will be effective), that is, with the p_i in arithmetic progression. Kemp's (1987a) weapon defense model led to the assumption that the p_i form a two-parameter geometric progression, that is, to a *log-linear relationship for the probabilities* of the form

$$\ln p_i = \ln C + (i - 1) \ln Q, \qquad i = 1, 2, \ldots, n. \qquad\qquad (3.105)$$

The pgf becomes

$$G_P(z) = \prod_{i=0}^{n-1} [1 + C \, Q^i (z - 1)]$$

$$= {}_1\Phi_0[Q^{-n}; -; Q, Q^n C(1 - z)], \qquad (3.106)$$

which can be expanded using Heine's theorem; see Chapter 1, Section A12, and Kemp (1987a). As n becomes large the distribution tends to an Euler distribution; see Chapter 4, Section 12.6.

An alternative two-parameter relationship between the p_i is $p_i = cq^{i-1}/(1 + cq^{i-1})$, i.e.

$$\ln\{p_i/(1 - p_i)\} = \ln c + (i - 1)\ln q, \qquad i = 1, 2, \ldots, n; \qquad (3.107)$$

this is a *log-linear-odds relationship*. It was adopted by Kemp and Kemp (1991) in their study of plausible non-binomial models for Weldon's dice data and for multiple-channel production processes such as plastic intrusion moulding. The log-linear-odds model also arises from a stationary stochastic process; see Kemp and Newton (1990) who were interested in a situation involving a dichotomy between parasites on hosts with and without open wounds resulting from previous parasite attacks. The pgf is

$$G_P(z) = \prod_{i=0}^{n-1} [(1 + cq^i z)/(1 + cq^i)]$$

$$= \frac{{}_1\Phi_0[q^{-n}; -; q, -cq^n z]}{{}_1\Phi_0[q^{-n}; -; q, -cq^n]}$$

$$= K \sum_{x=0}^{n} \frac{(1 - q^n) \cdots (1 - q^{n-x+1})}{(1 - q) \cdots (1 - q^x)} q^{x(x-1)/2} (cz)^x \qquad (3.108)$$

from Heine's theorem, where K is a normalizing constant.

Kemp and Kemp (1991) developed an alternative two-parameter Poisson trials model for the dice data, based on the hypothesis of one dud die. The pgf is now $(1 - P + Pz)(1 - p + pz)^{n-1}$, where p is the probability of success for the fair dice and P is the probability of success for the dud die. An interesting feature of Kemp and Kemp's four sets of calculations for dice data is the closeness of the fits given by the log-linear-odds and one-dud-die models. They commented that "the closeness of the results when fitting these models suggests that other Poisson trial models would give very similar results and that, unless one is interested in the parameters $\{p_i\}$ per se, the extra effort involved in estimation for more complicated Poisson trials models is unlikely to be worthwhile."

The *Lexian sampling scheme* is a different variant of binomial sampling. It was introduced by Lexis (1877), who was dissatisfied with the then commonly held and often erroneous assumption of homogeneity in sampling. In the context of dice throwing, let p_{ij} again be the probability of success for die i on throw j, but now assume that $p_{ij} = p_j$ (constant) for all n dice on throw j, with independence between all of the tosses. This corresponds to the assumption that for any particular throw the dice all have the same degree of bias, with the degree of bias varying from throw to throw. The pgf for the number of successes per throw is now

$$G_L(z) = \sum_{j=1}^{k} \frac{(1 - p_j + p_j z)^n}{k}, \tag{3.109}$$

with mean

$$\mu = \sum_{j=1}^{k} \frac{np_j}{k} = n\bar{p}, \qquad \text{say,} \tag{3.110}$$

and second factorial moment $\sum_{j=1}^{k} n(n-1)p_j^2/k$ (obtained by differentiating the pgf). Hence the variance is

$$\mu_2 = \sum_{j=1}^{k} \frac{n(n-1)p_j^2}{k} + n\bar{p} - n^2\bar{p}^2 = n\bar{p}(1 - \bar{p}) + n(n - 1) \sum_{j=1}^{k} \frac{n(n-1)(p_j - \bar{p})^2}{k}$$

$$= n\bar{p}(1 - \bar{p}) + n(n - 1)\sigma_b^2, \tag{3.111}$$

where σ_b^2 is the between-throws variance of the p_{ij}; see, e.g., Aitken (1945). The variance now exceeds the variance of a binomial variable with parameters $n, \bar{p}$; the excess increases with $n(n - 1)$. The outcome of the model is of course a *mixed binomial distribution*; see Chapter 8, Section 2.4 for situations where the p_j for the different throws are predetermined, and Chapter 8, Section 3.3 for situations where the p_j can be regarded as a sample from some population. Stuart and Ord (1987, pp. 165 and 171) have commented that the Lexian model is also equivalent to a special case of cluster sampling.

For the *Coolidge scheme* the p_{ij} are assumed to vary both within and between throws. The pgf for the number of successes per throw is now

$$G_C(z) = k^{-1} \sum_{j=1}^{k} \prod_{i=1}^{n} (1 - p_{ij} + p_{ij}z); \tag{3.112}$$

the mean is

$$\mu = \sum_{i=1}^{n} \sum_{j=1}^{k} p_{ij}k^{-1}, \tag{3.113}$$

and the variance is

$$\mu_2 = n\bar{p}(1 - \bar{p}) + n^2 \sum_{j=1}^{k}(p_j - \bar{p})^2 - \sum_{i=1}^{n}\sum_{j=1}^{k}(p_{ij} - \bar{p})^2 k^{-1}, \qquad (3.114)$$

which will often be very similar to the variance for the Lexian scheme; see Aitken (1945, pp. 53–54). The difficulty with this model is the impossibility of estimating the excessively large number of parameters from a data set.

Ottestad (1943) seems to have been unaware of the work of Coolidge (1921) and Aitken (1939, 1st edition) and gave the name *Poisson-Lexis* to the Coolidge scheme. He referred to previous work by Charlier (1920) and Arne Fisher (1936). For all three situations (Poisson, Lexis, and Poisson-Lexis) he examined the cases (1) n fixed from sample to sample and (2) n varying from sample to sample.

The Lexis ratio provides a test for the homogeneity of binomial sampling; Zabell (1983) provides relevant references.

12.3 Binomial-Binomial Lagrangian Distributions

The general form of a Lagrangian distribution was described in Chapter 2, Section 5.2. Let $g(z) = (q + pz)^m$, where $q = 1 - p$ and $mp < 1$, and suppose that $z = u \cdot g(z)$ such that $u = 0$ for $z = 0$ and $u = 1$ for $z = 1$. Then $f(z) = (q' + p'z)^n$, where $q' = 1 - p'$, expanded as a power series in u, is the pgf for the binomial-binomial Lagrangian distributions of Consul and Shenton (1972). We have

$$\Pr[X = 0] = f(0) = (q')^n,$$

$$\Pr[X = x] = \frac{1}{x!}\left[\frac{d^{x-1}}{dz^{x-1}}\left\{(q + pz)^{mx}np'(q' + p'z)^{n-1}\right\}\right]_{t=0}$$

$$= \frac{n}{x}(q')^n(pq^{m-1})^x \sum_{j=0}^{k}\binom{n-1}{j}\binom{mx}{x-j-1}\left(\frac{p'q}{pq'}\right)^{j-1}$$

$$= \frac{n}{mx+1}\binom{mx+1}{x}(q')^n\left(\frac{p'q}{pq'}\right)(pq^{m-1})^x$$

$$\times {}_2F_1[1 - x, 1 - n; mx - x + 2; \frac{p'q}{pq'}], \quad x = 1, 2, 3, \ldots, (3.115)$$

where $k = \min(x - 1, n - 1)$.

When $g(z) = 1$, the outcome is the usual binomial distribution if $0 < p' < 1$ and n is a positive integer; however, $g(z) = 1$ gives the negative binomial distribution if $q' = 1 + P$, $0 < P$, $n = -k < 0$; see Chapter 5. The special case with $p = p'$ (and hence $q = q'$) yields the "generalized negative binomial" distribution of Jain and Consul (1971); see Chapter 5, Section 12.3.

Consul and Shenton (1972) studied a number of other special cases of (3.115), including

1. the "binomial-delta" distribution with $g(z) = (q + pz)^m$, $f(z) = z^n$, $mp < 1$, and pmf

$$\Pr[X = x] = \frac{n}{x}\binom{mx}{x - n}p^{x-n}q^{n+mx-x}, \qquad \text{for } x \geq n; \qquad (3.116)$$

2. the "binomial-Poisson" distribution with $g(z) = (q + pz)^m$, $f(z) = e^{M(z-1)}$, $mp < 1$, and pmf

$$\Pr[X = x] = e^{-M}\frac{(Mq^m)^x}{x!}{}_2F_0[1 - x, -mx; \frac{p}{Mq}], \qquad \text{for } x \geq 0; \qquad (3.117)$$

3. the "binomial-negative-binomial" distribution with $g(z) = (q + pz)^m$, $f(z) = (Q - Pz)^{-k}$, $mp < 1$, and pmf

$$\Pr[X = x]$$
$$= \frac{\Gamma(k + x)}{x!\Gamma(x)}Q^{-k}\left(\frac{Pq^m}{Q}\right)^x {}_2F_1[1 - x, -mx; 1 - x - k; \frac{-pQ}{qP}],$$
$$\text{for } x \geq 0; \qquad (3.118)$$

4. the "Poisson-delta" distribution with $g(z) = e^{\theta(z-1)}$, $f(z) = z^n$, $\theta < 1$, and pmf

$$\Pr[X = x] = \frac{n}{x}\frac{e^{-\theta x}(\theta x)^{x-n}}{(x - n)}, \qquad \text{for } x \geq n; \qquad (3.119)$$

5. the "Poisson-Poisson" distribution with $g(z) = e^{\theta(z-1)}$, $f(z) = e^{M(z-1)}$, $\theta < 1$, and pmf

$$\Pr[X = x] = M(M + \theta x)^{x-1}e^{-(M+\theta x)}/x!, \qquad \text{for } x \geq 0; \qquad (3.120)$$

6. the "Poisson-binomial" distribution with $g(z) = e^{\theta(z-1)}$, $f(z) = (q + pz)^n$, $\theta < 1$, and pmf

$$\Pr[X = 0] = q^n,$$
$$\Pr[X = x] = \frac{(\theta x)^{x-1}}{x!}e^{-\theta x}npq^{n-1}{}_2F_0[1-x, 1-n; \frac{p}{\theta qx}],$$
$$\text{for } x \geq 1; \qquad (3.121)$$

7. the "Poisson-negative-binomial" distribution with $g(z) = e^{\theta(z-1)}$, $f(z) = (Q - Pz)^{-k}$, $\theta < 1$, and pmf

$$\Pr[X = 0] = Q^{-k}$$

$$\Pr[X = x] = \frac{(\theta x)^{x-1}}{x!} e^{-\theta x} kPQ^{-k-1} {}_2F_0[1-x, 1+k; \frac{-P}{\theta Qx}],$$

for $x \geq 1$;   (3.122)

8. the "negative-binomial-delta" distribution with $g(z) = (Q - Pz)^{-k}$, $f(z) = z^n$, $kP < 1$, and pmf

$$\Pr[X = x] = \frac{n}{x} \frac{\Gamma(kx + x - 1)}{(x - n)! \Gamma(kx)} \left(\frac{P}{Q}\right)^{x-n} Q^{-kx}, \qquad \text{for } x \geq n; (3.123)$$

9. the "negative-binomial-Poisson" distribution with $g(z) = (Q - Pz)^{-k}$, $f(z) = e^{M(z-1)}$, $kP < 1$, and pmf

$$\Pr[X = x] = \frac{e^{-M} M^x}{x!} Q^{-kx} {}_2F_0[1 - x, kx; -; \frac{-P}{MQ}], \qquad \text{for } x \geq 0;$$

(3.124)

10. the "negative-binomial-binomial" distribution with $g(z) = (Q - Pz)^{-k}$, $f(z) = (q + pz)^n$, $kP < 1$, and pmf

$$\Pr[X = 0] = q^n,$$

$$\Pr[X = x]$$

$$= npq^{n-1} \frac{\Gamma(kx+x-1)}{x! \Gamma(kx)} \left(\frac{P}{Q}\right)^{x-1} Q^{-kx} {}_2F_1[1-x, 1-n; 2-x-kx; \frac{-pQ}{Pq}],$$

for $x \geq 1$;   (3.125)

11. the "negative-binomial-negative-binomial" distribution with $g(z) = (Q - Pz)^{-k}$, $f(z) = (Q' - P'z)^{-M}$, $kP < 1$, and pmf

$$\Pr[X = 0] = (Q')^{-M},$$

$$\Pr[X = x]$$

$$= (Q')^{-M} \left(\frac{P'}{Q'Q^k}\right)^x \frac{\Gamma(M + x)}{x! \Gamma(M)} {}_2F_1[1-x, kx; 1-M-x; \frac{PQ'}{P'Q}],$$

for $x \geq 1$.   (3.126)

When $P = P'$ (and hence $Q = Q'$), the pmf for the negative-binomial-negative-binomial distribution simplifies to

$$\Pr[X = x] = \frac{M}{(M+kx+x)} \frac{\Gamma(kx+M+x+1)}{x!\,\Gamma(M+kx+1)} \left(\frac{P}{Q}\right)^x Q^{-(M+kx)},$$

$$\text{for } x \geq 0. \qquad (3.127)$$

Let $g_1, g_2, g_3, \ldots$ and $f_1, f_2, f_3, \ldots$ be the cumulants for the two constituent distributions with pgf's $g(z)$ and $f(z)$. Consul and Shenton (1972) were able to show that in terms of these cumulants the first four moments of the distribution (3.115) are

$$\mu_1' = f_1/(1 - g_1),$$

$$\mu_2 = \frac{f_2}{(1 - g_1)^2} + \frac{f_1 g_2}{(1 - g_1)^3},$$

$$\mu_3 = \frac{f_3}{(1 - g_1)^3} + \frac{f_1 g_3 + 3 f_2 g_2}{(1 - g_1)^4} + \frac{3 f_1 g_2^2}{(1 - g_1)^5},$$

$$\mu_4 = 3\mu_2^2 + \frac{\mu_2}{(1 - g_1)^4}\{15 g_2^2 + 4 g_3 (1 - g_1)\}$$

$$+ \frac{f_4(1 - g_1) + 6 f_3 g_2 + f_1 g_4}{(1 - g_1)^5} + \frac{6 f_1 g_2 g_3^2}{(1 - g_1)^6}. \qquad (3.128)$$

Consul and Shenton (1973) have investigated the relationship of these distributions to queueing theory, and they have also examined the asymptotically limiting forms. These distributions should not be confused with the quasi-binomial I and II distributions of Consul and Mittal (1975), Fazal (1976), Mishra and Sinha (1981), and Lingappiah (1987). The binomial-binomial distributions belong to the Gould family, whereas the quasi-binomial distributions belong to the Abel family; see Charalambides (1990), also Chapter 2, Section 5.3.

12.4 Weighted Binomial Distributions

Consider the situation where an event that occurs is not necessarily included in a sample, though it has a certain probability of being recorded. Let X be a rv with pmf p_x, and suppose that when the event $X = x$ occurs the probability of ascertaining it is $w(x)$. The pmf of the recorded distribution is then

$$\Pr[X = x] = w(x)p_x / \sum_x w(x)p_x. \qquad (3.129)$$

Let $\sum_x w(x)p_x z^x = E[w(x)z^x]$; then the pgf is

$$G(z) = \frac{E[w(x)z^x]}{E[w(x)]}. \tag{3.130}$$

The concept of a distribution weighted in this manner has been developed by Rao. Two important papers are Rao (1965) on the earlier theory of weighted distributions and Rao's (1985) comprehensive review of their applications. Patil, Rao, and Zelen (1986) have produced an extensive computerized bibliography.

Patil, Rao, and Ratnaparkhi (1986) suggested ten types of weight functions thought to be useful in scientific work; these include

$$w(x) = x^a, \qquad 0 < \alpha \leq 1,$$

$$w(x) = x(x-1)\ldots(x-\alpha+1),$$

$$w(x) = \theta^x,$$

and

$$w(x) = \alpha x + \beta.$$

When $w(x) = x$ the *ascertained distribution* is said to be *size-biased*. Patil and Rao (1978) showed that if $B(n,p)$, $P(\lambda)$ and $NB(k,p)$ denote the binomial, Poisson and negative binomial distributions, then their size-biased forms are $1+B(n-1,p)$, $1+P(\lambda)$ and $1+NB(k+1,p)$, respectively, and gave similar results for other distributions. A number of further results, with particular emphasis on models, mixtures of distributions, and form-invariance under size-biased sampling, with appropriate references, are in Patil and Rao (1978) and Patil, Rao, and Ratnaparkhi (1986); see also Patil, Rao, and Zelen (1988).

Kocherlakota and Kocherlakota (1990) have concentrated on weighted binomial distributions. Let $B_n(x;p) = \binom{n}{x}p^x(1-p)^{n-x}$, $x = 0,1,\ldots,n$; then

$$G(z) = \frac{\sum w(x)B_n(x;p)z^x}{\sum w(x)B_n(x;p)} \tag{3.131}$$

denotes the pgf of a weighted binomial distribution. We have

$$\frac{\partial^r}{\partial z^r}G(z) = \frac{n!p^r}{(n-r)!}\frac{\sum w(x+r)B_{n-r}(x;p)z^x}{\sum w(x+r)B_{n-r}(x;p)}, \tag{3.132}$$

and setting $z = 1$ in this expression gives the rth factorial moment. Moreover the only pgf satisfying

$$\frac{\partial^r}{\partial z^r}G(z) = \frac{n!p^r}{(n-r)!}\frac{\sum w(x+r)\Pr[X=x]z^x}{\sum w(x+r)\Pr[X=x]} \tag{3.133}$$

is $G(z) = (1-p+pz)^n$; that is, the relationship (3.132) characterizes the binomial distribution. Kocherlakota and Kocherlakota (1990) have studied estimation for weighted binomial distributions, both when $w(x)$ is specified and when $w(x) = x^\alpha$, α unknown, and have investigated tests of hypotheses when $w(x) = x^\alpha$ for changes in α and also for changes in p. They have also examined goodness-of-fit tests and tests for competing models, with numerical illustrations.

12.5 Pseudo-Binomial Variables

The pgf $(1 - p + pz)^m = (1 - Qz)^m/(1 - Q)^m$ is only valid when either (1) $0 < p < 1$ (that is, $Q < 0$) and m is a positive integer (the binomial distribution), or (2) $p < 0$ (that is, $0 < Q < 1$) and $m < 0$ (the negative binomial distribution; see Chapter 5). No other valid combination of values of p and m is possible. Clearly convolutions of two binomials, or of two negative binomials, or of a binomial with a negative binomial, produce valid distributions (the convolution of two distributions was defined in Chapter 1, Section B9).

The concept of a binomial pseudo-variable, that is, an entity having a non-valid pgf of the form $(1 - Qz)^m/(1 - Q)^m$, was introduced by Kemp (1979). She examined conditions under which convolutions involving Bernoulli or geometric varables with pseudo-Bernoulli variables could give valid distributions, and found that there are two possibilities,

1. $g(z) = (1 - Q)(1 - Q_1 z)(1 - Qz)^{-1}(1 - Q_1)^{-1}$, where $0 < Q_1 < Q < 1$;
2. $g(z) = (1 - Q)(1 - Q_1)(1 - Qz)^{-1}(1 - Q_1 z)^{-1}$, where $0 < -Q_1 < Q < 1$.

Kemp related the resultant distributions to various stochastic processes such as the Galton-Watson branching process, the nonhomogeneous birth-and-death process with one initial individual, and the queue-size distribution for the M/M/1 queue with arrivals in pairs.

Kemp showed that in both cases the outcome is infinitely divisible and hence

$$g(z) = (1 - Q)^\gamma(1 - Q_1 z)^\gamma(1 - Qz)^{-\gamma}(1 - Q_1)^{-\gamma},$$

$$\text{with} \quad 0 < Q_1 < Q < 1,$$

and $\qquad g(z) = (1 - Q)^\gamma(1 - Q_1)^\gamma(1 - Qz)^{-\gamma}(1 - Q_1 z)^{-\gamma},$

$$\text{with} \quad 0 < -Q_1 < Q < 1,$$

are valid pgf's, provided that $\gamma > 0$. This result enabled Kemp to give six sets of conditions on Q_1, Q_2, U_1, U_2 under which

$$G(z) = \left(\frac{1 - Q_1 z}{1 - Q_1}\right)^{U_1} \left(\frac{1 - Q_2 z}{1 - Q_2}\right)^{U_2} \tag{3.134}$$

gives a valid nondegenerate distribution on the nonnegative integers. Distributions with pgf's of this form include the distribution for the homogeneous birth-death-and-immigration process with n initial members, Sankaran's (1970) Poisson-Lindley distribution, Bhattacharya's (1966) "generalized binomial" distribution for accident proneness, Plunkett and Jain's (1975) Gegenbauer distribution, Phillips' (1978) distribution of scores for two-person games, and Ong and Lee's (1986) generalized noncentral negative binomial distribution. (The special case $U_1 = U_2$ is examined in Chapter 11, Section 7, under the name Gegenbauer distribution.)

Kemp gave a number of formulas for the probabilities for (3.134) and related these to previous expressions given by Bailey (1964) and McKendrick (1926); see also Irwin (1963). A useful recurrence relationship is

$$(x + 1)\Pr[X = x + 1] = \{(x - U_1)Q_1 + (x - U_2)Q_2\}\Pr[X = x]$$
$$+\{U_1 + U_2 - x + 1\}Q_1Q_2 \Pr[X = x - 1] \quad (3.135)$$

for $x > 0$, with $\Pr[X = -1] = 0$ and $\Pr[X = 0] = (1 - Q_1)^{-U_1}(1 - Q_2)^{-U_1}$. The moment properties can be obtained via the cumulants (these are the sums of the cumulants for the two component distributions, even when they are not valid distributions). An even easier approach is via the factorial cumulant generating function; from

$$\ln G(1 + t) = \sum_{i=1,2} U_i \ln(1 + p_i t) \quad (3.136)$$

where $p_i = Q_i/(Q_i - 1)$, $i = 1, 2$, we have

$$\kappa_{[r]} = (-1)^{r-1}(r - 1)!\{U_1 p_1^r + U_2 p_2^r\}. \quad (3.137)$$

12.6 Correlated Binomial Variables

Many data sets concerning the sex of sibs within families exhibit underdispersion compared with the binomial distribution. Correlation between the sexes of adjacent sibs has been reported by a number of authors [see e.g. Greenberg and White (1965)], and consequently Markov chain models that lead to underdispersion (also overdispersion) have been proposed. Edwards (1960) suggested a Markov chain with $p_{mm} = p + rq$, $p_{mf} = q - rq$, $p_{fm} = p - rp$, $p_{ff} = q + rp$, where p_{mf} is the transition probability for a boy followed by a girl. The correlation between successive births is then r, and the number of boys in n successive (single) births has the pgf

$$G(z) = [pz, \quad q] \begin{bmatrix} (p + rq)z, & q - rq \\ (p - rp)z, & q + rp \end{bmatrix}^{n-1} \begin{bmatrix} 1 \\ 1 \end{bmatrix}. \quad (3.138)$$

Other types of models for sibling data that lead to underdispersion are discussed, with references, in Brooks, James, and Gray (1991).

Consider now the sum of k identically distributed but not independent random variables. Suppose that they have a symmetric joint distribution with no second-order or higher-order "interactions." Two possibilities have been suggested, depending on whether an "additive" or a "multiplicative" definition of "interaction" is adopted; the advantages and disadvantages of the two schemes were discussed by Darroch (1974).

The additive model was put forward almost simultaneously by Kupper and Haseman (1978) and Altham (1978). Let $\Pr[X_j = i] = p_i$ for $j = 1, \ldots, k$, $i = 0, 1$, and let $\Pr[X_j = i_1, X_{j'} = i_2] = p_{i_1 i_2}$. Suppose also that $Z_k = X_1 + \cdots X_k$. Then

$$\frac{\Pr[X_1 = i_1, \ldots, X_k = i_k]}{p_{i_1} \cdots p_{i_k}} = \left(\sum_{1 \le a < b \le k} \frac{p_{i_a i_b}}{p_{i_a} p_{i_b}} \right) - \frac{k(k-1)}{2} + 1, \qquad (3.139)$$

for $i_j = 0, 1$ with $j = 1, \ldots, k$; this gives a symmetric joint distribution defined by the two parameters p_1 and p_{11}, since $p_{00} + p_{01} = p_0 = 1 - p_1$ and $p_{01} = p_{10}$. Taking $1 - \alpha = p_{10}/(p_0 p_1)$, Altham deduced that

$$\Pr[Z_k = j] =$$
$$\binom{k}{j} p_1^j (1 - p_1)^{k-j} \left\{ \frac{\alpha}{2} \left(\frac{j(j-1)}{p_1} + \frac{(k-j)(k-j-1)}{1 - p_1} \right) - \frac{\alpha k(k-1)}{2} + 1 \right\};$$
$$(3.140)$$

she showed that for certain values of α the distribution (3.140) is a mixture of three binomial distributions and fitted the distribution to some data using moment estimation. The case $\alpha = 0$ corresponds to independence between the X_i's and so gives a binomial distribution, whereas $\alpha > 0$ (< 0) corresponds to positive (negative) pairwise association. The constraints on α required in order that (3.140) is a valid pmf do not allow very strong pairwise association.

For the multiplicative model the assumption is that

$$\Pr[X_1 = i_1, \ldots, X_k = i_k] = K \prod_{1 \le a < b \le k} \phi_{i_a i_b}, \qquad (3.141)$$

whence

$$\Pr[Z_k = j] = K \binom{k}{j} \phi_{00}^{(k-j)(k-j-1)/2} \phi_{01}^{(j-k)j} \phi_{11}^{j(j-k)/2}; \qquad (3.142)$$

see Altham's paper. Taking $\theta = \phi_{01}/(\phi_{00}\phi_{11})^{1/2}$ and $p = \phi_{11}^{(k-1)/2}/(\phi_{11}^{(k-1)/2} + \phi_{00}^{(k-1)/2})$ gives

$$\Pr[Z_k = j] = C'\binom{k}{j}p^j(1-p)^{k-j}\theta^{j(k-j)}, \qquad (3.143)$$

where C' is a normalizing constant. This is an alternative two-parameter generalization of the binomial distribution, to which it reduces when $\theta = 1$. For $\theta > 1$ there is negative association producing a strongly unimodal distribution that is more peaked than the binomial; for $\theta < 1$ the association is positive, giving a flatter distribution.

The usefulness of the two distributions for toxicological experiments with laboratory animals, compared with that of the binomial distribution, has been discussed by Haseman and Kupper (1979); see also Makuch, Stephens and Escobar (1989). The beta-binomial is overdispersed in relation to a binomial distribution; see Chapter 6, Section 2.2. An extended beta-binomial distribution that permits underdispersion (correponding to negative association) has been discussed in relation to toxicological data by Prentice (1986). A more general modification of the binomial distribution that includes Altham's additive model as a special case has been proposed by Ng (1989). Rudolfer (1990) has compared his $\{0,1\}$-state Markov chain model of extra-binomial variation to Altham's additive and multiplicative models and to the beta-binomial model.

Paul's (1985) three-parameter, beta-correlated binomial distribution was modified in Paul (1987) in order to overcome certain theoretical difficulties associated with the earlier version of the distribution.

CHAPTER 4

Poisson Distribution

1 DEFINITION

A random variable X is said to have a Poisson distribution with parameter θ if

$$\Pr[X = x] = \frac{e^{-\theta}\theta^x}{x!}, \qquad x = 0, 1, 2, \cdots, \ \theta > 0. \tag{4.1}$$

The characteristic function is $\exp\{\theta(e^{it} - 1)\}$, and the pgf is

$$G(z) = e^{\theta(z-1)} = {}_0F_0[\ ;\ ;\theta z]/{}_0F_0[\ ;\ ;\theta] = {}_0F_0[\ ;\ ;\theta(z-1)]. \tag{4.2}$$

The distribution is a power series distribution with infinite nonnegative integer support. It belongs to the exponential family of distributions, and is both a Kemp hypergeometric probability distribution and a Kemp hypergeometric factorial moment distribution.

The mean and variance are $\mu = \mu_2 = \theta$.

2 HISTORICAL REMARKS AND GENESIS

Poisson (1837, Secs. 73, pp. 189–190, and 81, pp. 205–207) published the following derivation of the distribution that bears his name. He approached the distribution by considering the limit of a sequence of binomial distributions with

$$p_{x,N} = \Pr[X = x] = \begin{cases} \binom{N}{x}p^x(1-p)^{N-x} & \text{for } x = 0, 1, \ldots, N, \\ 0 & \text{for } x > N, \end{cases}$$

151

in which N tends to infinity and p tends to zero, while Np remains finite and equal to θ. It can be established by direct analysis that

$$\lim_{\substack{N \to \infty \\ Np=\theta}} \sum_w p_{x,N} = \sum_w \frac{e^{-\theta}\theta^x}{x!}, \tag{4.3}$$

where $\sum_w$ denotes summation over any (finite or infinite) subset w of the nonnegative integers $0, 1, 2, \ldots$.

The result had been given previously by de Moivre (1711) in *De Mensura Sortis*, p. 219. Bortkiewicz (1898) considered circumstances in which Poisson's distribution might arise. From the point of view of Poisson's own approach, these are situations where, *in addition* to the requirements of independence of trials and consistency of probability from trial to trial, the number of trials must be very large while the probability of occurrence of the outcome under observation must be small. Although Bortkiewicz called this the *Law of Small Numbers* there is no need for $\theta = Np$ to be "small". It is the largeness of N and the smallness of p that are important. In Bortkiewicz's (1898) book, one of the "outcomes" considered was the number of deaths from kicks by horses, per annum, in Prussian Army Corps. Here was a situation where the probability of death from this cause was small, while the number of soldiers exposed to risk (in any one Corps) is large. Whether the conditions of independence and constant probability are satisfied is doubtful. However, the data available to Bortkiewicz were quite satisfactorily fitted by Poisson distributions, and have been very widely quoted as an example of the applicability of this distribution. The data have been discussed in detail by Quine and Seneta (1987). Bortkiewicz also obtained satisfactory fits for fatal accident data and suicide data, including data on numbers of suicides per year for Prussian boys under ten years old (also for girls).

Bortkiewicz gave tables of the probabilities, and he obtained many properties of the distribution, such as difference and differential equations for the probabilities, and the moments (derived as limiting forms of the moments of the binomial distribution).

Thiele (1889) gave a completely different derivation. He showed that for the distribution with cumulants $\kappa_r = ba^r$, "the observed values are $0, a, 2a, 3a, \ldots$ and the relative frequency of ra is $b^r/r!$"; he considered that the distribution "is perhaps superior to the binomial as a representative of some skew laws of error."

Charlier (1905a) considered the binomial sampling scheme with pgf $\prod_{i=0}^{n}(1 - p_i + p_i z)$ and hence showed that the success probabilities p_i do not need to be constant for the Poisson limit to hold.

Charlier (1905b) derived the distribution from the differential-difference equations

$$\frac{dp_0}{dt} = -p_0, \quad \frac{dp_r}{dt} = -p_r + p_{r-1}, \quad r \geq 1, \tag{4.4}$$

where the probabilities are functions of time. Bateman (1910) and McKendrick (1914) also gave this model (the pure birth process with constant birth rate).

"Student" (W. S. Gosset) (1907) used the Poisson distribution to represent, to a first approximation, the number of particles falling in a small area A when a large number of such areas are spread at random over a surface large in comparison with A.

The Poisson distribution may also arise for events occurring "randomly and independently" in time. If it be supposed that the future lifetime of an item of equipment is independent of its present age (Chapter 18, Section 1), then the lifetime can be represented by a random variable T with pdf of form

$$p_T(t) = \tau^{-1} \exp(-t/\tau), \qquad t \geq 0, \ \tau \geq 0,$$

that is, by an exponential random variable. The expected value of T is τ (Chapter 18, Section 4). Now imagine a situation in which each item is replaced by another item with exactly the same lifetime distribution. Then the distribution of the number of failures (complete lifetimes) X in a period of length t is given by

$$\Pr[X = x] = \Pr[T_1 + T_2 + \cdots + T_{x-1} > t] - \Pr[T_1 + T_2 + \cdots + T_x > t], \quad (4.5)$$

where the lengths of successive lifetimes are denoted by $T_1, T_2, \ldots$, etc., and $x \geq 1$. From the relationship between the χ^2 and Poisson distributions (see Section 12.2 of this chapter), it follows that

$$\Pr[X = x] = \Pr[\chi^2_{2(x+1)} > 2t/\tau] - \Pr[\chi^2_{2x} > 2t/\tau]$$

$$= \sum_{j=0}^{x} e^{-t/\tau} \frac{(t/\tau)^j}{j!} - \sum_{j=0}^{x-1} e^{-t/\tau} \frac{(t/\tau)^j}{j!}$$

$$= e^{-t/\tau} \frac{(t/\tau)^x}{x!}, \qquad (4.6)$$

as in (4.1), with θ replaced by t/τ. Hence X has a Poisson distribution with parameter t/τ.

This mode of genesis underlies the use of the Poisson distribution to represent variations in the number of particles ("rays") emitted by a radioactive source in forced periods of time. Rutherford and Geiger (1910) gave some numerical data that were fitted well by a Poisson distribution; see also Rutherford et al. (1930).

A unification of the then existing theory was given by Bortkiewicz (1915), who utilized the gap distribution concept; see Haight (1967). The Poisson distribution can be characterized by the renewal counting process with exponential inter-arrival times.

Lévy (1937b, pp. 173-174) initiated the axiomatic approach. The Poisson distribution is the counting distribution for a Poisson process. This is a stochastic point process satisfying the following conditions: whatever the number of points in the interval $(0, t)$, the probability that in the interval $(t, t + \delta t)$ a point occurs is $\theta \delta t + o(\delta t)$, and the probability that more than one point occurs is $o(\delta t)$; see e.g. Feller (1968, p. 447).

Parzen (1962, p. 118) adopted a more rigorous approach. He showed that the Poisson process $X(t)$ satisfies the following five axioms:

Axiom 0: $X(0) = 0$.

Axiom 1: $X(t)$ has independent increments; that is, for all t_i such that $t_0 < t_1 < \ldots < t_n$, the random variables $X(t_i) - X(t_{i-1})$, $i = 1, 2, \ldots, n$, are independent.

Axiom 2: For any $t > 0$, $\quad 0 < \Pr[X(t) > 0] < 1$.

Axiom 3: For any $t > 0$,

$$\lim_{h \to 0} \frac{\Pr[X(t + h) - X(t) \geq 2]}{\Pr[X(t + h) - X(t) = 1]} = 0.$$

Axiom 4: $X(t)$ has stationary increments; that is, for points $t_i > t_j \geq 0$ (and $h > 0$), the random variables $X(t_i) - X(t_j)$ and $X(t_i + h) - X(t_j + h)$ are equidistributed.

There are several proofs that Axioms 0 to 4 imply that there exists a constant θ such that

$$\Pr[X(t) = x] = \frac{e^{-\theta t}(\theta t)^x}{x!}.$$

Modification of these axioms leads to more general stochastic processes; for example, replacing Axiom 4 by

$$\lim_{h \to 0} \frac{1 - \Pr[X(t + h) - X(t) = 0]}{h} = \lambda(t)$$

leads to the nonhomogeneous process [Parzen (1962, p. 125)]. In his monograph on the Poisson distribution, Haight (1967) discussed other axiomatic approaches such as that of Fisz and Urbanik (1956) who gave a sufficient condition for a process satisfying Axioms 0–2 to be a Poisson process. Haight also described a number of special probability models including, for example, the model of Hurwitz and Kac (1944). Johnson and Kotz (1977) focussed on urn models related to the Poisson distribution.

Since

$$\Pr[X > x + y | X > x] = \Pr[X > y]$$

iff X is exponentially distributed [Feller (1957, Ch. 17); Parzen (1962)], it follows from the exponential distribution of interarrival times for a Poisson process that the distribution of the time between two Poisson events is surprisingly the same as the distribution of the time between an arbitrary point in time and the next Poisson event.

A Poisson distribution is the outcome of an aspect of maximum disorder that arises in information theory; see Renyi (1964). A Poisson arrangement of points, obtained by thoroughly shuffling other points, was studied by Maruyama (1955) and Watanabe (1956); Haight (1967) gave further references.

Kreweras (1979) has described two unusual modes of genesis for the Poisson distribution with $\theta = 1$. Consider all $n!$ permutations of the integers $1, 2, \ldots, n$ and let $u(n, x)$ be the number of these permutations in which there are x pairs of integers in their natural order. For example, with $n = 5$, the permutation 12543 has $x = 1$, 12534 has $x = 2$, and 12345 has $x = 4$. Then

$$\lim_{n \to \infty} \left\{ \frac{u(n, x)}{n!} \right\} = \frac{e^{-1}}{x!}. \tag{4.7}$$

That is, the distribution of the number of pairs of integers in natural order (termed by Kreweras the *regularity* of the permutation) tends to a Poisson distribution with $\theta = 1$.

His other mode of genesis involves the consideration of partitions of the first $2n$ integers into pairs of integers. The total number of these partitions is

$$1 \times 3 \times 5 \times \ldots \times (2n - 1) = \frac{2^{-n}(2n)!}{n!};$$

the proportion of them containing exactly x pairs of two consecutive integers also tends to $e^{-1}/x!$ as $x \to \infty$.

The Poisson distribution is a limiting form for the binomial distribution (see Chapter 3, Section 6.1) and for many other distributions. In particular, it arises as

$$\lim_{\substack{k \to \infty \\ kp = \theta}} (1 + p - ps)^{-k} = e^{\theta(z-1)}, \tag{4.8}$$

that is, as a limiting form of the negative binomial distribution; see Chapter 5, Section 12.1. A realization of this relationship would have done much to defuse the controversy aroused by Whittaker (1914) regarding the intrinsic merits of the Poisson and negative binomial models.

Widdra (1972) obtained a Poisson distribution as the limiting distribution (as $n \to \infty$) of the number of successes in n *partially dependent* trials. He showed that if Y_{gh} are indicator variables with $Y_{ih}Y_{jh} = 0$ if $i \neq j$, where $i, j = 1, \ldots, m$ and $h = 1, \ldots, n$, but otherwise Y_{gh} and Y_{ij} are independent with constant success probability $E[Y_{gh}] = p$ for all g, h, then

$$\sum_{i=1}^{m}\sum_{h=1}^{n}\sum_{\substack{k=1 \\ h<k}}^{n} Y_{ih}Y_{ik}$$

has a limiting Poisson distribution, as $m, n \to \infty$, with $mn(n-1)p^2/2 = \theta$.

Important results by Chen (1975), on convergence to the Poisson distribution for the number of successes in n dependent trials using only the first two moments, have been reexamined by Arratia, Goldstein, and Gordon (1989, 1990). A simple proof concerning Poisson approximations to power series distributions, depending only on the continuity of probability generating functions, appears in Pérez-Abreu (1991).

Serfling (1977) has obtained reliability bounds for k-out-of-n systems based on Poisson approximations.

In recent years, the Poisson distribution has been used to a markedly increasing extent. Like the binomial distribution, the Poisson distribution often serves as a standard from which to measure departures, even when it is not itself an adequate representation of the real situation.

The handbook by Haight (1967) on the Poisson distribution contains much interesting information on the Poisson and related distributions.

3   MOMENTS

The simplicity of the Poisson pgf, $G(z) = \exp\{\theta(z-1)\}$, leads to simple expressions for the cumulant generating function, factorial moment generating function, and factorial cumulant generating function.

The cumulant generating function is $\theta(e^t - 1)$, so

$$\kappa_r = \theta \qquad \text{for all } r \geq 1. \tag{4.9}$$

The fmgf is $e^{\theta t}$, whence

$$\mu'_{[r]} = \theta^r \qquad \text{for all } r \geq 1, \tag{4.10}$$

and the fcgf is θt, whence

$$\kappa_{[1]} = \theta, \quad \kappa_{[r]} = 0 \qquad \text{for all } r > 1. \tag{4.11}$$

From the relationship between factorial moments and moments about the origin, we have

$$\mu'_r = \sum_{j=0}^{r} \frac{\theta^j}{j!} \Delta^j 0^r = \sum_{j=0}^{r} S(r,j)\theta^j, \qquad r = 1, 2, \ldots,$$

$$= \theta\mu'_{r-1} + \theta\frac{\partial\mu'_{r-1}}{\partial\theta}, \qquad r = 2, 3, \ldots, \tag{4.12}$$

where $\Delta^j 0^r$ is a difference of zero and $S(r, j)$ is a Stirling number of the second kind; see Chapter 1, Section A3.

The uncorrected moment generating function is

$$E[e^{tX}] = \exp[\theta(e^t - 1)], \tag{4.13}$$

and the central moment generating function is

$$E[e^{t(X-\mu)}] = E[e^{t(X-\theta)}] = e^{\theta(e^t - 1 - t)}; \tag{4.14}$$

the latter yields the following recurrence relationship for the moments about the mean

$$\mu_{r+1} = r\theta\mu_{r-1} + \theta\frac{\partial\mu_r}{\partial\theta}, \qquad r = 2, 3, \ldots \tag{4.15}$$

[Riordan (1937)]. Also Kendall (1943) showed that

$$\mu_r = \theta\sum_{j=0}^{r-2}\binom{r-1}{j}\mu_j, \qquad r = 2, 3, \ldots. \tag{4.16}$$

Hence

$$\mu = \kappa_1 = \theta,$$
$$\mu_2 = \kappa_2 = \theta,$$
$$\mu_3 = \kappa_3 = \theta,$$
$$\mu_4 = 3\theta^2 + \theta,$$
$$\mu_5 = 10\theta^2 + \theta, \qquad \text{etc.} \tag{4.17}$$

A Poisson random variable with *parameter* θ is sometimes said to have a Poisson distribution with *expected value* θ.

The moment ratios $\sqrt{\beta_1}$ and β_2 are given by

$$\sqrt{\beta_1} = \theta^{-1/2}, \qquad \beta_2 = 3 + \theta^{-1}. \tag{4.18}$$

Note that the Poisson distribution has the properties:

$$E[X] = \text{Var}(X) \quad \text{and} \quad \beta_2 - \beta_1 - 3 = 0.$$

The coefficient of variation is $\mu_2^{1/2}/\mu = \theta^{-1/2}$. The index of dispersion is $\mu_2/\mu = 1$; this is widely used in ecology as a standard against which to measure clustering (overdispersion with $\mu_2 > \mu$) or repulsion (underdispersion with $\mu_2 < \mu$).

The mean deviation of the Poisson distribution is

$$\nu_1 = E[|X - \theta|] = 2e^{-\theta}\theta^{[\theta]+1}/[\theta]! \tag{4.19}$$

(where $[\cdot]$ denotes the integer part); this was shown by Ramasubban (1958), who also gave an expression for the mean difference. Crow (1958) also discussed the mean deviation, and he showed that the rth inverse (ascending) factorial moment is

$$\mu_{-[r]} = E[\{(X+1)^{[r]}\}^{-1}] = e^{-\theta} \sum_{j=0}^{\infty} \frac{\theta^j}{(j+r)!}$$

$$= \theta^{-r}\left(1 - e^{-\theta}\sum_{j=0}^{r-1}\frac{\theta^j}{j!}\right). \qquad (4.20)$$

For the Poisson distribution

$$\frac{\text{mean deviation}}{\text{standard deviation}} = \frac{2\theta^{[\theta]+1/2}}{[\theta]!e^{\theta}}$$

$$\doteq \sqrt{\frac{2}{\pi}}\left(\frac{\theta}{[\theta]}\right)^{[\theta]+1/2}\exp\left[-(\theta-[\theta]) - \frac{1}{12[\theta]}\right], \quad (4.21)$$

where $[\cdot]$ denotes the integer part. As θ increases, the ratio oscillates; it tends to $\sqrt{2/\pi} \doteq 0.798$.

Haight (1967) quoted a number of formulas relating to the incomplete moments.

4 PROPERTIES

From (4.1)

$$\frac{\Pr[X=x+1]}{\Pr[X=x]} = \frac{\theta}{x+1}, \qquad x = 0, 1, \ldots, \qquad (4.22)$$

whence it follows that $\Pr[X=x]$ increases with x to a maximum at $x = [\theta]$ (or to two equal maxima at $x = \theta - 1$ and $x = \theta$ if θ is an integer) and thereafter decreases as x increases.

Note that equation (4.19) can be written as

$$\nu_1 = 2\theta \max_x \Pr[X=x].$$

Moreover $\Pr[X=x]$ increases monotonically with θ for fixed x if $\theta \leq x$, and decreases monotonically if $\theta \geq x$.

The probabilities are log-concave; that is,

$$\{\Pr[X=x+1]\}^2 > \Pr[X=x]\Pr[X=x+2]. \qquad (4.23)$$

They do not satisfy the log-convexity condition that is a <u>sufficient</u> condition for infinite divisibility; nevertheless,

$$e^{\theta(s-1)} = \left[e^{(\theta/n)(s-1)}\right]^n \tag{4.24}$$

for any positive integer n, and so the distribution is *infinitely divisible*.

Hadley and Whitin (1961) and Said (1958) listed a considerable number of formulas involving Poisson probabilities that might be of use in engineering and operations research. A few are listed here; others appear in various parts of this chapter. We use the notations

$$w(x, \theta) = \frac{e^{-\theta} \theta^x}{x!} \quad \text{and} \quad Q(x, \theta) = \sum_{j=x}^{\infty} w(j, \theta).$$

Then from (4.1) and (4.22)

$$xw(x, \theta) = \theta w(x - 1, \theta),$$

$$\sum_{j=0}^{x} w(j, \theta_1)w(x - j, \theta_2) = w(x, \theta_1 + \theta_2),$$

$$\sum_{j=x}^{\infty} jw(j, \theta) = \theta Q(x - 1, \theta),$$

$$xQ(x + 1, \theta) = xQ(x, \theta) - \theta w(x - 1, \theta),$$

$$\sum_{j=x}^{\infty} Q(j, \theta) = \theta Q(x - 1, \theta) + (1 - x)Q(x, \theta),$$

$$\sum_{j=0}^{x} Q(j, \theta) = \theta[1 - Q(x, \theta)] - xQ(x + 1, \theta),$$

$$\sum_{j=x}^{\infty} j^2 w(j, \theta) = \theta Q(x - 1, \theta) + \theta^2 Q(x - 2, \theta),$$

$$\sum_{j=x}^{\infty} jQ(j, \theta) = \frac{1}{2}\theta^2 Q(x - 2, \theta) + \theta Q(x - 1, \theta) - \frac{1}{2}x(x - 1)Q(x, \theta),$$

$$\sum_{j=0}^{x} jQ(j, \theta) = \frac{1}{2}\theta^2[1 - Q(x - 1, \theta)] + \theta[1 - Q(x, \theta)] + \frac{1}{2}x(x + 1)Q(x + 1, \theta),$$

$$\frac{\partial w(x, \theta)}{\partial \theta} = w(x - 1, \theta) - w(x, \theta) \tag{4.25}$$

[see also Haight (1967)].

The cumulative distribution function can be expressed in terms of the incomplete gamma function (see Chapter 1, Section A5) as

$$\sum_{j=0}^{x} \Pr[X = j] = \frac{\Gamma(x + 1, \theta)}{\Gamma(x)},$$

$$\sum_{j=x+1}^{\infty} \Pr[X = j] = \frac{\gamma(x + 1, \theta)}{\Gamma(x)}. \tag{4.26}$$

This gives a very important relationship between the Poisson and χ^2 distributions; see Sections 5 and 12.2 of this chapter. The cdf is a nonincreasing function of θ for fixed x. Also

$$\sum_{j=0}^{x} w(j, \theta) \geq \sum_{j=0}^{x-1} w(j, \theta - 1) \quad \text{if } x \leq \theta - 1, \tag{4.27}$$

$$\sum_{j=0}^{x} w(j, \theta) \leq \sum_{j=0}^{x-1} w(j, \theta - 1) \quad \text{if } x \geq \theta \tag{4.28}$$

[see Anderson and Samuels (1967)].

Kemp and Kemp (1990) have considered the behavior of the sums of two adjacent probabilities. Crow and Gardner (1959) have studied the sums of r adjacent probabilities. Cheng (1949) has studied the median of the Poisson distribution; it lies between $[\theta - 1]$ and $[\theta + 1]$, where $[\cdot]$ denotes the integer part; see also Lidstone (1942). Teicher (1955) gave results for $\sum_{j=0}^{[\theta]} w(j, \theta)$; see also Kemp (1988a) and Section 5 of this chapter.

Because the distribution is log-concave, $1/r_x > 1/r_{x+1}$ where r_x is the hazard function (failure rate); the distribution therefore has an increasing failure rate. The entropy is

$$\theta \ln \theta - \theta - e^{-\theta} \sum_{j \geq 0} \frac{\theta^j \{\ln(j!)\}}{j!}. \tag{4.29}$$

If X_1 and X_2 are independent variables, each having a Poisson distribution, with expected values θ_1 and θ_2, respectively, then $X_1 + X_2$ has a Poisson distribution with expected value $\theta_1 + \theta_2$. This follows from (4.2) above. The distribution of the mean for samples of size n from a Poisson(θ) distribution therefore has pmf

$$p_{\bar{x}} = \frac{e^{-n\theta}(n\theta)^{\sum x}}{(\sum x)!}, \quad \text{where} \quad \bar{x} = 0, 1/n, 2/n, \ldots. \tag{4.30}$$

The distribution of the difference $X_1 - X_2$ cannot be expressed in so simple a form. It will be discussed in Section 12.3 of this chapter.

The *conditional* distribution of X_1, given that $X_1 + X_2 = n$, is a binomial with parameters $n, \theta_1/(\theta_1 + \theta_2)$. If $X_1, \ldots, X_k$ is a sample of k independent observations from a Poisson(θ) distribution, then the conditional distribution of $X_1, \ldots, X_k$, given that $\sum_{j=1}^{k} X_j = N$, is multinomial with N trials and k equally likely categories.

Approximate formulas for the mean and variance of the median of a sample of n observations from a Poisson distribution were derived by Abdel-Aty (1954).

Steutel and Thiemann (1989b) have used their results concerning exceedance times for a gamma process to obtain approximations for the means and variances of the *order statistics* for the Poisson distribution. They obtained the following exact results for samples of size 2:

$$E[X_{2;2}] = \mu + \mu e^{-2\mu}\{I_0(2\mu) + I_1(2\mu)\}, \tag{4.31}$$

$$\mathrm{Var}(X_{2;2}) = \mu + \mu^2 e^{-4\mu}\{I_0(2\mu) + I_1(2\mu)\}^2 + \mu e^{-2\mu}I_0(2\mu), \tag{4.32}$$

where $X_{1;n} \leq X_{2;n} \leq \ldots \leq X_{n;n}$ are the order statistics for a sample of size n, and $I_j(\cdot)$ denotes a modified Bessel function of the first kind (Chapter 1, Section A5). They also gave approximate formulas for $E[X_{j;n}]$ and $\mathrm{Var}(X_{j;n})$ for large μ.

The limiting form of the Poisson distribution as θ becomes large is normal; the standardized variable

$$Y = \frac{X - \theta}{\theta^{1/2}}$$

approaches a standard normal distribution. This is useful for giving an approximation for the probability tail when θ is moderately large, and for obtaining approximate confidence intervals; see Section 7.3 of this chapter.

Results on sampling moments for random samples from a Poisson distribution have been obtained by Gart and Pettigrew (1970). For example,

$$E\left[k_j \,\Big|\, \sum x_i = X\right] = \bar{x}, \qquad j = 1, 2, \ldots, \tag{4.33}$$

where k_j is the jth k-statistic (see Chapter 1, Section B6). They found moreover that

$$\mathrm{Var}(k_j|X) = E\left[\sum_{r=1}^{j} A_{rj}X(X-1)\ldots(X-r+1)/n^r\right], \tag{4.34}$$

where the A_{rj}'s can be found from the sampling cumulants of the k-statistics as given in Stuart and Ord (1987, Ch. 12). In particular

$$\text{Var}(k_2|X) = \frac{2X(X-1)}{n^2(n-1)}, \tag{4.35}$$

$$\text{Var}(k_3|X) = \frac{6X(X-1)}{n^2(n-1)}\left(3 + \frac{X-2}{n-2}\right), \tag{4.36}$$

$$\text{Var}(k_4|X) = \frac{2X(X-1)}{n^2(n-1)}$$

$$\times \left(49 + \frac{108(X-2)}{n-2} + \frac{12(n+1)(X-2)(X-3)}{n(n-2)(n-3)}\right). \tag{4.37}$$

5 APPROXIMATIONS, BOUNDS, AND TRANSFORMATIONS

The relationship between the Poisson and χ^2 distributions (see Section 12.2) implies that approximations to the cumulative distribution function of the central χ^2 distribution can also be used as approximations to Poisson probabilities, and vice versa. Thus if X has distribution (4.1), then

$$\Pr[X \le x] = \Pr[\chi^2_{2(x+1)} > 2\theta]. \tag{4.38}$$

The Wilson-Hilferty approximation to the χ^2 distribution (Chapter 17) yields

$$\Pr[X \le x] \doteq (2\pi)^{-1/2} \int_z^\infty e^{-u^2/2}\, du, \tag{4.39}$$

where

$$z = 3\left[\left(\frac{\theta}{x+1}\right)^{1/3} - 1 + \frac{1}{9(x+1)}\right](x+1)^{1/2}.$$

As $\theta \to \infty$, the standardized Poisson distribution tends to the unit normal distribution. Modifying the crude approximation "$Z = (X - \theta)/\sqrt{\theta}$ is unit-normally distributed", as in the Cornish-Fisher expansion (see Chapter 17), leads to $Z - (1/3)(Z^2-1)\theta^{-1/2} + (1/36)(7Z^3-Z)\theta^{-1} - \ldots$ being unit-normally distributed.

A formal Edgeworth expansion gives

$$\Pr(X \le x) = \Phi(z) - \phi(z)\left[\frac{z^2-1}{6\theta^{1/2}} + \frac{z^5 - 7z^3 + 3z}{72\theta}\right.$$

$$\left. + \frac{5z^8 - 95z^6 + 384z^4 - 129z^2 - 123}{6480\theta^{3/2}}\right] + O(\theta^{-2}), \tag{4.40}$$

where $z = (x - \theta + 1/2)\theta^{-1/2}$; see, for example, Matsunawa (1986).

Peizer and Pratt's (1968) very accurate approximation is $\Pr[X \leq x] \doteq \Phi(z)$, where

$$z = \left(x - \theta + \frac{2}{3} + \frac{\epsilon}{x+1}\right) \left\{1 + T\left(\frac{x+1/2}{\theta}\right)\right\}^{1/2} \theta^{-1/2},$$

$T(y) = (1 - y^2 + 2y \ln y)(1-y)^{-2}$, $T(1) = 0$, and $\epsilon = 0$ for simplicity or $\epsilon = 0.02$ for more accuracy. Molenaar (1970a) suggested taking $\epsilon = 0.022$. Peizer and Pratt discussed the error of this approximation.

A variance-stabilization transformation is often wanted. A very simple solution is to take $2\sqrt{X}$. This is approximately normally distributed with expected value $2\sqrt{\theta}$ and variance 1. An improvement, suggested by Anscombe (1948), is to use $2\sqrt{X + \frac{3}{8}}$. Freeman and Tukey (1950) have suggested the transformed variable

$$Y = \sqrt{X} + \sqrt{X+1}. \tag{4.41}$$

Tables of this quantity, to two decimal places, for $x = 0(1)50$, have been given by Mosteller and Youtz (1961).

Molenaar (1970a, b, 1973) has made a detailed study of these and other approximations. As simple approximations, with "reasonable" accuracy, he advises

$$P = \sum_{j=0}^{x} e^{-\theta} \frac{\theta^j}{j!} \doteq \begin{cases} \Phi\{2(x+3/4)^{1/2} - 2\theta^{1/2}\} & \text{for } 0.06 < P < 0.94, \\ \Phi\{2(x+1)^{1/2} - 2\theta^{1/2}\} & \text{for tails.} \end{cases} \tag{4.42}$$

Individual Poisson probabilities may be approximated by the formula

$$\Pr[X = x] = (2\pi)^{-1/2} \int_{K_-}^{K_+} e^{-u^2/2} du, \tag{4.43}$$

where $K_- = (x - \theta - 1/2)\theta^{-1/2}$ and $K_+ = (x - \theta + 1/2)\theta^{-1/2}$.

When x is large, Stirling's expansion for $\Gamma(x+1) = x!$ has often been used; this gives

$$\Pr[X = x] \approx \frac{e^{x-\theta}}{\sqrt{2\pi x}} \left(\frac{\theta}{x}\right)^x \left\{1 + \frac{1}{12x} + \frac{1}{288x^2} - \frac{139}{51840x^3}\right.$$

$$\left. - \frac{571}{2488320x^4} + \dots\right\}^{-1}. \tag{4.44}$$

Kemp (1988b, 1989) derived a J-fraction approximation [Hart (1968)] from Stirling's expansion:

$$\Pr[X = x] \approx \frac{e^{x-\theta}}{\sqrt{2\pi x}} \left(\frac{\theta}{x}\right)^x \left[1 - \frac{1/12}{x + 1/24 + 293/(8640x)}\right]. \tag{4.45}$$

When $\theta = r$, where r is an integer, this gives the easily computed approximation for the modal probability

$$\Pr[X = r] \approx (2\pi r)^{-1/2} \left[1 - \frac{1}{u + 1/2 + 293/(60u)} \right], \qquad (4.46)$$

where $u = 12r$. A similar expression for the modal cumulative probability for integer θ is

$$\sum_{j=0}^{r} \Pr[X = r] \approx 0.5 + 2(2\pi r)^{-1/2} 3^{-1} \left[1 - \frac{a}{u + b + c/(u + d)} \right], \qquad (4.47)$$

where $a = 23/15$, $b = 15/14$, $c = 30557/4508$, and $d = 138134432/105880005$ [Kemp (1988b)]. Kemp gave parallel formulas for the modal probability and modal cumulative probability when $\theta = r + 0.5$, r an integer. She also gave adjustments to these formulas for $\theta = r + \alpha$ and $\theta = r + 0.5 + \beta$, where α and β are fractional. The accuracy of these approximations is discussed in Section 6.

Coming now to bounds, which have the advantage over approximations that the sign of the error is known, we have first the inequality [Teicher (1955)]

$$\Pr[x \leq \theta] \geq e^{-1}. \qquad (4.48)$$

(If θ is an integer the right-hand side can be replaced by $1/2$.) Bohman (1963) has given the following inequalities:

$$\Pr[X \leq x] \leq (2\pi)^{-1/2} \int_{-\infty}^{(x+1-\theta)/\sqrt{\theta}} e^{-u^2/2} du, \qquad (4.49)$$

$$\Pr[X \leq x] \geq [\Gamma(\theta + 1)]^{-1} \int_0^x t^{\theta} e^{-t} dt. \qquad (4.50)$$

Samuels (1965) gave these very simple bounds:

$$\Pr[X \leq x - 1] > 1 - \Pr[X \leq x] \qquad \text{if } x \geq \theta, \qquad (4.51)$$

$$\Pr[X \leq x] \geq \exp\{-\lambda/(x + 1)\} \qquad \text{if } x + 1 \geq \theta. \qquad (4.52)$$

Devroye (1986, p. 508) has suggested a number of upper bounds for the "normalized log" probabilities

$$q_j = \ln \Pr[X = m + j] + \ln(m!) + \theta - m \ln \theta, \qquad (4.53)$$

where $m = [\theta]$ ($[\cdot]$ denotes the integer part).

Approximations for the distribution of the range for a Poisson distribution, based on the work of Johnson and Young (1960) on the multinomial distribution, have been investigated numerically by Bennett and Nakamura (1970).

Pettigrew and Mohler (1967) have used the relationship with the multinomial distribution to devise a quick verification test for the Poisson distribution; see Section 7.1.

6 COMPUTATION AND TABLES

The calculation of an individual Poisson probability involves $x!$. Direct multiplication, with $x! = 1 \times 2 \times \cdots \times (x-1)x$, can be used for low values of x. However, $x!$ increases with x very rapidly, necessitating the use of an approximation such as Stirling's expansion or Kemp's (1989) J-fraction (see the previous section).

The customary method of calculation for a complete set of Poisson probabilities, or for the cumulative probabilities, is via the recurrence relation

$$\Pr[X = 0] = e^{-\theta},$$

$$\Pr[X = x] = \frac{\theta}{x} \Pr[X = x - 1], \qquad x \geq 1. \tag{4.54}$$

The problem of build-up errors through recursive calculations does not seem to be serious. Kemp and Kemp have studied the accuracy of individual and cumulative probabilities when computed recursively

1. from the origin using

$$\Pr[X = x] = \frac{\theta}{x} \Pr[X = x - 1],$$

2. from the origin using

$$\ln \Pr[X = x] = \ln \Pr[X = x - 1] + \ln \theta - \ln x,$$

3. from the mode using Kemp's (1988b) modal approximations (see the previous section).

Their results were reported briefly in Kemp and Kemp (1991). Kemp's approximations were found to differ from the true values of the modal and cumulative probabilities by less than 10^{-7} when $\theta \geq 10.5$, and by less than 10^{-10} for $\theta \geq 30.5$.

Fox and Glynn (1988) have put forward an algorithm to compute a complete set of individual Poisson probabilities, with truncation error in the two tails rigorously bounded from above and the remaining probabilities rigorously bounded from below; the algorithm is designed to avoid underflow or overflow.

The first detailed published tables for the Poisson distribution are contained in Molina's (1942) volume, where six-place tables for the Poisson probabilities with parameter

$$\theta = 0.001(0.001)0.01(0.01)0.3(0.1)1(1)5$$

and for the values of $x = 0(1)\infty$ are given. Cumulative probabilities are also given for the same values of x and θ.

Pearson and Hartley (1976) gave six-place values of $\Pr[X = x]$ for $\theta = 0.1(0.1)15.0$ and $x = 0(1)\infty$, and they presented the probabilities $\Pr[X \leq x]$ for

$$\theta = 0.0005(0.0005)0.005(0.005)0.05(0.05)1(0.1)5(0.25)10(0.5)20(1)60$$

and $x = 1(1)35$. Kitagawa's (1952) tables are more detailed. He presented the probabilities $\Pr[X = x]$ to eight decimal places for $x = 0(1)\infty$ and $\theta = 0.001(0.001)1.000$; for the same range of x with $\theta = 1.01(0.01)5.00$, to eight places; and for the same range of x with $\theta = 5.01(0.01)10.00$, to seven places. In Janko's (1958) tables, six-place values of $\Pr[X = x]$ are given with $\theta = 0.1(0.1)15.0(1)16(2)30$ for $x = 1(1)\infty$, and similarly for the cumulative probabilities.

Hald and Kousgaard (1967) published values of θ satisfying the equation

$$e^{-\theta} \sum_{j=0}^{c} \frac{\theta^j}{j!} = P$$

to four significant figures, for $P = 0.001, 0.005, 0.01, 0.025, 0.05, 0.1, 0.2, 0.8, 0.9, 0.95, 0.975, 0.99, 0.995, 0.999$ and $c = 0(1)50$.

The *Tables of the Incomplete Gamma Function Ratio* by Khamis and Rudert (1965) may also be used to obtain Poisson probabilities, for distributions with

$$\theta = 0.00005(0.00005)0.0005(0.0005)0.005(0.005)0.5(0.025)3.0$$

$$(0.05)8.0(0.25)33.0(0.5)83.0(1)125.$$

Values are given to ten decimal places.

Table 2 on pages 100 to 101 of the first edition of this book gave details of some other tables.

7 ESTIMATION

7.1 Model Selection

The Poisson distribution provides the simplest model for discrete data with infinite support. This has led to the development of a number of graphical tests for its suitability as a model given an empirical data set.

Dubey (1966b) suggested plotting f_r/f_{r+1} against r, where f_r is an observed frequency, $r = 0, 1, \ldots$. For a Poisson population this plot should be a straight line with both intersect and slope equal to $1/\theta$.

Ord (1967a, 1972) found that $u_r = rf_r/f_{r-1}$ is a better diagnostic. When plotted against r this should give a straight line with $u_r = c_0 + c_1 r$ for a number of discrete distributions; see Chapter 2, Section 3.2. For the Poisson distribution the expected relationship is $u_r = \theta$. Ord suggested an "intuitively reasonable smoothing procedure" using $v_r = 0.5(u_r + u_{r+1})$ instead of u_r. Gart (1970) pointed out that this relationship ($u_r = \theta$) also holds for a truncated Poisson distribution.

Grimm (1970) has discussed the construction of Poisson probability paper, using the property that $X^{1/2}$ is approximately normally distributed with mean $\theta^{1/2}$ and variance 0.25; see Section 5. His Poisson probability paper uses contours of $\Pr[X \leq c]$, plotted on a cumulative normal scale, against θ on a square-root scale, for $c = 0, 1, \ldots$. When cumulative relative frequencies are marked on the intersections of the horizontal cumulative probability lines and the contours for the appropriate values of c, then the marked points should lie on a vertical line. Grimm discussed a further use of his paper for the graphical determination of $100(1 - \alpha)\%$ confidence intervals.

Hoaglin, Mosteller and Tukey (1985) have developed a "Poissonness" plot based on

$$\ln\{r!\,\Pr[X = r]\} = -\theta + r \ln \theta.$$

A plot of $\ln\{r!f_r/N\}$, where $\sum_{j \geq 0} f_j = N$, against r should therefore have intercept $(-\theta)$ and slope $\ln \theta$. They suggested a method of "levelling" the Poisson plot by taking an assumed value θ_0 of θ. A plot of the count parameter $\ln\{r!f_r/N\} + \theta_0 - r \ln \theta_0$ against r should then have intercept $\theta_0 - \theta$ and slope $\ln(\theta/\theta_0)$. A good choice of θ_0 (e.g., $\bar{x}$) should give a nearly horizontal plot. Hoaglin et al. suggested a smoothing procedure for f_r and the superimposition of approximate significance levels for the count metameter. They considered their procedure to be more resistant against outliers and less prone to bias than a method based on the ratios of successive observed frequencies. However, their method lacks the simplicity of Ord's.

Rapid graphical tests for a number of small samples from a Poisson distribution (with possibly different values of θ) have been devised by Pettigrew and Mohler (1967), using the following result of Johnson and Young (1960). The approximate percentiles R_p of the conditional distribution of R, the range of a Poisson sample of size k, given the mean $\bar{x}$, are given by $R_p = w_p(\bar{x})^{1/2}$; w_p are the percentiles of the distribution of W, where W is the range of k independent unit normal variates. If R is plotted on an arithmetic scale, and $\bar{x}$ on a square-root scale, then the theoretical percentile lines form rays through the origin, and the points for the various samples should lie within a wedge-shaped area. Alternatively, if logarithmic paper is used, then the percentile lines are parallel, since

$$\ln R_p = \ln w_p + 0.5 \ln \bar{x},$$

and the plotted points should lie within them.

7.2 Point Estimation

Only one parameter θ is used in defining a Poisson distribution. Hence there is only the need to estimate a single parameter, though different functions of this parameter may be estimated in different circumstances.

Given n independent rv's $X_1, X_2, \ldots, X_n$, each with distribution (4.1), the MLE of θ is

$$\hat{\theta} = \frac{1}{n} \sum_{j=1}^{n} X_j. \tag{4.55}$$

This is of course the first-moment estimate of θ. $\sum_{j=1}^{n} X_j$ has a Poisson distribution with parameter $n\theta$; it is a complete sufficient statistic for θ. The variance of $\hat{\theta}$ is θ/n. This is equal to the Cramèr-Rao bound; $\hat{\theta}$ is the MVUE of θ.

When all the data are available, the MLE $(\hat{\theta})$ of θ is so easily calculated that it has, in practice, almost always been used. However, when some of the data are omitted, or are inaccurately observed, the situation is less clear-cut. The most important special situation of this kind is that in which the zero class is not observed. A random variable so obtained has the *positive Poisson* distribution; this will be discussed in Section 10.1, together with appropriate methods of estimation for θ.

The problem of estimating the mean of a Poisson distribution in the presence of a nuisance parameter has been examined by Yip (1988). Ahmed (1991) has considered the estimation of θ given samples from two possibly identical Poisson populations.

An intuitive estimator for $e^{-\theta}$, the "probability of the zero class," is $\exp(-\hat{\theta})$. This is a biased estimator — its expected value is $\exp[-\theta(1-e^{-1/n})]$. The MVUE of $e^{-\theta}$ is

$$T = \left(1 - \frac{1}{n}\right)^{n\hat{\theta}}; \tag{4.56}$$

the variance of this estimator is $e^{-2\theta}(e^{\theta/n} - 1)$. The mean square error of T is less than that of $\exp(-\hat{\theta})$ for $e^{-\theta} < 0.45$ [Johnson (1951)]. (Of course, if there is doubt whether the distribution of each X_j is really Poisson, it is safer to use the proportion of the $X's$ that are equal to zero as an estimator of the probability of the zero class.)

The estimation of the Poisson probability (4.1), for general x, has been considered by Barton (1961) and Glasser (1962). An unbiased estimator of

$e^{-\theta}\theta^x/x!$ is the random variable Y, defined by

$$Y = \begin{cases} 1 & \text{if } X_1 = x, \\ 0 & \text{if } X_1 \neq x. \end{cases} \tag{4.57}$$

The MVUE of $e^{-\theta}\theta^x/x!$ is

$$
\begin{aligned}
E[Y|\hat{\theta}] &= \Pr[Y = 1|\hat{\theta}] \\
&= \Pr[X_1 = x|\hat{\theta}] \\
&= \Pr\left[(X_1 = x) \cap \left(\sum_{j=2}^{n} X_j = n\hat{\theta} - x \right) \right] \Big/ \Pr\left[\sum_{j=1}^{n} X_j = n\hat{\theta} \right] \\
&= \frac{(e^{-\theta}\theta^x/x!)\{e^{-\theta(n-1)}[(n-1)\theta]^{n\hat{\theta}-x}/(n\hat{\theta}-x)!\}}{e^{-n\theta}(n\theta)^{n\hat{\theta}}/(n\hat{\theta})!} \\
&= \binom{n\hat{\theta}}{x} \left(\frac{1}{n} \right)^x \left(1 - \frac{1}{n} \right)^{n\hat{\theta}-x}. \tag{4.58}
\end{aligned}
$$

The MVUE of $\sum(e^{-\theta}\theta^x/x!)$ is

$$\sum \binom{n\hat{\theta}}{x} \left(\frac{1}{n} \right)^x \left(1 - \frac{1}{n} \right)^{n\hat{\theta}-x},$$

where $\sum$ denotes summation with respect to x over any (finite or infinite) subset of the nonnegative integers.

Estimation of the *ratio* of expected values θ_1, θ_2 of two Poisson distributions based on observed values of independent random variables that have such distributions has been described by Chapman (1952b). He showed that there is no unbiased estimator of the ratio θ_1/θ_2 with finite variance but that the estimator $x_1/(x_2 + 1)$ (in an obvious notation) is "almost unbiased."

The rationale of Bayesian inference for the Poisson distribution has been explained in detail by Irony (1992).

Empirical Bayes estimation for the Poisson distribution has been investigated by a number of authors. Robbins (1956), also Good (1953), developed an EB method for estimating θ that is known as the frequency-ratio method. Here x is the current observation, there are n past observations, and $f_n(x)$ denotes the number of past observations having the value x; this gives $(x + 1)f_n(x + 1)/\{1 + f_n(x)\}$ as the EB estimator of θ. The frequency ratio estimator, graphed as a function of x, is far from smooth, however. Maritz (1969) reviewed and extended various smoothing procedures that had previously been proposed, and he made a careful assessment of their comparative performance. A thorough and up-to-date account of EB point estimation for the Poisson distribution, with applications, is in Maritz and Lwin (1989).

Sadooghi-Alvandi (1990) has examined the estimation of θ from a sample of n observations using a LINEX loss function of the form

$$L(\theta_B, \theta) = b\{e^{a(\theta_B - \theta)} - a(\theta_B - \theta) - 1\}, \qquad b \geq 0, \ a \neq 0 \qquad (4.59)$$

introduced by Varian (1975), see also Zellner (1986). Given a gamma prior distribution with pdf $f(\theta|\gamma, \lambda) = \lambda^\gamma \theta^{\gamma-1} e^{-\lambda\theta}/\Gamma(\gamma)$, where $0 \leq \theta < \infty$ and $\gamma > 0$, $\lambda > 0$, then, provided that $\gamma + n + a \geq 0$, the unique Bayesian estimate of θ, relative to (4.59), is

$$\theta_B = \left\{\frac{n}{a}\ln\left(\frac{\lambda + n + a}{\lambda + n}\right)\right\}\left(\bar{x} + \frac{\gamma}{n}\right). \qquad (4.60)$$

Bayesian methods for the simultaneous estimation of the means of several independent Poisson distributions have also attracted attention. Clevenson and Zidek (1975) proposed a generalized Bayes method where the loss function is the sum of the component standardized squared error losses. They showed that the method is minimax, and also (both numerically and asymptotically) that it is better than using the sample means as estimators of the corresponding population means. Peng (1975) has investigated the use of the aggregate squared-error loss function $\sum_i(\theta_i^* - \theta_i)^2$. This approach has been extended by Hudson and Tsui (1981); see also Hwang (1982) and Tsui and Press (1982). Simultaneous estimation of Poisson means under entropy loss has been studied by Ghosh and Yang (1988) and Yang (1990).

7.3 Confidence Intervals

Since the Poisson distribution is a discrete distribution, it is not possible to construct confidence intervals for θ with an exactly specified confidence coefficient of say, $100(1 - \alpha)\%$. *Approximate* $100(1 - \alpha)\%$ confidence limits for θ given an observed value of X, where X has the distribution (4.1), are obtained by solving the equations

$$\exp(-\theta_L)\sum_{j=x}^{\infty}\frac{\theta_L^j}{j!} = \frac{\alpha}{2}, \qquad (4.61)$$

$$\exp(-\theta_U)\sum_{j=0}^{x}\frac{\theta_U^j}{j!} = \frac{\alpha}{2}, \qquad (4.62)$$

for θ_L, θ_U, respectively, and using the interval (θ_L, θ_U).

From the relationship between the Poisson and χ^2 distributions (Section 12.2), the equations (4.61) and (4.62) can be written as

$$\theta_L = 0.5\chi^2_{2x,\alpha/2}, \tag{4.63}$$

$$\theta_U = 0.5\chi^2_{2(x+1),1-\alpha/2}. \tag{4.64}$$

So the values of θ_L and θ_U can be found by interpolation (with respect to the number of degrees of freedom) in tables of percentage points of the central χ^2 distribution.

If θ is expected to be fairly large, say greater than 15, a normal approximation to the Poisson distribution might be used. Then

$$\Pr[|X - \theta| < u_{\alpha/2}\sqrt{\theta}\,|\theta] = 1 - \alpha, \tag{4.65}$$

where

$$(2\pi)^{-1/2} \int_{u_{\alpha/2}}^{\infty} e^{-u^2/2}du = \frac{\alpha}{2}.$$

From (4.65) $\Pr[\theta^2 - \theta\{2X + u^2_{\alpha/2}\} + X^2 < 0|\theta] = 1 - \alpha,$
that is,

$$\Pr\left[X + 0.5u^2_{\alpha/2} - u_{\alpha/2}(X + 0.25u^2_{\alpha/2})^{1/2}\right.$$
$$\left. < \theta < X + 0.5u^2_{\alpha/2} + u_{\alpha/2}(X + 0.25u^2_{\alpha/2})^{1/2}|\theta\right] \doteq (1 - \alpha).$$

The limits

$$X + 0.5u^2_{\alpha/2} \pm u_{\alpha/2}\sqrt{X + 0.25u^2_{\alpha/2}}$$

thus enclose a confidence interval for θ with confidence coefficient approximately equal to $100(1 - \alpha)\%$.

Molenaar (1970b, 1973) made recommendations concerning approximations for confidence limits for θ; he also cited other relevant references. Values of approximate 95% confidence limits for θ, given an observed value of X, are given by Mantel (1962). The "limit factors" shown must be *multiplied* by the observed value of X to obtain limits for θ. There is a similar table published by the Society of Actuaries (1951).

Tables for θ_L and θ_U are contained in Pearson and Hartley (1976). Crow and Gardner (1959) constructed a modified version of these tables, so that (1) the confidence belt is as narrow as possible, *measured in the direction of the observed variable*, and (2) among such narrowest belts, it has the smallest possible upper confidence limits. Condition 2 reduces, in particular, the width of confidence intervals corresponding to small values of the observed variable.

Let X_1 and X_2 be two independent Poisson rv's with parameters θ_1 and θ_2, respectively. Let $F_{m,n}(X)$ represent the distribution function of the ratio of two independent chi-square variables χ^2_{2m}/χ^2_{2n} with $2m$ and $2n$ degrees of freedom. Bol'shev (1965) has shown that for any α, where $0 \le \alpha < 1$, the solution of the equation

$$F_{x_1, x_2+1}(X) = \alpha$$

satisfies the inequality

$$\inf_{\theta_1, \theta_2} \Pr\left[X < \frac{\theta_1}{\theta_2}\right] \ge 1 - \alpha.$$

Thus X can be taken to be the lower bound of a confidence interval for θ_1/θ_2 with minimal confidence coefficient $(1 - \alpha)$. Similarly the solution of

$$F_{x_1+1, x_2}(Y) = 1 - \alpha$$

satisfies the inequality

$$\inf_{\theta_1, \theta_2} \Pr\left[Y > \frac{\theta_1}{\theta_2}\right] \ge 1 - \alpha,$$

and thus Y can be taken to be the upper bound of a confidence interval for θ_1/θ_2 with minimal confidence coefficient $(1 - \alpha)$. These results are useful for testing the hypothesis $\theta_1/\theta_2 < c$, where c is some constant (and similarly the hypothesis $\theta_1/\theta_2 > c$).

Chapman (1952b) used the fact that the conditional distribution of X_1, given $X_1+X_2=x$, is binomial with parameters x and $\theta_1/(\theta_1 + \theta_2)$; see Section 8. Forming approximate confidence intervals for $\theta_1/(\theta_1 + \theta_2) = (1 + \theta_2/\theta_1)^{-1}$ is then equivalent to doing the same for θ_1/θ_2.

Casella and Robert (1989) were critical of the method of randomized Poisson confidence intervals of Stevens (1957) and Blyth and Hutchinson (1961), where random noise is added to the data in order to obtain exact intervals. They introduced instead the concept of refined confidence intervals; this is a numerical procedure whereby the endpoints of a confidence interval are shifted in order to produce the smallest possible interval. Their paper is particularly concerned with confidence intervals for the Poisson distribution. They also considered certain other discrete distributions. Their method becomes very computer intensive.

7.4 Model Verification

If a data set comes from a Poisson distribution its variance should be equal to its mean. This can be tested by the Index of Dispersion test, with $(N-1)s^2/\bar{x}$ having approximately a χ^2 distribution with $N - 1$ degrees of freedom. Conditions under which the variance test is optimal for testing the homogeneity of a complete Poisson sample have been examined by Neyman and Scott

(1966), Potthoff and Whittinghill (1966), and Moran (1973); see also Gart (1974).

Tests for a fully specified Poisson distribution, based on the empirical distribution function, are discussed by Stephens (1986). The best known of these is Pearson's χ^2 goodness-of-fit test; see e.g. Tallis (1983) for details and further references. In particular, Stephens discusses Kolmogorov-Smirnov goodness-of-fit tests; see also Wood and Altavela (1978).

Verification tests for a homogeneous Poisson process are customarily based on the exponentiality of the interarrival times.

8 CHARACTERIZATIONS

There has been much work concerning characterizations of the Poisson distribution. Work prior to 1974 is summarized in Kotz (1974).

Raikov (1938) showed that if X_1 and X_2 are independent random variables and X_1+X_2 has a Poisson distribution, then X_1 and X_2 must each have Poisson distributions. (A similar property also holds for the sum of any number of independent Poisson random variables.)

Moran (1952) discovered a fundamental property of the Poisson distribution. If X_1 and X_2 are independent nonnegative integer-valued rv's, such that the conditional distribution of X_1 given the total $X_1 + X_2$ is a binomial distribution with a common parameter p for all given values of $X_1 + X_2$, and if there exists at least one integer i such that $\Pr[X_1 = i] > 0$, $\Pr[X_2 = i] > 0$, then X_1 and X_2 are both Poisson rv's. Chatterji (1963) showed that if X_1 and X_2 are independent nonnegative integer-valued random variables, and if

$$\Pr[X_1 = x_1 | X_1 + X_2 = x] = \binom{x}{x_1} p_x^{x_1}(1 - p_x)^{x-x_1}, \quad x_1 = 0, 1, \ldots, x, \quad (4.66)$$

then it follows that

1. p_x does not depend on x, but equals a constant p for all values of x,
2. X_1 and X_2 each have Poisson distributions with parameters in the ratio $p : (1 - p)$.

This characterization has been extended by Bol'shev (1965) to n variables $X_1, X_2, \ldots, X_n$. The condition is that the conditional distribution of $X_1, X_2, \ldots, X_n$, given $\sum X_i$, be multinomial; see also Volodin (1965). Bol'shev suggested that this property might be used to generate Poisson-distributed random variables for a number of Poisson distributions, using only a single original Poisson variate (with a large expected value) that is split up according to a multinomial with fixed cell probabilities; Brown and Bromberg (1984) have developed this idea further. Patil and Seshadri (1964) showed that Moran's characterization is a particular case of a more general characterization for a pair of independent random variables X_1 and X_2 with specified conditional distribution of X_1 given $X_1 + X_2$. Further extensions to the Moran characterization have been given by Janardan and Rao (1982) and Alzaid, Rao, and Shanbhag (1986). For comments on a suggested extension to truncated rv's see Panaretos (1983b).

Lukacs (1956) showed that if $X_1, X_2, \ldots, X_n$ is a random sample from some distribution, then the distribution from which the sample is taken is Poisson with parameter θ iff the statistic $n\bar{x}$ has a Poisson distribution with parameter $n\theta$.

Suppose that $X_1, X_2, \ldots, X_n$ is a random sample from a distribution, and suppose that r and s are two positive integers. Assume also that the $(r+s)$th moment of the distribution exists, that the cdf is zero for $x < 0$, and that the cdf is greater than zero for $x \geq 0$. Then Lukacs (1965) showed that the distribution is Poisson iff the k-statistic $k_{r+s} - k_r$ has a constant regression on $k_1 = \bar{x}$, that is, iff $E[k_{r+s} - k_r | k_1] = 0$.

Rao and Rubin (1964) obtained the following characterization of the Poisson distribution. If X is a discrete rv taking only nonnegative integer values and the conditional distribution of Y given $X = x$ is binomial with parameters x, p (p not depending on x), then the distribution of X is Poisson iff

$$\Pr[Y = k | Y = X] = \Pr[Y = k | Y \neq X].$$

Rao (1965) gave the following physical basis for the model described above:

X represents a "naturally" occurring quantity, which is observed in such a way that some of the components of X may not be counted.
Y represents the value remaining (and actually observed) after this "destructive process."

That is to say, suppose that an original observation is distributed according to a Poisson distribution with parameter θ, and that the probability that the original observation n is reduced to r due to a destructive process is

$$\binom{n}{r} \pi^r (1 - \pi)^{n-r}, \qquad 0 \leq \pi \leq 1.$$

If Y denotes the resultant rv, then

$$\Pr[Y = r] = \Pr[Y = r | \text{undamaged}] = \Pr[Y = r | \text{damaged}]$$

$$= \frac{e^{-\theta\pi}(\theta\pi)^r}{r!}; \qquad (4.67)$$

furthermore condition (4.67) characterizes the Poisson distribution.

Rao and Rubin's characterization has generated a lot of attention. For instance, Srivastava and Srivastava (1970) showed that if the original observations have a Poisson distribution, and if condition (4.67) is satisfied, then the destructive process is binomial.

Let X be a discrete rv with support $0, 1, \ldots$; suppose also that if a random observation $X = n$ is obtained, then n Bernoulli trials with probability p of success are performed. Let Y and Z be the resultant numbers of successes and failures, respectively. Then Srivastava (1971) showed that X has a Poisson distribution iff Y and Z are independent.

An alternative, simpler, proof of the Rao-Rubin theorem was given by Wang (1970). Wang also introduced the idea of binomial *splitting*, that is, compounding X with the binomial distribution

$$\binom{n}{r} \pi^r (1 - \pi)^{n-r}, \qquad 1 \le \pi \le 1,$$

and the similarly defined notion of binomial *expanding*. He showed that the Poisson distribution is the only one that remains invariant under binomial splitting or expanding.

The Rao-Rubin characterization was extended to a pair of independent Poisson variables (and also to a multivariate set of Poisson variables) by Talwalker (1970). This extension was conjectured independently by Srivastava and Srivastava (1970) as follows:

Let $X = \binom{X_1}{X_2}$ be a non–degenerate, discrete random vector such that X_1, X_2 take nonnegative integer values. Let

$$\binom{n_1}{r_1} \pi_1^{r_1} \phi_1^{n_1 - r_1} \quad \text{and} \quad \binom{n_2}{r_2} \pi_2^{r_2} \phi_2^{n_2 - r_2}$$

be the independent probabilities that the observations n_1, n_2 on X_1, X_2 are reduced to r_1, r_2, respectively, during the destructive process. Let $Y = \binom{Y_1}{Y_2}$ denote the resultant random vector where Y_1 and Y_2 take the values $0, 1, 2, \ldots,$; let $r = \binom{r_1}{r_2}$ denote the vector with elements r_1 and r_2 . Then

$$\begin{aligned}
\Pr[Y = r] &= \Pr[Y = r | \text{damaged}] \\
&= \Pr[Y = r | x_1 \text{damaged}, x_2 \text{undamaged}] \\
&= \Pr[Y = r | x_1 \text{undamaged}, x_2 \text{damaged}] \\
&= \Pr[Y = r | \text{undamaged}]
\end{aligned}$$

iff X has the double Poisson distribution with pgf $\exp\{\theta_1(z_1 - 1) + \theta_2(z_2 - 1)\}$.

An elementary (and elegant) proof was given by Shanbhag (1974); see also Aczel (1972) and van der Vaart (1972). Aczel (1975) studied the problem under weaker conditions. Note that the simpler proof by Wang (1970) of the Rao-Rubin characterization holds also in the multivariate case.

Characterization of the binomial, Poisson and negative binomial trio of distributions via the generalized Rao-Rubin condition has been studied by Talwalker (1975, 1980) and Rao et al. (1980). The generalized Rao-Rubin condition is

$$\Pr[Y = k | \text{probability of survival is } \pi]$$
$$= \Pr[Y = k | \text{undamaged, probability of survival is } \pi'],$$

where Y is the resultant rv with support $0, 1, \ldots$.
Krishnaji (1974) obtained the following result: If $Y = X_1 + X_2$ and

$$\Pr[X_1 = x_1 | Y = y] = \binom{y}{x_1} p^{x_1}(1 - p)^{y - x_1},$$

then $E[X_1 | X_2] = \theta p$ if and only if Y has a Poisson distribution with expected value θ.

Another variant of the Rao-Rubin theorem has been proved by Shanbhag and Clark (1972):
Let X have a power series distribution with pmf

$$\Pr[X = x] = \frac{a_x \theta^x}{A(\theta)}, \qquad x = 0, 1, \ldots, \tag{4.68}$$

and assume that the conditional distribution of Y given that $X = n$ has pmf $\Pr[Y = r | X = n] = g(r; n)$, $r = 0, 1, \ldots, n$, with mean and variance $n\pi$ and $n\pi(1 - \pi)$, where π does not depend on θ. Then

$$E[Y] = E[Y | Y = x] \quad \text{and} \quad \text{Var}(Y) = \text{Var}(Y | Y = x)$$

iff $\{P_x\}$ is Poisson and $g(n; n) = \pi^n$. (Note that if $g(r; n)$ is binomial, then its mean and variance are $n\pi$ and $n\pi(1 - \pi)$, respectively.)

Further results relating to the Rao-Rubin characterization have been obtained by Shanbhag (1974), Shanbhag and Rajamannar (1974), Srivastava and Singh (1975), Shanbhag and Panaretos (1979), Gupta (1981), and Kourouklis (1986).

Korwar (1975) established that if X and Y are nonnegative integer-valued rv's such that, for all x,

$$\Pr[Y = y | X = x] = \binom{x}{y} p^y (1 - p)^{x - y}, \qquad y = 0, 1, \ldots, x,$$

where p is some constant such that $0 < p < 1$, not dependent on x, then X

has a Poisson distribution iff

$$\Pr[Y = y] = \Pr[Y = y | X = x] = \Pr[Y = y | X > Y]$$

(cf. (4.67)).

The investigations into empirical Bayes estimation for the Poisson distribution by Maritz (1969) have led to a different kind of characterization theorem. Shanbhag and Clark (1972) showed that if X has a power series distribution (4.68), and if θ is a rv with pdf

$$f(\theta) = k \sum_{j=0}^{\infty} b_i \theta^i,$$

where $b_i \geq 0$, $0 < \theta < c$, and c may be infinite, then $\beta(x) = E[\theta | X = x]$ is linear iff $f(\theta)$ is exponential and X has a Poisson distribution.

Characterizations based on properties of sample statistics were also developed by Kharshikar (1970) and by Shanbhag (1970a). Kharshikar showed that if X belongs to a power series family, then the condition $E[s^2/\bar{x}|\bar{x} > 0] = 1$ (where $\bar{x}$ and s^2 are the sample mean and variance statistics) characterizes a Poisson distribution. Shanbhag proved a more general version of this theorem involving a quadratic form in the sample observations. The modification of Shanbhag's result by Wang (1972) was later shown by Shanbhag (1973) to need amendment.

Another sort of characterization (one based on a Bhattacharya covariance matrix) was obtained by Shanbhag (1972). This one would seem to have limited practical use, as it is valid also for the binomial and negative binomial distributions.

Daboni (1959) has given the following characterization in terms of mixtures of binomial distributions. Suppose that X is a random variable distributed as a mixture of binomial distributions with parameters N, p. Then X and $(N - X)$ are independent iff N has a Poisson distribution.

Boswell and Patil (1973) showed that the following four statements about a discrete random variable $X(\theta)$, $\theta \geq 0$, $X(0) \equiv 0$, are equivalent:

1. There exists a real a such that $\Pr[X(\theta) \geq a] = 1$, and $\partial P_x/\partial\theta = P_{x-1} - P_x$, $\theta > 0$.
2. $\sum_{r=0}^{x} \partial P_r/\partial\theta = -P_x$, $\theta > 0$.
3. For any function $f(X(\theta))$ that has finite expectation,

$$\frac{\partial E[f(X(\theta))]}{\partial\theta} = E[f(X(\theta) + 1) - f(X(\theta))].$$

4. $X(\theta)$ has a Poisson distribution, $\theta > 0$.

Boswell and Patil also gave a more general formulation that allows $X(\theta)$ to be either discrete or continuous.

Patil and Ratnaparkhi (1977) established that if X has a power series distribution, then a necessary and sufficient condition for X to have a Poisson distribution is $\mu'_{[2]} = \mu^2$, where $\mu'_{[2]}$ is the second factorial moment.

Samaniego (1976) termed $Y = X_1 + X_2$ a "convoluted Poisson variable" if X_1 and X_2 are independent and X_1 has a Poisson distribution. He obtained characterizations of convoluted Poisson distributions.

The well-known characterization in terms of the conditional (multinomial) distribution of independent variables $X_1, \ldots, X_n$ given $\sum_{j=1}^{n} X_j = z$ (see (4.66) et seq.) has been extended to zero-truncated X variables by Singh (1978). A characterization based on a multivariate splitting model was obtained by Rao and Srivastava (1979); another characterization based on this model was given by Ratnaparkhi (1981).

The following new type of characterization is due to Rao and Sreehari (1987). A nonnegative integer-valued rv X has a Poisson distribution iff

$$\sup \left(\frac{\text{Var}(h(X))}{\text{Var}[X] E[\{h(X+1) - h(X)\}^2]} \right) = 1, \qquad (4.69)$$

where the supremum is taken over all real-valued functions $h(\cdot)$ such that $E[\{h(X+1) - h(X)\}^2]$ is finite.

Characterizations of the Poisson distribution based on discrete analogues of the Cramèr-Wold and Skitovich-Darmois theorems have been derived by McKenzie (1991).

Diaconis and Zabell (1991) have pointed out that the Poisson distribution is characterized by the identity

$$E[\theta f(X+1)] = E[X f(X)]$$

for every bounded function $f(\cdot)$ on the integers. For the background to this identity, see Stein (1986, Ch. 9).

The considerable amount of effort put into the study of characterizations for the Poisson distribution has concentrated largely on a limited range of types of characterizations. Haight (1972) attempted to unify certain isolated results via Svensson's (1969) theorem. This states that if X has a distribution on the nonnegative integers, then there exists a two-dimensional rv (X, Y) with the property that the pgf of $(X|X + Y)$ is of the form

$$\phi(z) = (1 - p + pz)^n$$

if and only if

$$\pi_{X+Y}(z) = \pi_X\{(z - 1 + p)/p\},$$

where π_Y is the pgf of Y. Haight saw that this theorem permits the characterization of certain distributions of the sum of two rv's, omitting in general the assumption of independence. For instance, Chatterji's characterization theorem (4.66) is a straightforward consequence of Svensson's theorem.

In a lengthy and detailed account of characterizations of the *Poisson process*, Galambos and Kotz (1978) gave a number of new results; they related these where possible to the work of previous authors. Their researches were especially concerned with characterizations based on age and residual life, on rarefactions of renewal processes, on geometric compounding, and on damage models, and were motivated by characterizations of the exponential distribution. A characterization of the Poisson process by properties of conditional moments has been given by Bryc (1987) and extended by Wesolowski (1988).

9 APPLICATIONS

The Poisson distribution has been described as playing a "similar role with respect to discrete distributions to that of the normal for absolutely continuous distributions," Douglas (1980). It is used (1) as an approximation to the binomial and other distributions, (2) when events occur randomly in time or space, (3) in certain models for the analysis of contingency tables, and (4) in the empirical treatment of count data.

In many elementary textbooks it is introduced as a limit, and hence as an approximation, for the binomial distribution where the occurrence of an event is rare and there are many trials. This approach underlies the use of the Poisson distribution in quality control, for the number of defective items per batch; see e.g. Walsh (1955), van der Waerden (1960), and Chatfield (1983). The Poisson is also a limiting form for the hypergeometric distribution (Chapter 6, Section 5) and hence provides an approximation for sampling without replacement. Feller (1968, p. 59) pointed out the relevance of the Poisson limiting form for Boltzmann-Maxwell statistics in quantum statistics and in the theory of photographic plates.

Early writers on the analysis of quadrat data [e.g., Greig-Smith (1964)] justified the use of the Poisson distribution for such data via the relationship of the Poisson distribution with the binomial distribution [see also Seber (1982b, p. 24)]. Quadrat data have been collected extensively in ecology, geology, geography, and urban studies. A more fundamental explanation for the widespread applicability of the Poisson distribution for counts per unit of space or volume is the Poisson process [see e.g. Cliff and Ord (1981)].

The Poisson process also has great importance concerning counts of events per unit of time, particularly in queueing theory. Here the intervals between successive events have independent identical exponential distributions, and consequently the number of events in a specified time interval has a Poisson distribution; see Section 2 of this chapter. Consider, for example, the problem of finding the number of telephone channels in use (or customers waiting in line, etc.) at any one time. Suppose that there are "infinitely many" (actually, a large number of) channels available, that the holding time (length of call) T for each call has the exponential distribution with pdf

$$p(t) = \phi^{-1}e^{-t/\phi}, \qquad \phi > 0,\ t > 0,$$

and that incoming calls arrive at times following a Poisson process, with the average number of calls per unit time equal to θ. Then the function $P_N(\tau)$, representing the probability that exactly N channels are being used at time τ after no channels are in use, satisfies the differential equations

$$P_0'(\tau) = -\theta P_0(\tau) + \phi^{-1}P_1(\tau),$$

$$P_N'(\tau) = -(\theta + N\phi^{-1})P_N(\tau) + \theta P_{N-1}(\tau) + (N+1)\phi^{-1}P_{N+1}(\tau), \quad N \geq 1.$$

$$(4.70)$$

The "steady state" probabilities $(\lim_{\tau \to \infty} P_N(\tau) = P_N)$ satisfy the equations

$$\theta P_0 = \phi^{-1}P_1,$$

$$(\theta + N\phi^{-1})P_N = \theta P_{N-1} + (N+1)\phi^{-1}P_{N+1}. \qquad (4.71)$$

On solving these equations, it is found that

$$P_N = \frac{e^{-\theta\phi}(\theta\phi)^N}{N!}, \qquad N = 0, 1, \ldots, \qquad (4.72)$$

which is (4.1) with θ replaced by $\theta\phi$; see e.g. Gnedenko and Kovalenko (1989).

There are many excellent texts on stochastic processes that devote considerable attention to the Poisson process. Doob (1953) mentions its use for molecular and stellar distributions, while Parzen (1962) gives applications to particle counters, birth processes, renewal processes, and shot noise. Taylor and Karlin's (1984) introductory text is markedly application-oriented, with applications in, inter alia, engineering, biosciences, medicine, risk theory, commerce, and demography. The Poisson process for entities in space has been discussed in depth by Ripley (1981) and Stoyan, Kendall, and Mecke (1987).

Poisson regression, that is, the analysis of the relationship between an observed count and a set of explanatory variables, is described by Koch, Atkinson, and Stokes (1986) in a very clear introductory article. Its use in the analysis of binary data is discussed by Cox and Snell (1989); see especially pp. 146–147 for the interrelationships between Poisson and multinomial models. This relationship is examined by Sandland and Cormack (1984), Cormack (1989), and Cormack and Jupp (1991) in relation to their work on the log-linear analysis of capture-recapture data. Its role in the analysis of frequencies when summarizing data sets is described lucidly by Bishop, Fienberg, and Holland (1975). Fienberg (1982) gave helpful pointers to the literature about the treatment of cross-classified categorical data in general. Applications of Poisson regression include bioassay, counts of colonies of bacteria or viruses for varying dilutions and/or experimental conditions, equipment malfunctions for varying operational conditions, cancer incidence, and mortality and morbidity statistics. Applications to economic problems are described by Hausman et al. (1984). Lee (1986) has cited a number of other references; he was particularly concerned with testing a Poisson model against other discrete models, including the negative binomial.

In the empirical treatment of count data the Poisson distribution is often used as a yardstick to assess the degree and nature of nonrandomness. Mixed Poisson distributions and Poisson-cluster distributions have been extensively developed as tools for dealing with nonhomogeneous and clumped data; see Chapters 8 and 9.

Chapter 7 in Haight (1967) gives a multitude of references concerning uses of the Poisson distribution, categorized by industry, agriculture and ecology, biology, medicine, telephony, accidents, commerce, queueing theory, sociology and demography, traffic flow theory, military, particle counting, and miscellaneous.

Recent applications of the Poisson distribution include, for example, its use for sister chromatid exchanges in the study of DNA breakage and reunion [Margolin et al. (1986)]. Goldstein (1990) has applied it to DNA sequence matching. The annual volumes of the *Current Index to Statistics* and the quarterly issues of *Statistical Theory and Method Abstracts* give leads to other recent uses, both applied and theoretical.

10 TRUNCATED AND MISRECORDED POISSON DISTRIBUTIONS

10.1 Left-Truncation

The commonest form of truncation is the omission of the zero class; this occurs if the observational apparatus becomes active only when at least one event occurs. The pmf of the corresponding truncated Poisson distribution is

$$\Pr[X = x] = (1 - e^{-\theta})^{-1} e^{-\theta} \theta^x / x! = (e^{\theta} - 1)^{-1} \theta^x / x!, \quad x = 1, 2, \ldots. \quad (4.73)$$

This is usually called the *positive Poisson* distribution. Cohen (1960a) called it a *conditional Poisson* distribution.

The rth factorial moment of X is

$$E[X(X - 1) \ldots (X - r + 1)] = (1 - e^{-\theta})^{-1} \theta^r. \quad (4.74)$$

The mean and variance are

$$E[X] = \frac{\theta}{1 - e^{-\theta}}, \tag{4.75}$$

$$\text{Var}[X] = \frac{\theta}{1 - e^{-\theta}} - \frac{\theta^2 e^{-\theta}}{(1 - e^{-\theta})^2}. \tag{4.76}$$

The expected value of X^{-1} is

$$E[X^{-1}] = \frac{1}{e^{\theta} - 1} \sum_{j=1}^{\infty} \frac{\theta^j}{j! j} \tag{4.77}$$

[Grab and Savage (1954)]. Tiku (1964) showed that

$$E[X^{-1}] \doteq [(\theta - 1)(1 - e^{-\theta})]^{-1} \tag{4.78}$$

for sufficiently large θ. Grab and Savage (with addenda in the same volume) gave tables of $E[X^{-1}]$ to five decimal places for $\theta = 0.01, 0.05(0.05)$ $1.0(0.1)2.0(0.2)5.0(0.5)7(1)10(2)20$.

The maximum likelihood estimator $\hat{\theta}$ of θ, given observed values of n independent random variables $X_1, X_2, \ldots, X_n$ each having the same positive Poisson distribution, satisfies the equation

$$\bar{x} = n^{-1} \sum_{j=1}^{n} x_j = \frac{\hat{\theta}}{1 - e^{-\hat{\theta}}}. \tag{4.79}$$

Equation (4.79) may be solved numerically; the process is quite straightforward. David and Johnson (1952) gave tables of the function $\hat{\theta}/(1 - e^{-\hat{\theta}})$. Although Irwin (1959) has derived an explicit expression for the solution of (4.79), namely

$$\hat{\theta} = \bar{x} - \sum_{j=1}^{\infty} \frac{j^{j-1}}{j!} (\bar{x} e^{-\bar{x}})^j, \tag{4.80}$$

the methods suggested above are usually easier to apply. Kemp and Kemp (1988) have shown that

$$\frac{6(\bar{x} - 1)}{\bar{x} + 2} < \hat{\theta} < \bar{x}\{1 - (e^{\bar{x}} - \bar{x})^{-1}\}. \tag{4.81}$$

For large values of n, the variance of the ML estimator $\hat{\theta}$ is given by

$$\text{Var}(\hat{\theta}) \approx \theta(1 - e^{-\theta})^2(1 - e^{-\theta} - \theta e^{-\theta})^{-1} n^{-1}. \tag{4.82}$$

David and Johnson (1952) have suggested an estimator based on the sample moments

$$T_r = \frac{1}{n} \sum_{j=1}^{n} x_j^r, \qquad r = 1, 2, \dots.$$

Their estimator is

$$\theta^\dagger = T_2 T_1^{-1} - 1, \tag{4.83}$$

and its variance is approximately

$$(\theta + 2)(1 - e^{-\theta})n^{-1}.$$

Its asymptotic efficiency (relative to the ML estimator) reaches a minimum of about 70% between $\theta = 2.5$ and $\theta = 3$. The efficiency increases to 100% as θ increases [David and Johnson (1952)].

Tate and Goen (1958) have shown that the MVUE of θ is

$$\theta^* = \frac{n\bar{x}S(n\bar{x}-1,n)}{S(n\bar{x},n)} = \frac{\bar{x}\{1 - S(n\bar{x}-1,n-1)\}}{S(n\bar{x},n)}, \tag{4.84}$$

where $S(a,b)$ is the Stirling number of the second kind defined in Chapter 1, Section A3. There are tables of the multiplier of $\bar{x}$ in Tate and Goen (1958). For large n the multiplier is approximately $\{1 - (1 - n^{-1})^{n\bar{x}-1}\}$.

Let n_0 denote the number of missing values with $X = 0$. David and Johnson (1952) posed the following question: "Are we to regard the truncated sample as one of a sequence of samples of equal size (i.e., n constant), or as the result of truncation of complete samples of the same size (i.e., $n + n_0$ constant)?" The previous discussion has been based on the fixed n case. McKendrick (1926) estimated θ on the alternative assumption of fixed $N = n + n_0$. For Poisson variables the first two factorial moments are θ and θ^2. McKendrick's method uses the first two observed factorial moments to obtain an initial estimate of n_0. The estimated mean and the expected zero frequency are then used to estimate θ, and hence n_0, iteratively. The method is of considerable historical interest and is not difficult to apply, but is now rarely used. Assuming the same model (fixed $N = n + n_0$), Dahiya and Gross (1971) obtained a conditional MLE of N, and hence estimated n_0. Blumenthal, Dahiya, and Gross (1978) derived the unconditional MLE of θ, and also a modified MLE which appeared better in certain respects.

If the first r_1 values $(0, 1, \dots, (r_1 - 1))$ are omitted, then we have a *left-truncated Poisson distribution* with pmf

$$\Pr[X = x] = \frac{e^{-\theta}\theta^x}{x!} \left[1 - e^{-\theta} \sum_{j=0}^{r_1-1} \frac{\theta^j}{j!} \right]^{-1}, \qquad x = r_1, r_1 + 1, \dots. \tag{4.85}$$

The MLE, $\hat{\theta}$, of θ, given values of n independent rv's $X_1, X_2, \ldots, X_n$, each distributed as in (4.85), satisfies the equation

$$\bar{x} = \left[\hat{\theta} - e^{\hat{\theta}} \sum_{j=1}^{r_1-1} \frac{\hat{\theta}^j}{(j-1)!} \right] \times \left[1 - e^{\hat{\theta}} \sum_{j=0}^{r_1-1} \left(\frac{\hat{\theta}}{j!} \right)^j \right]^{-1}. \qquad (4.86)$$

As an initial value for use in the iterative solution of (4.86), the simple estimator

$$\theta^* = n^{-1} \sum_j (x_j | x_j > r_1) \qquad (4.87)$$

that was proposed by Moore (1954) might be used. For the case $r_1 = 1$ (the positive Poisson) this becomes

$$\theta^* = n^{-1} \sum_j (x_j | x_j > 1), \qquad (4.88)$$

with variance

$$\mathrm{Var}\,\theta^* = \theta \{ 1 + \theta (e^\theta - 1)^{-1} \} n^{-1} \qquad (4.89)$$

[Plackett (1953)], and its efficiency is quite high (over 90%).

Rider (1953) constructed an estimator for θ for the left-truncated Poisson distribution based on the uncorrected sample moments. Minimum variance unbiased estimators have been given by Tate and Goen (1958). Maximum likelihood estimators based on censored samples from truncated Poisson distributions have been discussed by Murakami (1961).

Rao and Rubin (1964) gave a characterization for the left-truncated Poisson distribution analogous to their characterization for the complete Poisson distribution in Section 8 of this chapter.

10.2 Right-Truncation and Double-Truncation

Right-truncation (omission of values exceeding a specified value, r_2) can occur if the counting mechanism is unable to deal with large numbers. Often the *existence* of these high values is known, even if their exact magnitude is not; in this case, if there are n' values greater than r_2, then the MLE, $\hat{\theta}$, satisfies the equation

$$\sum_j (x_j | x_j \le r_2) = \hat{\theta} \left[N - n' \left[1 + \frac{\hat{\theta}^{r_2}}{r_2!} \left\{ \sum_{j=r_2+1}^{\infty} \frac{\hat{\theta}^j}{j!} \right\}^{-1} \right] \right] \qquad (4.90)$$

[Tippett (1932)]. When this information is not available, then it is appropriate to use the *right-truncated Poisson distribution* with pmf

$$\Pr[X = x] = \frac{\theta^x}{x!} \left[\sum_{j=0}^{r_2} \frac{\theta^j}{j!} \right]^{-1}, \qquad x = 0, 1, \ldots, r_2. \qquad (4.91)$$

If $X_1, X_2, \ldots, X_n$ are n independent rv's each having this pmf, then the MLE, $\hat{\theta}$, of θ satisfies the following equation:

$$\sum_{j=0}^{r_2} \frac{(\bar{x} - j)\hat{\theta}^j}{j!} = 0. \qquad (4.92)$$

The solution of this equation for $\hat{\theta}$ is a function of $\bar{x}$ alone, and not of n. Cohen (1961) has provided tables from which $\hat{\theta}$ can be obtained, given r_2 and $\bar{x}$. When the tables are insufficient, a solution may be obtained by interpolation. Equation (4.92) can be rewritten in the form

$$\bar{x} = \left[\hat{\theta} \sum_{j=0}^{r_2-1} \{ e^{-\hat{\theta}} \hat{\theta}^j / j! \} \right] \bigg/ \left[\sum_{j=0}^{r_2} e^{-\hat{\theta}} \hat{\theta}^j / j! \right], \qquad (4.93)$$

and the right-hand side evaluated either numerically or by the use of tables of the cumulative probabilities; see Section 6 of this chapter.

The asymptotic variance of $\hat{\theta}$ is $\theta \psi(\theta) n^{-1}$, where

$$\psi(\theta) = \frac{\left[\sum_{j=0}^{r_2} (\theta^j / j!) \right]^2}{\left\{ \sum_{j=0}^{r_2-1} (\theta^j / j!) \right\} \left\{ \sum_{j=0}^{r_2-1} (\theta^j / j!) + \theta^{r_2+1} / r_2! \right\} - (\theta^{r_2+1} / r_2!) \sum_{j=0}^{r_2} (\theta^j / j!)}. \qquad (4.94)$$

Moore (1954) suggested the simple estimator (analogous to (4.87))

$$\theta^* = \sum_j \frac{x_j}{m}, \qquad (4.95)$$

where m is the number of values of x that are less than $(r_2 - 1)$; this is an unbiased estimator of θ.

The right-truncated Poisson arises in telephony when there are just n lines, and any calls that arrive when all n lines are busy are not held. This gives the loss $M/M/n$ queueing system with

$$\Pr[X = x] = \Pr[X = 0] \frac{\theta^x}{x!}, \qquad 0 \le x \le n, \qquad (4.96)$$

where $\Pr[X = n]$ is the proportion of time for which the system is fully occupied, that is, the proportion of lost calls. The expression for $\Pr[X = n]$ is known as *Erlang's loss formula*.

The doubly-truncated Poisson distribution [Cohen (1954)] is the distribution of a rv for which

$$\Pr[X = x] = \frac{\theta^x}{x!} \left[\sum_{j=r_1}^{r_2} \frac{\theta^j}{j!} \right]^{-1}, \qquad x = r_1, r_1 + 1, \ldots, r_2, \quad 0 < r_1 < r_2. \quad (4.97)$$

If $X_1, X_2, \ldots, X_n$ are independent rv's, each with pmf (4.97), the MLE, $\hat{\theta}$, of θ satisfies the equation

$$\bar{x} = \left[\sum_{j=r_1}^{r_2} \{j\hat{\theta}^j/j!\} \right] \bigg/ \left[\sum_{j=r_1}^{r_2} \hat{\theta}^j/j! \right]. \qquad (4.98)$$

This can be solved in the same manner as (4.92). Moore (1954) has suggested a statistic analogous to (4.95) that is an unbiased estimator of θ.

Doss (1963) has compared the efficiency of the ML estimators of θ for the untruncated and for the doubly-truncated Poisson distributions. He showed that the variance of $\hat{\theta}$ for untruncated data is always less than the variance of $\hat{\theta}$ for doubly-truncated data. Comparisons of this kind are relevant when it is possible to control the method of observation to produce distributions either of type (4.1) or of type (4.97).

When all values of x that are greater than a certain value K are recorded as that value, the resultant distribution is

$$\Pr[X = x] = \frac{e^{-\theta}\theta^x}{x!}, \qquad x < K,$$

$$\Pr[X = K] = \sum_{j=K}^{\infty} \frac{e^{-\theta}\theta^j}{j!}. \qquad (4.99)$$

Newell (1965) has applied this distribution to the number of hospital beds occupied when K is the total number of beds available.

10.3 Misrecorded Poisson Distributions

These distributions attempt to take into account errors in recording a variable which in reality does have a Poisson distribution. Suppose that the zero-class alone is misrecorded. Then the pmf is

$$\Pr[X = 0] = \omega + (1 - \omega)e^{-\theta},$$

$$\Pr[X = x] = (1 - \omega)\frac{e^{-\theta}\theta^x}{x!}, \qquad x \geq 1. \qquad (4.100)$$

When $0 < \omega < 1$ (that is, when there is over-reporting) the distribution is known as the *Poisson-with-zeroes*, or alternatively as the *zero-inflated Poisson distribution*. For the *zero-deflated Poisson distribution* (that is, when there is under-reporting) a necessary condition on ω is

$$(1 - e^\theta)^{-1} < \omega < 0.$$

These are zero-modified Poisson distributions; see Chapter 8, Section 2.2.

The distribution defined by the equations

$$\Pr[X = 0] = e^{-\theta}(1 + \theta\lambda),$$
$$\Pr[X = 1] = \theta e^{-\theta}(1 - \lambda),$$
$$\Pr[X = x] = \frac{\theta^x e^{-\theta}}{x!}, \qquad x \geq 2, \tag{4.101}$$

corresponds to a situation in which values from a Poisson distribution are recorded correctly, except that, when the true value is 1, there is a probability λ that it will be recorded as zero. Cohen (1960c) obtained the following formulas for ML estimators of θ and λ, respectively:

$$\hat{\theta} = 0.5[\{\bar{x} - 1 + f_0\} + \{(\bar{x} - 1 + f_0)^2 + 4(\bar{x} - f_1)\}^{1/2}], \tag{4.102}$$
$$\hat{\lambda} = (f_0 - f_1\hat{\theta}^{-1})(f_0 + f_1)^{-1}, \tag{4.103}$$

where $\bar{x}$ is the sample mean, and f_0 and f_1 are the observed <u>relative</u> frequencies of zero and unity. The asymptotic variances of $\hat{\theta}$ and $\hat{\lambda}$ are, respectively,

$$\theta(1 + \theta)(1 + \theta - e^{-\theta})^{-1}n^{-1}$$

and

$$(1 + \lambda\theta - \lambda e^{-\theta})(1 - \lambda)(\theta e^{-\theta})^{-1}(1 + \theta - e^{-\theta})^{-1}n^{-1},$$

where n denotes the sample size. The asymptotic correlation between the estimators is

$$[(1 - \lambda)e^{-\theta}(1 - \theta)^{-1}(1 + \lambda\theta - \lambda e^{-\theta})^{-1}]^{1/2}.$$

Shah and Venkataraman (1962) gave formulas for the moments of the distribution; they also constructed estimators of θ and λ based on the sample moments.

Cohen (1959) considered the case where a true value of $(c+1)$ is sometimes reported as c (with probability α); the pmf is

$$\Pr[X = x] = \frac{e^{-\theta}\theta^x}{x!}, \qquad x = 0, 1, \ldots, c - 1, c + 2, c + 3, \ldots,$$

$$\Pr[X = c] = \frac{e^{-\theta}\theta^c}{c!}\left(1 + \frac{\alpha\theta}{c + 1}\right),$$

$$\Pr[X = c + 1] = (1 - \alpha)\frac{e^{-\theta}\theta^{c+1}}{(c + 1)!}. \tag{4.104}$$

Cohen gave formulas for the maximum likelihood estimators $\hat{\theta}$ and $\hat{\alpha}$, and also expressions for their asymptotic variances and covariance. In further papers Cohen (1960b, d) has considered other similarly misrecorded Poisson distributions.

11 POISSON-STOPPED-SUM DISTRIBUTIONS

Poisson-stopped-sum distributions have pgf's of the form

$$G(z) = e^{\theta\{g(z)-1\}}; \tag{4.105}$$

they arise as the distribution of the sum of a Poisson number of iid rv's with pgf $g(z)$. The term *Poisson-stopped-sum* was introduced by Godambe and Patil (1975) and used by Douglas (1980) in his book on contagious distributions.

Because of their infinite divisibility (see Chapter 9, Section 3) these distributions have very great importance in discrete distribution theory. They are known by a number of other names. Feller (1943) used the term *generalized Poisson*; this usage is common, though subject to unfortunate ambiguity. Galliher et al. (1959) and Kemp (1967a) called them *stuttering Poisson*; Thyrion (1960) called them *Poisson par grappes*. The name *compound Poisson* was used by Feller (1950, 1957, 1968) and Lloyd (1980). The term "compound" is particularly confusing because it is also widely used for mixed Poisson distributions, as in the first edition of this book.

While certain very common distributions, such as the negative binomial (see Chapter 5), are both Poisson-stopped-sum distributions *and* mixed Poisson distributions, there are others that belong to only one of these two families of distributions.

The term clustered Poisson distribution is also used. Suppose that the number of egg masses per plant laid by an insect has a Poisson distribution, that the number of larvae hatching per egg mass has pgf $g(z)$, and that egg masses hatch independently. Then the total number of larvae per plant has a Poisson-stopped-sum distribution.

These distributions have also been called composed Poisson distributions. Let

$$g(z) = a_0 + a_1 z + a_2 z^2 + \dots, \quad a_i \geq 0, \ i = 0, 1, 2, \dots.$$

Then the Poisson-stopped-sum distribution has pgf

$$G(z) = e^{\lambda a_1(z-1)} e^{\lambda a_2(z^2-1)} e^{\lambda a_3(z^3-1)} \dots; \tag{4.106}$$

thus the distribution can also be interpreted as the convolution of a Poisson singlet distribution with a Poisson doublet distribution, a Poisson triplet distribution, and so on; in other words, it is the distribution of $X_1 + X_2 + X_3 + \ldots$, where X_1 has a Poisson distribution and X_r, $r = 2, 3 \ldots$ has a Poisson distribution on $x = 0, r, 2r, \ldots$. A European term for "convolution" has been "composition."

Clearly, if $g(z)$ is a pgf, then $a_i \geq 0$ for all i, and $G(z)$ is infinitely divisible. Conversely, if $G(z)$ is infinitely divisible, then $g(z)$ is a valid pgf; see, e.g., Feller (1968, p. 290).

If $g(z)$ is the pgf of a rv with finite support, then the Poisson-stopped sum of such rv's has been called an *extended Poisson of order k* [Hirano and Aki (1987)]; when $g(z)$ is a discrete uniform distribution with support $1, 2, \ldots, k$, the outcome is the *Poisson of order k* [Phillipou (1983)]; see Chapter 10, Section 6.

The best-known Poisson-stopped-sum distribution is the negative binomial; here $g(z)$ is the logarithmic distribution (and has infinite support). Chapter 5 is devoted to this very important distribution and to its special form, the geometric.

The *Lagrangian-Poisson distribution* has received much attention; it is a Poisson-stopped sum of Borel-Tanner rv's. In his book Consul (1989) called it the "generalized Poisson distribution" and gave many earlier references.

The general theory of Poisson-stopped-sum distributions is given in Chapter 9, where many special forms such as the Hermite, Neyman Type A, Pólya-Aeppli, and Lagrangian-Poisson are discussed in detail.

12 OTHER RELATED DISTRIBUTIONS

12.1 The Normal Distribution

The limiting distribution of the standardized Poisson variable $(X - \theta)\theta^{-1/2}$, where X has distribution (4.1), is a unit normal distribution. That is,

$$\lim_{\theta \to \infty} \Pr[\alpha < (X - \theta)\theta^{-1/2} < \beta] = (2\pi)^{-1/2} \int_\alpha^\beta e^{-u^2/2} du. \qquad (4.107)$$

12.2 The Gamma Distribution

The following important formal relation between the Poisson and gamma (and hence chi-square) distributions has already been noted in Sections 2 and 5 of this chapter. If Y has pdf

$$f(y) = \frac{1}{(\alpha - 1)!} y^{\alpha - 1} e^{-y}, \qquad 0 < y,$$

and α is a positive integer, then

$$\Pr[Y > y] = \frac{1}{(\alpha - 1)!} \int_y^\infty u^{\alpha-1} e^{-u} du$$

$$= \frac{1}{(\alpha - 1)!} y^{\alpha-1} e^{-y} + \frac{1}{(\alpha - 2)!} \int_y^\infty u^{\alpha-2} e^{-u} du \qquad \text{(integrating by parts)}$$

$$= \frac{1}{(\alpha - 1)!} y^{\alpha-1} e^{-y} + \frac{1}{(\alpha - 2)!} y^{\alpha-2} e^{-y} + \cdots + y e^{-y} + e^{-y}$$

$$= \Pr[X < \alpha], \qquad (4.108)$$

where X is a Poisson rv with parameter y. If $\alpha = \nu/2$, where ν is an even integer, then

$$\Pr[X < \nu/2] = \Pr[\chi_\nu^2 > 2y].$$

12.3   Sums and Differences of Poisson Variates

If X_i, $i = 1, 2, \ldots$, are independent Poisson rv's with parameters θ_i, then $\sum_i X_i$ has a Poisson distribution with parameter $\sum_i \theta_i$; this is the reproductive property of the Poisson distribution.

The sum of n iid positive (i.e., zero-truncated) Poisson rv's with parameter θ was studied by Tate and Goen (1958); the distribution is known as the *Stirling distribution of the second kind*. The pgf is

$$G(z) = \frac{(e^{\theta z} - 1)^n}{(e^\theta - 1)^n} \qquad (4.109)$$

and the pmf is

$$\Pr[Y = y] = \frac{n! S(y,n) \theta^y}{(e^\theta - 1)^n y!}, \qquad y = n, n+1, \ldots, \qquad (4.110)$$

where $S(y,n)$ is the Stirling number of the second kind. The mean and variance are

$$\mu = n\theta(1 - e^{-\theta})^{-1} \quad \text{and} \quad \mu_2 = \mu(1 + \theta - \mu/n),$$

respectively. The distribution is a generalized power series distribution, and therefore the MLE of θ is given by

$$\bar{x} = n\hat{\theta}(1 - e^{-\hat{\theta}})^{-1}. \qquad (4.111)$$

Tate and Goen considered minimum variance unbiased estimation of θ; see also Ahuja (1971a). Ahuja and Ennking (1972) obtained a recurrence relationship for the pmf, and also an explicit expression for the cdf as a sum of a linear combination of incomplete gamma functions. Tate and Goen, and Ahuja, also considered sums of iid Poisson rv's left-truncated at $c > 1$. Cacoullos (1975) has developed *multiparameter Stirling distributions of the second kind* in his study of the sum of iid Poisson rv's left-truncated at differing, unknown points.

Gart (1974) studied the exact moments of the statistic for the variance homogeneity test for samples from the zero-truncated Poisson distribution, and gave a number of references to earlier work on the distribution.

The distribution of the difference between two independent random variables, each having a Poisson distribution, has attracted some attention. Strackee and van der Gon (1962) state, "In a steady state the number of light quanta, emitted or absorbed in a definite time, is distributed according to a Poisson distribution. In view thereof, the physical limit of perceptible contrast in vision can be studied in terms of the difference between two independent variates each following a Poisson distribution." The distribution of differences may also be relevant when a physical effect is estimated as the difference between two counts, one when a "cause" is acting, and the other a "control" to estimate the "background effect."

Irwin (1937) studied the case when the two variables X_1 and X_2 each have the same expected value θ. Evidently for $y \geq 0$,

$$\Pr[X_1 - X_2 = y] = e^{-2\theta} \sum_{j=y}^{\infty} \theta^{j+(j-y)} [j!(j-y)!]^{-1}$$

$$= e^{-2\theta} I_{y/2}(2\theta), \tag{4.112}$$

where $I(\cdot)$ is a modified Bessel function of the first kind; see Chapter 1, Section A5. (By symmetry $\Pr[X_1 - X_2 = y] = \Pr[X_2 - X_1 = y]$.) In this particular case all the cumulants of odd order are zero, and all those of even order are equal to 2θ. The pgf is $\exp\{\theta(z + z^{-1} - 2)\}$ and the approach to normality is rapid.

Skellam (1946), de Castro (1952), and Prekopa (1952) discussed the problem when $E[X_1] = \theta_1 \neq E[X_2] = \theta_2$. In this case, for $y > 0$,

$$\Pr[X_1 - X_2 = y] = e^{-(\theta_1+\theta_2)} \sum_{j=y}^{\infty} \theta_1^j \theta_2^{j-y} [j!(j-y)!]^{-1}$$

$$= e^{-(\theta_1+\theta_2)} \left(\frac{\theta_1}{\theta_2}\right)^{y/2} I_{y/2}(2\sqrt{\theta_1\theta_2}) \tag{4.113}$$

(note the relationship to the noncentral χ^2 distribution, Chapter 28). The tail probabilities can be formulated in terms of the tail probabilities of the non-

Table 4.1 Combinations of Values of θ_1 and θ_2 in Strakee and van der Gon's Tables of $\Pr[X_1 - X_2 = y]$

θ_1	$\frac{1}{4}$	1	4	$\frac{1}{2}$	1	2	4	8	1	2	4	2	4	8	4	8
θ_2	$\frac{1}{4}$	$\frac{1}{4}$	$\frac{1}{4}$	$\frac{1}{2}$	$\frac{1}{2}$	$\frac{1}{2}$	$\frac{1}{2}$	$\frac{1}{2}$	1	1	1	2	2	2	4	8

central chi-square distribution, providing a generalization of the relationship between the Poisson and chi-square distributions [Johnson (1959)].

Strakee and van der Gon (1962) gave tables of the cumulative probability $\Pr[X_1 - X_2 = y]$ to four decimal places for the combinations of values of θ_1 and θ_2 given in Table 4.1.

Their tables also show the differences between the normal approximations [see Fisz (1953)]

$$\Pr[X_1 - X_2 \leq y] \doteq [2\pi(\theta_1 + \theta_2)]^{-1/2} \int_{-\infty}^{y+1/2} \exp\left[-\frac{(\theta_1 - \theta_2 - y)^2}{2(\theta_1 + \theta_2)}\right] dy, \quad (4.114)$$

and the tabulated values. Romani (1956) showed that all the odd cumulants of $X_1 - X_2$ are equal to $\theta_1 - \theta_2$, and that all the even cumulants are equal to $\theta_1 + \theta_2$. He also discussed the properties of the ML estimator of $E[X_1 - X_2] = \theta_1 - \theta_2$. Katti (1960) studied $E|X_1 - X_2|$.

Ratcliffe (1964) carried out a Monte Carlo experiment to assess the accuracy with which the distribution of $(X_1 - X_2)(X_1 + X_2)^{-1/2}$ is represented by a unit normal distribution for the special case $\theta_1 = \theta_2 = \theta$.

Consider now $X = Y + Z$, where Y is a Poisson rv with parameter θ, and $Z \geq 0$ is a discrete rv independent of Y. Samaniego (1976) discussed the applicability of such a model to various situations, such as signal + noise processes and misclassification. He examined ML estimation of θ when the distribution of Z is known, and he related this to Crow and Gardner's (1959) concept of confidence belts (see Section 7.3 of this chapter). Huang and Fung's (1989) *intervened truncated Poisson distribution* arises when Z has a truncated Poisson distribution (left, right, or double truncation). The resultant distribution can be used to model situations where the Poisson parameter is changed by an "intervention" at some point during the observational period. Moment and ML estimation equations were derived by Huang and Fung.

12.4 Hyper-Poisson Distributions

Staff (1964, 1967) defined the *displaced Poisson distribution* as a left-truncated Poisson distribution that has been "displaced" by subtraction of a constant so that the lowest value that is taken by the variable is zero. Thus the left-truncated distribution

$$\Pr[Y = y] = \frac{\theta^y}{y!} \left\{ \sum_{j=r+1}^{\infty} \frac{\theta^j}{j!} \right\}^{-1}, \qquad y = r+1, r+2, \ldots,$$

corresponds to the displaced distribution

$$\Pr[X = x] = \frac{\theta^{x+r+1}}{(x+r+1)!} \left\{ \sum_{j=r+1}^{\infty} \frac{\theta^j}{j!} \right\}^{-1}, \qquad x = 0, 1, 2, \ldots. \qquad (4.115)$$

Staff also obtained the distribution via the recurrence relation

$$\Pr[X = x+1] = \frac{\theta}{r+x} \Pr[X = x], \qquad x = 1, 2, \ldots.$$

If r is known, estimation of θ is straightforward. Staff also considered the situation in which both r and θ have to be estimated. The simplest formulas that he obtained were for estimators based on the sample mean $\bar{x}$, standard deviation (s) and proportion (f_0) in the zero class. They are

$$r^* = (s^2 - \bar{x})[1 - f_0(1+\bar{x})]^{-1}, \qquad (4.116)$$

$$\lambda^* = \bar{x} + r^*(1 - f_0) \qquad (4.117)$$

(for the displaced Poisson distribution r is necessarily a positive integer, though the restriction to integers is not necessary for the hyper-Poisson distribution, see below). If r is less than 8, the efficiency of r^* relative to the MLE exceeds 70% (e.g., if $r = 2$ then the efficiency of r^* exceeds 89%).

By stating its pgf in the form

$$G(z) = \frac{{}_1F_1[1; r; \theta z]}{{}_1F_1[1; r; \theta]}, \qquad r \text{ an integer}, \qquad (4.118)$$

we see immediately that the displaced Poisson is a special case of the more general *hyper-Poisson distribution* with pgf

$$G(z) = \frac{{}_1F_1[1; \lambda; \theta z]}{{}_1F_1[1; \lambda; \theta]}, \qquad (4.119)$$

λ nonnegative real. Bardwell and Crow (1964) termed the distribution *sub-Poisson* for $\lambda < 1$, and *super-Poisson* for $\lambda > 1$. It is easy to obtain the recurrence relation

$$(\lambda + x) \Pr[X = x+1] = \theta \Pr[X = x], \qquad x = 1, 2, \ldots. \qquad (4.120)$$

Summation of both sides of the equation gives

$$\mu = (\theta + 1 - \lambda) - (1 - \lambda)\Pr[X = 0]$$
$$= (\theta + 1 - \lambda) - (1 - \lambda)(_1F_1[1; \lambda; \theta])^{-1}. \tag{4.121}$$

The distribution is a power series distribution (Chapter 2, Section 2) belonging to the generalized hypergeometric family (Chapter 2, Section 4.1); this enabled Kemp (1968a,b) to obtain recurrence relationships of the form

$$\mu'_{[i+2]} = (\theta - \lambda - i)\mu'_{[i+1]} + \theta(i + 1)\mu'_{[i]}. \tag{4.122}$$

It follows that all moments of the distribution are finite, and that the variance is

$$\text{Var}(X) = (\theta - \lambda + 1 - \mu)\mu + \theta. \tag{4.123}$$

Barton (1966) commented that the distribution can be regarded as a mixed Poisson distribution, where the Poisson parameter has a truncated Pearson Type III distribution (Chapter 8, Section 3.2(10)). Kemp (1968a) showed, more generally, that mixing a hyper-Poisson distribution with parameters λ and θ can yield a hyper-Poisson with parameters $(\lambda + \eta)$ and θ, where $\eta > 0$.

Bardwell and Crow (1964) described various methods of estimating λ and θ, given values $x_1, x_2, \cdots, x_n$, of n independent rv's each having the distribution (4.119).

First, suppose that λ is known. Since (4.119) can be regarded as a power series distribution (Chapter 2, Section 2), the MLE, $\hat{\theta}$, is obtained by equating the expected value (4.121) to the sample arithmetic mean $\bar{x}$, giving

$$\bar{x} = \hat{\theta} + (1 - \lambda)\{1 - (_1F_1[1; \lambda; \hat{\theta}])^{-1}\}, \tag{4.124}$$

that is,

$$_1F_1[1; \lambda; \hat{\theta}] = \frac{1 - \lambda}{1 - \lambda - \bar{x} + \hat{\theta}}. \tag{4.125}$$

The variance of $\hat{\theta}$ is approximately $\theta^2[n\text{Var}(X)]^{-1}$ where $\text{Var}(X)$ is as given in (4.123).

The uniformly minimum variance unbiased estimator of θ is *approximately* $\lambda\bar{x}$ when n is large. Crow and Bardwell (1965) obtained the values

$$0 \qquad \text{when } \bar{x} = 0$$
$$\lambda n^{-1} \qquad \text{when } \bar{x} = n^{-1},$$
$$2\lambda n^{-1}[1 + n^{-1}(\lambda - 1)(\lambda + 1)^{-1}]^{-1} \quad \text{when } \bar{x} = 2n^{-1}.$$

Second, when both λ and θ are unknown, the equations satisfied by the ML estimators $\hat{\lambda}$ and $\hat{\theta}$ are

$$_1F_1[1; \hat{\lambda}; \hat{\theta}] = \frac{1 - \hat{\lambda}}{1 - \hat{\lambda} - \bar{x} + \hat{\theta}} \tag{4.126}$$

(cf. (4.125)) and

$$\sum_{j=1}^{\infty} \hat{\theta}^j (\hat{\lambda}^{[j]})^{-1} S_j = {}_1F_1[1; \hat{\lambda}; \hat{\theta}] n^{-1} \sum_{i=1}^{n} S_{x_i}, \tag{4.127}$$

where $S_j = \sum_{i=1}^{j} (\lambda + i - 1)^{-1}$.

Crow and Bardwell (1965) also described some other estimators. All of them seem to have rather low efficiencies, so we will only mention a pair, θ^* and λ^*, that use the proportion of zeroes (f_0) and the first two sample moments $\bar{x}$ and $m_2' = n^{-1} \sum_{j=1}^{n} x_j^2$. They are

$$\theta^* = [(1 - f_0)m_2' - \bar{x}^2][1 - f_0(\bar{x} + 1)]^{-1}, \tag{4.128}$$

$$\lambda^* = 1 + \frac{(m_2' - \bar{x}^2) - \bar{x}}{1 - f_0(\bar{x} + 1)}. \tag{4.129}$$

Some queueing theory, with a hyper-Poisson distribution of arrivals, has been worked out by Nisida (1962).

The hyper-Poisson is a special case of Hall (1966) and Bhattacharya's (1966) *confluent hypergeometric distribution* with pgf

$$G(z) = \frac{_1F_1[a; \lambda; \theta z]}{_1F_1[a; \lambda; \theta]}.$$

Also it is a member of the extended Katz family of distributions; see Chapter 2, Section 3.1. Inference for this family has been studied by Gurland and Tripathi (1975) and Tripathi and Gurland (1977, 1979).

12.5 Grouped Poisson Distributions

In Section 2 it was shown that if the times between successive events have a common exponential distribution, then the total number of events in a fixed time T has a Poisson distribution. Morlat (1952) constructed a "generalization of the Poisson law" by supposing that the common distribution of the times (t) is a gamma distribution with origin at zero (Chapter 17). If the common pdf is

$$[\Gamma(\alpha)]^{-1} t^{\alpha-1} e^{-t}, \qquad t \geq 0,$$

then the probability of x events in time T is

$$\Pr[X = x] = [\Gamma(x\alpha)]^{-1} \int_0^T \left\{ 1 - \frac{y^\alpha \Gamma(x\alpha)}{\Gamma((x+1)\alpha)} \right\} y^{x\alpha-1} e^{-y} dy. \qquad (4.130)$$

If α is an integer, we have the remarkable formula

$$\Pr[X = x] = \sum_{j=x\alpha}^{(x+1)\alpha-1} \frac{e^{-T} T^j}{j!}, \qquad (4.131)$$

showing that the probabilities are given by grouping together an α number of successive Poisson probabilities.

The "generalized Poisson" distributions of Gold (1957) and Gerstenkorn (1962) [see also Godwin (1967)] have pmf's formed from cumulative Poisson probabilities. We have

$$\Pr[X = x] = q^x e^{-qu} \left(\sum_{j=x}^{\infty} \frac{u^j}{j!} \right) c^{-1}, \qquad x = 0, 1, 2, \ldots, \qquad (4.132)$$

where $u > 0$, c is an appropriate normalizing constant, and either (1) $q > 0$, $q \neq 1$, or (2) $q = 1$.
In case 1 the pgf is

$$G(z) = \left(\frac{1-q}{1-qz} \right) \left(\frac{qze^{u(qz-1)} - 1}{qe^{u(q-1)} - 1} \right); \qquad (4.133)$$

in case 2, the pgf is

$$G(z) = \frac{ze^{u(z-1)} - 1}{(1+u)(z-1)} \qquad (4.134)$$

and now the probabilities are the normalized tail-probabilities of the Poisson distribution.

These "generalized Poisson" distributions are related, but are not identical, to the *burnt-fingers distribution* of Greenwood and Yule (1920), McKendrick (1926), Arbous and Kerrich (1951), and Irwin (1953) (see Chapter 2, Section 4.2). The burnt-fingers distribution arises if the first event occurs in time $(t, t + \partial t)$ with probability $(a\partial t + o(\partial t))$, while subsequent events occur with probability $(b\partial t + o(\partial t))$. The probabilities are

$$\Pr[X = 0] = e^{-at},$$

$$\Pr[X = x] = e^{-bt} \frac{a}{b} \left(\frac{b}{b-a} \right)^x \sum_{j=x}^{\infty} \frac{(b-a)^j}{j!}, \qquad x = 1, 2, \ldots, \qquad (4.135)$$

and the pgf is

$$g(z) = \frac{aze^{b(z-1)} + e^{-a}(b-a)(z-1)}{a-b+bz}.$$ (4.136)

12.6 Heine and Euler Distributions

The Heine and Euler distributions were first studied by Benkherouf and Bather (1988) as prior distributions for stopping time strategies when sequentially drilling for oil. Kemp (1992a, b) has shown that they are members of a family of q-series analogues of the Poisson distribution with pmf satisfying

$$\Pr[X = x] = \Pr[X = 0]\alpha^x/\{(1-q)(1-q^2)\cdots(1-q^x)\}, \qquad x = 1,2,\ldots;$$ (4.137)

the determination of $\Pr[X = 0]$ is discussed below. The pgf's of these distributions are expressible as

$$G(z) = \frac{{}_0\Phi_0[\ ;\ ;q,\alpha z]}{{}_0\Phi_0[\ ;\ ;q,\alpha]},$$ (4.138)

using basic hypergeometric functions (q-hypergeometric series); see Chapter 1, Section A12. For the Heine distribution $1 < q$ and $\alpha < 0$, whereas for the Euler distribution $0 < q < 1$ and $0 < \alpha < 1$. The Poisson distribution with parameter θ is the limiting form for both distributions when $q \to 1$ and $\alpha \to 0$ such that $\alpha/(1-q) = \theta$ remains constant. As $q \to \infty$, the Heine tends to a Bernoulli distribution, and as $q \to 0$, the Euler tends to a geometric distribution.

The disjoint restrictions on their parameters yield different summation formulas for their pgf's. Using Heine's theorem (Chapter 1, Section A12), the pgf for the Heine distribution can be rewritten as

$$H(z) = \prod_{j\geq 0} \left(\frac{1+\beta Q^j z}{1+\beta Q^j} \right), \qquad \text{where } Q = q^{-1} \text{ and } \beta = -\alpha q^{-1}, \quad (4.139)$$

that is, where $0 < Q < 1$ and $0 < \beta$. A Heine random variable is therefore an infinite sum of independent Bernoulli rv's with changing (log-linear) odds; its mean and variance are

$$\mu = \prod_{j\geq 0} \left(\frac{\beta Q^j}{1+\beta Q^j} \right) \quad \text{and} \quad \mu_2 = \prod_{j\geq 0} \left(\frac{\beta Q^j}{(1+\beta Q^j)^2} \right),$$ (4.140)

and

$$\Pr[X = 0] = \prod_{j\geq 0} (1+\beta Q^j)^{-1}.$$ (4.141)

When β is restricted to $0 < \beta < 1$, an alternative formula for the mean is

$$\mu = \prod_{j \geq 0} \left\{ \frac{-(-\beta)^j}{(1 - Q^j)} \right\}. \tag{4.142}$$

For the Euler distribution, on the other hand, Heine's theorem gives

$$G(z) = \prod_{j \geq 0} \left(\frac{1 - \alpha q^j}{1 - \alpha q^j z} \right), \qquad 0 < q < 1, \; 1 < \alpha < 1, \tag{4.143}$$

showing that an Euler rv is an infinite sum of independent geometric rv's with changing (log-linear) probabilities of failure, that its mean and variance are

$$\mu = \prod_{j \geq 0} \left(\frac{\alpha q^j}{1 - \alpha q^j} \right) \quad \text{and} \quad \mu_2 = \prod_{j \geq 0} \left(\frac{\alpha q^j}{(1 - \alpha q^j)^2} \right), \tag{4.144}$$

and that

$$\Pr[X = 0] = \prod_{j \geq 0} (1 - \alpha q^j). \tag{4.145}$$

An alternative expression for the mean, valid for all α in $0 < \alpha < 1$, is

$$\mu = \prod_{j \geq 0} \left\{ \frac{\alpha q^j}{1 - q^j} \right\}. \tag{4.146}$$

Kemp (1992a) studied moments, cumulants, and other properties, as well as maximum likelihood and other methods of estimation, for both distributions. Both have infinite support, are unimodal, and have increasing failure rates. The Euler distribution is overdispersed and infinitely divisible, whereas the Heine distribution is underdispersed and not infinitely divisible. Kemp (1992a) found, moreover, that $-1 < q < 0$, $0 < \alpha < 1$, gives rise to a third member of this family, and she called it the pseudo-Euler distribution; it is infinitely divisible but, unlike the others, it can be multimodal.

Kemp (1992b) investigated steady-state Markov chain models for the Heine and Euler distributions; these include current-age models for discrete renewal processes, success-runs processes with nonzero probabilities that a trial is abandoned, Foster processes, and equilibrium random walks corresponding to M/M/1 queues with elective customer behavior.

CHAPTER 5

Negative Binomial Distribution

1 DEFINITION

Many different models give rise to the negative binomial distribution, and consequently there is a variety of definitions in the literature. The two main dichotomies are (1) between parameterizations and (2) between points of support.

Formally, the negative binomial distribution can be defined in terms of the expansion of the negative binomial expression $(Q - P)^{-k}$, where $Q = 1 + P$, $P > 0$, and k is positive real; the $(x + 1)$th term in the expansion yields $\Pr[X = x]$. This is analogous to the definition of the binomial distribution in terms of the binomial expression $(\pi + \omega)^n$, where $\omega = 1 - \pi$, $0 < \pi < 1$, and n is a positive integer.

Thus the *negative binomial distribution* with parameters k, P, is the distribution of the random variable X for which

$$\Pr[X = x] = \binom{k + x - 1}{k - 1} \left(\frac{P}{Q}\right)^x \left(1 - \frac{P}{Q}\right)^k, \qquad x = 0, 1, 2, \ldots, \qquad (5.1)$$

where $Q = 1 + P$, $P > 0$, and $k > 0$. Unlike the binomial distribution, here there is a nonzero probability for X to take any specified nonnegative integer value as in the case of the Poisson distribution.

The pgf is

$$G(z) = (1 + P - Pz)^{-k} \qquad (5.2)$$

$$= {}_1F_0[k; ; P(z - 1)] \qquad (5.3)$$

$$= \frac{{}_1F_0[k; ; Pz/(1 + P)]}{{}_1F_0[k; ; P/(1 + P)]}. \qquad (5.4)$$

The corresponding cf is $(1 + P - Pe^{it})^{-k}$. The mean and variance are

$$\mu = kP \quad \text{and} \quad \mu_2 = kP(1 + P). \qquad (5.5)$$

This parameterization (but with the symbol p instead of P) is the one introduced by Fisher (1941).

Other early writers adopted different parameterizations. Jeffreys (1941) had $b = P/(1+P)$, $\rho = kP$, giving the pgf $[(1 - bz)/(1 - b)]^{\rho - \rho/b}$, and $\mu = \rho$, $\mu_2 = \rho/(1 - b)$. Anscombe (1950) used the form $\alpha = k$, $\lambda = kP$, giving the pgf $(1 + \lambda/\alpha - \lambda z/\alpha)^{-\alpha}$, and $\mu = \lambda$, $\mu_2 = \lambda(1 + \lambda/\alpha)$.

Evans (1953) took $a = P$, $m = kP$, giving the pgf $(1 + a - az)^{-m/a}$, and $\mu = m$, $\mu_2 = m(1+a)$. This parameterization has been popular in the ecological literature. Some writers, for instance Patil et al. (1984), have called this the Pólya–Eggenberger distribution, as it arises as a limiting form of Eggenberger and Pólya's (1923) urn model distribution. Other authors, notably Johnson and Kotz (1977) and Berg (1988b), call the (nonlimiting) urn model distribution the Pólya–Eggenberger distribution; see Chapter 6, Section 2.4. For both distributions "Pólya–Eggenberger" is quite often abbreviated to "Pólya".

A further parameterization that has gained wide favor is $q = P/(1 + P)$, $k = k$, giving

$$G(z) = \left(\frac{1 - q}{1 - qz} \right)^k, \tag{5.6}$$

$$\Pr[X = x] = \binom{k + x - 1}{k - 1} q^x (1 - q)^k, \qquad x = 0, 1, 2, \ldots, \tag{5.7}$$

and $\mu = kq/(1 - q)$, $\mu_2 = kq/(1 - q)^2$. Sometimes $\lambda = P/(1 + P)$ is used in order to avoid confusion with the binomial parameter q.

Clearly k need not be an integer. When k is an integer, the distribution is sometimes called the *Pascal distribution* [Pascal (1679)]. The name "Pascal distribution" is, however, more often applied to the distribution shifted k units from the origin, i.e. with support $k, k + 1, \ldots$; this is also called the binomial waiting-time distribution.

The geometric is the special case $k = 1$ of the negative binomial distribution.

Kemp (1967a) has summarized four commonly encountered types of pgf for the negative binomial and geometric distributions as follows:

Negative binomial   Geometric

1. $p^k(1 - qz)^{-k}$ $p(1 - qz)^{-1}$ $\left. \right\}$ $\left\{ \begin{array}{l} p + q = 1 \\ 0 < p < 1 \end{array} \right.$

2. $p^k z^k(1 - qz)^{-k}$ $pz(1 - qz)^{-1}$

3. $(Q - Pz)^{-k}$ $(Q - Pz)^{-1}$ $\left. \right\}$ $\left\{ \begin{array}{l} Q = 1 + P, P > 0 \\ (Q = 1/p, P = q/p, \\ \text{i.e., } p = 1/Q, q = P/Q) \end{array} \right.$

4. $z^k(Q - Pz)^{-k}$ $z(Q - Pz)^{-1}$

In cases (1) and (3), k is positive real, the support is $0, 1, 2, \ldots$ and the distribution is a power series distribution. For cases (2) and (4), k is necessarily a positive integer; the distribution has support $k, k+1, k+2, \ldots$ and the distribution is a generalized power series distribution (for GPSD, see Chapter 2, Section 2.1).

Case (1) shows that the distribution is a generalized hypergeometric probability distribution (with argument parameter q; cf. (5.4)), while case (3) shows that it is a generalized hypergeometric factorial moment distribution (with argument parameter P; cf. (5.3)). Other families to which the negative binomial belongs are the exponential family (provided k is fixed), the Katz, and the Ord families.

Bartko (1961) has written a good review article on various aspects of the negative binomial distribution.

2 GEOMETRIC DISTRIBUTION

In the special case $k = 1$ the pmf is

$$\Pr[X = j] = Q^{-1} \left(\frac{P}{Q} \right)^{j}$$

$$= pq^{j}, \qquad j = 0, 1, 2, \ldots. \tag{5.8}$$

These values are in geometric progression and so this distribution is called a *geometric distribution*; sometimes it is called a *Furry distribution* [Furry (1937)].

Its properties can be obtained from those of the negative binomial, as the special case $k = 1$.

The geometric distribution also possesses a property similar to the "non–aging" (or "Markovian") property of the exponential distribution (Chapter 18, Section 2). This is

$$\Pr[X = x + j | X \geq j] = \frac{Q^{-1}(P/Q)^{x+j}}{(P/Q)^{j}} = Q^{-1} \left(\frac{P}{Q} \right)^{x} = \Pr[X = x]. \tag{5.9}$$

This property characterizes the geometric distribution (among all distributions restricted to the nonnegative integers), just as the corresponding property characterizes the exponential distribution. The distribution is commonly said to be a discrete analogue of the exponential distribution.

The geometric distribution may be extended to cover the case of a variable taking values $\theta_0, \theta_0 + \delta, \theta_0 + 2\delta, \ldots$ ($\delta > 0$). Then, in place of (5.8), we have

$$\Pr[X = \theta_0 + j\delta] = Q^{-1} \left(\frac{P}{Q} \right)^{j}. \tag{5.10}$$

The characterization summarized in (5.9) also applies to this distribution with X replaced by $(\theta_0 + X\delta)$, and k replaced by $(\theta_0 + j\delta)$.

Other characterizations are described in Section 9.1 of this chapter.

Another special property of the geometric distribution is that if a mixture of negative binomial distributions (as in (5.1)) is formed by supposing k to have the geometric distribution

$$\Pr[k = j] = (Q')^{-1}\left(\frac{P'}{Q'}\right)^{j-1}, \qquad j = 1, 2, \ldots, \tag{5.11}$$

then the resultant mixture distribution is also a geometric distribution of the form (5.8) with Q replaced by $QQ' - P'$ [Magistad (1961)].

The geometric, like the negative binomial distribution, is infinitely divisible; see Chapter 9, Section 3, for a definition of infinite divisibility.

The Shannon (first–order) entropy of the geometric distribution is $P\log_2 P - Q\log_2 Q$. The second–order entropy is $\log_2(1 + 2P)$.

Margolin and Winokur (1967) have obtained formulas for the moments of the order statistics for the geometric distribution; see also Kabe (1969). They have tabulated values of the mean and variance of the rth–order statistic in a sample of size n for $n = 1, 5(5)20$, $r = 1(1)5(5)20$, $1 \le r \le n$, $q = 0.25(0.25)0.75$. The figures are given to two decimal places. Steutel and Thiemann's (1989a) expressions for the order statistics were derived using the independence of the integer and fractional parts of exponentially distributed rv's. The computation of the order statistics from the geometric distribution has also been studied by Adatia (1991). Adatia also obtained an explicit formula for the expected value of the product of two such order statistics.

Order statistics from a continuous distribution form a Markov chain, but this is not in general true for discrete distributions. At first it was thought that exceptionally the Markov property holds for the geometric distribution [Gupta and Gupta (1981)]; later this was disproved by Nagaraja (1982) and Arnold et al. (1984). Nagaraja's (1990) lucid survey article on order statistics from discrete distributions documents a number of characterizations of the geometric distribution based on its order statistics; see Section 9.1 of this chapter.

For computer generation of geometric random variates, see Chapter 1, Section C5.

Estimation of the parameter of the geometric distribution is particularly straightforward; because it is a power series distribution the first–moment equation is also the ML equation. Hence

$$\hat{P} = \bar{x}. \tag{5.12}$$

A moment-type estimator for the geometric distribution with either or both tails truncated has been obtained by Kapadia and Thomasson (1975), who compared its efficiency with that of the ML estimator. Estimation for the geometric distribution with unknown P and unknown location parameter has been studied by Klotz (1970) (ML estimation), Iwase (1986) (MVU estimation), and Yanagimoto (1988) (conditional ML estimation). Vit (1974) has examined tests for homogeneity.

If $X_1, X_2, \ldots, X_k$ are random variables each with the geometric distribution with pmf

$$\Pr[X = x] = Q^{-1} \left(\frac{P}{Q} \right)^x, \qquad x = 0, 1, 2, \ldots,$$

then $\sum_{i=1}^{k} X_i$ is a negative binomial variable with parameters k and P; see Section 5 of this chapter. Using this fact, Clemans (1959) has constructed charts from which confidence intervals for P, given $k^{-1} \sum_{i=1}^{k} x_i$, can be read off.

Applications of the geometric distribution include runs of one species with respect to another in transects through plant populations [Pielou (1962, 1963)], a ticket control problem [Jagers (1973)], a surveillance system for congenital malformations [Chen (1978)], and estimation of animal abundance [Seber (1982b)]. Mann et al. (1974) treated applications in reliability theory.

The distribution is used in Markov chain models, for example, in meteorological models of weather cycles and precipitation amounts [Gabriel and Neumann (1962)]. Many other applications in queueing theory and applied stochastic models are discussed by Taylor and Karlin (1984) and Bhat (1984). Daniels (1961) has investigated the representation of a discrete distribution as a mixture of geometric distributions and has applied this to busy–period distributions in equilibrium queueing systems. Sandland (1974) has put forward a building society membership scheme and a length of tenure scheme as models for the truncated geometric distribution with support $0, 1, \ldots, n - 1$.

3 HISTORICAL REMARKS AND GENESIS

Special forms of the negative binomial distribution were discussed by Pascal (1679). A derivation as the distribution of the number of tosses of a coin necessessary to achieve a fixed number of heads was published by Montmort (1714) in his solution of the Problem of Points; see Todhunter (1865, p. 97). A very clear interpretation of the pmf as a density function was given by Galloway (1839, pp. 37–38) in his discussion of the Problem of Points. Let X be the random variable representing the number of independent trials necessary to obtain k occurrences of an event that has a constant probability of occurring at each trial. Then

$$\Pr[X = k + j] = \binom{k + j - 1}{k - 1} p^k (1 - p)^j, \qquad j = 1, 2, \ldots; \qquad (5.13)$$

that is, X has a negative binomial distribution (case (2) in Kemp's list in the previous section).

Meyer (1879, p. 204) obtained the pmf as the probability of exactly j male births in a birth sequence containing a fixed number of female births; he assumed a known constant probability of a male birth. He also gave the cdf in a form that we now recognize as the upper tail of an F–distribution (equivalent to an incomplete Beta function; see Section 6).

Student (1907) found empirically that certain haemocytometer data could be fitted well by a negative binomial distribution. Whittaker (1914) continued this approach. Unfortunately she did not realize that the Poisson distribution is a limiting form for both the binomial and also the negative binomial distributions (see Section 12.1), and she aroused considerable controversy concerning the relative merits of the Poisson and the negative binomial.

Greenwood and Yule (1920) derived the following relationship between the Poisson and negative binomial distributions. Suppose that we have a mixture of Poisson distributions, such that the expected values θ of the Poisson distributions vary according to a gamma distribution with pdf

$$f(\theta) = \{\beta^\alpha \Gamma(\alpha)\}^{-1} \theta^{\alpha-1} \exp(-\theta/\beta), \qquad \theta > 0,\ \alpha > 0,\ \beta > 0.$$

Then

$$\Pr[X = x] = \{\beta^\alpha \Gamma(\alpha)\}^{-1} \int_0^\infty \theta^{\alpha-1} e^{-\theta/\beta} (\theta^x e^{-\theta}/x!)\, d\theta$$

$$= \binom{\alpha + x - 1}{\alpha - 1} \left(\frac{\beta}{\beta+1}\right)^x \left(\frac{1}{\beta+1}\right)^\alpha. \tag{5.14}$$

So X has a negative binomial distribution with parameters α and β. This type of model was used to represent "accident proneness" by Greenwood and Yule. The parameter θ represents the expected number of accidents for an individual. This is assumed to vary from individual to individual.

Another important derivation is that of Lüders (1934); see also Quenouille (1949). Here the negative binomial arises as the distribution of the sum of N independent random variables each having the same logarithmic distribution (Chapter 7), where N has a Poisson distribution. Thyrion (1960) called this an *Arfwedson process*. Boswell and Patil (1970) termed it a Poisson sum (Poisson–stopped sum) of logarithmic rv's. Let $Y = X_1 + X_2 + \cdots + X_N$, where the X_i are iid logarithmic rv's with pgf $\ln(1 - \theta z)/\ln(1 - \theta)$. Assume also that N is a Poisson rv (with parameter λ) which is independent of the X_i. Then the pgf of Y is

$$\exp\left[\lambda\left\{\frac{\ln(1 - \theta z)}{\ln(1 - \theta)} - 1\right\}\right] = \left(\frac{1 - \theta}{1 - \theta z}\right)^{-\lambda/\ln(1-\theta)}; \tag{5.15}$$

see Chapter 7, Section 2.

The negative binomial as a limiting form for Pólya and Eggenberger's urn model was mentioned in Section 1. Consider a random sample of n balls from an urn containing Np white balls and $N(1 - p)$ black balls. Suppose that after each draw the drawn ball is replaced together with $c = N\beta$ others of the same color. Let the number of white balls in the sample be X. Then

$$\Pr[X = x] = \binom{n}{x} \left(\frac{p}{\beta}\right)^{[x]} \left(\frac{q}{\beta}\right)^{[n-x]} \Bigg/ \left(\frac{1}{\beta}\right)^{[n]}, \qquad (5.16)$$

where $a^{[x]} = a(a + 1)\ldots(a + x - 1)$; see Chapter 6, Section 2.4. The limiting form as $n \to \infty$, $p \to 0$, $\beta \to 0$ such that $np \to \eta k$, $n\beta \to \eta$ is negative binomial with pmf (5.1), where $P = \eta$; see Eggenberger and Pólya (1923, 1928).

This limiting form of (5.16) has been called a "Pólya" distribution by, for instance, Gnedenko (1961), Arley and Buch (1950), and Hald (1952). On the other hand, Bosch (1963) called (5.16) a "Pólya" distribution.

Patil and Joshi (1968) termed the negative binomial a "Pólya–Eggenberger", and the distribution (5.16) simply a "Pólya" distribution. Proofs of the limiting form appear in Bosch (1963), Lundberg (1940), Feller (1968, p. 480), and Boswell and Patil (1970). Thompson (1954) showed that a negative binomial distribution can also be obtained (approximately) from a modified form of Neyman's contagious distribution model (Chapter 9, Section 6).

Distribution (5.16) arises also as a beta mixture of binomial distributions [Skellam (1948)]; see Chapter 6, Section 2.2, and Chapter 8, Section 3.3. Boswell and Patil (1970) derived the negative binomial as a limiting form of this mixture of binomials.

Feller (1957, p. 253) has pointed out that the negative binomial can be regarded as a convolution of a fixed number of geometric distributions; here, as for the inverse sampling model (5.13), the exponent k is necessarily an integer. Maritz (1952) considered how the negative binomial could arise from the addition of a set of correlated Poisson rv's. Kemp (1968a) showed that weighting a negative binomial with parameters k and $P = q/p$, using the weight function (sampling chance) $a_x = \alpha^x$ gives another negative binomial, but with parameters k and $\alpha q/(1 - \alpha q)$.

Bhattacharya (1966) obtained the negative binomial by mixing his confluent hypergeometric distributions with a "generalized exponential" distribution; the pgf of the outcome is

$$\int_0^\infty \frac{{}_1F_1(a; b; \theta z)}{{}_1F_1(a; b; \theta)} \times \frac{c^a(c + 1)^{b-a}\theta^{b-1}e^{-(c+1)\theta}{}_1F_1(a; b; \theta)d\theta}{\Gamma(b)}$$

$$= \left(1 - \frac{1}{c + 1}\right)^a \Bigg/ \left(1 - \frac{z}{c + 1}\right)^a. \qquad (5.17)$$

Bhattacharya showed that the "generalized exponential" mixing distribution is unique by virtue of the uniqueness of the Mellin transform (Chapter 1, Section A10), and he applied his results to the theory of accident proneness in the case where $a = 1$ and the number of accidents sustained by an individual has a sub–Poisson distribution (Chapter 4, Section 12.4).

The result of mixing negative binomials with constant exponent parameter k, using a beta distribution with parameters c and $k - c$, where $k > c > 0$, is another negative binomial distribution with exponent parameter c; we have

$$\int_0^1 [(1 + P\theta - P\theta z)^{-k}] \frac{\theta^{c-1}(1 - \theta)^{k-c-1} d\theta}{B(c, k - c)} = (1 + P - Pz)^{-c}. \quad (5.18)$$

Mixing Katti (1966) Type H_2 distributions (Chapter 6, Section 11) using a particular beta distribution can also yield a negative binomial distribution:

$$\int_0^1 {}_2F_1[k, a; b; P\theta(z - 1)] \frac{\theta^{b-1}(1 - \theta)^{a-b-1} d\theta}{B(b, a - b)}$$

$$= {}_3F_2[k, a, b; b, a; P(z - 1)]$$

$$= (1 + P - Pz)^{-k}. \quad (5.19)$$

Also a gamma mixture of Poisson$\wedge$Beta distributions (Chapter 8, Section 3.2(8)) can give rise to a negative binomial:

$$\int_0^\infty {}_1F_1[a; a + b; P\theta(z - 1)] \frac{e^{-\theta}\theta^{a+b-1} d\theta}{\Gamma(a + b)}$$

$$= {}_2F_1[a, a + b; a + b; P(z - 1)]$$

$$= (1 + P - Pz)^{-a}. \quad (5.20)$$

These results are special cases of those in Chapter 8, Section 3.4; see also Kemp (1968a).

The negative binomial arises also from several well–known stochastic processes. The time–homogeneous birth–and–immigration process with zero initial population was first obtained by McKendrick (1914); the equivalence of the distributions arising from this process, from Greenwood and Yule's model as a gamma mixture of Poisson distributions, and from Lüders and Quenouille's Poisson–stopped sum of logarithmic distributions model, was discussed by Irwin (1941). The nonhomogeneous process with zero initial population known as the *Pólya process* was developed by Lundberg (1940) in the context of risk theory. Other stochastic processes that lead to the negative binomial include the simple birth process with nonzero initial population size [Yule (1925) and Furry (1937)], Kendall's (1948) nonhomogeneous birth–and–death process with zero death rate, and the simple birth–death–and–immigration process with zero initial population of Kendall (1949).

The geometric distribution is the equilibrium distribution of queue length for the M/M/1 queue, whilst the negative binomial is the equilibrium queue–length distribution for the M/M/1 queue with a particular form of balking; see Haight (1957) and also Bhat (1984). The negative binomial can also be obtained as the equilibrium solution for a particular type of Markov chain known as a Foster process [Foster (1952)].

4 MOMENTS

From the probability generating function $(1 + P - Pz)^{-k}$, it follows that the factorial moment generating function is $(1 - Pt)^{-k}$, and so

$$\mu'_{[r]} = \frac{(k + r - 1)!}{(k - 1)!} P^r, \qquad r = 1, 2, \ldots. \tag{5.21}$$

Also the factorial cumulant generating function is $-k \ln(1 - Pt)$, whence

$$\kappa_{[r]} = k(r - 1)! P^r, \qquad r = 1, 2, \ldots. \tag{5.22}$$

The relationship with the binomial pgf is readily apparent — replacing N by $(-k)$ and π by $(-P)$ in the well–known formulas for the moment properties of the binomial distribution gives the corresponding formulas for the negative binomial distribution. In particular

$$\mu_r = kPQ \sum_{j=0}^{r-2} \binom{r-1}{j} \mu_j + P \sum_{j=0}^{r-2} \binom{r-1}{j} \mu_{j+1},$$

$$\mu = \kappa_1 = kP \qquad\qquad = \frac{kq}{p},$$

$$\mu_2 = \kappa_2 = kP(1 + P) \qquad = \frac{kq}{p^2},$$

$$\mu_3 = \kappa_3 = kP(1 + P)(1 + 2P) \qquad = \frac{kq(1 + q)}{p^3},$$

$$\mu_4 = 3k^2 P^2 (1 + P)^2 + kP(1 + P)(1 + 6P + 6P^2)$$

$$= \frac{3k^2 q^2}{p^4} + \frac{kq(p^2 + 6q)}{p^4}, \tag{5.23}$$

and

$$\sqrt{\beta_1} = \frac{1 + 2P}{\{kP(1 + P)\}^{1/2}} = \frac{1 + q}{\sqrt{kq}},$$

$$\beta_2 = 3 + \frac{(1 + 6P + 6P^2)}{kP(1 + P)} = 3 + \frac{p^2 + 6q}{kq}, \qquad (5.24)$$

where $p = Q^{-1}$ and $q = PQ^{-1}$.

The alternative notation, with the pgf in the form $p^k(1 - qz)^{-k}$, shows that the uncorrected moment generating function is $p^k(1 - qe^t)^{-k}$, the central moment generating function is $e^{-kqt/p}p^k(1 - qe^t)^{-k}$, whence

$$\mu_{r+1} = q\frac{\partial \mu_r}{\partial q} + \frac{rkq}{p^2}\mu_{r-1},$$

and the cumulant generating function is $k \ln p - k \ln(1 - qe^t)$.

Because this is a power series distribution with series parameter q, the cumulants satisfy

$$\kappa_{r+1} = q\frac{\partial \kappa_r}{\partial q}, \qquad r = 1, 2, \dots. \qquad (5.25)$$

The distribution is overdispersed (variance greater than the mean), with an index of dispersion equal to $p^{-1} = 1 + P$. The coefficient of variation is $(kq)^{-1/2} = \{(1 + P)/(kP)\}^{1/2}$.

The mean deviation is

$$\nu_1 = \frac{2m(k + m - 1)!P^m}{m!(k - 1)!Q^{m+k-1}} = \frac{2m(k + m - 1)!p^{k-1}q^m}{m!(k - 1)!}, \qquad (5.26)$$

where $m = [kP] + 1$ (that is, m is the smallest integer greater than the mean μ); see Kamat (1965).

5 PROPERTIES

From the relationship

$$\frac{\Pr[X = x + 1]}{\Pr[X = x]} = \frac{(k + x)P}{(x + 1)Q}, \qquad (5.27)$$

it can be seen that

$$\Pr[X = x + 1] < \Pr[X = x] \qquad \text{if } x > kP - Q,$$

and that $\qquad \Pr[X = x] \geq \Pr[X = x - 1] \qquad \text{if } x \leq kP - P. \qquad (5.28)$

So when $(k-1)P$ is not an integer, there is a single mode at $[(k-1)P]$, where $[\cdot]$ denotes the integer part. When $(k-1)P$ is an integer, then there are two equal modes at $X = (k-1)P$ and $X = kP - Q$. If $kP < Q$, the mode is at $X = 0$.

For fixed values of x and k the probabilities increase monotonically with P; for fixed x and P they increase monotonically with k.

When $k < 1$, we have $p_x p_{x+2}/p_{x+1}^2 > 1$ (where $p_x = \Pr[X = x]$), and therefore the probabilities are log–convex; when $k > 1$, we have $p_x p_{x+2}/p_{x+1}^2 < 1$ and so now the probabilities are log–concave. Although the probabilities satisfy the log–convexity condition that is a sufficient condition for infinite divisibility only when $k < 1$, nevertheless the distribution is a Poisson–stopped sum of logarithmic rv's, and so is infinitely divisible for all values of k.

The log–convexity/log–concavity properties imply that the distribution has a decreasing hazard (failure) rate for $k < 1$ and an increasing hazard rate for $k > 1$. For $k = 1$ the failure rate is constant. This is the no–memory (Markovian) property of the geometric distribution; see Sections 2 and 9.1 of this chapter.

If X_1 and X_2 are independent variables, each having a negative binomial distribution with the same series parameter q but with possibly different power parameters k_1 and k_2, then $X_1 + X_2$ also has a negative binomial distribution; its pgf is

$$(1 + P - Pz)^{-k_1-k_2} = p^{k_1+k_2}(1 - qz)^{-k_1-k_2}. \tag{5.29}$$

As k tends to infinity and P to zero, with kP remaining fixed ($kP = \theta$), the right–hand side of (5.1) tends to the value $e^{-\theta}\theta^k/k!$, corresponding to a Poisson distribution with expected value θ.

Young (1970) has studied the moments of the order statistics for the negative binomial distribution and has tabulated $E[X_{(r)}]$ for a sample of size n to two decimal places, for all r, when $n = 2(1)8$, $p = 0.01, 0.1(0.2)0.7$. He showed that when $p = Q^{-1}$ is close to unity, there is a good gamma approximation and Gupta's (1960a) tables for gamma order statistics can be used.

Pessin (1961, 1965) has noted that as $Q \to \infty$ with k constant, the standardized negative binomial tends to a gamma distribution.

6 APPROXIMATIONS AND TRANSFORMATIONS

The sum of a number of negative binomial terms can be expressed in terms of an incomplete beta function ratio, and hence as a sum of binomial terms. We have

$$\sum_{j=r}^{\infty} \Pr[X=j] = \frac{(k+r-1)!p^k q^r}{(k-1)!\,r!}\left\{1 + \frac{(k+r)q}{(r+1)} + \cdots\right\}$$

$$= \frac{(k+r-1)!p^k q^r}{(k-1)!\,r!}{}_2F_1[1, k+r; r+1; q]$$

$$= \frac{(k+r-1)!q^r}{(k-1)!\,r!}{}_2F_1[r, 1-k; r+1; q]$$

$$= \frac{B_q(r,k)}{B(r,k)} = I_q(r,k). \tag{5.30}$$

Therefore
$$\Pr[X \le r] = 1 - I_q(r+1, k)$$
$$= I_p(k, r+1)$$
$$= \Pr[Y \ge k], \tag{5.31}$$

where Y is a binomial rv with pgf $(q + pz)^{k+r}$. This formula has been re-discovered on many occasions. Patil (1963) gives a list of references; see also Morris (1963). Approximations for binomial distributions (already discussed in Chapter 3, Section 6) can thereby be applied to negative binomial distributions.

Bartko (1966) has studied five different approximations for cumulative negative binomial probabilities. Their accuracy is similar to that of approximations for the binomial distribution. The two most useful approximations in Bartko's opinion are

1. a corrected (Gram-Charlier) Poisson approximation

$$\Pr[X \le x] = e^{-kP}\sum_{j=0}^{x}\frac{(kP)^j}{j!} - \frac{(x-kP)}{2(1+P)}e^{-kP}\frac{(kP)^x}{x!}, \tag{5.32}$$

2. the Camp-Paulson approximation (see Chapter 26)

$$\Pr[X \le x] = \frac{1}{\sqrt{2\pi}}\int_{-\infty}^{K} e^{-u^2/2}du,$$

where

$$K = \frac{1}{3}\left[\frac{9x+8}{x+1} - \frac{(9k-1)\{kP/(x+1)\}^{1/3}}{k}\right]\left[\frac{\{kP/(x+1)\}^{2/3}}{k} - \frac{1}{x+1}\right]^{-1/2}. \tag{5.33}$$

Table 5.1a   Maximum Absolute Errors for the Corrected Gram-Charlier Poisson Approximation to the Negative Binomial Distribution

$(1+P)^{-1}$	5	10	25	50	100
			k		
0.05	0.309				
0.10	0.232	0.221			
0.20	0.146	0.138	0.131		
0.30	0.096	0.090	0.085	0.082	
0.40	0.062	0.058	0.054	0.053	
0.50	0.039	0.036	0.034	0.033	0.032
0.60	0.023	0.022	0.020	0.019	0.018
0.70	0.013	0.011	0.010	0.010	0.010
0.80	0.005	0.005	0.004	0.004	0.004
0.90	0.001	0.001	0.001	0.001	0.001
0.95	0.0001	0.0002	0.0003	0.0002	0.0003

Adapted from Bartko (1966).

Of these (5.33) is remarkably accurate, but it is much more complicated than (5.32). Tables 5.1a and 5.1b (taken from Bartko's paper) show the maximum absolute error in individual probabilities using (5.32) and (5.33), respectively.

Peizer and Pratt (1968) and Pratt (1968) have obtained extremely accurate normal approximations of the form

$$\Pr[X \le r] = \int_{-\infty}^{z_i} (2\pi)^{-1/2} e^{-x^2/2} dx,$$

where $\quad z_i = d_i \left\{ \dfrac{1 + p\, g(a) + q\, g(b)}{(r + k + 1/6)pq} \right\}^{1/2}, \quad i = 1, 2,$ (5.34)

$$a = \frac{r + 0.5}{(r + k)q},$$

$$b = \frac{k - 0.5}{(r + k)p},$$

$$g(u) = \frac{1 + u}{1 - u} + \frac{2u \ln u}{(1 - u)^2},$$

and $\quad d_1 = (r + 2/3)p - (s - 1/3)q,$

$$d_2 = d_1 + 0.02 \left(\frac{p}{r + 1} - \frac{q}{k} + \frac{p - 0.5}{r + k + 1} \right).$$

Peizer and Pratt provided a table of $g(u), 0 \le u \le 1$, and evidence concerning the remarkable accuracy of the two approximations.

Table 5.1b   Maximum Absolute Errors for the Camp-Paulson Approximation to the Negative Binomial Distribution

$(1+P)^{-1}$	k				
	5	10	25	50	100
0.05	0.001	0.001	0.0002		
0.10	0.001	0.001	0.0002	0.0001	
0.20	0.001	0.001	0.0002	0.0001	
0.30	0.001	0.001	0.0001	0.0001	
0.40	0.001	0.001	0.0001	0.0001	0.00002
0.50	0.001	0.001	0.0001	0.0001	0.00003
0.60	0.001	0.001	0.0002	0.0001	0.00004
0.70	0.001	0.001	0.0003	0.0001	0.0001
0.80	0.002	0.002	0.0006	0.0003	0.0001
0.90	0.004	0.002	0.002	0.0007	0.0004
0.95	0.003	0.005	0.003	0.002	0.001

Adapted from Bartko (1966).

We mention also Guenther's (1972) approximation based on the incomplete gamma function; see also Binns (1974). Guenther's approximation is

$$\Pr[X \leq r] \doteq \Pr\left[\chi^2_{2kq} \leq (2r+1)p\right], \tag{5.35}$$

where χ^2_{2kq} is a chi-squared variable; see Chapter 17. Best and Gipps (1974) presented evidence that

$$\Pr[X \leq r] \doteq \Pr\left[\chi^2_{8kq/(q+1)^2} \leq \frac{\{4r+2+4kq/(1+q)\}p}{(1+q)}\right] \tag{5.36}$$

provides a considerable improvement over (5.35). Tables of the incomplete gamma function can be used for these approximations. Guenther gave references concerning such tables.

A transformation that approximately normalizes, and also approximately equalizes the variance, merits special discussion. The formulas $E[X] = kP$, $\text{Var}(X) = kP(1+P)$, suggest the transformation

$$Y_1 = \sqrt{k}\sinh^{-1}\sqrt{\frac{X}{k}}, \tag{5.37}$$

with Y_1 approximately distributed as a standard normal variable. This transformation was applied by Beall (1942) to certain entomological data. More detailed investigations by Anscombe (1948) indicated that the transformation

$$Y_2 = \sqrt{k - 0.5}\, \sinh^{-1} \sqrt{\frac{X + 3/8}{k - 3/4}} \qquad (5.38)$$

is preferable; see also Laubscher (1961).

7   COMPUTATION AND TABLES

Computation of individual negative binomial probabilities can be reduced to the calculation of the corresponding binomial probabilities by the use of the relationship between the tails of the binomial and negative binomial distributions; see the previous section. For low values of r the probabilities can also be computed by recursion from $\Pr[X = 0]$ using

$$\Pr[X = r + 1] = \frac{(k + r)q}{r + 1}\, \Pr[X = r]. \qquad (5.39)$$

For higher values of r Stirling's expansion (Chapter 1, Section A2) can be used for the gamma functions in the expression

$$\Pr[X = r + 1] = \frac{\Gamma(k + r)p^k q^r \Gamma(r + 1)\Gamma(k)}{;}$$

this gives

$$\ln \Pr[X = r] \doteq (k - 1)\ln\left\{\frac{(k + r)p}{k}\right\} + (r + 0.5)\ln\left\{\frac{(k + r)q}{r}\right\}$$
$$-0.5\ln\left\{\frac{2\pi kq}{p^2}\right\} - \frac{1}{12k} - \frac{k}{12r(k + r)}. \qquad (5.40)$$

Cumulative negative binomial probabilities can be computed from cumulative binomial probabilities or by summation of individual probabilities; alternatively they can be approximated using appropriate formulae from the previous section.

However, for fractional values of k, and for convenience in looking up sequences of values, direct tables can be useful. Values of $\Pr[X = r]$ for

$$Q^{-1} = 0.05 \qquad\qquad k = 0.1(0.1)0.5\,,$$

$$Q^{-1} = 0.10 \qquad\qquad k = 0.1(0.1)1.0\,,$$

$$Q^{-1} = 0.12(0.02)0.20 \quad k = 0.1(0.1)2.5\,,$$

$$Q^{-1} = 0.22(0.02)0.40 \quad k = 0.1(0.1)2.5(0.5)5.0\,,$$

$$Q^{-1} = 0.42(0.02)0.60 \quad k = 0.1(0.1)2.5(0.5)10.0\,,$$

$$Q^{-1} = 0.62(0.02)0.80 \quad k = 0.2(0.2)5.0(1)20,$$

and a few other pairs of values of Q^{-1} and k, are given to six decimal places in the tables of Williamson and Bretherton (1963).

Grimm (1962) gave values of individual probabilities and of the cumulative distribution function to five decimal places for

$$kP = 0.1(0.1)1.0(0.2)4.0(0.5)10.0,$$

$$Q = 1.2, 1.5, 2.0(1)5.$$

Brown (1965) gave values to four decimal places of the same quantities for

$$kP = 0.25(0.25)1.0(1)10,$$

$$Q = 1.5(0.5)5.0(1)7.$$

Taguti (1952) gave minimum values of r for which

$$\sum_{j=0}^{r} (j!)^{-1} h(h+d) \dots (h + (j-1)d)(1+d)^{-(h/d)-j} \geq \alpha$$

for $\alpha = 0.95, 0.99$. These are (approximate) percentage points of negative binomial distributions with $k = h/d$, $P = d$.

8 ESTIMATION

8.1 Model Selection

Early graphical methods for identifying whether or not a negative binomial model is appropriate for a particular type of data were based on ratios of factorial moments [Ottestad (1939)], or probability-ratio cumulants [Gurland (1965)], or ratios of factorial cumulants [Hinz and Gurland (1967)]. Ord's method of plotting $u_r = rf_r/f_{r-1}$ against r (where f_r is an observed frequency) gives an upward-sloping straight line, $u_r \doteq (k+r-1)p$; see Ord (1967a, 1972) and Tripathi and Gurland (1979). Grimm's (1970) method, and the methods of Hoaglin, Mosteller, and Tukey (1985) can also be used; see Chapter 4, Section 7.1.

8.2 *P* Unknown

Consider the total number of trials $k+X$ needed in order to obtain k successes when the probability of a success is p (the inverse sampling model). Then the minimum variance unbiased estimator of p $(= (1 + P)^{-1})$, based on a single observation x of X, is

$$p^\circ = \frac{k - 1}{k + x - 1}, \tag{5.41}$$

and the Cramér-Rao lower bound on its variance is

$$\text{Var}(p^\circ) \geq \frac{p^2 q}{k}. \tag{5.42}$$

Best (1974) stated that

$$\text{Var}(p^\circ) = p^2 \sum_{r=1}^{\infty} \binom{k + r - 1}{r}^{-1} q^r; \tag{5.43}$$

Mikulski and Smith (1976) showed that

$$\text{Var}(p^\circ) \leq \frac{p^2 q}{k - p + 2}. \tag{5.44}$$

These bounds on the variance of p° have been sharpened by Ray and Sahai (1978) and Sahai and Buhrman (1979).

Given a sample of observations from a negative binomial distribution with power parameter k, then consideration of the pgf in the form $G(z) = p^k(1 - qz)^{-k}$ shows that the distribution is a power series distribution, and hence the ML equation for q is the first-moment equation $\bar{x} = k\hat{q}/(1 - \hat{q})$; thus

$$\hat{q} = \frac{\bar{x}}{k + \bar{x}}. \tag{5.45}$$

Roy and Mitra (1957) showed that the uniform minimum variance unbiased estimator of $P = q/p$ is $\tilde{\theta}/(1 - \tilde{\theta})$, where

$$\tilde{\theta} = \sum_x x f_x \Big/ \Big\{ \sum_x (k + x) f_x - 1 \Big\}, \tag{5.46}$$

and the f_x are the observed frequencies. The UMVUE of μ is $\sum x f_x / (k \sum f_x)$ and the UMVUE of μ_2 is $\sum x f_x \sum (x + 1) f_x / \{n(n + 1)\}$ [Guttman (1958)].

Irony (1990) has commented that the steps needed to make Bayesian inferences about q parallel those needed for Bayesian inferences about the binomial parameter p.

Maynard and Chow (1972) have constructed an approximate Pitman-type "close" estimator of P for small sample sizes. Scheaffer (1976) has studied methods for obtaining confidence intervals for $p = 1 - q$. Gerrard and Cook (1972) and Binns (1975) have considered sequential estimation of the mean $kq/(1 - q)$ when k is known.

8.3 Both Parameters Unknown

We turn now to the situation where both parameters are unknown. Because of the variability of the sample variance of the negative binomial distribution, underdispersed samples with the sample variance less than the sample mean $(s^2 < \bar{x})$ will occasionally be encountered, even when a negative binomial model is appropriate. However, whenever this occurs, the appropriateness of the model should be examined; see, e.g., Clark and Perry (1989).

Method of Maximum Likelihood
The *maximum likelihood* estimators satisfy the equations

$$\hat{k}\hat{P} = \bar{x}, \tag{5.47}$$

$$\ln(1 + \hat{P}) = \sum_{j=1}^{\infty} \{(\hat{k} + j - 1)^{-1} \sum_{i=j}^{\infty} f_j\}, \tag{5.48}$$

where f_j is an observed frequency; see Fisher (1941), Bliss and Fisher (1953), and Wise (1946). Iteration is required for the solution of these equations. It is important to realize that iteration may be very slow if the initial estimates are poor. Rapid explicit methods that can provide good initial estimates have therefore been studied extensively. Ross and Preece (1985) have advocated the use of the maximum likelihood program, MLP, of Ross (1980).

Method of Moments
The simplest way to estimate the parameters is by the method of moments (MM), that is, by equating the sample mean $\bar{x}$ and sample variance s^2 to the corresponding population values. Thus, if $x_1, x_2, \ldots, x_n$ are n observed values (supposed independent), we calculate the solutions $\tilde{k}, \tilde{P}$ of the equations

$$\tilde{k}\tilde{P} = \bar{x} \quad \text{and} \quad \tilde{k}\tilde{P}(1 + \tilde{P}) = s^2;$$

this gives

$$\tilde{P} = \frac{s^2}{\bar{x}} - 1, \tag{5.49}$$

$$\tilde{k} = \frac{\bar{x}^2}{s^2 - \bar{x}}. \tag{5.50}$$

Method of Mean-and-Zero-Frequency
In place of equation (5.48) an equation obtained by equating the observed and expected numbers of zero values may be used. This equation is

$$f_0 = (1 + P^\dagger)^{-k^\dagger},$$

where f_0 is the number of zero values. Combining this equation with the equation $k^\dagger P^\dagger = \bar{x}$ gives

$$\frac{P^\dagger}{\ln(1 + P^\dagger)} = -\frac{\bar{x}}{\ln f_0}. \tag{5.51}$$

The Digamma Method
For large values of k and P, Anscombe (1948, 1950) suggested an iterative method using the transformation

$$Y = 2\sinh^{-1}\sqrt{\frac{X + 3/8}{k - 3/4}} \qquad \text{(see Section 6).}$$

The variance of Y is approximately $\psi'(k)$, where $\psi(\cdot)$ and $\psi'(\cdot)$ are the digamma and trigamma functions; see Chapter 1, Section A2. Starting with a trial value k_0 of k, values of y corresponding to the observed values of X are calculated. The variance s_y^2 of these y's is then calculated, and a new value k_1 is obtained by solving the equation

$$s_y^2 = \psi'(k_1).$$

The process is continued until $k_{i+1} \doteq k_i \doteq k^\ddagger$ within a desired degree of accuracy. The method is now rarely used.

Generalized Minimum Chi-squared Method
Gurland (1965) and Gurland and Tripathi (1975) have put forward a method based on the solution of linear equations involving functions of the moments and/or frequencies; see also Katti and Gurland (1962a). Gurland (1965) and Hinz and Gurland (1967) concluded that estimators based on the factorial cumulants and a certain function of the zero-frequency have good efficiency relative to maximum likelihood.

Anscombe (1950) gave a graph from which the efficiencies of certain methods can be judged. He found that, although the method of moments and the mean-and-zero-frequency method each have high efficiency over some part of the parameter space, there are regions where both have low efficiency. Shenton and Meyers (1965) made an extensive study of the then current estimation methods, with particular interest in the bias of the moment estimator of k. Bowman and Shenton (1965, 1966) have obtained asymptotic formulas for the variances, covariances and biases of the moment and ML estimators.

Pieters et al. (1977) made small-sample comparisons of various methods using simulation. Willson, Folks and Young (1984) extended this work by considering not only the bias but also the standard deviation and the mean square error of the MM and ML estimators, and by comparing these to a proposed multistage estimation procedure. They used 18 combinations of the parameters, and samples of size 50 and 100. In her comment on their work, Bowman (1984) pointed out the riskiness in depending on small samples when estimating k, and questioned the choice of $n = 5$ as the initial sample size for the multistage procedure.

Kemp (1986) sought to explain Anscombe's findings regarding the efficiencies of certain methods by showing that the solution to an approximation to the ML equations that is valid for a certain region of the parameter space can give useful approximations to the ML estimators over that region. Kemp and Kemp (1988) went further with this approach by using (5.47) and an approximation to the digamma function implicit in (5.48) to obtain the equations

$$k^* = \frac{\bar{x}\bar{x}_w(\eta - 1)}{\bar{x}_w - \bar{x}\eta},$$

$$P^* = \frac{\bar{x}}{k^*}, \tag{5.52}$$

where $\bar{x}_w = \sum_x x f_x \eta^x$ is a weighted mean of the sample distribution, and

$$\eta = \frac{\tilde{k} + 1}{\tilde{k} + 2} \qquad \text{for} \quad \bar{x} \le 2,$$

$$= \frac{\tilde{k} + \bar{x}/2}{\tilde{k} + 1 + \bar{x}/2} \qquad \text{for} \quad \bar{x} > 2, \tag{5.53}$$

where $\tilde{k}$ is the moment estimate of k. A large simulation study confirmed that these explicit estimators have high efficiency relative to ML for nearly all of the parameter space.

Anraku and Yanagimoto (1990) have adopted the parameterization

$$G(z) = (1 + \theta\mu - \theta\mu z)^{-1/\theta};$$

this is obtained by setting $P = \theta\mu$, $k = 1/\theta$ in (5.1) where μ is the population mean. In this parameterization the index of dispersion is $\mu_2/\mu = 1 + \theta\mu$. The authors' focus of attention is the estimation of θ (the "dispersion parameter" in their paper) conditional on a knowledge of μ. They commented that "$\bar{x}$ is a reasonable estimator of μ irrespective of the estimator of θ," and obtained the conditional likelihood

$$L_c(x;\theta) = \sum_{i=1}^{n}\sum_{j=1}^{x_i} \frac{j-1}{1+\theta(j-1)} - \sum_{i=1}^{t} \frac{i-1}{n+\theta(i-1)}, \qquad (5.54)$$

where $t = \sum X_i$, when $s^2 > \bar{x}$. This is maximized by their conditional ML estimator, θ_c. For $s^2 \leq \bar{x}$ their estimator is defined to be zero. Anraku and Yanagimoto have investigated the biases, etc., associated with θ_c, compared with the unconditional ML estimator and the moment estimator of θ, both theoretically and via simulation.

Clark and Perry (1989) have advocated the use of $\theta = 1/k$ instead of k because this "avoids problems caused by infinite values of $\hat{k}$ when $s^2 = \bar{x}$; also confidence intervals for θ are continuous and usually more symmetric than those for k, which may be discontinuous." Clark and Perry's maximum quasi-likelihood estimator (MQLE), $\hat{\theta}$ of θ, is obtained by solving iteratively

$$\sum_{i=1}^{n}\left[\hat{\theta}^{-2}\ln\left(\frac{1+2m}{1+\hat{\theta}x_i}\right) - \frac{x_i}{1+\hat{\theta}x_i} + \frac{1+6x_i}{2(\hat{\theta}+6+6\hat{\theta}x_i)}\right] = \frac{n}{2(\hat{\theta}+6)}, \qquad (5.55)$$

where m is taken to be $\hat{\mu}$, and x_i, $i = 1, 2, \ldots, n$, are the observations. The authors indicate how a 95% profile likelihood confidence interval for θ can be constructed. Their simulation results show that MQL estimation performed slightly better than estimation by the method of moments (MM), except when the sample size is not more than 20. Piegorsch (1990) carried out further simulation studies, showing that MQL appears to be slightly less biased than ML or MM; he recommends the use of MQL, provided that the sample size is greater than 20, and that θ is "not very small."

Maximum-likelihood fitting of a negative binomial distribution to coarsely grouped data has been described by O'Carroll (1962).

8.4 Data Sets with a Common Parameter

The estimation of an (assumed) common value of k from data from a number of negative binomial distributions was discussed briefly by Anscombe (1950). A much more detailed discussion was given by Bliss and Owen (1958). Suppose that in a sequence of n samples of sizes n_i, $i = 1, 2, \ldots, n$, the observed means and standard deviations are denoted by $\bar{x}_i$, and s_i^2, respectively, where $i = 1, 2, \ldots, n$. Then the moment estimators of k are

$$\tilde{k}_i = \frac{\bar{x}_i^2}{s_i^2 - \bar{x}_i}, \qquad i = 1, 2, \ldots, n,$$

and their approximate variances are

$$\frac{2k(k+1)(Q_i/P_i)^2}{n_i}, \qquad i = 1, 2, \ldots, n.$$

As a first approximation, the weights $w_i = n_i(\tilde{P}_i/\tilde{Q}_i)^2$ may be used, giving

$$k^{(1)} = \sum_{i=1}^{n} w_i \tilde{k}_i \Bigg/ \sum_{i=1}^{n} w_i .$$

Using this value, $k^{(1)}$, new estimates of P_i and new weights can be calculated; these can then be used to obtain an improved estimate of k.

A standard method for testing the hypothesis of a common value of k was provided by Bliss and Fisher (1953).

Testing for a common value of P has also been investigated by Meelis (1974), who distinguished three situations:

1. P known to the experimenter.
2. P unknown to the experimenter, and the sample sizes for the n samples all equal.
3. P unknown to the experimenter, and the sample sizes not necessarily equal.

Freeman (1980) has studied methods based on minimum chi-square for fitting two parameter discrete distributions to many data sets with one common parameter. He discussed fitting negative binomials with a common value of k, and also with a common value of P; he examined the meanings of these two models with particular reference to Taylor's power law [Taylor (1961)].

9 CHARACTERIZATIONS

9.1 Geometric Distribution

The geometric distribution is characterized by the Markovian property

$$\Pr[X = x + y | X \geq y] = \Pr[X = x]; \qquad (5.56)$$

see Section 2 of this chapter. This is a discrete analogue of the Markovian property of the exponential distribution [Hawkins and Kotz (1976)]. Many other characterizations of the geometric are similarly analogues of exponential characterizations.

Shanbhag (1970b) showed that if X is a discrete positive integer-valued random variable, then

$$\Pr[X > a + b | X > a] = \Pr[X > [a + b] - [a]] \qquad (5.57)$$

for all a and b iff X has a geometric distribution. Similarly X has a geometric distribution iff

$$E[X|X > y] = E[X] + [y] + 1 \qquad (5.58)$$

for all y where $[y]$ denotes the integer part of y.

It is also possible to characterize any geometric distribution by the distribution of the difference between two independent random variables having the same geometric distribution, Puri (1966). Puri also showed that if the common distribution has parameter P, then the absolute difference distribution can be constructed as the distribution of the sum of two independent random variables, one distributed binomially with parameters $n = 1$, $p = P(1+2P)^{-1}$ and the other distributed geometrically with parameter P.

The independence of the difference $X_1 - X_2$ and $\min(X_1, X_2)$ for two independent random variables underlies another type of characterization. Ferguson (1964, 1965) and Crawford (1966) found that $\min(X_1, X_2)$ and $X_1 - X_2$ are independent random variables iff X_1 and X_2 are either both exponential or both geometric rv's with the same location and scale parameters; see also Srivastava (1974).

Srivastava (1965) obtained an extended form of this characterization, applicable to n independent discrete random variables $X_1, X_2, \ldots, X_n$. The essential condition is that $\min(X_1, X_2, \ldots, X_n)$ and $\sum_{i=1}^{n} [X_i - \min(X_1, X_2, \ldots, X_n)]$ are mutually independent.

Srivastava (1974) showed furthermore that if X_1 and X_2 have independent nonnegative discrete distributions, then

$$\Pr[U = j, V = l] = \Pr[U = j]\Pr[V = l] \qquad (5.59)$$

(where $U = \min(X_1, X_2)$ and $V = X_1 - X_2$) for all j and $l = 1, 2$, iff X_1 and X_2 have geometric distributions with possibly different parameters.

Gupta's (1970) result based on a triplet of order statistics is as follows (in the discrete case): Let $X_1, X_2, \ldots, X_n$ be independent and identically distributed discrete rv's with order statistics

$$X_{(1)} \leq X_{(2)} \leq \ldots \leq X_{(n)}.$$

Then the independence of $X_{(i)}$ and $X_{(k)} - X_{(j)}$ for one triplet i, j, k, $1 \leq i < j < k \leq n$ provides a characterization of the geometric distribution. The independence of $X_{(1)}$ and $\sum_{j=2}^{k} \{X_{(j)} - X_{(1)}\}$ also provides a characterization [Srivastava (1974)].

Suppose now that X has an arbitrary nonnegative discrete distribution satisfying the condition that for some interval $(\delta_1, \delta_2]$ we have $\Pr[\delta_1 < X \leq \delta_2] = 0$, where $0 \leq \delta_1 < \delta_2 < \infty$. Puri and Rubin (1970) proved that if X_1 and X_2 independently have the same distribution as X, then X and the absolute difference $|X_1 - X_2|$ have the same distribution iff X has the pmf

$$\Pr[X = 0] = \alpha,$$

$$\Pr[X = x\tau] = 2\alpha(1 - \alpha)(1 - 2\alpha)^{x-1}, \qquad x = 1, 2, \ldots, \qquad (5.60)$$

where $\tau > 0$, and either $\alpha = 1$ or $0 < \alpha \leq 0.5$.

Puri's (1973) characterization theorem concerning the sum of two independent nonnegative discrete rv's X_1 and X_2 states that

$$\Pr[X_1 \leq x] - \Pr[X_1 + X_2 \leq x] = a\Pr[X_1 + X_2 = x], \qquad x = 0, 1, 2, \ldots, \quad (5.61)$$

where $a > 0$, iff

$$\Pr[X_2 = x] = \left(\frac{1}{1 + a}\right)\left(\frac{a}{1 + a}\right)^x, \qquad x = 0, 1, 2, \ldots, \qquad (5.62)$$

i.e. iff X_2 has a geometric distribution.

The analogue of Puri and Rubin's (1972) characterization of mixtures of exponential rv's supposes that $(X_1, X_2, \ldots, X_n)$ is a vector of n nonnegative discrete rv's such that

$$\Pr[X_i > x_i | i = 1, 2, \ldots, n] = \beta\Pr[X_i = x_i | i = 1, 2, \ldots, n], \qquad (5.63)$$

where $\beta > 0$ and $x_i = 0, 1, 2, \ldots$. Then the only distributions of $X_1, X_2, \ldots, X_n$ satisfying this condition are mixtures of geometric distributions.

Uppuluri, Feder and Shenton (1967) obtained a characterization based on the observation that the geometric distribution is obtained as the limiting form of the sequence $\{Y_n\}$ defined by the stochastic model

$$Y_n = 1 + V_n Y_{n-1}, \qquad (5.64)$$

where V_i are iid with $\Pr[V_i = 0] = a$ and $\Pr[V_i = 1] = 1 - a$, with $0 < a < 1$; see also Paulson and Uppuluri (1972).

Lukacs (1965) gave characterizations based on moment properties. Let $X_1, X_2, \ldots, X_n$ be a random sample from a distribution with finite variance, and put $S_j = \sum_{i=1}^{n}(X_i)^j, j = 1, 2$. Then the distribution is geometric iff

$$T = \left(\frac{n+1}{n-1}\right)S_2 - \left(\frac{2}{n-1}\right)S_1^2 - S_1 \qquad (5.65)$$

has zero regression on S_1. Lukacs also showed that $0 < \kappa_1/\kappa_2 < 1$, $\kappa_1/\kappa_2 = 1/Q$, $\kappa_2 = \kappa_1^2 + \kappa_1$ characterizes the geometric distribution with characteristic function $\phi(t) = (Q - Pe^{it})^{-1}$ and mean and variance $\mu = \kappa_1 = kP$ and $\mu_2 = \kappa_2 = kP(1 + P)$.

Kagan, Linnik and Rao (1973) showed that Patil and Seshadri's (1964) general result for discrete distributions implies the following characterization:

If the conditional distribution of X_1, given $X_1 + X_2$, has a discrete uniform distribution for all values of the total $X_1 + X_2$, then X_1 and X_2 both have the same geometric distribution.

Dallas (1974) found that if X has a nondegenerate distribution on the nonnegative integers, with $\Pr[X = k] \neq 0$ for all $k = 0, 1, 2, \ldots$, then X has a geometric distribution iff $\operatorname{Var}(X|X > c) = d$ where d is constant $(= kP(1+P))$, for $c = -1, 0, 1, \ldots$. This is a "non-aging" property (cf. (5.56)).

Srivastava (1979) and Ahsanullah and Holland (1984) have also given two characterizations of the geometric distribution based on properties of record value distributions. Arnold (1980) has given further characterizations in terms of order statistics. Yet more characterizations have been proved by Shaked (1974), Shanbhag (1974), Chong (1977), Nagaraja and Srivastava (1987), Nagaraja (1988), Wesolowski (1989), and Khalil, Dimitrov, and Dion (1991). Nagaraja (1990) has included in his paper on order statistics for discrete distributions a comprehensive review of characterizations for the geometric distribution based (1) on the independence of certain functions of its order statistics and (2) on its distributional properties.

9.2 Negative Binomial Distribution

Only a few characterizations have been obtained for the negative binomial distribution. Patil and Seshadri's (1964) general result implies that if the conditional distribution of X_1, given $X_1 + X_2$, is negative hypergeometric with parameters m and n for all values of the total $X_1 + X_2$, then X_1 and X_2 both have negative binomial distributions, with parameters (m, θ) and (n, θ) respectively [Kagan, Linnik and Rao (1973)].

Meredith (1971) has made the following comment on Ottestadt's (1944) derivation of the negative binomial as a mixed Poisson distribution by assuming that the regression of Θ (the mixing variable) on X is linear, without making any assumption about the form of $f(\theta)$: Because the mixing distribution of a mixed Poisson is identifiable [Feller (1943)], it follows that the regression of Θ on X is linear iff Θ has a gamma distribution and the mixed Poisson distribution is negative binomial.

Some of the other modes of genesis in Section 3 can similarly be made the basis of characterizations. For example, a Poisson-stopped sum of a distribution on the positive integers is a negative binomial iff the distribution on the positive integers is logarithmic.

10 APPLICATIONS

The negative binomial distribution has become increasingly popular as a more flexible alternative to the Poisson distribution, especially when it is doubtful whether the strict requirements, particularly independence, for a Poisson distribution will be satisfied.

Negative binomial distributions have been found to provide useful representations in many fields. Many researchers, including Arbous and Kerrich (1951), Greenwood and Yule (1920), and Kemp (1970), have applied it to accident statistics. Furry (1937) and Kendall (1949) have shown its applicability in birth-and-death processes. It has been found useful for psychological data by Sichel (1951), as a lag distribution for time series in economics by Solow (1960), and in market research and consumer expenditure by Chatfield, Ehrenberg and Goodhardt (1966), Chatfield (1975), and Goodhardt, Ehrenberg and Chatfield (1984). Medical and military applications have been described by Chew (1964) and by Bennett and Birch (1964). Burrell and Cane (1982) have used negative binomial models for lending-library data.

Bliss and Fisher (1953) successfully fitted the distribution to a large number of biometrical data sets. When Martin and Katti (1965) fitted the negative binomial and certain other distributions to 35 ecological data sets, they found that the negative binomial and the Neyman type A have very wide applicability. Elliott's (1979) manual highlights its usefulness for analysing samples of freshwater fauna. Wilson and Room (1983), Binns (1986), and Perry (1984) have used the negative binomial for modelling entomological data.

The distribution has been used to model family size by Rao et al. (1973). Janardan and Schaeffer (1981) considered it as a possible alternative to the logarithmic distribution for the number of different compounds identified in water samples. The following three very diverse applications were examined by Clark and Perry (1989): cell-centres in grid squares [see also Crick and Lawrence (1975) and Diggle (1983)], red mites on apple leaves [see also Bliss and Fisher (1953)], and counts of cycles to failure of worsted yarn [see also Box and Cox (1964)].

Boswell and Patil (1970) gave a useful account of some dozen processes leading to the negative binomial distribution; these and others are given in Section 3. Physical applications involving queueing theory and other stochastic processes were described by Bhat (1984) and by Taylor and Karlin (1984). Autoregressive moving-average processes with geometric and negative binomial marginal distributions were discussed by McKenzie (1986).

There is a problem however. If the negative binomial is found empirically to give a good fit for a particular kind of data, then the experimenter may still have to decide how to interpret the fit in terms of the many possible modes of genesis of the distribution. In particular, a good fit does not, on its own, distinguish between the heterogeneous (mixed) Poisson model and the Poisson-stopped sum model.

There are of course situations where a good fit is not obtainable with the negative binomial distribution, and in such cases it is usual to consider the possibility of a mixture of distributions or a contagious distribution (possibly with more than two parameters); see Chapters 8 and 9.

11 TRUNCATED NEGATIVE BINOMIAL DISTRIBUTIONS

In the most common form of truncation, the zeroes are not recorded, giving the pmf

$$\Pr[X = x] = (1 - Q^{-k})^{-1}\binom{k + x - 1}{k - 1}\left(\frac{P}{Q}\right)^{x}\left(1 - \frac{P}{Q}\right)^{k}, \qquad x = 1, 2, \dots,$$
$$(5.66)$$

where $Q = 1 + P$. This situation occurs in applications of the negative binomial such as the number of offspring per family, the number of claims per claimant, and the number of occupants per car. Other examples have been given by Sampford (1955) and Brass (1958).

Boswell and Patil (1970) have given a genesis of the *zero–truncated negative binomial distribution* as a distribution for the sizes of groups (a group–size distribution). They have also shown that it can be derived as a mixture of zero–truncated Poisson distributions, since (amending a typo in the original)

$$\binom{k + x - 1}{x}\left(\frac{1}{1 + \theta}\right)^{x}\left(\frac{\theta}{1 + \theta}\right)^{k}\Bigg/\left\{1 - \left(\frac{\theta}{1 + \theta}\right)^{k}\right\}$$

$$= \binom{k + x - 1}{x}\left\{\frac{\theta^{k}(1 + \theta)^{-x-k}}{1 - \{\theta/(1 + \theta)\}^{k}}\right\}$$

$$= \int_{0}^{\infty}\frac{e^{-\lambda}\lambda^{x}}{x!(1 - e^{-\lambda})} \times ce^{-\lambda\theta}\lambda^{k-1}(1 - e^{-\lambda})d\lambda, \qquad x = 1, 2, \dots, \quad (5.67)$$

where $c^{-1} = \Gamma(k)\{\theta^{-k} - (1 + \theta)^{-k}\}$ and $\theta = 1/P$.

The moments about zero are easily calculated as

$$\mu_{r}' = (1 - Q^{-k})^{-1}m_{r}', \qquad (5.68)$$

where m_{r}' is the corresponding uncorrected moment of the untruncated negative binomial distribution. Thus

$$\mu = E[X] = kP(1 - Q^{-k})^{-1}$$
$$E[X^{2}] = (kPQ + k^{2}P^{2})(1 - Q^{-k})^{-1},$$

whence
$$(5.69)$$

$$\mu_{2} = \frac{(kPQ + k^{2}P^{2})}{(1 - Q^{-k})} - \frac{k^{2}P^{2}}{(1 - Q^{-k})^{2}}. \qquad (5.70)$$

Rider (1962b) has given tables of $E[X^{-1}]$ and of $\mathrm{Var}(X^{-1})$ to five decimal places for $P = 0.01, 0.05(0.05)1.00(1)5$, $k = 1(1)10$. Govindarajulu (1962a) proposed the approximations

$$E[X^{-1}] \doteq (kP - Q)^{-1},$$
$$\mathrm{Var}(X^{-1}) \doteq Q(kP - Q)^{-2}(kP - 2Q)^{-1}. \tag{5.71}$$

The maximum likelihood estimators of k and Q are given by

$$0 = \sum_{x \geq 1} f_x \left[\frac{1}{\hat{k}} + \frac{1}{\hat{k}+1} + \cdots + \frac{1}{\hat{k}+x-1} - \frac{\ln \hat{Q}}{1 - \hat{Q}^{-\hat{k}}} \right],$$

$$0 = \sum_{x \geq 1} f_x \left[\frac{x}{1 - \hat{Q}} - \frac{\hat{k}}{1 - \hat{Q}^{-\hat{k}}} \right], \tag{5.72}$$

where f_x is the observed frequency of the observation x; see David and Johnson (1952) who also gave expressions for the variances and covariance of the ML estimates.

Solution of the maximum likelihood equations requires iteration. Wyshak (1974) has published a computer algorithm to solve the ML equations and hence obtain estimates of k and p $(= 1/(1 + P))$ and their variances.

David and Johnson (1952) found that the moment equations (obtained by equating the sample and expected values of the mean and of the variance) also do not have an explicit solution. They considered introducing a third equation (obtained by equating the sample and observed third moments), with which simple explicit solutions could be obtained. These estimators are, however, very inefficient, and the authors recommended using ML methods instead. Rider (1955) also derived estimators based on the moments, and Shah (1961) gave formulas for their asymptotic variances and covariances.

Sampford (1955) proposed a trial–and–error method for solving the moment equations. He found that for all combinations of the values $k = 0.5, 1(1)5$ and $kP = 0.5, 1, 2, 5$, the efficiency of the moment estimators is at least 56%, rising to at least 85%, for $k \geq 3$, and at least 91% for $k = 5$.

Brass (1957) proposed using f_1 together with $\bar{x}$ and s^2. This yields the formulas

$$\tilde{Q} = s^2(\bar{x})^{-1}(1 - f_1)^{-1},$$
$$\tilde{k} = (\bar{x} - \tilde{q}f_1)(\tilde{Q} - 1)^{-1}. \tag{5.73}$$

He showed that the efficiency (as measured by the ratio of the generalized variances) of his method is greater than that of Sampford's moment method when $k \leq 5$, and not much less when $k > 5$. Brass also suggested a modification of the ML equations, replacing $\hat{k}\hat{P}\hat{Q}^{-(\hat{k}+1)}/(1-\hat{Q}^{-\hat{k}})$ by f_1, which seemed to give very little reduction in efficiency except for $k < 1$.

Pichon et al. (1976) proposed the use of only $\bar{x}$ and f_1, but without investigating properties of the method.

Schenzle (1979) has studied the asymptotic efficiencies of Sampford's trial-and-error method (using $\bar{x}$ and s^2), the method of Brass (using $\bar{x}$, s^2 and f_1), and the method of Pichon et al. (using $\bar{x}$ and f_1). He found that Pichon et al. had correctly conjectured that their method gives slightly better efficiency than the other two for small values of k. However, for small values of k (e.g., $k < 1$) and values of μ in the range $2 \le \mu \le 8$, he considered that the efficiencies of all three methods become so low that none of these simple methods is applicable and ML estimation should be used instead.

Minimum variance unbiased estimation has been investigated by Cacoullos and Charalambides (1975).

Ahuja (1971b) has studied the n-fold convolution of the zero–truncated negative binomial distribution. He observed that it is a special case of the generalized power series distribution with series function

$$f(\theta) = \left\{ (1 - \theta)^{-k} - 1 \right\}^n ,$$

where $\theta = P/Q$, and hence showed that the pmf is

$$\Pr[X = x] = \sum_{i=1}^{n} (-1)^{n-i} \binom{n}{i} \binom{x + ki - 1}{x} \frac{\theta^x}{\{(1 - \theta)^{-k} - 1\}^n} \qquad (5.74)$$

for $x = n, n + 1, \ldots$. He also obtained an expression for the cdf in terms of a linear combination of incomplete beta functions. Ahuja and Enneking (1974) have studied the n-fold convolution of the negative binomial truncated by omission of $x = 0, 1, \ldots, c$ where c is a positive integer. Charalambides (1977b) explored the use of associated Lah numbers and generalized Lah numbers for these two types of convolutions.

Saleh and Rahim (1972) gave a formula for the convolution of n truncated negative binomial variables as an example of their general method for investigating convolutions of truncated discrete distributions.

Hamdan (1975) has obtained an expression for the correlation between the numbers of two types of children when family size is assumed to be negative binomial but the number of fertile childless families is unknown; he has applied his results to the data of Reed and Reed (1965).

The *displaced negative binomial distribution* is obtained by truncation of the first r probabilities of a negative binomial distribution. Shah (1971) showed that it has pmf

$$\Pr[X = x] = \frac{(k + x - 1)! \theta^x / x!}{\sum_{x \ge r} (k + x - 1)! \theta^x / x!}, \qquad x = r, r + 1, \ldots, \qquad (5.75)$$

and investigated ML and other methods of estimation.

12 OTHER RELATED DISTRIBUTIONS

12.1 Limiting Forms

The negative binomial distribution is a limiting form of the hypergeometric–type distributions of Chapter 6, Section 2.4, with pgf

$$G(z) = \frac{{}_2F_1[-n, -a; b - n + 1; z]}{{}_2F_1[-n, -a; b - n + 1; 1]},$$

1. when $a < 0$, $b < 0$, and n is a positive integer (Type IIA hypergeometric),
   if $n \to \infty$, $b \to -\infty$, and $n/(n - b) \to q$, $0 < q < 1$,
   that is, if $n \to \infty$, $b \to -\infty$, and $-n/b \to P > 0$.
2. when $a + b < n < 0 < a$, and a is a positive integer (Type IIIA hypergeometric),
   if $a \to \infty$, $b \to -\infty$, and $-a/b \to q$, $0 < q < 1$,
   that is, if $a \to \infty$, $b \to -\infty$, and $-a/(a + b) \to P > 0$.
3. when $a < 0$, $n < 0$, $0 < a + b + 1$ (Type IV hypergeometric),
   if $a \to -\infty$, $b \to \infty$, and $-a/b \to q$, $0 < q < 1$,
   that is, if $a \to -\infty$, $b \to \infty$, and $-a/(a + b) \to P > 0$.

In turn the logarithmic distribution of Chapter 7 with parameter $\alpha = q$ is the limiting form of the zero–truncated negative binomial distribution as $k \to 0$.

Relationships among the negative binomial, Poisson and binomial distributions have been mentioned in Chapters 2, 3 and 4. It is convenient to make some comments here concerning these relationships. Each of the three distributions has a pgf of the form

$$[(1 + \gamma) - \gamma z]^{-m}.$$

For the negative binomial, $\gamma > 0$, $m > 0$, while for the binomial, $-1 < \gamma < 0$, $m < 0$ an integer. The Poisson distribution corresponds to the limiting intermediate case, where $\gamma \to 0$, $m \to \infty$, with $m\gamma = \theta$, θ fixed. That is to say, the Poisson distribution with parameter θ is the limiting form of the negative binomial distribution with pgf $(1 - q)^k/(1 - qz)^k$ as $k \to \infty$, $q \to 0$, $kq \to \theta$.

The most obvious distinction between the three distributions is in the value of the ratio of the variance to the mean. This is less than unity for the binomial, equal to unity for the Poisson, and greater than unity for the negative binomial distribution. Sampling variation may yield values of this ratio that do not accord with the underlying distribution; nevertheless, the value of the ratio of the sample variance to the sample mean is a good guide as to which of these three distributions is appropriate when fitting a given data set, *given that one of them is to be used* (see also the remarks in Section 8.1 of this chapter).

It is interesting to note that for all Poisson distributions, the point (β_1, β_2) lies on the straight line $\beta_2 - \beta_1 - 3 = 0$. On the other hand, for the binomial distribution with parameters n and p we have $\beta_2 - \beta_1 - 3 = -2/n$; for the negative binomial distribution with parameters k and P we have $\beta_2 - \beta_1 - 3 = 2/k$.

12.2 Engen's Extended Negative Binomial Model

An extension of the negative binomial distribution has been proposed by Engen (1974, 1978) in connection with a model for species frequency data. Suppose that each member of a population of individuals (elements) belongs to one and only one of a number of species (classes) $C_1, C_2, \ldots, C_s$ where s is the number of species (possibly infinite). Let R_j, $j = 1, \ldots, N$ be the number of species represented by exactly j individuals in a sample of individuals, and suppose that sampling is such that

$$E[R_j] \propto \frac{w^k \Gamma(k+j)(1-w)^j}{\Gamma(k+1)j!}, \qquad j = 0, 1, 2, \ldots, \quad 0 < w < 1,$$

$$\propto w^k (k+1)(k+2) \cdots (k+j-1)(1-w)^j/j!. \qquad (5.76)$$

The number of species represented by no individuals is unascertainable. For a given total number of observed species the conditional distribution of species frequency therefore has the pgf

$$
\begin{aligned}
G(z) &= \frac{\sum_{j\geq 1} \alpha w^k (k+1) \cdots (k+j-1)(1-w)^j z^j/j!}{\sum_{j\geq 1} \alpha w^k (k+1) \cdots (k+j-1)(1-w)^j/j!} \\[2mm]
&= \frac{\{1 - (1-w)z\}^{-k} - 1}{\{1 - (1-w)\}^{-k} - 1} \\[2mm]
&= z \frac{{}_2F_1[k+1, 1, 2; (1-w)z]}{{}_2F_1[k+1, 1, 2; (1-w)]}.
\end{aligned}
\qquad (5.77)
$$

This is a zero–truncated negative binomial distribution (Section 11 of this chapter); it is also closely related to an extended hypergeometric distribution (Chapter 6, Section 11). Willmot (1988a) has commented that it belongs to Sundt and Jewell's family of distributions (Chapter 2, Section 3.1).

As $k \to 0$ Engen's model tends to Fisher's logarithmic series model (see Chapter 7, Section 2); the corresponding conditional (one–parameter) pgf is $g(z) = \ln\{1 - (1-w)z\}/\ln w$. However, the form of (5.76) led Engen to remark that "it appears that the natural lower bound for k seems to be -1 (Engen, 1974) and not zero as claimed by Fisher [et al.] (1943)." Values of k in the range $-1 < k < 0$ do indeed give a valid distribution. Its mean and variance are

$$\mu = \frac{k(1-w)}{w(1-w^k)} \quad \text{and} \quad \mu_2 = \frac{k(1-w)}{w^2(1-w^k)} \left[1 - \frac{k(1-w)w^k}{(1-w^k)}\right]. \tag{5.78}$$

12.3 Lagrangian "Generalized Negative Binomial Distribution"

The binomial-binomial Lagrangian distribution of Consul and Shenton (1972) has already been documented in Chapter 3, Section 12.3 (see also Chapter 2, Section 5.2). The special case of the Lagrangian binomial-binomial distribution with $p = p'$ (and hence with $q = q'$) had previously been introduced into the literature by Jain and Consul (1971); it is included in this chapter because they called it the *"generalized negative binomial distribution"*, and this name has continued to be used. When $p = p'$ (and $q = q'$), the pmf (3.115) simplifies to

$$\Pr[X = x] = \frac{n}{(n+mx)} \binom{n+mx}{x} p^x q^{n+mx-x}, \qquad \text{for } x \geq 0, \tag{5.79}$$

where $x = 0, 1, \ldots$, and the restrictions on the parameters are $0 < p < 1$, $n > 0$, $p < mp < 1$. The first three moments are

$$\mu = np(1-mp)^{-1},$$

$$\mu_2 = npq(1-mp)^{-3},$$

$$\mu_3 = \mu_2 \left[\frac{3mpq}{(1-mp)^2} + \frac{4q-1}{1-mp}\right] - \frac{2npq^2}{(1-mp)^4}. \tag{5.80}$$

The distribution has received much attention; see for instance Gupta (1974) and Kumar and Consul (1979). Normal and inverse Gaussian limiting forms were examined by Consul and Shenton (1973). Characterization theorems appear in Jain and Consul (1971), Consul (1974), and Consul and Gupta (1980). The negative moments have been investigated by Kumar and Consul (1979). Jain and Consul (1971) fitted the distribution to data using estimation by the method of moments, while minimum variance unbiased estimation has been investigated by Kumar and Consul (1980) and Consul and Famoye (1989). A stochastic urn model for the distribution was devised by Famoye and Consul (1989). The difference of two such distributions is the subject of Consul (1989). In Chapter 2, Section 5.3, we commented that the distribution belongs to Charalambides' (1986a) family of Gould distributions.

12.4 Weighted Negative Binomial Distributions

The theory of *weighted negative binomial distributions* is analogous to that for weighted binomial distributions (see Chapter 3, Section 12.4).

An interesting variant is that of Bissell (1972a, b); see also Scheaffer and Leavenworth (1976). These authors studied estimation for the negative binomial distribution when the sampling units vary in size; they were interested in counts of flaws in strips of cloth where the strips are of varying length. This required a form of weighting according to the size of the sampling unit. Bissell showed that the way that the weighting factor enters into the pmf depends on the appropriate negative binomial model; it is not the same for the gamma mixture of Poissons model as for the model corresponding to a Poisson-stopped sum of logarithmic variables.

12.5 Convolutions Involving Negative Binomial Variates

Convolutions of distributions were defined in Chapter 1, Section B9. The convolution of a Poisson with a negative binomial distribution (the *Poisson*negative binomial distribution*) appeared in a little-known paper by Lüders (1934) as his *Formel II distribution*.

We consider first Lüders' Formel I distribution. He obtained this as a convolution of Poisson singlet, doublet, triplet, etc., variables, with parameters

$$\lambda_1 = \frac{\beta}{1+\gamma} \ , \ \lambda_2 = \frac{\beta\gamma}{2(1+\gamma)^2} \ , \ \lambda_3 = \frac{\beta\gamma^2}{3(1+\gamma)^3} \ , \ \dots \ , \ \lambda_r = \frac{\beta\gamma^{r-1}}{r(1+\gamma)^r} \ , \ \dots,$$

respectively. The pgf for this convolution is then

$$G(z) = \exp\left[\frac{\beta}{\gamma}\sum_{r\geq 1}\left(\frac{\gamma}{1+\gamma}\right)^r\frac{(z^r-1)}{r}\right] = (1+\gamma-\gamma z)^{-\beta/\gamma}. \qquad (5.81)$$

Lüders' Formel I distribution is therefore a negative binomial distribution.

Lüders Formel II distribution is the outcome when the Poisson singlet, doublet, triplet, etc. parameters are

$$\lambda_1 = \lambda \ , \ \lambda_2 = \frac{b}{2} \ , \ \lambda_3 = \frac{bq}{3} \ , \ \dots \ , \ \lambda_r = \frac{bq^{r-2}}{r} \ , \ \dots,$$

respectively. This is a three-parameter distribution. Lüders derived a general formula for the probabilities and expressions for the first three moments; he used the latter to fit the distribution by the method of moments to the well-known haemocytometer data of "Student" (1907).

The pgf can be shown to be

$$G(z) = \exp\left[\left(\lambda-\frac{b}{q}\right)(z-1)+\frac{b}{q^2}\sum_{r\geq 1}\frac{q^r(z^r-1)}{r}\right]$$

$$= e^{(\lambda-b/q)(z-1)}(1-q)^{b/q^2}(1-qz)^{-b/q^2}, \qquad (5.82)$$

that is, the convolution of a Poisson pgf with a negative binomial pgf. Using the convolution property the pmf is found to be

$$\Pr[X = x] = \frac{e^{-\theta} \theta^x (1 - q)^k}{(k - 1)! x!} \sum_{j=0}^{x} \binom{x}{j} \left(\frac{q}{\theta}\right)^j, \qquad x = 0, 1, \ldots, \qquad (5.83)$$

where $\theta = \lambda - b/q$ and $k = b/q^2$. The cumulants are the sums of Poisson and negative binomial cumulants.

We note that the *Poisson*binomial distribution* (a convolution of a Poisson and a binomial distribution) arises in stochastic process theory as the equilibrium solution for a simple immigration-death process [Cox and Miller (1965, p. 168)]; this mode of genesis can be reinterpreted as the equilibrium solution for Palm's trunking problem concerning a telephone exchange with an infinite number of channels (an M/M/∞ queue) [Feller (1957, pp. 414, 435)]. The properties of the distribution do not seem to have received attention per se; however, they are analogous to those for the Formel II distribution.

The Poisson*negative binomial distribution is known in the actuarial literature as the *Delaporte distribution*, where it has arisen as a three-parameter mixture of Poisson distributions; see Chapter 8, Section 3.2. It has been put forward as an alternative to the more usual assumption of a two-parameter gamma mixture (i.e., the negative binomial distribution) in the theory of insurance claims; see Delaporte (1959) and also Willmot (1989) and Willmot and Sundt (1989) who have provided useful bibliographies.

The mixing distribution is assumed to have the density function

$$f(\lambda) = \frac{\beta^\alpha}{\Gamma(\alpha)} (\lambda - \gamma)^{\alpha - 1} e^{\beta(\gamma - \lambda)}, \qquad \alpha, \beta > 0, \ \lambda > \gamma \geq 0, \qquad (5.84)$$

yielding a mixed Poisson distribution with pgf

$$G(z) = \int_{\gamma}^{\infty} e^{\lambda(z-1)} f(\lambda) d\lambda$$

$$= e^{\gamma(z-1)} \left(\frac{\beta + 1}{\beta} - \frac{z}{\beta}\right)^{-\alpha}. \qquad (5.85)$$

Willmot (1989) has examined its tail behavior; certain asymptotic results are given in Willmot and Sundt (1989).

Ong and Lee(1979) derived their *noncentral negative binomial distribution* as a mixture of negative binomial variables with pmf $\binom{t+x-1}{x}(1 - a)^t a^x$, $x = 0, 1, \ldots$, where T is a rv, $T = Y + v$, v is constant, and Y is a Poisson rv with parameter λ. The resultant mixture has the pmf

$$\Pr[X = x] = \sum_{y=0}^{\infty} \frac{e^{-\lambda} \lambda^y}{y!} \binom{y + v + x - 1}{x} (1 - a)^{y+v} a^x$$

$$= e^{-\lambda}(1-a)^v a^x \binom{v+x-1}{x} {}_1F_1[v+x; v; \lambda(1-a)] \quad (5.86)$$

and pgf

$$G(z) = \left(\frac{1-a}{1-az}\right)^v \exp\left\{\lambda\left(\frac{1-a}{1-az}-1\right)\right\}; \quad (5.87)$$

this is the pgf for the convolution of a negative binomial rv with a Pólya-Aeppli rv (see Chapter 9, Section 7, for the Pólya-Aeppli distribution). The cumulants of the distribution are therefore the sums of the component Poisson and Pólya-Aeppli cumulants.

Ong and Lee (1979) showed that the distribution arises also as a mixture of Poisson distributions, using a continuous mixing distribution that involves a Bessel function and is related to the noncentral chi-square distribution of Chapter 28. This mode of genesis has applications to neural counting mechanisms [see McGill (1967)]. It also has applications to photon counting [see Teich and McGill (1976)]. Ong and Lee gave a useful recurrence relation for the probabilities. They also obtained a characterization and demonstrated various estimation procedures.

Ong and Lee's (1986) generalization of their noncentral negative binomial distribution is the convolution of two binomial-type distributions; see Kemp (1979) and Chapter 3, Section 12.5. Ong and Lee obtained formulations via mixing processes and gave examples of fits of this distribution to data using moment estimation.

12.6 Pascal-Poisson Distribution

The *Pascal-Poisson distribution* was first derived as a negative binomial-stopped sum of Poisson distributions (the *negative binomial$\bigvee$Poisson distribution*) by Kocherlakota Subrahmaniam (1966) as a limiting form of the more general contagious distribution arising from a larvae survival model that is described in Chapter 9, Section 9. (Stopped-sum distributions are defined in Chapter 9, Section 1.)

The pgf of the Pascal-Poisson distribution is

$$G(z) = {}_1F_0[a;\ ; \frac{\mu}{a\phi}\{\exp(\phi(z-1))-1\}]$$

$$= \left[1 + \frac{\mu}{a\phi} - \frac{\mu}{a\phi}e^{\phi(z-1)}\right]^{-a}, \quad (5.88)$$

where a, ϕ and μ are all positive. The form of the pgf (5.88) shows, by Gurland's theorem (Chapter 8, Section 3.1), that the Pascal-Poisson distribution is both a negative binomial-stopped sum of Poisson variables and also a Poisson mixture of negative binomial distributions. It is a three-parameter distribution; it was re-examined in a paper by Kathleen Subrahmaniam (1978).

Subrahmaniam (1966) gave a recursion formula for the probabilities and showed that the mean and variance are

$$\mu \quad \text{and} \quad \mu_2 = \mu \left(1 + \phi + \frac{\mu}{a} \right). \tag{5.89}$$

He fitted the distribution to Beall and Rescia's (1953) data sets by an ad hoc method with $a = 1, 2$.

Subrahmaniam (1978) set $\phi = c$ and obtained the following formulas for the probabilities:

$$\Pr[X = 0] = \left[1 - \frac{\mu}{ac} \{ \exp(-c) - 1 \} \right]^{-a}, \tag{5.90}$$

$$\Pr[X = x] = \left(\frac{ac}{ac + \mu} \right)^a \frac{c^x}{\Gamma(a)\Gamma(x+1)} \sum_{j=1}^{\infty} \frac{\Gamma(a+j)}{\Gamma(j+1)} \left(\frac{\mu}{ac + \mu} \right)^j j^x e^{-jc}, \tag{5.91}$$

for $x = 1, 2, \ldots.$ She concentrated on estimation for the distribution assuming that a is known, and gave moment and also ML estimators for μ and c; she derived expressions for their variances. The relative efficiency of moment estimation was found to decrease as μ increases (a and c fixed), and to decrease as c increases (μ and a fixed). Maximum likelihood estimation proved much superior. The Pascal-Poisson distribution gave a somewhat better fit than a negative binomial to data on molecular evolutionary events in a study on amino acids.

12.7 A Riff-Shuffle Distribution

Consider two packs of cards, A and B, each containing m cards. At each trial a card is taken independently with probability p from pack A and probability $q = 1 - p$ from pack B. Let X be the rv that represents the number of cards chosen from one pack when all the m cards of the other pack have been chosen. That is to say, trials continue until one of the two packs is exhausted; the number of cards remaining in the other pack is then $m - X$.

Uppuluri and Blot (1970) [see also Lingappaiah (1987)] showed that the appropriate pmf is

$$\Pr[X = x] = \binom{m + x - 1}{x} (p^m q^x + q^m p^x), \qquad x = 0, 1, \ldots, m. \tag{5.92}$$

(We have $\sum_{x=0}^{m} \Pr[X = x] = 1$, since

$$\sum_{x=0}^{m} \binom{m + x - 1}{x} p^m q^x = I_p(m, m),$$

where $I_p(m, m)$ is an incomplete beta function, and $I_p(m, m) + I_q(m, m) = 1$.)
The distribution can be regarded as a mixture of two tail-truncated negative
binomial distributions with parameters (m, p) and (m, q), where $p + q = 1$.

From Uppuluri and Blot's graphical analysis of the distribution it would
appear to be unimodal. It is very skew with a long tail to the left when
$p = 0.5$, nearly symmetrical when $0.3 < p < 0.4$ and when $0.6 < p < 0.7$, and
becomes very skew with a long tail to the right as p approaches 0 or 1 (the
properties of the distribution are symmetric in p and $q = 1 - p$).

Uppuluri and Blot gave expressions for the moment generating function,
and for the mean and variance in terms of incomplete beta functions; they
found moreover that as m becomes large,

$$\mu \approx m - 2\sqrt{\frac{m}{\pi}} \qquad \text{when } p = 0.5 = q,$$

$$\approx \frac{mq}{p} \qquad \text{when } p \neq 0.5, \tag{5.93}$$

$$\mu_2 \approx 2m - \frac{4m}{\pi} - 2\sqrt{\frac{m}{\pi}} \qquad \text{when } p = 0.5 = q,$$

$$\approx \frac{mq}{p^2} \qquad \text{when } p \neq 0.5. \tag{5.94}$$

They showed also that as m becomes large $(X - mq/p)/(mq/p^2)^{1/2}$ tends
to normality, and they considered applications to baseball series, the Banach
matchbox problem, and to the genetic code problem.

Lingappaiah (1987) has investigated Bayesian estimation of the parameter
p.

CHAPTER 6

Hypergeometric Distributions

1 DEFINITION

Consider the classical situation in which a hypergeometric distribution arises. Suppose that an urn contains N balls, of which Np are white and $N - Np$ are black. If a sample of n balls were to be drawn at random from the urn with each drawn ball being replaced immediately, then the probability of drawing a white ball would be constant $(= p)$ and the distribution of the number X of white balls drawn would be binomial with parameters n, p. However, if a sample of n balls is drawn at random from an urn with Np white balls and $N - Np$ black balls,*without replacing any balls in the urn at any stage*, then the probability that the number X of white balls in the sample of n balls is equal to x is

$$\Pr[X = x] = \binom{Np}{x}\binom{N - Np}{n - x} \Big/ \binom{N}{n} \tag{6.1}$$

$$= \binom{n}{x}\binom{N - n}{Np - x} \Big/ \binom{N}{Np}, \tag{6.2}$$

where $\max(0, n - N + Np) \leq x \leq \min(n, Np)$. This defines a *hypergeometric distribution* with parameters n, Np, N, all positive; the re-expression of (6.1) as (6.2) shows that the distribution is unaltered when n and Np are interchanged. Note that the pmf of a hypergeometric-type distribution can always be manipulated into the form $\binom{a}{b}\binom{c}{d}/\binom{a+c}{b+d}$.

The reason for the name "hypergeometric" is that the probabilities are the successive terms in the expansion of

$$\frac{(N - n)!(N - Np)!}{N!(N - Np - n)!} \, {}_2F_1[-n, -Np; N - Np - n + 1; 1],$$

where

$$2F_1[\alpha, \beta; \gamma; z] = 1 + \frac{\alpha\beta}{\gamma} \cdot \frac{z}{1!} + \frac{\alpha(\alpha + 1)\beta(\beta + 1)}{\gamma(\gamma + 1)} \cdot \frac{z^2}{2!} + \cdots$$

is a Gaussian hypergeometric series (Chapter 1, Section A6). The pgf for the classical hypergeometric distribution is therefore

$$G(z) = \frac{2F_1[-n, -Np; N - Np - n + 1; z]}{2F_1[-n, -Np; N - Np - n + 1; 1]}; \tag{6.3}$$

this can be restated as

$$G(z) = 2F_1[-n, -Np; -N; 1 - z], \tag{6.4}$$

using a result from the theory of terminating Gaussian hypergeometric series. The cf is

$$\frac{2F_1[-n, -Np; N - Np - n + 1; e^{it}]}{2F_1[-n, -Np; N - Np - n + 1; 1]}, \tag{6.5}$$

and the mean and variance are

$$\mu = \mathrm{E}[X] = np \quad \text{and} \quad \mu_2 = \mathrm{Var}(X) = np(1 - p)(N - n)/(N - 1), \tag{6.6}$$

(further moment formulas are given in Section 3 of this chapter).

The classical hypergeometric distribution, like the negative (inverse) hypergeometric (i.e., the beta-binomial) and also the beta-negative-binomial distribution (Sections 2.2 and 2.3 of this chapter, respectively), is a member of Ord's (1967a) difference equation system of discrete distributions (see Section 2.4 of this chapter, also Chapter 2, Section 3.2). However, none of the three hypergeometric-type distributions is a power series distribution (Chapter 2, Section 2).

From (6.3), the classical hypergeometric distribution is a Kemp generalized hypergeometric probability distribution (GHPD) (Chapter 2, Section 4.1); (6.4) shows that it is also a Kemp generalized hypergeometric factorial moment distribution (GHFD) (Chapter 2, Section 4.2).

2 HISTORICAL REMARKS AND GENESIS

2.1 Classical Hypergeometric Distribution

De Moivre's (1711, p. 236) solution to a generalization of Huyghen's fourth proposed problem gave the probability of drawing x white and $c - x$ black counters from a collection containing a white and b black counters (without replacements) as the ratio of

$$\frac{b}{1} \times \frac{b-1}{2} \times \cdots \times \frac{b-c+x+1}{c-x} \times \frac{a}{1} \times \frac{a-1}{x} \times \cdots \times \frac{a-x+1}{x}$$

to

$$\frac{a+b}{1} \times \frac{a+b-1}{2} \times \cdots \times \frac{a+b-c+1}{c};$$

this gives the hypergeometric probability mass function with $a = Np$, $b = N - Np$, and $c = n$. The multivariate hypergeometric pmf was obtained by Simpson in 1740 [see Todhunter (1865, p. 206)], but little attention was given to the univariate distribution until Cournot (1843, pp. 43, 68, 69) applied it to matters concerning conscription, absent parliamentary representatives, and the selection of deputations and juries.

The properties of the distribution were investigated in depth by Pearson (1895, 1899, 1924), who was interested in developing the system of continuous distributions that now bears his name via limiting forms of discrete distributions. Important further properties of the hypergeometric distribution were obtained by Romanovsky (1925).

2.2 Negative (Inverse) Hypergeometric Distribution: Hypergeometric Waiting-Time Distribution, Beta-Binomial Distribution

Consider now the following result that seems to have been derived for the first time by Condorcet in 1785 [see Todhunter (1865, p. 383)].

Let A and B be two mutually exclusive events that have already occurred v and w times respectively in $v+w$ trials. Let $n = p+q$. Then the probability that in the next n trials events A and B will happen p and q times respectively is

$$\frac{(p+q)!}{p!q!} \int_0^1 x^{v+p}(1-x)^{w+q} dx \Bigg/ \int_0^1 x^v(1-x)^w dx$$

$$= \frac{(p+q)!(v+p)!(w+q)!(v+w+1)!}{p!q!(v+p+w+q+1)!v!w!}$$

$$= \frac{n!(-v-w-n-2)!(-v-1)!(-w-1)!}{p!(n-p)!(-v-1-p)!(-w-n-1+p)!(-v-w-2)!}$$

$$= \binom{n}{p}\binom{-v-w-n-2}{-v-1-p} \Bigg/ \binom{-v-w-2}{-v-1}, \tag{6.7}$$

$$= \frac{(-v-1)!(-w-1)!n!(-v-w-2-n)!}{p!(-v-1-p)!(n-p)!(-w-1-n+p)!(-v-w-2)!}$$

$$= \binom{-v-1}{p}\binom{-w-1}{n-p} \Big/ \binom{-v-w-2}{n}, \qquad p = 0,1,\dots,n; \quad (6.8)$$

see also Pearson (1907).

This is a hypergeometric distribution with parameters $n = n$, $Np = -v-1$, and $N = -v-w-2$. Because N is negative, it is known as the *negative hyper-geometric distribution*. It is also known (for reasons that will become apparent) as the *inverse hypergeometric distribution* or *hypergeometric waiting-time distribution*, as the *beta-binomial distribution*, and as the *Markov-Pólya distribution*.

Its pgf is

$$G(z) = \frac{{}_2F_1[-n,v+1;-w-n;z]}{{}_2F_1[-n,v+1;-w-n;1]} \qquad (6.9)$$

$$= {}_2F_1[-n,v+1;v+w+2;1-z], \qquad (6.10)$$

and the mean and variance are

$$\mu = \frac{n(v+1)}{v+w+2},$$

$$\mu_2 = \frac{n(v+1)(w+1)(v+w+n+2)}{(v+w+2)^2(v+w+3)}. \qquad (6.11)$$

(Higher moments are discussed in Section 3 of this chapter.)

Condorcet's derivation assumes that sampling takes place from an infinite population. Prevost and Lhuilier in 1799 recognized that an equivalent expression is obtained when two samples are taken in succession from a finite population, without any replacements; see Todhunter (1865, p. 454).

The derivation of the *distribution of the number of exceedances* is mathematically similar to Condorcet's result; see Gumbel and von Schelling (1950) and Sarkadi (1957b). Gumbel (1958) credits this to Thomas (1948). Consider two independent random samples of sizes m and n, drawn from a population in which a measured character has a continuous distribution. The number of *exceedances* X is defined as the number (out of n) of the observed values in the second sample that exceed the rth largest of the m values in the first sample. The probability that in n future trials there will be x values exceeding the rth largest value in m past trials is

$$\Pr[X=x] = \binom{m}{r} r \binom{n}{x}(m+n)^{-1} \Big/ \binom{m+n-1}{r+x-1}. \qquad (6.12)$$

This can be manipulated in the same way as (6.7) and (6.8) to give

$$\Pr[X=x] = \binom{n}{x}\binom{-m-n-1}{-r-x}\bigg/\binom{-m-1}{-r}$$
$$= \binom{-r}{x}\binom{-m+r-1}{n-x}\bigg/\binom{-m-1}{n}, \qquad x = 0, 1, \ldots, n. \quad (6.13)$$

The pgf is

$$G(z) = \frac{{}_2F_1[-n,r;-n-m+r;z]}{{}_2F_1[-n,r;-n-m+r;1]}. \qquad (6.14)$$

Irwin (1954) pointed out that sampling for a sample of fixed size n from an urn with Np white balls and $(N - Np)$ black balls as in Section 1, but with replacement together with an additional similarly colored ball after each ball is drawn, also gives rise to the negative hypergeometric distribution. This is a particular case of Pólya-type sampling (see Section 2.4 of this chapter).

Suppose now that sampling without replacement (as described in Section 1 of this chapter) is continued until k white balls are obtained ($0 < k \leq Np$). The distribution of the number of draws that are required is known as the *inverse hypergeometric distribution* or *hypergeometric waiting-time distribution*. (The model is analogous to the inverse binomial sampling model for the negative binomial distribution; however, the range of possible values for a negative hypergeometric distribution is finite because there is not an infinitude of black balls that might be drawn.) For this distribution

$$\Pr[X=x] = \binom{Np}{k-1}(Np-k+1)\binom{N-Np}{x-k}(N-x+1)^{-1}\bigg/\binom{N}{x-1} \quad (6.15)$$
$$= \binom{x-1}{x-k}\binom{N-x}{N-Np-x+k}\bigg/\binom{N}{N-Np}$$
$$= \binom{N-Np}{x-k}\binom{-N-1}{-x}\bigg/\binom{-Np-1}{-k}$$
$$= \binom{-k}{x-k}\binom{k-Np-1}{N-Np-x+k}\bigg/\binom{-Np-1}{N-Np}, \qquad (6.16)$$

where $x = k, k+1, \ldots, k+N-Np$, by manipulation of the factorials. Comparison with (6.7) shows that this is a negative hypergeometric distribution shifted k units away from the origin. The term "inverse hypergeometric distribution" can refer either to the total number of draws, as above, or to the number of unsuccessful draws, as in Kemp and Kemp (1956a) and Sarkadi (1957a). Bol'shev (1964) related the inverse sampling model to a two-dimensional random walk.

Yet another derivation, closely related mathematically to Condorcet's, gives the distribution as a mixture of binomial distributions, with the binomial parameter p having a beta distribution:

$$\Pr[X = x] = \int_0^1 \frac{n!}{x!(n-x)!} p^x (1-p)^{n-x} \times \frac{p^{\alpha-1}(1-p)^{\beta-1}dp}{B(\alpha,\beta)} \quad (6.17)$$

$$= \binom{n}{x}\binom{-\alpha-\beta-n}{-\alpha-x} \Big/ \binom{-\alpha-\beta}{-\alpha}$$

$$= \binom{-\alpha}{x}\binom{-\beta}{n-x} \Big/ \binom{-\alpha-\beta}{n}, \quad (6.18)$$

where $x = 0, 1, \ldots, n$. The pgf is

$$G(z)\frac{{}_2F_1[-n,\alpha;-\beta-n+1;z]}{{}_2F_1[-n,\alpha;-\beta-n+1;1]}. \quad (6.19)$$

This model has been obtained, and subsequently applied, in many different fields by a number of research workers (see Section 9.2 of this chapter). When $\alpha = \beta = 1$ the outcome is the discrete rectangular distribution (see Section 10.1 of this chapter).

Guenther (1975) has written a helpful review paper concerning the negative (inverse) hypergeometric distribution.

2.3 Beta-Negative-Binomial Distribution: Beta-Pascal Distribution, Generalized Waring Distribution

The *beta-negative-binomial distribution* was obtained analogously to the beta-binomial distribution by Kemp and Kemp (1956a), who commented that it arises both as a beta mixture of negative binomial distributions with the pgf $(1-\lambda)^k/(1-\lambda z)^k$ and as an F-distribution mixture of the negative binomial distribution with pgf $(1 + P - Pz)^{-k}$. We have

$$G(z) = \int_0^1 \left(\frac{1-\lambda}{1-\lambda z}\right)^k \times \frac{\lambda^{\ell-1}(1-\lambda)^{m-1}d\lambda}{B(\ell,m)}$$

$$= \int_0^1 (1 + P - Pz)^{-k} \times \frac{P^{l-1}(1+P)^{-\ell-m}dP}{B(\ell,m)}$$

$$= \frac{{}_2F_1[k,\ell;k+\ell+m;z]}{{}_2F_1[k,\ell;k+\ell+m;1]}. \quad (6.20)$$

The probabilities are

$$\Pr[X = x] = \binom{-k}{x}\binom{m+k-1}{-\ell-x} \Big/ \binom{m-1}{-\ell}$$

$$= \binom{-\ell}{x}\binom{\ell + m - 1}{-k - x} \bigg/ \binom{m - 1}{-k}, \qquad x = 0, 1, \ldots . \quad (6.21)$$

The mean and variance are

$$\mu = \frac{k\ell}{m - 1},$$

$$\mu_2 = \frac{k\ell(m + k - 1)(m + \ell - 1)}{(m - 1)^2(m - 2)}. \qquad (6.22)$$

The moments exist only for $r < m$, however; see Section 3 of this chapter.

Another name for the distribution is the *inverse Markov-Pólya distribution*. Kemp and Kemp (1956a) pointed out that this distribution also arises by inverse sampling from a Pólya urn *with additional replacements*; see the next subsection.

Unlike the classical hypergeometric and the negative hypergeometric distributions the support of the beta-negative-binomial-distribution is infinite. Also, unlike those distributions, the beta-negative-binomial is not a Kemp GHF distribution. It is, however, a Kemp GHP distribution,

We note in passing that a beta mixture of negative binomial distributions with pgf $(1 + P - Pz)^{-k}$ gives rise to a distribution with a much more complicated expression for its pmf; its pgf is

$$H(z) = \int_0^1 (1 + P - Pz)^{-k} \times \frac{P^{\ell-1}(1 - P)^{m-1}dP}{B(\ell, m)}$$

$$= {}_2F_1[k, \ell; \ell + m; z - 1], \qquad (6.23)$$

and hence it is a Kemp GHF distribution although it is not a Kemp GHP distribution; see Kemp and Kemp (1975) and Chapter 2, Section 4.2. It does not belong to the class of Pólya distributions discussed in the next section.

The term "Beta-Pascal" is often applied to a shifted form of the distribution (6.20) with support $k, k + 1, \ldots$; see e.g. Raiffer and Schlaifer (1961, pp. 238, 270), Dubey (1966a). Here k is necessarily an integer.

The unshifted distribution (6.20) (with support $0, 1, \ldots$) has been studied in considerable detail by Irwin (1963, 1968, 1975a, b, c) and Xekalaki (1981, 1983a, b, c, d) under the name *generalized Waring distribution*. Irwin (1963) developed it from the following generalization of Waring's expansion (see Section 10.4 of this chapter):

$$\frac{(c - a - 1)!}{(c - a + k - 1)!} = \frac{(c - 1)!}{(c + k - 1)!} \left[1 + \frac{ak}{(c + k)} + \frac{a(a + 1)k(k + 1)}{(c + k)(c + k + 1)2!} + \cdots \right]$$

$$= \frac{(c - 1)!}{(c + k - 1)!} {}_2F_1[a, k; c + k; 1].$$

Irwin's procedure of setting $\Pr[X = x]$ proportional to the $(x + 1)$th term in this series is equivalent (as he realized) to adopting the pgf

$$G(z) = \frac{{}_2F_1[a, k; c + k; z]}{{}_2F_1[a, k; c + k; 1]}, \tag{6.24}$$

where $k = -n$, $a = -Np$ and $c + k = N - Np - n + 1$ gives the classical hypergeometric distribution. But note that Irwin's restrictions on the parameters are $c > a > 0, k > 0$.

Irwin used the factorial moments

$$\mu'_{[r]} = \frac{(a + r - 1)!(k + r - 1)!(c - a - r - 1)!}{(a - 1)!(k - 1)!(c - a - 1)!} \tag{6.25}$$

(which can be obtained by successive differentiation of the pgf; see Chapter 1, Section B9) to obtain

$$\mu = \frac{ak}{c - a - 1},$$

$$\mu_2 = \frac{ak(c - a + k - 1)(c - 1)}{(c - a - 1)^2(c - a - 2)} \tag{6.26}$$

(cf. (6.6), (6.11), and (6.22)).

Further properties of the distribution, relationships to its Pearson-type continuous analogues, tail-length behavior, and parameter estimation are the subjects of Irwin (1968, 1975a, b, c).

Xekalaki (1981) has written an anthology of results concerning models, and some related characterizations, for the generalized Waring distribution. She has studied urn models, mixture models, conditionality models, and STER models. In Xekalaki (1983a) she studied infinite divisibility, completeness, and regression properties of the distribution. In Xekalaki (1985) she showed that the distribution can be determined uniquely from a knowledge of certain conditional distributions and some appropriately chosen regression functions. Applications of the distribution are given in Section 9.3.

Special cases of the distribution are the Yule and Waring distributions; see Sections 10.3 and 10.4.

2.4 Pólya Distributions: Generalized Hypergeometric Distributions

The urn models described earlier in this chapter are all particular cases of the Pólya urn model. This was put forward by Eggenberger and Pólya (1923) as a model for contagious distributions, that is, for situations where the occurrence of an event has an aftereffect; see also Jordan (1927) and Eggenberger and Pólya (1928).

Suppose that a finite urn initially contains w white balls and b black balls, and that balls are withdrawn one at a time, with immediate replacement together with c balls of a similar color. Then the probability that x white balls are drawn in a sample of n withdrawals is

$$
\begin{aligned}
\Pr[X = x] &= \binom{n}{x} \frac{w(w+c)\ldots\{w+(x-1)c\}b(b+c)\ldots\{b+(n-x-1)c\}}{(w+b)(w+b+c)\ldots\{w+b+(n-1)c\}} \\
&= \binom{n}{x} \frac{B(x+w/c, n-x+b/c)}{B(w/c, b/c)} \\
&= \binom{n}{x} \binom{-n-(w+b)/c}{-x-w/c} \Big/ \binom{-(w+b)/c}{-w/c} \\
&= \binom{-w/c}{x}\binom{-b/c}{n-x} \Big/ \binom{-(w+b)/c}{n} .
\end{aligned} \tag{6.27}
$$

Other ways of expressing the probabilities are discussed in Bosch (1963).
    The pgf is

$$
G(z) = \frac{{}_2F_1[-n, w/c; -n+1-b/c; z]}{{}_2F_1[-n, w/c; -n+1-b/c; 1]}. \tag{6.28}
$$

Pólya (1930) pointed out the following particular cases: if c is positive, then success and failure are both contagious; if $c = 0$, then events are independent (the classical binomial situation); while if c is negative, then each withdrawal creates a reversal of fortune. When c is negative such that w/c is a negative integer (e.g. $c = -1$), the outcome is the classical hypergeometric distribution, whereas when c is positive such that w/c is a positive integer (e.g. $c = +1$), the negative hypergeometric distribution is the result. Inverse sampling for a fixed number of white balls leads to the inverse (negative) hypergeometric when w/c is a negative integer, and to a beta-negative-binomial distribution when w/c is a positive integer.
    In

$$
\Pr[X = x] = \binom{a}{x}\binom{b}{n-x} \Big/ \binom{a+b}{n} \tag{6.29}
$$

it is clearly not essential that all the parameters n, a, b are positive; in fact with certain restrictions we can take any two of them negative and the remaining one positive, and we still obtain a probability mass function. The conditions under which (6.29) provides an honest distribution, with n, a and b taking real values, have been investigated by Davies (1933, 1934), Noack (1950), and Kemp and Kemp (1956a). Such distributions were termed *generalized hypergeometric distributions* by Kemp and Kemp. They have pgf's of the form

$$
G(z) = \frac{{}_2F_1[-n, -a; b-n+1; z]}{{}_2F_1[-n, -a; b-n+1; 1]}, \tag{6.30}
$$

Table 6.1 Conditions for the Existence of Type I, II, III, and IV Generalized Hypergeometric Distributions

Type	Conditions	Support
Type IA(i)	$n - b - 1 < 0$; n an integer; $0 \le n - 1 < a$	$x = 0, 1, \ldots, n$
Type IA(ii)	$n - b - 1 < 0$; a an integer; $0 \le a - 1 < n$	$x = 0, 1, \ldots, a$
Type IB	$n - b - 1 < 0$; $J < a < J+1$; $J < n < J+1$	$x = 0, 1, \ldots$
Type IIA	$a < 0 < n$; n an integer; $b < 0$; $b \ne -1$	$x = 0, 1, \ldots, n$
Type IIB	$a < 0 < a + b + 1$; $J < n < J+1$; $J < n - b - 1 < J+1$	$x = 0, 1, \ldots$
Type IIIA	$n < 0 < a$; a an integer; $b < n - a$; $b \ne n - a - 1$;	$x = 0, 1, \ldots, a$
Type IIIB	$n < 0 < a + b + 1$; $J < a < J+1$; $J < n - b - 1 < J+1$	$x = 0, 1, \ldots$
Type IV	$a < 0$; $n < 0$; $0 < a + b + 1$	$x = 0, 1, \ldots$

Note: J denotes a nonnegative integer (the same integer for any one type of distribution.)

and form a subset of Kemp's (1968a, b) wider class of GHP distributions with pgf's of the form

$$G(z) = \frac{{}_pF_q[a_1, a_2, \ldots, a_p; b_1, b_2, \ldots, b_q; \lambda z]}{{}_pF_q[a_1, a_2, \ldots, a_p; b_1, b_2, \ldots, b_q; \lambda]};$$

see Chapter 2, Section 4.1.

Among the class of distributions with pmf's of the form (6.29) there will be some for which $\Pr[X = x] = 0$ if x exceeds some integer r. This will be so if a or n is a positive integer. Kemp and Kemp introduced the convention that $\Pr[X = x] = 0$ for all $x \ge r + 1$ if $\Pr[X = r + 1] = 0$. They used the definition

$$\frac{\alpha!}{(\alpha + \beta)!} = (-1)^\beta (-\alpha)(-\alpha - 1) \ldots (-\alpha - \beta - 1)$$

$$= \frac{(-1)^\beta (-\alpha - \beta - 1)!}{(-\alpha - 1)!} \qquad (6.31)$$

when $\alpha < 0$ and $\beta < 0$ with β an integer; see Chapter 1, Section A1. Moreover, when both α and β are negative integers

$$\frac{\alpha!}{(\alpha + \beta)!} = \lim_{\varepsilon \to 0} \frac{\Gamma(\alpha + \varepsilon + 1)}{\Gamma(\alpha + \varepsilon + \beta + 1)} = \frac{(-1)^\beta (-\alpha - \beta - 1)!}{(-\alpha - 1)!}$$

as before. They also imposed the restriction $\Pr[X = 0] \ne 0$ in order to ease the derivation of the moment and other properties of the distributions.

With these conventions, they distinguished four main types of distribution corresponding to (6.29), divided into subtypes as in Table 6.1.

The classical hypergeometric distribution belongs to Type IA(i) or IA(ii) with n, a and b all integers. The negative (inverse) hypergeometric distribution belongs to Type IIA or IIIA with n, a and b again all integers. A sort of

Table 6.2   Relations between Types of Hypergeometric Distributions

Kemp and Kemp	Name	Support	Ord	Shimizu
IA(i), IA(ii)	Classical hypergeometric	Finite	I(a)	A1
IIA, IIIA	Negative (inverse) hypergeometric (Beta-binomial)	Finite	I(b)	A2
IB	—	Infinite	I(e)	B1
IIB, IIIB	—	Infinite	I(e)	B2
IV	Beta-negative-binomial	Infinite	VI	B3

dualism can be noted between the following types: Types IA(i) and IA(ii); Types IIA and IIIA; and Types IIB and IIIB (making the substitutions $a \leftrightarrow n$ and $a + b - n \leftrightarrow b$). No meaningful models have been obtained for Types IB, IIB or IIIB. The beta-negative-binomial is Type IV.

Sarkadi (1957a) extended the class of distributions corresponding to (6.29) by including the cases $b = -1$, $b = n - a - 1$ that were excluded from Types IIA and IIIA, respectively, by Kemp and Kemp. He pointed out that the sum of the probabilities over the ranges $0 \leq x \leq n$ and $0 \leq x \leq a$, respectively, is equal to unity in both cases, and so (6.29) defines a proper distribution. By changing Kemp and Kemp's definition of $\alpha!/(\alpha + \beta)!$, Shimizu (1968) and Sibuya and Shimizu (1981) were able to include further distributions with support $[m_1, m_2]$, $[m_1, \infty)$, $[-m_2, -m_1]$, $(-\infty, -m_1]$, where m_1 and m_2 are positive integers. Their new types are distributions of the form $\pm X \pm k$, where X is a rv of one of the types in Table 6.1, and k is an integer. Table 6.2 gives the broad relationships between Kemp and Kemp (1956a), Ord (1967b), and Shimizu (1968) hypergeometric-type distributions.

Review articles concerning generalized hypergeometric distributions are those of Guenther (1983) and Sibuya (1983).

Kemp and Kemp's (1975) paper was concerned with models for generalized hypergeometric distributions. Besides urn models and models for contagion, it gave models (1) based on equilibrium stochastic processes, (2) STER models, (3) conditionality models, (4) weighting models, and (5) mixing models. It showed that

1. An equilibrium time-homogeneous stochastic process with birth and death rates λ_i and μ_i, such that $\lambda_{i-1}/\mu_i = (a_1 + i - 1)/(b + i - 1)$, can yield a Type IIA/IIIA or a Type IV distribution, by a suitable choice of parameters. Similarly

$$\lambda_{i-1}/\mu_i = (a_1 + i - 1)(a_2 + i - 1)/\{(b + i - 1)i\}$$

can lead to any one of Types IA, IIA/IIIA, or IV, by a suitable choice of parameters.

2. STER distributions arise in connection with an inventory decision problem. If demand is a discrete rv with pgf $G(z) = \sum_{i \geq 0} p_i z^i$, then the corresponding STER distribution has probabilities that are Sums successively Truncated from the Expectation of the Reciprocal of the demand variable, giving the STER pgf

$$H(z) = (1-z)^{-1}(1-p_0)^{-1} \int_z^1 \frac{\{G(z) - p_0\}dz}{z}; \qquad (6.32)$$

see Bissinger (1965) and Chapter 11, Section 12. Kemp and Kemp found that a Type IIA/IIIA demand distribution with support $1, 2, \ldots, \min(n, a)$ can give rise to a STER distribution that is also Type IIA/IIIA; see also Kemp and Kemp (1969b).

3. Let X and Y be iid discrete rv's. If X and Y are both binomial, with parameters (n, p) and (m, p), then the conditional distribution of $X|X + Y$ is hypergeometric (Type IA). If X and Y are both negative binomial, with parameters (u, λ) and (v, λ), then $X|X + Y$ has a negative hypergeometric (Type IIA/IIB) distribution [Kemp (1968a)]. If

$$G_X(z) = \frac{{}_1F_1[-n; c; -\lambda z]}{{}_1F_1[-n; c; \lambda]}$$

and $G_Y(z) = \exp \lambda(z - 1)$, then the distributions of $X|X+Y$ and $Y|X+ Y$ are both Type IA. Discrete Bessel distributions for X and Y can also lead to a Type IA distribution for $X|X + Y$. In a parallel manner, binomial distributions with parameters (n, p) and $(m, 1 - p)$ for X and Y lead to a Type IA distribution for $X|Y - X$. Kemp and Kemp (1975) gave further models of this kind. See Kemp (1968a) for the general theory.

4. Weighting models give rise to distributions that have been modified by the method of ascertainment. When the weights (sampling chances) w_x are proportional to the the value of the observation (i.e. to x), the distribution with pgf ${}_2F_1[a_1, a_2; b; z]/{}_2F_1[a_1, a_2; b; 1]$ is ascertained as the distribution with pgf

$$G(z) = \frac{z_2 F_1[a_1 + 1, a_2 + 1; b + 1; z]}{{}_2F_1[a_1 + 1, a_2 + 1; b + 1; 1]};$$

if w_x is proportional to $x!/(x - k)!$, then the same initial distribution is ascertained as the distribution with pgf

$$G(z) = \frac{z^k {}_2F_1[a_1 + k, a_2 + k; b + k; z]}{{}_2F_1[a_1 + k, a_2 + k; b + k; 1]}$$

[Kemp (1968a)].

5. Kemp and Kemp pointed out that a Beta mixture of extended beta-binomial distributions can, under certain circumstances, give rise to a beta-binomial distribution; see also Chapter 8, Section 3.4. Two possibilities are as follows:

$$G_1(z) = \int_0^1 {}_2F_1[-n, a; c; y(z-1)] \frac{y^{c-1}(1-y)^{d-c-1} dy}{B(c, d-c)}$$

$$= {}_2F_1[-n, a; d; z-1]; \tag{6.33}$$

$$G_2(z) = \int_0^1 {}_2F_1[-n, d; b; y(z-1)] \frac{y^{c-1}(1-y)^{d-c-1} dy}{B(c, d-c)}$$

$$= {}_2F_1[-n, c; b; z-1]. \tag{6.34}$$

They also commented that a Type IIA/IIIA distribution can be obtained as a Gamma mixture of restricted Laplace-Haag distributions, and they pointed out that a Type IV distribution can be derived as a mixture of Poisson distributions.

3 MOMENTS

The moment properties of the hypergeometric distribution can be obtained from the factorial moments, and indeed exist only when the factorial moments exist. The general form for the rth factorial moment (if it exists) for the distribution with pgf (6.30) is

$$\mu'_{[r]} = \frac{n!a!(a+b-r)!}{(n-r)!(a-r)!(a+b)!}, \tag{6.35}$$

and so

$$_2F_1[-a, -n; -a-b; -t] \tag{6.36}$$

can be treated as the fmgf for the factorial moments.
The moments exist for

Type IA(i)	Always (zero if $r > n$)
Type IA(ii)	Always (zero if $r > a$)
Type IB	When $r < a+b+1$
Type IIA	Always (zero if $r > n$)
Type IIB	Never
Type IIIA	Always (zero if $r > a$)
Type IIIB	Never
Type IV	When $r < a+b+1$.

In other words,

$\mu'_{[r]}$ is finite for all r for Types IA(i), IA(ii), IIA and IIIA,

$\mu'_{[r]}$ is finite for $r < a + b + 1$ for Types IB and IV.

Provided that the specified moment exists, it is straightforward (though tedious) to show via the factorial moments that

$$E[X] = \mu = \frac{na}{a+b},$$

$$\text{Var}(X) = \mu_2 = \frac{nab(a+b-n)}{(a+b)^2(a+b-1)},$$

$$\mu_3 = \frac{\mu_2(b-a)(a+b-2n)}{(a+b)(a+b-2)},$$

$$\mu_4 = \frac{\mu_2}{(a+b-2)(a+b-3)}\left\{(a+b)(a+b+1-6n) + 3ab(n-2)\right.$$

$$\left. + 6n^2 + \frac{3abn(6-n)}{a+b} - \frac{18abn^2}{(a+b)^2}\right\}. \tag{6.37}$$

The moment ratios are

$$\sqrt{\beta_1} = \left[\frac{(a+b-1)}{abn(a+b-n)}\right]^{1/2}\frac{(b-a)(a+b-2n)}{(a+b-2)}, \tag{6.38}$$

$$\beta_2 = \frac{(a+b)^2(a+b-1)}{nab(a+b-n)(a+b-2)(a+b-3)}\left\{(a+b)(a+b+1-6n)\right.$$

$$\left. +3ab(n-2) + 6n^2 + \frac{3abn(6-n)}{a+b} - \frac{18abn^2}{(a+b)^2}\right\} + 3. \tag{6.39}$$

As $a \to \infty$, $b \to \infty$ such that $a/(a+b) = p$ (constant), the moment properties tend to those of the binomial distribution, provided that n is a positive integer. When n is negative and $a/(a+b)$ tends to λ, $\lambda < 0$, the moment properties tend to those of the negative binomial distribution; see Table 6.3 on page 256.

An alternative approach to the moment properties is via the differential equation for the moment generating function. The pgf is $G(z) = K_2F_1[-n, -a; b-n+1; z]$, where K is a normalizing constant. This satisfies

$$\theta(\theta + b - n)G(z) = z(\theta - n)(\theta - a)G(z), \tag{6.40}$$

where θ is the differential operator $z\,d/dz$. The umgf is $G(e^t)$; from the relationship between the θ and $D = d/dz$ operators (Chapter 1, Section A4) it follows that $G(e^t)$ satisfies

$$D(D + b - n)G(e^t) = e^t(D - n)(D - a)G(e^t), \qquad (6.41)$$

where D is the differential operator d/dt. The cmgf is $M(t) = e^{-\mu t}G(e^t)$, and this satisfies

$$(D + \mu)(D + \mu + b - n)M(t) = e^t(D + \mu - n)(D + \mu - a)M(t). \qquad (6.42)$$

Identifying the coefficients of t^0, t^1, t^2 and t^3 in (6.42) gives expressions for the first four central moments that are equivalent to (6.37). Higher moments may be obtained similarly. This is essentially the method of Pearson (1899).

Lessing (1973) has shown that the uncorrected moments can be obtained from the following expression for the umgf:

$$G(e^t) = \frac{(a + b - n)!}{(a + b)!} \frac{\partial^n}{\partial y^n} \left[(1 + ye^t)^a (1 + y)^b \right]_{y=0}. \qquad (6.43)$$

Janardan (1973b) commented that this result is a special case of the expression obtained by Janardan and Patil (1972) for the umgf of the s-dimensional multivariate hypergeometric distribution and that similar results hold for the other hypergeometric-type distributions in Section 2.

The following finite difference relation holds among the central moments $\{\mu_j\}$:

$$(a + b)\mu_{r+1} = \{(1 + E)^r - E^r\}[\mu_2 + \alpha\mu_1 + \beta\mu_0], \qquad (6.44)$$

where E is the displacement operator (i.e., $E^p(\mu_s) \equiv \mu_{s+p}$),

$$\alpha = -a + \frac{n(a - b)}{a + b},$$

$$\beta = \frac{nab(a + b - n)}{(a + b)^2},$$

and $\mu_0 = 1$, $\mu_1 = 0$ [Pearson (1924)]. A somewhat similar relationship holds among the incomplete moments. For more details about incomplete moments, see Ayyanger (1934).

The mean deviation is

$$\nu_1 = E[|X - na/(a + b)|]$$

$$= \frac{2m(b - n + m)}{(a + b)} \binom{a}{m}\binom{b}{n - m} \bigg/ \binom{a + b}{n}, \qquad (6.45)$$

where m is the greatest integer not exceeding $\mu + 1$ [Kamat (1965)].

Matuszewski (1962) and Chahine (1965) have studied the ascending factorial moments of the negative (inverse) hypergeometric distribution.

The classical hypergeometric distribution is Type IA(i)/IA(ii) with n, a and b all positive integers; its factorial moments are finite and nonzero for $r \leq \min(n, a)$, and zero otherwise. We have $n = n$, $a = Np$, $b = N - Np$, and so

$$E[X] = \mu = np,$$

$$\text{Var}(X) = \mu_2 = \frac{np(1 - p)(N - n)}{N - 1}, \qquad \text{etc.} \qquad (6.46)$$

The negative (inverse) hypergeometric distribution is Type IIA with n, $-a$ and $-b$ all positive integers; here the factorial moments are finite and nonzero for $r \leq n$, and zero otherwise. Setting $n = n$, $-a = v+1$, $-b = w+1$ as in (6.8) gives expressions (6.11) for the mean and variance. For inverse sampling, as in (6.15), with support $k, k+1, \ldots, k+N-Np$, we have $n = N - Np$, $-a = k$, $-b = Np - k + 1$ and

$$\mu = k + \frac{na}{a + b} = k + \frac{(N - Np)k}{Np + 1} = \frac{k(N + 1)}{Np + 1},$$

$$\mu_2 = \frac{k(N - Np)(Np + 1 - k)(N + 1)}{(Np + 1)^2(Np + 2)}, \qquad \text{etc.} \qquad (6.47)$$

For the beta-binomial model the integer restrictions can be relaxed to n a positive integer, $-a = \alpha$ and $-b = \beta$, both positive real. The moments are now

$$\mu = \frac{n\alpha}{\alpha + \beta},$$

$$\mu_2 = \frac{n\alpha\beta(\alpha + \beta + n)}{(\alpha + \beta)^2(\alpha + \beta + 1)}, \qquad \text{etc.} \qquad (6.48)$$

(By the duality relationship between Types IIA and IIIA, these Type IIA distributions are also Type IIIA with an appropriate transformation of parameters.)

The beta-negative-binomial distribution (generalized Waring distribution) is Type IV, with $-n = k$, $-a = \ell$, $b = \ell + m - 1$, all positive real. The rth moment exists only for $r < m$. Subject to their existence,

$$\mu = \frac{k\ell}{m - 1},$$

$$\mu_2 = \frac{k\ell(\ell + m - 1)(m + k - 1)}{(m - 1)^2(m - 2)}, \qquad \text{etc.} \qquad (6.49)$$

For the beta-Pascal distribution with the same parameters, but support k, $k+1, \ldots$, where k is a positive integer, we have

$$\mu = k + \frac{k\ell}{m-1} = \frac{k(\ell + m - 1)}{m-1},$$

$$\mu_2 = \frac{k\ell(\ell + m - 1)(m + k - 1)}{(m-1)^2(m-2)}, \qquad \text{etc.} \qquad (6.50)$$

4 PROPERTIES

Let

$$f(x|n, a, b) = \Pr[X = x] = \binom{a}{x}\binom{b}{n-x} \Big/ \binom{a+b}{n}$$

and

$$F(x|n, a, b) = \sum_j \Pr[X = j] = \sum_j \binom{a}{j}\binom{b}{n-j} \Big/ \binom{a+b}{n}, \qquad (6.51)$$

where the range of summation for j is $\max(0, n - b) \le j \le x$ (or $0 \le j \le \min(x, n, a)$ when n and a are positive). Then the following probability relationships hold:

$$f(x + 1|n, a, b) = \frac{(a - x)(n - x)}{(x + 1)(b - n + x + 1)} f(x|n, a, b), \qquad (6.52)$$

$$f(x|n, a + 1, b - 1) = \frac{(a + 1)(b - n + x)}{(a + 1 - x)b} f(x|n, a, b), \qquad (6.53)$$

$$f(x|n + 1, a, b) = \frac{(b - n + x)(n + 1)}{(n + 1 - x)(a + b - n)} f(x|n, a, b), \qquad (6.54)$$

$$f(x|n, a, b + 1) = \frac{(a + b - n + 1)(b + 1)}{(b - n + x + 1)(a + b + 1)} f(x|n, a, b). \qquad (6.55)$$

Also

$$f(x|n, a, b) = f(n - x|n, b, a) \qquad (6.56)$$

$$= f(a - x|a + b - n, a, b) \qquad (6.57)$$

$$= f(b - n + x|a + b - n, b, a); \qquad (6.58)$$

see Lieberman and Owen (1961). Furthermore

$$F(x|n, a, b) = 1 - F(n - x - 1|n, b, a) \qquad (6.59)$$

$$= F(b - n + x|a + b - n, b, a) \qquad (6.60)$$

$$= 1 - F(a - x - 1|a + b - n, a, b). \qquad (6.61)$$

Raiffa and Schlaifer (1961) obtained relationships between the tails of the hypergeometric, beta-binomial (negative hypergeometric) and beta-negative-binomial distributions. These authors used a different notation from Lieberman and Owen (1961). Let $F_h(\cdot)$ and $G_h(\cdot)$ denote the lower and upper tails of a classical hypergeometric distribution; then

$$G_h(k|n, \ell+m-1, k+m-1)$$

$$= \sum_{x \geq k} \binom{k+m-1}{x}\binom{\ell+n-k}{n-x} \bigg/ \binom{\ell+m+n-1}{n}$$

$$= \sum_{x \leq n-k-1} \binom{\ell+n-k}{x}\binom{k+m-1}{n-x} \bigg/ \binom{\ell+m+n-1}{n}$$

$$= F_h(n-k-1|n, \ell+m-1, \ell+n-k). \tag{6.62}$$

Furthermore, let

$$G_{\beta b}(k|m, \ell+m, n) = \sum_{x \geq k} \binom{-m}{x}\binom{-\ell}{n-x} \bigg/ \binom{-\ell-m}{n}$$

$$= \sum_{x \geq k} \int_0^1 \binom{n}{x} p^x (1-p)^{n-x} \times \frac{p^{m-1}(1-p)^{\ell-1}dp}{B(\ell,m)} \tag{6.63}$$

be the upper tail of a beta-binomial (negative hypergeometric) distribution. Also let

$$F_{\beta Pa}(n|m, \ell+m, k)$$

$$= \sum_{x \leq n-k} \binom{-k}{x}\binom{k+m-1}{-\ell-x} \bigg/ \binom{m-1}{-\ell}$$

$$= \sum_{x \leq n-k} \int_0^1 \binom{k+x-1}{x}(1-\lambda)^k \lambda^x \times \frac{\lambda^{\ell-1}(1-\lambda)^{m-1}d\lambda}{B(\ell,m)} \tag{6.64}$$

denote the lower tail of a beta-negative-binomial (beta-Pascal) distribution. Then from the relationship between the tails of a binomial and a negative binomial distribution (Chapter 5 Section 6), Raiffa and Schlaifer proved that

$$G_{\beta b}(k|m, \ell+m, n) = F_{\beta Pa}(n|m, \ell+m, k) \tag{6.65}$$

and hence that

$$G_{\beta b}(k|m, \ell+m, n) = F_{\beta Pa}(n|m, \ell+m, k)$$

$$= G_h(k|n, \ell + m - 1, k + m - 1) \tag{6.66}$$

$$= F_h(n - k - 1|n, \ell + m - 1, \ell + n - k), \tag{6.67}$$

by a probabilistic argument; see Raiffa and Schlaifer (1961, pp. 238-239).

From (6.52), $\Pr[X = x + 1]$ is greater or less than $\Pr[X = x]$ according as

$$\frac{(a - x)(n - x)}{(x + 1)(b - n + x + 1)} \gtrless 1, \tag{6.68}$$

that is, according as

$$x \lessgtr \frac{(n + 1)(a + 1)}{(a + b + 2)} - 1. \tag{6.69}$$

Let $c = (n + 1)(a + 1)/(a + b + 2)$. Then $\Pr[X = x]$ increases with x, reaching a maximum at the greatest integer that does not exceed c, and then decreases. The mode of the distribution is therefore at $[c]$, where $[\cdot]$ denotes the integer part. If c is an integer, then there are two equal maxima at $c - 1$ and c. (Note that if a and b are large, then the mode is very close to the mean, since $\mu = na/(a + b)$.)

The classical hypergeometric distribution is known to have a monotone likelihood ratio in x for known values of n and $a + b$ [Ferguson (1967)].

The classical hypergeometric distribution tends to a Poisson distribution with mean μ as $n \to \infty$, $(a + b) \to \infty$ such that $na/(a + b) = \mu$, μ constant. Feller (1957), Nicholson (1956), Molenaar (1970a), and Lieberman and Owen (1961) have examined conditions under which it tends to a normal distribution.

As $a \to \infty$, $b \to \infty$ such that $a/(a + b) = p$, p constant, $0 < p < 1$, a Type IA(i) distribution tends to a binomial distribution with parameters n, p.

If a is a positive integer, then as $n \to \infty$, $b \to \infty$ such that $n/b = p$, p constant, $0 < p < 1$, a Type IA(ii) distribution tends to a binomial distribution with parameters a, p.

As a and b both tend to $-\infty$ in such a way that $a/(a + b) = p$, p constant, $0 < p < 1$, a Type IIA distribution tends to a binomial distribution with parameters n, p.

As $a \to \infty$, $b \to -\infty$ such that $a/b = -\lambda$, λ constant, $0 < \lambda < 1$, a Type IIIA distribution tends to a negative binomial distribution with pgf $(1 - \lambda)^k/(1 - \lambda z)^k$, where $k = -n$.

As $a \to -\infty$, $b \to \infty$ such that $a/b = -\lambda$, λ constant, $0 < \lambda < 1$, a Type IV distribution similarly tends to a negative binomial distribution.

From the duality relationship between Type IIA and Type IIIA, a Type IIA distribution can also tend to a negative binomial, and a Type IIIA can tend to a positive binomial distribution.

A comparison between hypergeometric Type IA(i), IIA, IIIA, IV, binomial, Poisson, and negative binomial distributions [from Kemp and Kemp (1956a)] is presented in Table 6.3; see also Table 3.1 in Chapter 3, Section 10.

Table 6.3 Comparison of Hypergeometric Type IA(ii), IIA, IIIA, IV, Binomial, Poisson, and Negative Binomial Distributions

$$\Pr[X = x]$$

x	IA(ii)	Binomial	IIA	Poisson	IIIA	Negative Binomial	IV
0	0.076	0.107	0.137	0.135	0.123	0.162	0.197
1	0.265	0.269	0.266	0.271	0.265	0.269	0.267
2	0.348	0.302	0.270	0.271	0.284	0.247	0.220
3	0.222	0.201	0.184	0.180	0.195	0.164	0.144
4	0.075	0.088	0.093	0.090	0.093	0.089	0.083
5	0.013	0.026	0.036	0.036	0.032	0.042	0.044
6	0.001	0.006	0.011	0.012	0.007	0.017	0.023
7	0.000	0.001	0.003	0.003	0.001	0.007	0.011
8	0.000	0.000	0.000	0.001	0.000	0.002	0.005
9	—	0.000	0.000	0.000	—	0.001	0.003
10	—	0.000	0.000	0.000	—	0.000	0.001
11	—	—	—	0.000	—	0.000	0.001
≥ 12	—	—	—	0.000	—	0.000	0.001

Note: For each distribution the mean is 2, $|n| = 10$, $|a/(a + b)| = 0.2$, and $|a + b| = 40$.
A dash means that the probability is zero.
0.000 means that the probability is less than 0.0005.
Adapted from Kemp and Kemp (1956a).

5 APPROXIMATIONS AND BOUNDS

There is a considerable variety of approximations to the individual probabilities, and also to cumulative sums of probabilities, for the classical hypergeometric distribution. Many of these are based on the approximation of the hypergeometric distribution (6.1) by a binomial distribution with parameters n, p.

Sródka (1963) obtained the following very good bounds on the probabilities:

$$\binom{n}{x} \left(\frac{Np - x}{N}\right)^x \left(\frac{N - Np - n + x}{N}\right)^{n-x} \left(1 + \frac{6n^2 - 6n - 1}{12N}\right) < \Pr[X = x]$$

$$< \binom{n}{x} p^x (1 - p)^{n-x} \left(1 - \frac{n}{N}\right)^{-n} \left(1 + \frac{6n^2 + 6n - 1}{12N}\right)^{-1}. \tag{6.70}$$

For sufficiently large N this can be simplified to

$$\binom{n}{x} \left(\frac{Np - x}{N}\right)^x \left(\frac{N - Np - n + x}{N}\right)^{n-x}$$

$$< \Pr[X = x] < \binom{n}{x} p^x (1 - p)^{n-x} \left(1 - \frac{n}{N}\right)^{-n}. \tag{6.71}$$

It is often adequate to use the simple binomial approximation

$$\Pr[X = x] \doteq \binom{n}{x} p^x (1-p)^{n-x} \tag{6.72}$$

when $n < 0.1N$.

There is a marked improvement if n and p are replaced by n^* and p^*, where

$$p^* = \frac{(n-1) + (N-n)p}{N-1},$$

$$n^* = \frac{np}{p^*}, \tag{6.73}$$

that is, if $n^* p^*$ and $n^* p^* (1 - p^*)$ are set equal to the theoretical mean and variance of the hypergeometric distribution [Sandiford (1960)].

Greater accuracy still may be obtained by using the following modification suggested by Ord (1968a):

$$\Pr[X = x] \doteq \binom{n}{x} p^x (1-p)^{n-x} \left[1 + \frac{x(1-2p) + np^2 - (x-np)^2}{2Np(1-p)} \right]. \tag{6.74}$$

This can be reformulated as

$$\Pr[X = x] \doteq \binom{n}{x} p^x (1-p)^{n-x} \left[1 + \frac{(2x-np)(n-1)}{2N(1-p)} - \frac{x(x-1)}{2Np(1-p)} \right]. \tag{6.75}$$

Burr (1973) showed that

$$\Pr[X = x] = \binom{n}{x} p^x (1-p)^{n-x} \left[1 + \frac{\{x - (x-np)^2\}}{2Np} + O\left(\frac{1}{N^2 p^2} \right) \right], \tag{6.76}$$

and that for $n > Np$ a closer approximation is obtained by interchanging the roles of n and Np. Burr's approximation is equivalent to

$$\Pr[X = x] \doteq \binom{n}{x} p^x (1-p)^{n-x} \left[1 + \frac{(2x-np)n}{2N} - \frac{x(x-1)}{2Np} \right]. \tag{6.77}$$

Ma (1982) independently derived an approximation for $n \le Np$ that is equivalent to Ord's approximation. He also showed that when $n > Np$, interchanging the roles of n and Np gives a better approximation. He drew the conclusion from his numerical comparisons with Burr's approximations that "in most practical cases the effects of these approximations are almost uniformly better than Burr's."

Since, as already noted, the hypergeometic distribution is unchanged by interchanging n and Np, it is clear that a binomial with parameters Np, n/N has a claim equal to that of a binomial with parameters n, p as an approximating distribution for (6.1). In addition the distribution of $n - x$ could be approximated by a binomial with parameters $N - Np$, n/N; similarly the distribution of $Np - x$ could be approximated by a binomial with parameters $N - n$, p. Brunk et al. (1968) have compared these approximations. Their investigations support the opinion of Lieberman and Owen (1961) that it is best to use the binomial with smallest power parameter, that is, $\min(n, Np, N - Np, N - n)$.

The following binomial-type approximation for the cumulative probabilities was obtained by Wise (1954):

$$\sum_{j=0}^{x} \Pr[X = j] \doteq \sum_{j=0}^{x} \binom{n}{j} w^j (1 - w)^{n-j}, \tag{6.78}$$

where $w = (Np - x/2)/(N - n/2 + 1/2)$; that is, the distribution (6.1) is approximated by a binomial distribution with parameters n, $(Np - x/2)/(N - n/2 - 1/2)$. An approximation of a similar form has been constructed by Bennett (1965). This is (for $Np \geq n$)

$$\sum_{j=0}^{x} \Pr[X = j] \doteq \sum_{j=0}^{x} \binom{n}{j} p^j (1 - p)^{n-j}$$
$$- \frac{n(n-1)p(1-p)}{2(N-1)} \left\{ \binom{n}{x+2} p^{x+2}(1-p)^{n-x-2} - \binom{n}{x+1} p^{x+1}(1-p)^{n-x-1} \right\} \tag{6.79}$$

(we can always arrange that $Np \geq n$).

Molenaar (1970a) found that the use of

$$p^{\ddagger} = \frac{Np - x/2}{N - (n-1)/2} - \frac{n(x - np - 1/2)}{6\{N - (n-1)/2\}^2} \tag{6.80}$$

gives very accurate results even when $n/N > 0.1$.

Uhlmann (1966) has made a systematic comparison between the hypergeometric distribution (with parameters n, Np, N $(0 < p < 1)$) and the binomial distribution (with parameters n, p). Denoting $\Pr[X \leq c]$ for the two distributions by $L_{N,n,c}(p)$ and $L_{n,c}(p)$, respectively, he showed that in general

$$
L_{N,n,c}(p) - L_{n,c}(p) \begin{cases} = 0 & \text{for } p = 0, \\ > 0 & \text{for } 0 < p \le c(n-1)^{-1}N(N+1)^{-1}, \\ < 0 & \text{for } c(n-1)^{-1}N(N+1)^{-1} + (N+1)^{-1} \le p < 1, \\ = 0 & \text{for } p = 1. \end{cases}
$$

$$(6.81)$$

If n is <u>odd</u> *and* $c = (n-1)/2$, then it is possible to eliminate the range of values

$$
c(n-1)^{-1}N(N+1)^{-1} < p < c(n-1)^{-1}N(N+1)^{-1} + (N+1)^{-1}, \qquad (6.82)
$$

where the sign of the difference is indeterminate, and replace (6.81) by

$$
L_{N,n,c}(p) - L_{n,c}(p) \begin{cases} > 0 & \text{for } 0 < p \le 0.5 \\ = 0 & \text{for } p = 0.5 \\ < 0 & \text{for } 0.5 < p < 1. \end{cases} \qquad (6.83)
$$

A more recent study of the relationship between hypergeometric and binomial pmf's is that of Ahrens (1987), who used simple majorizing functions (upper bounds) for their ratio. He found that the ratio of the hypergeometric to the binomial pmf can always be kept below $\sqrt{2}$ by a suitable choice of approximating binomial.

If the binomial approximation to the hypergeometric distribution can itself be approximated by a Poisson or normal approximation (see Chapter 3, Section 6.1), then there is a corresponding Poisson or normal approximation to the hypergeometric. Thus when p is small but n is large, the Poisson approximation

$$
\Pr[X = x] \doteq \frac{e^{-np}(np)^x}{x!} \qquad (6.84)
$$

may be used. Burr (1973) showed that

$$
\Pr[X = x] = e^{-np} \frac{(np)^x}{x!} \times \left[1 + \left(\frac{1}{2Np} + \frac{1}{2n} \right) \{ x - (x - np)^2 \} \right.
$$
$$
\left. + O\left(\frac{1}{x^2} + \frac{1}{n^2} \right) \right]. \qquad (6.85)
$$

The χ^2-test for association in a 2×2 contingency table uses a χ^2 approximation for the tail of a hypergeometric distribution; this is

$$
\Pr[X \le x] \doteq \Pr[\chi^2_{[1]} \ge T], \qquad (6.86)
$$

where $\quad T = \dfrac{(N-1)[x(N-n-Np+x)-(Np-x)(n-x)]^2}{Np(N-Np)n(N-n)}$

$$= \frac{(N-1)(x-np)^2}{p(1-p)n(N-n)}. \tag{6.87}$$

The relationship between the tails of a χ^2 distribution (Chapter 17) and a Poisson distribution means that this is a Poisson-type approximation for the cumulative hypergeometric probabilities. It can be improved by the use of Yates' correction, giving "the usual 1/2-corrected chi-statistic" of Ling and Pratt (1984). It is appropriate for n large provided that p is not unduly small.

From the relationship between the χ^2 distribution with one degree of freedom and the normal distribution, we have, when p is not small and n is large,

$$\Pr[X \leq x] \doteq (2\pi)^{-1/2} \int_{-\infty}^{y} \exp(-u^2/2)du, \tag{6.88}$$

with $\qquad\qquad y = \dfrac{(x-np+1/2)}{[(N-n)np(1-p)/(N-1)]^{1/2}}. \tag{6.89}$

Hemelrijk (1967) has reported that, unless the tail probability is less than about 0.07 <u>and</u> $Np + n \leq N/2$, some improvement is effected by replacing $(N-1)^{-1}$ by N^{-1} under the square root sign. A more refined normal approximation was proposed by Feller (1957) and Nicholson (1956).

Pearson (1906) approximated hypergeometric distributions by (continuous) Pearson type distributions. This work was continued by Davies (1933, 1934). The Pearson distributions that appeared most promising were Types VI or III. Bol'shev (1964) also proposed an approximation of this kind that gives good results for $N \geq 25$. Here the cumulative distribution function is approximated by an incomplete beta function ratio:

$$\Pr[X \leq x] \doteq I_{1-\xi}(\lambda - x + c, x - c + 1), \tag{6.90}$$

with

$$\xi = \frac{n+Np-1-2np}{N-2},$$

$$c = \frac{n(n-1)p(Np-1)}{(N-1)[(N-n)(1-p)+np-1]},$$

$$\lambda = \frac{(N-2)^2 np(N-n)(1-p)}{(N-1)[(N-n)(1-p)+np-1](n+Np-1-2np)}.$$

The approximation improves as $N \to \infty$.

Normal approximations would, however, seem to be the most successful. Ling and Pratt (1984) carried out an extensive empirical study of 12 normal and 3 binomial approximations for cumulative hypergeometric probabilities, including the following two relatively simple normal approximations of Molenaar (1970a, 1973):

$$\Pr[X \leq x] \doteq \Phi\{2[N-1]^{-1/2}[(x+1)^{1/2}(N-Np-n+x+1)^{1/2}$$
$$-(n-x)^{1/2}(Np-x)^{1/2}]\}, \tag{6.91}$$

$$\Pr[X \leq x] \doteq \Phi\{2N^{-1/2}[(x+0.75)^{1/2}(N-Np-n+x+0.75)^{1/2}$$
$$-(n-x-0.25)^{1/2}(Np-x-0.25)^{1/2}]\}. \tag{6.92}$$

The second of these is slightly better for $0.05 < \Pr[X \leq x] < 0.95$, and the first otherwise. Both are considerably more accurate than (6.88).

Ling and Pratt considered that binomial approximations are not appropriate as competitors to normal approximations because of the computational problems with the tails. The four normal approximations that they found best originated from an unpublished paper; this was submitted to the *Journal of the American Statistical Association* by D. B. Peizer in 1968 but was never revised nor resubmitted. Peizer's approximations are extremely good, but they are considerably more complicated than those above of Molenaar. The simplest of the Peizer approximations is

$$\Pr[X \leq x] \doteq \Phi\left(\frac{A'D' - B'C'}{|AD - BC|}\left\{2L\frac{2mnrsN'}{m'n'r's'N}\right\}^{1/2}\right), \tag{6.93}$$

where

$n = Np$	$n' = Np + 1/6$
$m = N - Np$	$m' = N - Np + 1/6$
$r = n$	$r' = n + 1/6$
$s = N - n$	$s' = N - n + 1/6$
$N = N$	$N' = N - 1/6$
$A = x + 1/2$	$A' = x + 2/3$
$B = Np - x - 1/2$	$B' = Np - x - 1/3$
$C = n - x - 1/2$	$C' = n - x - 1/3$
$D = N - Np - n + x + 1/2$	$D' = N - Np - n + x + 2/3.$

with equation number (6.94) at right.

Also $L = A \ln[AN/(nr)] + B \ln[BN/(ns)] + C \ln[CN/(mr)] + D \ln[DN/(ms)]$, with n, m, r, s, as above. See Ling and Pratt (1984) for details of the other approximations that they studied.

6 TABLES AND COMPUTATION

An extensive set of tables of individual and cumulative probabilities for the classical hypergeometric distribution has been prepared by Lieberman and Owen (1961). They give values for individual and cumulative probabilities to six decimal places for

$$N = 2(1)50(10)100, \quad Np = 1(1)N - 1, \quad n = 1(1)Np.$$

Less extensive tables were published earlier by Chung and DeLury (1950). Graphs based on hypergeometric probabilities were given by Clark and Koopmans (1959).

Guenther (1983) thought that the best way to evaluate $\Pr[X \leq x]$ is from the Lieberman and Owen tables, or by means of a packaged computer program. Computer algorithms have been provided by Freeman (1973) and Lund (1980). Lund's algorithm has been improved by Shea (1989) and by Berger (1991).

Guenther considered that if a packaged program is not available, it is relatively easy to write one; the first term, $\Pr[X = 0]$, can be computed as

$$\frac{(N - n) \cdots (N - n - Np + 1)}{N \cdots (N - Np + 1)}$$

by alternatively dividing and multiplying; the remaining terms can then be calculated recursively. There is, however, a problem with accuracy if $\Pr[X = 0]$ is very small.

Little serious attention seems to have been given to the use of Stirling's expansion for the computation of individual hypergeometric probabilities.

7 ESTIMATION

7.1 Classical Hypergeometric Distribution

In one of the most common situations, inspection sampling (see Section 9.1), there is a single observation of r defectives (successes!) in a sample of size n taken from a lot of size N. Both N and n are known, and the hypergeometric parameter Np denotes the number of defectives in a lot; an estimate of Np is required.

The maximum likelihood estimator $\widehat{Np}$ is the integer maximizing

$$\binom{\widehat{Np}}{r}\binom{N - \widehat{Np}}{n - r} \tag{6.95}$$

for the observed value r. From the relationship between successive probabilities, $\Pr[X = r|n, Np + 1, N] \gtrless \Pr[X = r|n, Np, N]$ according as $Np \lessgtr$

$n^{-1}r(N+1)-1$. Hence $\widehat{Np}$ is the greatest integer not exceeding $r(N+1)/n$; if $r(N+1)/n$ is an integer, then $\{r(N+1)/n\}-1$ and $r(N+1)/n$ both maximize the likelihood. The variance of $r(N+1)/n$ is, from (6.6),

$$\frac{(N+1)^2(N-n)p(1-p)}{n(N-1)}.$$

Neyman confidence intervals for Np have been tabulated extensively, notably by Chung and DeLury (1950) and in Owen (1962), and have been used widely. Steck and Zimmer (1968) outlined how these may be obtained; see also Guenther (1983). Steck and Zimmer also derived Bayes confidence intervals for Np based on various special cases of a Pólya prior distribution. They related these to Neyman confidence intervals. It would seem that Bayes confidence intervals are highly sensitive to choice of prior distribution.

A test of $Np = a$ against $Np = a_0$ is sometimes required. Guenther (1977) discussed hypothesis testing in this context, giving numerical examples.

In the simplest capture-recapture application (again see Section 9.1) we want to estimate the total size of a population N, with both $n_1 = Np$ (the number caught on the first occasion) and $n_2 = n$ (the number caught on the second occasion) known, given a single observation r. Here

$$\Pr[X=r|n_2,n_1,N+1] \gtrless \Pr[X=r|n_2,n_1,N] \qquad (6.96)$$

according as $N \lessgtr (n_1 n_2/r) - 1$. Hence the maximum likelihood estimator, $\hat{N}$, of N is the greatest integer not exceeding $n_1 n_2/r$; if $n_1 n_2/r$ is an integer, then $(n_1 n_2/r) - 1$ and $n_1 n_2/r$ both maximize the likelihood.

The properties of the estimator $\hat{N}$ have been discussed in detail by Chapman (1951). Usually $n_1 + n_2 \not> N$, in which case the moments of $\hat{N}$ are infinite. Because of the problems of bias and variability concerning $\hat{N}$, Chapman suggested instead the use of the estimator

$$N^* = \frac{(n_1+1)(n_2+1)}{(r+1)} - 1. \qquad (6.97)$$

He found that

$$E[N^* - N] = \frac{(n_1+1)(n_2+1)(N-n_1)!(N-n_2)!}{(N+1)!(N-n_1-n_2-1)!}, \qquad (6.98)$$

which is less than 1 when $N > 10^4$ and $n_1 n_2/N > 9.2$. The variance of N^* is approximately

$$N^2(m^{-1} + 2m^{-2} + 6m^{-3}), \qquad (6.99)$$

and its coefficient of variation is approximately $m^{-1/2}$, where $m = (n_1 n_2)/N$. Chapman concluded that "programmes in which the expected number of tagged members is much smaller than 10 may fail to give even the order of magnitude of the population correctly."

Robson and Regier (1964) discussed the choice of n_1 and n_2. Chapman (1948, 1951) showed how large-sample confidence intervals for N^* can be constructed; see also Seber (1982b).

The estimator

$$N^{**} = \frac{(n_1 + 2)(n_2 + 2)}{r + 2} \qquad (6.100)$$

has also been suggested for n_2 sufficiently large. Here

$$E(N^{**}) \doteqdot N(1 - m^{-1}),$$
$$\operatorname{Var}(N^{**}) \doteqdot N^2(m^{-1} - m^{-2} - m^{-3}). \qquad (6.101)$$

In epidemiological studies the estimation of a target population size N is quite often achieved by merging two lists of sizes n_1 and n_2; see Section 9.1. Here it is usual to have $n_1 + n_2 > N$, in which case $\hat{N}$ is unbiased, and

$$s^2 = \frac{(n_1 + 1)(n_2 + 1)(n_1 - r)(n_2 - r)}{(r + 1)^2(r + 2)}$$

(where r is the number of items in common on the two lists) is an unbiased estimator of $\operatorname{Var}(\hat{N})$ [Wittes (1972)].

7.2 Negative (Inverse) Hypergeometric Distribution

Consider the beta-binomial parameterization

$$\Pr[X = x] = \binom{-\alpha}{x}\binom{-\beta}{n - x} \Big/ \binom{-\alpha - \beta}{n} \qquad (6.102)$$

from Section 2.2, with $\alpha, \beta, n > 0$, n an integer. Moment and ML estimation procedures for the parameters α and β were devised by Skellam (1948) and Kemp and Kemp (1956b).

The moment estimators are obtained by setting

$$\bar{x} = \frac{\tilde{\alpha}n}{\tilde{\alpha} + \tilde{\beta}}, \qquad s^2 = \frac{n\tilde{\alpha}\tilde{\beta}(\tilde{\alpha} + \tilde{\beta} + n)}{(\tilde{\alpha} + \tilde{\beta})^2(\tilde{\alpha} + \tilde{\beta} + 1)}, \qquad (6.103)$$

that is,

$$\tilde{\alpha} = \frac{(n - \bar{x} - s^2/\bar{x})\bar{x}}{(s^2/\bar{x} + \bar{x}/n - 1)n},$$

$$\tilde{\beta} = \frac{(n - \bar{x} - s^2/\bar{x})(n - \bar{x})}{(s^2/\bar{x} + \bar{x}/n - 1)n}. \qquad (6.104)$$

Maximum likelihood estimation is reminiscent of ML estimation for the negative binomial distribution. Let the observed frequencies be f_x, $x =$

$0, 1, \ldots, n$, and set

$$A_x = f_{x+1} + f_{x+2} + \cdots + f_n,$$
$$B_x = f_0 + f_1 + \cdots + f_x;$$

then the total number of observations is $A_{-1} = B_n$.
The ML equations are

$$0 = F \equiv \sum_{x=0}^{n-1} \frac{A_x}{\hat{\alpha} + x} - \sum_{x=0}^{n-1} \frac{A_{-1}}{\hat{\alpha} + \hat{\beta} + x}$$

$$0 = G \equiv \sum_{x=0}^{n-1} \frac{B_x}{\hat{\beta} + x} - \sum_{x=0}^{n-1} \frac{A_{-1}}{\hat{\alpha} + \hat{\beta} + x}. \qquad (6.105)$$

Iteration is required for their solution. Given initial estimates α_1 and β_1 (e.g., the moment estimates), corresponding values of F_1 and G_1 can be computed; better estimates, α_2 and β_2, can then be obtained by solving the simultaneous linear equations

$$F_1 = (\alpha_2 - \alpha_1) \sum_{x=0}^{n-1} \frac{A_x}{(\alpha_1 + x)^2} - (\alpha_2 - \alpha_1 + \beta_2 - \beta_1) \sum_{x=0}^{n-1} \frac{A_{-1}}{(\alpha_1 + \beta_1 + x)^2}$$

$$G_1 = (\beta_2 - \beta_1) \sum_{x=0}^{n-1} \frac{B_x}{(\beta_1 + x)^2} - (\alpha_2 - \alpha_1 + \beta_2 - \beta_1) \sum_{x=0}^{n-1} \frac{A_{-1}}{(\alpha_1 + \beta_1 + x)^2}. \qquad (6.106)$$

The next cycle is then begun by calculating F_2 and G_2. Kemp and Kemp (1956b) reported that "on average five iterations were needed to stabilize the estimates to three decimal places."

Chatfield and Goodhardt (1970) put forward an estimation method based on the mean and zero frequency. For highly J-shaped distributions (e.g., distributions of numbers of items purchased) this method has good efficiency; however, it does require iteration.

Griffiths (1973) remarked that the ML estimators can be obtained by the use of a computer algorithm to maximize the log-likelihood. Williams (1975), like Griffiths, considered it advantageous to reparameterize, taking $\pi = \alpha/(\alpha + \beta)$ (the mean of the beta distribution) and $\theta = 1/(\alpha + \beta)$ (a shape parameter). Williams was hopeful that convergence of a likelihood maximization algorithm would be more rapid with this parameterization.

Qu et al. (1990) have pointed out that θ greater, equal, or less than zero gives the beta-binomial, binomial, and hypergeometric distributions, respectively. Hence ML estimation with this parameterization enables an appropriate distribution from within this group to be fitted to data without assuming one particular distribution. Qu et al. also proposed methods of testing $H_0 : \theta = 0$, using (1) a Wald statistic and (2) the likelihood ratio. He showed how the homogeneity of the parameters can be tested using the deviance.

Bowman, Kastenbaum, and Shenton (1992) have shown that the joint efficiency for the method of moments estimation of α and β is very high over much of the parameter space. Series were derived for the first four moments of the moment estimators; simulation approaches were used for validation.

Moment and maximum likelihood estimation for the case where all three parameters, α, β, and n, are unknown was discussed in outline by Kemp and Kemp (1956b).

7.3 Beta-Pascal Distribution

Dubey (1966a) studied estimation for the beta-Pascal distribution assuming k is known.

Given the beta-Pascal parameterization with

$$\Pr[X = y] = \int_0^1 \binom{y-1}{k-1} (1-\lambda)^k \lambda^{y-k} \times \frac{\lambda^{\ell-1}(1-\lambda)^{m-1} d\lambda}{B(\ell, m)}, \qquad y = k, k+1, \ldots,$$

$$(6.107)$$

the mean and variance are

$$\mu = k + \frac{k\ell}{(m-1)} = \frac{k(\ell + m - 1)}{(m-1)}, \qquad \text{provided that } m > 1,$$

$$\mu_2 = \frac{k\ell(k + m - 1)(\ell + m - 1)}{(m-1)^2(m-2)}, \qquad \text{provided that } m > 2 \qquad (6.108)$$

(note that the support is $k, k+1, \ldots$). Hence the moment estimates are

$$\tilde{m} = 2 + \frac{\bar{x}(\bar{x} - k)(k+1)}{(s^2 k - \bar{x}^2 + k\bar{x})},$$

$$\tilde{\ell} = \frac{(\tilde{m} - 1)(\bar{x} - k)}{k}. \qquad (6.109)$$

Dubey also discussed ML estimation assuming that k is known; there are close parallels with ML estimation for the Beta-binomial distribution. Irwin (1975b) described briefly ML estimation procedures for the three-parameter generalized Waring distribution.

8 CHARACTERIZATIONS

There are several characterizations for hypergeometric-type distributions.

Patil and Seshadri's (1964) very general result for discrete distributions has the following corollaries:

1. Iff the conditional distribution of X given $X + Y$ is hypergeometric with parameters a and b, then X and Y have binomial distributions with parameters of the form (a, θ) and (b, θ), respectively.
2. Iff the conditional distribution of X given $X + Y$ is negative hypergeometric with parameters α and β for all values of $X + Y$, then X and Y have negative binomial distributions with parameters of the form (α, θ) and (β, θ), respectively.

Further details are in Kagan, Linnik, and Rao (1973).

Consider now a family of $N + 1$ distributions indexed by $j = 0, 1, \ldots, N$, each supported on a subset of $\{0, 1, \ldots, n\}$, $n \leq N$. Skibinsky (1970) showed that this is the hypergeometric family with parameters N, n, j if and only if for each θ, $0 \leq \theta \leq 1$, the mixture of the family with binomial (N, θ) mixing distribution is the binomial (n, θ) distribution. Skibinsky restated this characterization as follows: Let $h_0, h_1, \ldots, H_N$ denote $N + 1$ functions on $\{0, 1, \ldots, n\}$; then

$$h_j(i) = \binom{j}{i} \binom{N - j}{n - i} \Big/ \binom{N}{n}, \tag{6.110}$$

$i = 0, 1, \ldots, n$, $j = 0, 1, \ldots, N$, if and only if the h_i are independent of θ and

$$\sum_{j=0}^{N} h_j \, b(j; N, \theta) = b(\cdot; n, \theta), \qquad 0 \leq \theta \leq 1, \tag{6.111}$$

where $b(\cdot; n\theta)$ is the binomial pmf with parameters n and θ. He reinterpreted this characterization as follows: Suppose that X has a binomial distribution with parameters N and θ and that Y is distributed on $0, 1, \ldots, n$; suppose also that h_j is defined as in (6.110). Then Y has a binomial distribution with parameters n and θ if the conditional distributions of $Y|X$ are specified by h_x. Conversely, if the conditional distributions of $Y|X$ are independent of θ, $0 \leq \theta \leq 1$, and Y is binomially distributed with parameters (n, θ), then the conditional distributions of $Y|X$ are specified by h_x.

Skibinsky noted the connection between his characterization and certain results of Mood (1943) and Hald (1960) concerning acceptance sampling. Neville and Kemp (1975) elaborated on this. Given a population of lots of fixed size N such that the distribution of the number of defectives per lot is binomial, suppose that a sample of size n is drawn without replacement from one of these lots and that the distribution of the number of defectives in a single sample is hypergeometric. Then the overall distribution of the number of defectives in such samples is binomial. In Hald's terminology, the binomial distribution is reproducible with respect to sampling without replacement.

Neville and Kemp (1975), also Janardan (1973a) in a departmental report, found that a similar characterization holds if the initial distribution of the number of defectives per lot is hypergeometric; the classical hypergeometric distribution is also reproducible with respect to sampling without replacement.

Patil and Ratnaparkhi (1977) obtained characterization results for the Poisson, binomial and negative binomial distributions, given a bivariate observation with the second component subject to damage, and linearity of regression of the first component on the second under the transition of the second component from the original to the damaged state. They described how these results can be extended to provide characterizations for the hypergeometric and negative hypergemetric distributions.

Qu et al. (1990) have proved the following characterization theorem for distributions with pmf of the form

$$\binom{n}{x} \prod_{i=0}^{x-1}(A + iS) \prod_{i=0}^{n-x-1}(B + iS) \Big/ \prod_{i=0}^{n-1}(A + B + iS)$$

$$= \binom{n}{x}\binom{-1/\theta - n}{-\pi/\theta - x} \Big/ \binom{-1/\theta}{-\pi/\theta}, \qquad (6.112)$$

where $\pi = A/(A + B)$ and $\theta = S/(A + B)$ (i.e., for Beta-binomial, binomial, and hypergeometric distributions, according as θ is greater, equal, or less than zero):

Let Y_j, $(j = 1, \ldots, n)$ be n binary variables taking the values 0 and 1. Then $Y = \sum_{j=1}^{n} Y_i$ has the above distribution iff the following conditions are satisfied:

1. The conditional expectation of Y_k, $k = 2, 3, \ldots, n$, with respect to $Y_1, \ldots, Y_{k-1}$ is a linear function of the observed values $Y_1, \ldots, Y_{k-1}$.
2. The expectation for each variable is the same; that is, $E[Y_j] = \pi$, $j = 1, \ldots, n$.
3. The correlation between each pair of variables Y_i and Y_j is the same and is equal to

$$\frac{E[(Y_i - E[Y_i])(Y_j - E[Y_j])]}{\sqrt{E[(Y_i - E[Y_i])^2]E[(Y_j - E[Y_j])^2]}} = \frac{\theta}{1 + \theta} = r \qquad (6.113)$$

(the interclass correlation), where $i = 1, \ldots, n$, $j > i$.

Xekalaki (1981) gave two characterizations for the generalized Waring distribution based on conditionality properties. Papageorgiou (1985) has obtained characterizations of the hypergeometric and negative hypergeometric distributions, using the linear regression of one random variable on another and the conditional distribution of the latter given the first variable; he suggested an application in genetics. Further characterizations of this kind are in Kyriakoussis and Papageorgiou (1991a).

9 APPLICATIONS

9.1 Classical Hypergeometric Distribution

Besides providing a fund of teaching examples, the hypergeometric distribution has a number of important practical applications.

In industrial quality control, lots of size N containing a proportion p of defectives are sampled using samples of fixed size n. The number of defectives X per sample is then a hypergeometric rv. If $x \leq c$ (the acceptance number), the lot is accepted; otherwise it is rejected. The design of suitable sampling plans requires the calculation of confidence intervals for Np, given c, N and n. Tables of these have been published by Chung and DeLury (1950) and Owen (1962). In many cases binomial or Poisson approximations to the hypergeometric distribution suffice.

Another useful application is in the estimation of the size of animal populations from "capture-recapture" data. This kind of application dates back at least to Peterson (1896), quoted by Chapman (1952a). Consider, for example, the estimation of the number N of fish in a pond. First a known number m of fish are netted, marked (tagged), and returned to the pond. A short time later, long enough to ensure (hopefully) random dispersion of the tagged fish but not long enough for natural changes to affect the population size, a sample of size n is taken from the pond, and the number X of tagged fish in the second sample is observed. The assumptions that the samples have fixed sizes m and n, and that the fish behave randomly and independently, lead to a hypergeometric distribution for X.

An alternative viewpoint is to regard the N organisms in the population as N independent trials each with the same probability of belonging to a given one of the four capture-recapture categories. This gives rise to a multinomial model where the probability of an observed pattern of captures and recaptures is proportional to the hypergeometric pmf [Cormack (1979)]. The feature that the sample sizes can be regarded as either fixed or random, with little change to the analysis of the data, occurs in other, more elaborate, capture-recapture sampling schemes, such as schemes with multiple recaptures at successive points of time and those with open (that is, changeable) populations rather than closed populations [Seber (1982a)].

There are variations on the situations leading to the basic hypergeometric capture-recapture model. As mentioned in Section 7.1, the estimation of a target population N in epidemiological studies (e.g., the number of anencephalics born in Boston Lying-In Hospital over several decades) can be achieved by counting the number of cases that appear on both of two lists of sizes n and m (a diagnostic registry list and a list of discharge summaries) [Wittes (1972)].

In opinion surveys a random sample of respondents of size n is drawn without replacement from a finite population of size N. From the proportion in the sample who answer a particular question positively, it is desired to estimate the proportion in the whole population who would answer positively.

A further important use of the hypergeometric distribution is in the analysis of 2×2 contingency tables with both sets of marginal frequencies fixed. The probability of a result as extreme as the observed result is the tail probability for a classical hypergeometric distribution. When testing whether or not the two dichotomies are independent (i.e., testing homogeneity), the test based on the computation of a hypergeometric cumulative probability is known as *Fisher's exact test*; see, e.g., Gibbons (1983). This test was suggested in the mid 1930's by Fisher, Irwin, and Yates; Gibbons has provided relevant references. This test can also be used in the case where only one set of marginal frequencies is fixed; here the test is conditional on the observed values of the unfixed marginal frequencies. Similarly it can be used when neither set of marginal frequencies is fixed.

Cochran (1954) recommended the use of Fisher's exact test, in preference to the approximate χ^2-test with continuity correction, (1) when $N \leq 20$, and (2) when $20 < N < 40$ and the smallest expected frequency is less than 5, where N is the total frequency. Tables for the test are provided in Pearson and Hartley (1976); see also Finney et al. (1963). Significance tests and confidence intervals for the odds ratio are described in Cox and Snell (1989) along with useful bibliographic notes.

The test has been used widely. Yates (1934) illustrated its use with data on breast- and bottle-fed children with normal and maloccluded teeth. Cox and Snell (1989) have examined data on physicians dichotomized as smokers/nonsmokers and as cancer patients/controls. We note that 2×2 tables can arise in two distinct contexts: one sample with two dichotomously observed variables (as in the two preceding examples); and two independent samples with one dichotomous variable. A sample of pregnant women who smoke and a second sample of pregnant nonsmokers, with both samples dichotomized by low/normal birthweight of infant, provides an example of the second context. A discussion of an application of Fisher's exact test in linguistics is given in the first edition of this book.

9.2 Negative (Inverse) Hypergeometric Distribution: Beta-Binomial Distribution

The waiting-time model for the inverse hypergeometric distribution has a number of practical uses. For instance, in inspection sampling, instead of taking a sample of fixed size n from a batch of items and then accepting the batch if the observed number of defectives is less than or equal to some predetermined value c (otherwise rejecting), a form of "curtailed sampling" can be adopted; see, e.g., Guenther (1969). Here items are drawn one at a time until either $c + 1$ defectives are observed (at which point the batch is rejected) or $n - c$ nondefectives are observed (and the batch is accepted). The number of observations required in order to reach a decision then has an inverse hypergeometric distribution.

The use of inverse hypergeometric sampling to estimate the size of a biological population has been studied by Bailey (1951) and Chapman (1952a). A sample of k individuals is caught, marked and released; afterward individuals are sampled one at a time until a predetermined number c of marked individuals are recaptured.

The negative hypergeometric as the distribution of the number of exceedances has been mentioned in Section 2.2.

Distribution-free prediction intervals can be constructed by taking two samples of sizes n_1 and n_2. Then the probability that there are w items in the second sample that are greater than the rth order statistic of the first sample is a negative hypergeometric tail probability. Guenther (1975) gave details.

Suppose now that $X_1, X_2, \ldots, X_{2s+1}$ and $Y_1, Y_2, \ldots, Y_m$ are independent random samples from continuous distributions with cdf's $F_1(x)$ and $F_2(x)$, respectively. Then the null hypothesis, H_0: $F_1(x) = F_2(x)$ for all x, can be tested using the statistic V equal to the number of observations in the second sample that are less than the sample median of the first sample; see Gart (1963) and also Mood (1950, pp. 395-398). The pmf of the statistic V is negative hypergeometric.

The Beta-binomial model for the distribution has been used very widely. Muench (1936, 1938) was interested in applications to medical trials. Skellam (1948) applied the model to the association of chromosomes and to traffic clusters. Barnard (1954), Hopkins (1955), and Hald (1960) recognized its importance in inspection sampling. Irwin (1954) and Griffiths (1973) used it for disease incidence. Kemp and Kemp (1956b) demonstrated its relevance for point quadrat data. Ishii and Hayakawa (1960) examined it as a model for the sex composition of families and for absences of students. Chatfield and Goodhardt (1970) have used it for analysing market purchases, Williams (1975) and Haseman and Kupper (1979) have employed it in toxicology, and Crowder (1978) has used it for seed germination data.

The distribution has also arisen in a Bayesian context from binomial sampling with a beta prior. An early discussion is in Pearson (1925). Guenther (1971) showed that the average cost per lot for the Hald linear cost model with a beta prior has a beta-binomial distribution. The beta-binomial distribution has itself been used as a prior distribution by Steck and Zimmer (1968).

9.3 Beta-Negative-Binomial Distribution: Beta-Pascal Distribution, Generalized Waring Distribution

In-depth studies of the application of the generalized Waring distribution to accident theory have been made by Irwin (1968) and Xekalaki (1983b). Irwin derived the distribution as the distribution of accidents in an accident-prone community exposed to varable risk, whereas Xekalaki's two derivations were based on a "contagion" hypothesis and on a "spells" hypothesis.

In a Bayesian context the beta-negative-binomial distribution can arise from negative binomial sampling with a beta prior.

10 SPECIAL CASES

10.1 Discrete Rectangular Distribution

The *discrete rectangular distribution* (sometimes called the *discrete uniform distribution*) is defined in its most general form by

$$\Pr[X = a + xh] = \frac{1}{n+1}, \qquad x = 0, 1, 2, \ldots, n. \qquad (6.114)$$

Various standard forms are in use. One that is frequently used is obtained by putting $a = 0$, $h = 1$, so that the values taken by X are $0, 1, \ldots, n$. Other forms can be obtained from this standard form by a linear transformation.

The pgf corresponding to $\Pr[X = x] = 1/(n+1), x = 0, 1, \ldots, n$ is

$$G(z) = (n+1)^{-1}(1 + z + z^2 + \cdots + z^n) \qquad (6.115)$$

$$= (n+1)^{-1}{}_2F_1[-n, 1; -n; z], \qquad (6.116)$$

showing that the distribution is a special case of the negative (inverse) hypergeometric (beta-binomial) distribution.

The pgf can also be stated as

$$G(z) = \frac{1 - z^{n+1}}{(n+1)(1-z)}; \qquad (6.117)$$

this result was known to de Moivre, who obtained the distribution of the sum of k such variables from the expansion of

$$\left(\frac{1 - z^{n+1}}{1 - z}\right)^k.$$

From (6.115) the umgf is

$$G(e^t) = \frac{1 + e^t + e^{2t} + \cdots + e^{nt}}{n+1}, \qquad (6.118)$$

whence

$$\mu_r' = \sum_{j=1}^{n} \frac{j^r}{n+1}. \qquad (6.119)$$

From (6.117) the cf is

$$G(e^{it}) = \frac{e^{(n+1)it} - 1}{(n+1)(e^{it} - 1)}, \tag{6.120}$$

and the cmgf is

$$e^{-\mu t} G(e^t) = \frac{e^{-nt/2}(e^{(n+1)t} - 1)}{(n+1)(e^t - 1)}$$

$$= \frac{\sinh(nt/2)}{(n+1)\sinh(t/2)}, \tag{6.121}$$

whence

$$\mu_{2r+1} = 0,$$

$$\mu_{2r} = (n+1)^{-1} \sum_{j=0}^{n} (j - n/2)^{2r}, \qquad r = 1, 2, \dots. \tag{6.122}$$

The factorial moment generating function is

$$G(1+t) = \frac{(1+t)^{n+1} - 1}{(n+1)t}$$

$$= \sum_{r=0}^{n} \frac{n!}{(n-r)!(r+1)} \times \frac{t^r}{r!}. \tag{6.123}$$

The first four moments are

$$\mu = \frac{n}{2},$$

$$\mu_2 = \frac{n(n+2)}{12},$$

$$\mu_3 = 0,$$

$$\mu_4 = \frac{n(n+2)(3n^2 + 6n - 4)}{240}. \tag{6.124}$$

The cumulants of the distribution are

$$\kappa_{2r+1} = 0,$$

$$\kappa_{2r} = (2r)^{-1}[(n+1)^{2r} - 1]B_{2r}, \qquad r = 1, 2, \dots, \tag{6.125}$$

where the B_{2r} are the Bernoulli numbers (Chapter 1, Section A9).

The distribution has no mode. The median is $n/2$ for n even, and it lies between $(n-1)/2$ and $(n+1)/2$ for n odd. The coefficient of variation is $\{(n+2)/(3n)\}^{1/2}$.

The hazard rate (failure rate) is increasing, with $r_x = 1/(n+1-x)$ [Patel (1973)].

The beta-binomial model for the discrete rectangular distribution involves a beta distribution with parameters $\alpha = 1$, $\beta = 1$, that is, a uniform distribution on the interval $[0, 1]$. The somewhat surprising implication is that a continuous uniform mixture of binomials gives rise to a discrete rectangular (discrete uniform) distribution.

The exceedance model is as follows: Whatever the distribution of Z, in a future random sample of size n of values of Z, the number of values that exceed a previously observed value has a rectangular distribution.

Irwin's (1954) urn model (a sample of fixed size with additional replacements) implies that the distribution arises if a fixed number of balls are drawn from an urn initially containing one white ball and one black ball, provided that each drawn ball is replaced together with another similar ball.

The inverse sampling model is often presented as a problem concerning keys. If a key ring contains one correct and N incorrect keys, then the number of unsuccessful keys that must be tried in order to find the correct key is a discrete rectangular rv, assuming that no key is tried more than once.

The largest order statistic T in a sample of size k from a discrete rectangular distribution with support $0, 1, \ldots, n$ is a complete and sufficient statistic for n. Also

$$T^* = \frac{T^{k+1} - (T-1)^{k+1}}{T^k - (T-1)^k} \tag{6.126}$$

is the UMVUE for n.

Patil and Seshadri's (1964) general characterization result for discrete distributions implies that iff the conditional distribution of X given $X+Y$ is discrete rectangular, then X and Y have identical geometric distributions.

10.2 Distribution of Leads in Coin Tossing

The *distribution of leads in coin tossing* [Chung and Feller (1949), Feller (1957, p. 77)] has the pmf

$$\Pr[X = x] = \frac{(2x)!(2n-2x)!2^{-2n}}{x!x!(n-x)!(n-x)!}, \qquad x = 0, 1, 2, \ldots, n. \tag{6.127}$$

Use of Legendre's duplication formula (Chapter 1, Section A2) gives

$$\Pr[X = x] = \frac{(-1/2)!(-1/2)!(-1)^n}{(-x-1/2)!(-n+x-1/2)!(n-x)!x!}$$

$$= \binom{n}{x}\binom{-n-1}{-1/2-x} \Big/ \binom{-1}{-1/2}$$

$$= \binom{-1/2}{x}\binom{-1/2}{n-x} \Big/ \binom{-1}{n}, \qquad x = 0,1,2,\ldots,n,$$

$$(6.128)$$

that is, the particular negative hypergeometric distribution with pgf

$$G(z) = \frac{(n-1/2)!}{n!(-1/2)!}{}_2F_1[-n,1/2;-n+1/2;z] \qquad (6.129)$$

[Kemp (1968b)]. This distribution also gives the solution to *Banach's matchbox problem*; see Feller (1957).

We find that

$$\frac{\Pr[X=x]-\Pr[X=x-1]}{\Pr[X=x-1]} = \frac{x-(n+1)/2}{x(n-x+1/2)}. \qquad (6.130)$$

Hence $\Pr[X=x] \gtrless \Pr[X=x-1]$ according as $x \gtrless (n+1)/2$; that is, the distribution is U-shaped.

The mean and variance are

$$\mu = \frac{n}{2}$$

$$\mu_2 = \frac{n(n+1)}{8}. \qquad (6.131)$$

The mean is the same as that for a discrete rectangular distribution with the same support, but the variance is always greater.

Feller (1957) has said, "The picturesque language of gambling should not detract from the general importance of the coin-tossing model. In fact, the model may serve as a first approximation to many more complicated chance-dependent processes in physics, economics, and learning-theory." He elaborated on this theme.

10.3 Yule Distribution

The *Yule distribution* was developed by Yule (1925) as a descriptor of the number of species of biological organisms per family, using a mixture of shifted geometric distributions with pmf

$$e^{-\beta u}(1-e^{-\beta u})^{y-1}, \qquad y = 1,2,\ldots, \qquad (6.132)$$

where u has an exponential distribution with pdf

$$f(u) = \theta^{-1}\exp(-u/\theta), \qquad \theta > 0, \quad 0 \le u. \qquad (6.133)$$

The resultant mixture distribution can be represented as

$$\text{Geometric}(e^{-\beta U}) \bigwedge_U \text{Exponential}(\theta);$$

the probabilities are

$$\Pr[Y = y] = \int_0^\infty e^{-\beta u}(1 - e^{-\beta u})^{y-1}\theta^{-1}e^{-u/\theta}\,du. \qquad (6.134)$$

Putting $v = e^{-\beta u}$ and $\rho = (\beta\theta)^{-1}$ (so that $\rho > 0$), this becomes

$$\Pr[Y = y] = \int_0^1 (1 - v)^{y-1}v^\rho\rho\,dv \qquad (6.135)$$

$$= \rho(\rho!)(y - 1)!/(y + \rho)!, \qquad y = 1, 2, \ldots. \qquad (6.136)$$

The pgf is

$$H(z) = \frac{\rho\,z}{\rho + 1}{}_2F_1[1, 1; \rho + 2; z], \qquad (6.137)$$

showing that the distribution is a special case of a beta-negative binomial distribution shifted to support $1, 2, \ldots$.

The integral (6.135) demonstrates that the distribution is also the outcome of a beta mixture of shifted geometric distributions with pmf $v(1 - v)^{y-1}$, $y = 1, 2, \ldots$; this was the mode of genesis underlying its use by Miller (1961) for the distribution of sizes of traffic clusters, and by Pielou (1962) for runs of plant species.

The mean and variance are

$$\mu = \frac{\rho}{\rho - 1} \qquad \text{provided that } \rho > 1,$$

$$\mu_2 = \frac{\rho^2}{(\rho - 1)^2(\rho - 2)} \qquad \text{provided that } \rho > 2. \qquad (6.138)$$

The name Yule has also been used for the distribution when it is shifted to the support $0, 1, \ldots$, that is, with pmf

$$\Pr[X = x] = \frac{\rho(\rho!)x!}{(x + \rho + 1)!}, \qquad x = 0, 1, \ldots. \qquad (6.139)$$

The corresponding pgf is

$$G(z) = \frac{\rho}{\rho + 1}{}_2F_1[1, 1; \rho + 2; z]; \qquad (6.140)$$

the mean and variance are now

$$\mu = \frac{1}{\rho - 1} \qquad \text{provided that } \rho > 1,$$

$$\mu_2 = \frac{\rho^2}{(\rho - 1)^2(\rho - 2)} \qquad \text{provided that } \rho > 2 \qquad (6.141)$$

(the variance is of course unchanged). More generally, the rth descending factorial moment for (6.139) is

$$\mu'_{[r]} = \frac{r!r!(\rho - r - 1)!}{(\rho - 1)!} \qquad \text{provided that } \rho > r, \qquad (6.142)$$

and the rth ascending factorial moment is

$$E[X(X+1)\ldots(X+r-1)] = \frac{r!\rho}{(\rho - r)(\rho - r + 1)} \qquad \text{provided that } \rho > r. \qquad (6.143)$$

The Yule distribution has been used to model word-frequency data by Simon (1955, 1960) and Haight (1966). Herdan (1961), however, was doubtful about the fit in the upper tail. Mandelbrot (1959) was especially critical about the use of values of ρ less than unity (i.e., the use of distributions with an infinite mean).

In his paper on natural law in the social sciences, Kendall (1961) applied the distribution to certain kinds of bibliographic data; see also Kendall (1960).

Haight (1966) has fitted the logarithmic, Yule and Borel (see Chapter 9, Section 11) distributions to four distributions of responses in psychological tests. In each case three methods of fitting were used:

1. Equating the sample and population means.
2. Equating the sample and population first frequencies.
3. Equating sample and population tail frequencies.

He found that generally the Yule distribution gave the best fit and the logarithmic distribution the worst fit. Furthermore, fitting by moments (method 1) generally gave a worse fit, for a given distribution, than did methods 2 and 3.

Xekalaki (1984) has shown how the distribution can arise in an econometric context. In Xekalaki (1983c) she applied it to an income under-reporting model and to an inventory control problem. The applications in the latter paper involved the following result: Let X and Y have discrete distributions on the nonnegative integers with the property that

$$\Pr[Y = y] = c^{-1} \sum_{x \geq y+1} \frac{\Pr[X = x]}{x}, \qquad (6.144)$$

where $c = 1 - \Pr[X = 0]$. Then $\Pr[Y = y] = \Pr[X = x]$ if and only if X has a Yule distribution with pmf (6.136).

The Yule distribution can be regarded as a discrete analogue of the Pareto distribution; see Xekalaki and Panaretos (1988). Also when $\rho = 1$, it can be regarded as an approximation to the Zeta distribution (Chapter 11, Section 20). Panaretos (1989a, b) has given an application to surname-frequency data.

Prasad (1957) has described a generalized form of the Yule distribution with the following pmf:

$$\Pr[Y = y] = \frac{2\lambda(\lambda + 1)}{(\lambda + y - 1)(\lambda + y)(\lambda + y + 1)}, \qquad y = 1, 2, \ldots, \qquad (6.145)$$

where $\lambda > 0$. The pgf is

$$H(z) = \frac{2z}{\lambda + 2} {}_2F_1[1, \lambda; \lambda + 3; z]. \qquad (6.146)$$

(Putting $\lambda = 1$, we obtain a Yule distribution with $\rho = 2$.) The expected value of Y is $1 + \lambda$; the second and higher moments are, however, infinite. The cdf is

$$\Pr[Y \le y] = 1 - \frac{\lambda(\lambda + 1)}{(\lambda + y)(\lambda + y + 1)}. \qquad (6.147)$$

Another generalization of the Yule distribution appears in Johnson and Kotz (1989).

10.4 Waring Distribution

The *Waring distribution* is a generalization of the Yule distribution that was developed by Irwin (1963), using the Waring expansion

$$\frac{1}{c - a} = \frac{1}{c} + \frac{a}{c(c + 1)} + \frac{a(a + 1)}{c(c + 1)(c + 2)} + \cdots; \qquad (6.148)$$

this converges for $c > a$. Taking $\Pr[X = x]$ proportional to the $x + 1$th term in the series gives

$$\Pr[X = x] = \frac{(c - a)(a + x - 1)! c!}{c(a - 1)!(c + x)!}, \qquad x = 0, 1, 2, \ldots; \qquad (6.149)$$

the correponding pgf is

$$G(z) = \frac{c - a}{c} {}_2F_1[1, a; c + 1; z]. \qquad (6.150)$$

The Yule is the special case $a = 1$.

The Waring is itself the special case $k = 1$ of the generalized Waring distribution; see Section 2.3; its moments and properties follow therefrom. In particular

$$\mu = \frac{a}{c-a-1} \qquad \text{provided that } c-a > 1,$$

$$\mu_2 = \frac{a(c-a)(c-1)}{(c-a-1)^2(c-a-2)} \qquad \text{provided that } c-a > 2. \quad (6.151)$$

Irwin commented that the series converges much more slowly than the geometric distribution, and said that "we can make the tail as long as we please, as c and $a \to 0$." He demonstrated how the distribution can be fitted, both by the mean-and-zero-frequency method and by maximum likelihood, using one of Kendall's (1961) bibliographic data sets.

The distribution arises as a special case of the beta-negative-binomial distribution in two ways. First,

$$G(z) = \int_0^1 \left(\frac{1-\lambda}{1-\lambda z}\right) \frac{\lambda^{a-1}(1-\lambda)^{c-a-1} \, d\lambda}{B(a, c-a)}, \quad (6.152)$$

showing that it is a mixture of geometric distributions [Miller (1961); Pielou (1962)].

Second,

$$G(z) = \int_0^1 \left(\frac{1-\lambda}{1-\lambda z}\right)^a \frac{(1-\lambda)^{c-a-1} \, d\lambda}{B(1, c-a)}; \quad (6.153)$$

this demonstrates that it is also a mixture of negative binomial distributions using a particular type of beta mixing distribution.

The hazard function of the Waring distribution has been investigated by Xekalaki (1983d).

Marlow's (1965) *factorial distribution* has the pgf

$$G(z) = \frac{{}_2F_1[1, \lambda - n + 1; \lambda + 1; z]}{{}_2F_1[1, \lambda - n + 1; \lambda + 1; 1]}, \quad (6.154)$$

where $n = 2, 3, 4, \ldots$, and $n - 1 < \lambda$, and hence it is a Waring distribution with $a = \lambda - n + 1$, $c = \lambda$.

11   EXTENDED HYPERGEOMETRIC DISTRIBUTIONS

The name *extended hypergeometric distribution* was given by Harkness (1965) to the conditional distribution of one of two binomial rv's given that their sum is fixed. The distribution had arisen in an investigation of the power function for the test of independence in a 2×2 contingency table. If X_i has parameters n_i and $p_i = 1 - q_i$ for $i = 1, 2$, then

$$\Pr[X_1 = x | X_1 + X_2 = m]$$

$$= \left[\sum_x \binom{n_1}{x} \binom{n_2}{m-x} \left(\frac{p_1 q_2}{q_1 p_2} \right)^x \right]^{-1} \binom{n_1}{x} \binom{n_2}{m-x} \left(\frac{p_1 q_2}{q_1 p_2} \right)^x \quad (6.155)$$

$$= \{ {}_2F_1[-n_1, -m; n_2 + 1 - m; \theta] \}^{-1} \binom{n_1}{x} \binom{n_2}{m-x} \theta^x \Big/ \binom{n_1 + n_2}{m}, \quad (6.156)$$

where $\theta = p_1 q_2 / (q_1 p_2)$ and $\max(0, m - n_2) \le x \le \min(n_1, m)$; the same limits apply to the summation in (6.155).

This is a four-parameter extension of the classical hypergeometric distribution of Section 2.1 above; the classical hypergeometric is of course the outcome when $p_1 = p_2$ (and $q_1 = q_2$). Unlike the classical hypergeometric, however, the distribution is a GPSD (Chapter 2, Section 2.1). The pgf is

$$G(z) = \frac{{}_2F_1[-n_1, -m; n_2 + 1 - m; \theta z]}{{}_2F_1[-n_1, -m; n_2 + 1 - m; \theta]}; \quad (6.157)$$

the distribution is therefore also GHPD, and its properties can be obtained as in Chapter 2, Section 4.1. For example, as $n_1 \to \infty$ and $n_2 \to \infty$ such that $n_1/(n_1 + n_2) = c$, c constant, it tends to a binomial distribution with parameters m and $\theta c/(\theta c + 1 - c)$. Hannan and Harkness (1963) derived a normal limiting form, and Harkness (1965) obtained a Poisson limiting form.

The rth factorial moment of the distribution is

$$\mu'_{[r]} = \frac{n_1! m! (n_2 - m)! \; {}_2F_1[r - n_1, r - m; r + n_2 + 1 - m; \theta]}{(n_1 - r)!(m - r)!(n_2 - m + r)! \; {}_2F_1[-n_1, -m; n_2 + 1 - m; \theta]}; \quad (6.158)$$

in particular

$$\mu = \frac{n_1 m \; {}_2F_1[1 - n_1, 1 - m; n_2 + 2 - m; \theta]}{(n_2 - m + 1) \; {}_2F_1[-n_1, -m; n_2 + 1 - m; \theta]}. \quad (6.159)$$

There are no simple explicit expressions for the moments, although there are recurrence relations such as

$$(1 - \theta)\mu'_2 = n_1 m \theta - \{ n_1 + n_2 - (n_1 + m)(1 - \theta) \} \mu. \quad (6.160)$$

Harkness (1965) studied maximum likelihood estimation for θ, assuming that the other three parameters n_1, n_2 and m are known. Let $y_1, y_2, \ldots, y_N$ be a sample of N independent observations from distribution (6.155). Then the MLE of θ satisfies the equation $\bar{y} = \mu$; this does not have an explicit solution. Using the "natural" estimator

$$\tilde{\theta} = \frac{\bar{y}(n_2 - m + \bar{y})}{(n_1 - \bar{y})(m - \bar{y})}, \tag{6.161}$$

Harkness obtained the following lower and upper bounds for $\hat{\theta}$:

1. For $\bar{y} \leq mn_1/(n_1 + n_2)$,

$$\tilde{\theta} \leq \hat{\theta} \leq \tilde{\theta} + \frac{\bar{y}\{mn_1 - (n_1 + n_2)\bar{y}\}}{mn_1(n_1 - \bar{y})(m - \bar{y})}.$$

2. For $\bar{y} \geq mn_1/(n_1 + n_2)$,

$$\tilde{\theta}\left[1 + \frac{(n_1 - \bar{y})(n_1 + n_2 - n_1 m - \bar{y})}{n_1(n_1 + n_2 - m)(n_1 - \bar{y})(m - \bar{y})}\right]^{-1} \leq \hat{\theta} \leq \tilde{\theta}.$$

Replacement of either (or both) of the binomial rv's for (6.155) by a negative binomial rv leads to other four-parameter GHP distributions that are extensions of the inverse (negative) hypergeometric and generalized Waring distributions in Sections 2.2 and 2.3 of this chapter. Morton (1991) has examined the use of an extended negative hypergeometric distribution in the analysis of data with extra-multinomial variation.

It is easy to construct ad hoc distributions with pgf's of the form $C\,_2F_1[a, b; c; \theta z]$. However, fitting a four-parameter distribution such as this may be troublesome — a small change in the estimate of one parameter may lead to big changes in the estimates of the other parameters. When a statistical model leads to a distribution of this kind, however, very often one or more of the parameters will be known (as in the Harkness model above). The lost-games distribution (Chapter 11, Section 10) is another example of a distribution with a pgf of this form that arises naturally and does not have the full complement of four parameters to be estimated.

Consider now a binomial distribution with parameters n and θp where p has a beta distribution on $(0, 1)$. This gives the following extension of the beta-binomial ($\equiv$ negative hypergeometric) distribution of Section 2.2 above:

$$G(z) = \int_0^1 (1 - \theta p + \theta pz)^n \times \frac{p^{\alpha-1}(1 - p)^{\beta-1}dp}{B(\alpha, \beta)}$$

$$= {}_2F_1[-n, \alpha; \alpha + \beta; \theta(1 - z)]; \tag{6.162}$$

this is a GHFD (Chapter 2, Section 4.2), and hence the pmf can be shown to be

$$
\Pr[X = x] = \binom{n}{x} \frac{\theta^x (\alpha + x - 1)!(\alpha + \beta - 1)!}{(\alpha - 1)!(\alpha + \beta + x - 1)!} \, {}_2F_1[-n + x, \alpha + x; \alpha + \beta + x; \theta],
$$

(6.163)

where $x = 0, 1, \ldots, n$.

Similarly, given a negative binomial distribution with pgf $(1 + \theta P - \theta P z)^{-k}$, then, if P has a beta distribution on $(0, 1)$, the resultant mixture has the pmf

$$
\Pr[X = x]
$$
$$
= \binom{k+x-1}{x} \frac{\theta^x (\alpha + x - 1)!(\alpha + \beta - 1)!}{(\alpha - 1)!(\alpha + \beta + x - 1)!} \, {}_2F_1[k + x, \alpha + x; \alpha + \beta + x; -\theta],
$$

(6.164)

where $x = 0, 1, \ldots$. This is not a direct extension of the beta-negative-binomial distribution. Instead it is an extension of distribution (6.23) which was mentioned in passing in Section 2.3 of this chapter; it is the type H_2 distribution of Gurland (1958) and Katti (1966).

12 OTHER RELATED DISTRIBUTIONS

The *positive hypergeometric distribution* is formed from the classical hypergeometric distribution by omitting the zero class. (If n exceeds $N - Np$, then there is no zero class.) If $n \leq N - Np$, then the pmf is

$$
\Pr[X = x] = \binom{Np}{x} \binom{N - Np}{n - x} \Big/ \left[\binom{N}{n} - \binom{N - Np}{n} \right],
$$
$$
x = 1, 2, \ldots, \min(n, Np).
$$

(6.165)

Govindarajulu (1962) has made a detailed study of the inverse moments of the distribution and has tabulated values of $E(X^{-r})$ for $r = 1, 2$.

The *noncentral hypergeometric distribution* is the name given by Wallenius (1963) to a distribution constructed by supposing that in sampling without replacement (as in Section 1 of this chapter), the probability of drawing a white ball, given that there are Np white and $(N - Np)$ black balls, is not p but $p/\{p + \theta(1 - p)\}$ with $\theta \neq 1$. The mathematical analysis following from this assumption is rather involved. Starting from the recurrence relationship

$$
\Pr[X = x | Np, n, N]
$$
$$
= \frac{p \Pr[X = x - 1 | Np - 1, n - 1, N - 1] + \theta(1 - p) \Pr[X = x | Np, n - 1, N - 1]}{p + \theta(1 - p)},
$$

(6.166)

Wallenius obtained the formula

$$\Pr[X = x] = \binom{Np}{x}\binom{N-Np}{n-x}\int_0^1 (1 - t^c)^x (1 - t^{\theta c})^{n-x}\,dt \qquad (6.167)$$

with $c = \{Np - x + \theta(N - Np - n + x)\}^{-1}$. He gave bounds for $\Pr[X = x]$, and he showed that, for n small compared with Np and $(N - Np)$, X is distributed approximately binomially with parameters n, $\{1 + \theta p/(1 - p)\}^{-1}$.

Janardan (1978) has developed a *generalized Markov-Pólya* (generalized negative hypergeometric) distribution based on a voting model. The pmf is

$$\Pr[X = k] = \binom{N}{k}\frac{J_k(a,c,t)J_{N-k}(b,c,t)}{J_N(a+b,c,t)}, \qquad (6.168)$$

where

$$J_k(a,c,t) = a(a + kt + c)(a + kt + 2c)\ldots(a + kt + (k-1)c).$$

Janardan gave recurrence relations for the probabilities as well as expressions for the mean and variance; he also considered certain special and limiting cases, and maximum likelihood estimation. A generalized inverse Markov-Pólya distribution was also introduced. The use of the generalized Markov-Pólya distribution as a random damage model was studied by B. R. Rao and Janardan (1984).

Consider the distribution of the number of items observed to be defective in samples from a finite population, when the detection of a defective is not certain. Johnson, Kotz, and Sorkin (1980) examined this problem in relation to audit sampling of financial accounts. If the probability of selecting y erroneous accounts is hypergeometric, that is, if

$$\Pr[Y = y]$$
$$= \binom{Np}{y}\binom{N-Np}{n-y}\Big/\binom{N}{n}, \qquad \max(0, n - N + Np) \le y \le \min(n, Np),$$

and if the conditional probability of detecting x among these y erroneous accounts is

$$\binom{y}{x}\pi^x(1 - \pi)^{y-x}, \qquad x = 0, 1, \ldots, y,$$

then the unconditional probability of detecting x erroneous accounts is

$$\Pr[X = x] = \sum_{y \ge x}\binom{Np}{y}\binom{N-Np}{n-y}\binom{N}{n}^{-1}\binom{y}{x}\pi^x(1 - \pi)^{y-x}, \qquad (6.169)$$

where $\max(0, n - N + Np) \leq x \leq \min(n, Np)$. Symbolically the distribution is

$$\mathrm{Bin}(Y, \pi) \bigwedge_Y \mathrm{Hypergeometric}(n, Np, N). \tag{6.170}$$

Distributions of this kind that arise in inspection sampling with imperfect inspection have been studied in depth by Johnson and Kotz and their co-workers; a full bibliography of their work is given in Johnson, Kotz, and Wu (1991). For an exposition of the types of distributions that can arise from particular screening procedures see Johnson and Kotz (1985). Johnson, Kotz, and Rodriguez (1985, 1986) have developed an *imperfect inspection hypergeometric distribution* that takes into account misclassification of nondefectives as well as defectives. This distribution can be represented symbolically as

$$\mathrm{Bin}(Y, p)^* \mathrm{Bin}(n - Y, p') \bigwedge_Y \mathrm{Hypergeometric}(n, D, N), \tag{6.171}$$

where (*) denotes the convolution operation; see Chapter 1, Section B9, and Chapter 3, Section 12.5.

The pmf is

$$\Pr[X = k | n, D, N; p, p']$$

$$= \binom{N}{n}^{-1} \sum_y \left[\binom{y}{w} \binom{n-y}{k-w} p^w (1-p)^{y-w} (p')^{k-w} (1-p')^{n-y-k+w} \right], \tag{6.172}$$

where $\max(0, n - N + D) \leq y \leq \min(n, D)$ and $\max(0, k - n + y) \leq w \leq \min(n, D)$. Perfect inspection corresponds to $p = 1$ and $p' = 0$. Johnson, Kotz, and Rodriguez provided tables of the pmf given k, the number of apparently nonconforming items, for $N = 100, 200, \infty$, and described how to interpolate for other values of N. The extension to double and multiple sampling schemes was investigated.

Hypergeometric, negative hypergeometric and generalized Waring (Pólya and inverse Pólya) distributions of order k were introduced by Panaretos and Xekalaki (1986a) using urn model derivations. Order k generalizations of the beta-binomial and beta-negative-binomial distributions were examined by Xekalaki, Panaretos, and Philippou (1987). New Pólya and inverse Pólya distributions of order k were obtained by Philippou, Tripsiannis, and Antzoulakis (1989) both by means of urn models and by beta-mixing of binomial and negative binomial distributions of order k. Godbole (1990b) has reconsidered the distributions of Panaretos and Xekalaki (1986a) and has also developed a waiting-time distribution of order k. Chapter 10, Section 6, deals with distributions of order k in greater depth.

The *intrinsic hypergeometric distribution* of Baldessari and Weber (1987) is the distribution of $\sum_{i=1}^N X_i$, where the X_i are dependent zero-one rv's.

CHAPTER 7

Logarithmic Distribution

1 DEFINITION

The random variable X has a *logarithmic distribution* if

$$\Pr[X = x] = \frac{a\theta^x}{x}, \qquad\qquad x = 1, 2, \ldots, \qquad (7.1)$$

$$= \frac{(x-1)\theta}{x}\Pr[X = x - 1], \qquad x = 2, 3, \ldots, \qquad (7.2)$$

where $0 < \theta < 1$ and $a = -[\ln(1-\theta)]^{-1}$. It is a one-parameter generalized power series distribution with infinite support on the positive integers.

The characteristic function is

$$\varphi(t) = \frac{\ln(1 - \theta e^{it})}{\ln(1 - \theta)}, \qquad (7.3)$$

and the pgf is

$$G(z) = \frac{\ln(1 - \theta z)}{\ln(1 - \theta)} = \frac{z\,{}_2F_1[1, 1; 2; \theta z]}{{}_2F_1[1, 1; 2; \theta]}. \qquad (7.4)$$

When shifted to the origin it is a Kemp generalized hypergeometric probability distribution; see Chapter 2, Section 4.1.

The mean and variance are

$$\mu = \frac{a\theta}{1 - \theta} \qquad (7.5)$$

$$\mu_2 = \frac{a\theta(1 - a\theta)}{(1 - \theta)^2} = \mu[(1 - \theta)^{-1} - \mu], \qquad (7.6)$$

where $a = -1/\ln(1 - \theta)$.

Confusion has sometimes arisen concerning the logarithmic distribution and the log-series distribution. The term *log-series distribution* appears in

285

the literature on species abundance data, in particular in Fisher, Corbet and Williams (1943) and Williams (1964), where a catch of individuals is found to contain S species and N individuals. If the expected *frequency* distribution of species represented r times is given by the terms of the series $-\alpha \ln(1-\theta)$, that is, $E[f_r] = \alpha \theta^r / r$, then $E[S] = -\alpha \ln(1-\theta)$, and $E[N] = \alpha \theta / (1-\theta)$. This is a two-parameter distribution; the parameter α is known as the index of diversity. Fisher's derivation of the log-series distribution was not well worded. Rao (1971), Boswell and Patil (1971), and Lo and Wani (1983) have endeavored to make his argument more rigorous; see also Wani (1978) concerning the interpretation of the parameters.

The name "logarithmic distribution" has also been used in the literature on floating-point arithmetic to denote the distribution of X where $\log_\beta X \pmod 1$ is uniformly distributed on $[0, 1]$; see, e.g., Bustoz et al. (1979).

2 HISTORICAL REMARKS AND GENESIS

In a little-known German paper, Lüders (1934) used the terms of a logarithmic series as the parameters for a convolution of Poisson, Poisson doublet, Poisson triplet, etc., distributions, thereby obtaining a negative binomial distribution. Quenouille (1949) reinterpreted this relationship, showing that the negative binomial arises as the sum of n independent logarithmic variables where n has a Poisson distribution; see Chapter 5, Section 3.

The log-series distribution has been used extensively by Williams [see, e.g., Williams (1947, 1964) and also Section 9 of this chapter]. This led to Fisher's derivation of the logarithmic distribution [Fisher et al. (1943)] as the limit as $k \to 0$ of the zero-truncated negative binomial distribution. A more rigorous proof is

$$\lim_{k \to 0} \frac{[(1-\theta)/(1-\theta z)]^k - (1-\theta)^k}{1 - (1-\theta)^k} = \lim_{k \to 0} \frac{(1-\theta z)^{-k} - 1}{(1-\theta)^{-k} - 1},$$

$$= \frac{\theta z + \theta^2 z^2 / 2 + \cdots}{\theta + \theta^2 / 2 + \cdots}; \tag{7.7}$$

see also Stuart and Ord (1987, p. 177). The formal mathematical relationship between Fisher's derivation and Quenouille's result was explained by Kemp (1978a).

An alternative model for the log-series distribution with two parameters was put forward by Anscombe (1950), who envisaged a sample of species from a finite but unknown number of species. He wrote "It is a multivariate distribution, consisting of a set of independent Poisson distributions with mean values αX, $\alpha X^2/2$, $\alpha X^3/3$, ... A 'sample' comprises one reading from each distribution." Boswell and Patil (1971) discussed this model at length and showed that it implies that the conditional distribution of the number of individuals behaves like the sum of S_0 independent logarithmic distributions (i.e. has a Stirling distribution of the first kind); see Section 11 of this chapter.

The negative binomial distribution is a gamma-mixed Poisson distribution (see Chapter 5, Section 3); this implies that there are two mixed Poisson models for the logarithmic distribution, corresponding to mixing after or mixing before truncation of the zero frequency. In the first case we have

$$\int_0^\infty \left\{ \frac{e^{\lambda z} - 1}{e^\lambda - 1} \right\} \frac{(e^\lambda - 1)e^{-\lambda/\theta} \, d\lambda}{(-\lambda)\log(1-\theta)} = \frac{\ln(1-\theta z)}{\ln(1-\theta)} \tag{7.8}$$

(this can be shown by expanding $e^{\lambda z} - 1$ as an infinite series in λz). In the second case we have

$$\int_\epsilon^\infty e^{\lambda(z-1)} \frac{e^{\lambda(\theta-1)/\theta} \, d\lambda}{\lambda E_1[\epsilon(1-\theta)/\theta]} = \frac{E_1[\epsilon/\theta - \epsilon z]}{E_1[\epsilon/\theta - \epsilon]}, \tag{7.9}$$

where $E_1[\omega]$ is the exponential integral

$$E_1[\omega] = \int_\omega^\infty e^{-t} t^{-1} dt = \Gamma(0, \omega); \tag{7.10}$$

the logarithmic distribution is obtained when the zero class is truncated and $\epsilon \to 0$. Such models were first suggested by Kendall (1948) in his paper on some modes of growth leading to the distribution.

The negative binomial distribution is the outcome of a number of stochastic processes, for example, the Yule-Furry process, the linear birth-death process, and the Pólya process. Kendall also studied the logarithmic distribution as a limiting form of such processes. He showed moreover that

$$\int_0^\infty \frac{z(1-v)}{(1-vz)} \frac{dv}{(v-1)\ln(1-\theta)} = \frac{\ln(1-\theta z)}{\ln(1-\theta)}, \tag{7.11}$$

that is, that the logarithmic distribution is a mixed shifted geometric distribution; see Chapter 1, Section C7 for a computer generation algorithm based on this property. Equivalently, the logarithmic distribution shifted to the support $0, 1, 2, \ldots$ is a mixed geometric distribution.

The steady-state birth-death process with state-dependent birth and death rates $\lambda_i = \lambda i$ and $\mu_i = \mu i$, $i \geq 1$, appears in Caraco (1979) in the context of animal group-size dynamics. For the steady-state distribution to be logarithmic it is necessary to assume that $\mu_i = \mu i$ for $i > 1$ and $\mu_i = 0$ when $i = 1$.

When Rao's damage process (Chapter 9, Section 2) is applied to the logarithmic distribution with parameter θ, the result is another logarithmic distribution with parameter $\theta p/(1-\theta+\theta p)$. Useful reviews of models for the logarithmic distributions are given by Nelson and David (1967), Boswell and Patil (1971), and Kemp (1981a). A functional equation for the pgf of the logarithmic distribution is presented in Panaretos (1987b).

3 MOMENTS

The rth factorial moment of the logarithmic distribution is

$$\mu'_{[r]} = a\theta^r \sum_{k=r}^{\infty} (k-1)(k-2)\ldots(k-r+1)\theta^{k-r}$$

$$= a\theta^r \frac{d^{r-1}}{d\theta^{r-1}} \left[\sum_{k=1}^{\infty} \theta^{k-1} \right]$$

$$= a\theta^r (r-1)!(1-\theta)^{-r}. \tag{7.12}$$

The moment generating function is

$$E[e^{tX}] = \frac{\ln(1-\theta e^t)}{\ln(1-\theta)}. \tag{7.13}$$

The first four uncorrected moments about the origin are

$$\mu'_1 = \mu = a\theta(1-\theta)^{-1},$$
$$\mu'_2 = a\theta^2(1-\theta)^{-2} + a\theta(1-\theta)^{-1} = a\theta(1-\theta)^{-2},$$
$$\mu'_3 = a\theta(1+\theta)(1-\theta)^{-3},$$
$$\mu'_4 = a\theta(1+4\theta+\theta^2)(1-\theta)^{-4}, \tag{7.14}$$

where $a = -1/\ln(1-\theta)$, and the corresponding central moments are

$$\mu_2 = a\theta(1-a\theta)(1-\theta)^{-2} = \mu'_1[(1-\theta)^{-1} - \mu'_1],$$
$$\mu_3 = a\theta(1+\theta-3a\theta+2a^2\theta^2)(1-\theta)^{-3},$$
$$\mu_4 = a\theta\{1+4\theta+\theta^2 - 4a\theta(1+\theta) + 6a^2\theta^2 - 3a^3\theta^3\}(1-\theta)^{-4}. \tag{7.15}$$

The coefficient of variation is

$$\frac{\mu_2^{1/2}}{\mu} = (a^{-1}\theta^{-1} - 1)^{1/2},$$

and the index of dispersion is $(1-a\theta)/(1-\theta) > 1$.

Table 7.1 (β_1, β_2) **Points of Logarithmic Distributions**

θ	β_1	β_2	θ	β_1	β_2
0.05	43.46	49.86	0.55	9.01	16.46
0.1	23.59	30.06	0.6	9.03	16.66
0.15	17.06	23.61	0.65	9.14	16.97
0.2	13.88	20.51	0.7	9.36	17.42
0.25	12.04	18.76	0.75	9.70	18.04
0.3	10.88	17.70	0.8	10.20	18.89
0.35	10.12	17.04	0.85	10.93	20.09
0.4	9.61	16.65	0.9	12.08	21.91
0.45	9.17	16.36	0.95	14.25	25.28
0.5	9.09	16.39			

The moment ratios $\beta_1 = \mu_3^2/\mu_2^3$ and $\beta_2 = \mu_4/\mu_2^2$ both tend to infinity as $\theta \to 0$ and as $\theta \to 1$, with

$$\lim_{\theta \to 0}(\beta_2/\beta_1) = 1 \quad \text{and} \quad \lim_{\theta \to 1}(\beta_2/\beta_1) = \frac{3}{2}.$$

The (β_1, β_2) points of the logarithmic distribution are given in Table 7.1. The moments about zero satisfy the relation

$$\mu'_{r+1} = \theta \frac{d\mu'_r}{d\theta} + \frac{a\theta}{1 - \theta}\mu'_r. \tag{7.16}$$

The central moments satisfy the relation

$$\mu_{r+1} = \theta \frac{d\mu_r}{d\theta} + r\mu_2\mu_{r-1}. \tag{7.17}$$

Since the logarithmic distribution is a power series distribution, it follows that the cumulants satisfy the recurrence relations [Khatri (1959)]

$$\kappa_r = \theta \frac{d\kappa_{r-1}}{d\theta}, \tag{7.18}$$

and that the factorial cumulants satisfy

$$\kappa_{[r]} = \theta \frac{d\kappa_{[r-1]}}{d\theta} - (r - 1)\kappa_{[r-1]}. \tag{7.19}$$

The mean deviation is

$$\nu_1 = 2a \sum_{k=1}^{[\mu]} \frac{(\mu - k)\theta^k}{k} = \frac{2a\theta\left\{\theta^{[\mu]} - \Pr[X > [\mu]]\right\}}{1 - \theta}, \tag{7.20}$$

where $\mu = a\theta(1 - \theta)^{-1}$ and $[\mu]$ denotes the integer part of μ; see Kamat (1965).

4 PROPERTIES

From (7.1)

$$\frac{\Pr[X = x + 1]}{\Pr[X = x]} = \frac{x\theta}{x + 1}, \qquad x = 1, 2, \ldots . \tag{7.21}$$

This ratio is less than 1 for all values of $x = 1, 2, \ldots$, since $\theta < 1$. Hence the value of $\Pr[X = x]$ decreases as x increases.

The failure rate r_x is given by

$$\frac{1}{r_x} = \sum_{i \geq x} \frac{\Pr[X = i]}{\Pr[X = x]}$$

$$= 1 + \frac{x\theta}{x + 1} + \frac{x\theta^2}{x + 2}$$

$$< \frac{1}{r_{x+1}}, \tag{7.22}$$

and hence the distribution has a decreasing failure rate [Patel (1973)].

Methods of approximating $\Pr[X \leq [\mu]]$ are described in Section 5. Approximation (7.25) leads to the following approximate formula for the median M of the distribution:

$$M \approx e^{-\gamma}(1 - \theta)^{-1/2} + 0.5 + e^{-2\gamma} \doteq 0.56146(1 - \theta)^{-1/2} + 0.81524, \tag{7.23}$$

where γ is Euler's constant; see Chapter 1, Section A2. This formula has been attributed to Grundy [see Williams (1964) and Gower (1961)] and it seems to give a good approximation if θ is not too small.

The distribution has a rather long positive tail. For large values of x the shape of the tail is similar to that of a geometric distribution (Chapter 5, Section 2) with parameter θ.

The entropy of the distribution is

$$a \sum_{j=1}^{\infty} \theta^j j^{-1} \log j - a\theta(1 - \theta)^{-1} \log \theta - \log a.$$

Sironomey (1962) has shown that this is an increasing function of θ. Sironomey also gave an expression for the entropy of an individual with respect to a species, for the log-series distribution.

Another property of the logarithmic distribution shifted to the support $0, 1, 2, \ldots$ is infinite divisibility, a result that was first proved by Katti (1967).

Table 7.2   Properties of the Logarithmic Distribution for Various Values of θ

θ	$\Pr[X = 1]$	$\Pr[X = 2]$	$\sum_{j \geq 10} \Pr[X = j]$	Mean
0.1	0.95	0.05	0.00	1.05
0.3	0.84	0.13	0.00	1.20
0.5	0.72	0.18	0.00	1.44
0.7	0.58	0.20	0.01	1.94
0.8	0.50	0.20	0.03	2.49
0.85	0.45	0.19	0.05	2.99
0.90	0.39	0.18	0.09	3.91
0.95	0.32	0.15	0.19	6.34
0.99	0.21	0.11	0.40	21.50
0.995	0.19	0.09	0.47	37.56
0.999	0.14	0.07	0.59	144.62
0.9999	0.11	0.05	0.69	1085.63

From Kemp (1981b).

Infinite divisibility follows from the log-convexity of the probabilities [see Steutel (1970)]. The shifted logarithmic distribution arises therefore from a Poisson distribution of clusters of various sizes. Explicit formulas for the cluster-size probabilities were given by Kemp (1978a).

Table 7.2 from Kemp (1981b) summarizes various properties of the logarithmic distribution.

5   APPROXIMATIONS AND BOUNDS

The cumulative probability

$$\Pr[X \leq x] = \frac{-\sum_{j=1}^{x} \theta^j/j}{\ln(1-\theta)} = 1 + \frac{\sum_{j=x+1}^{\infty} \theta^j/j}{\ln(1-\theta)}$$

can be approximated using

$$\sum_{j=x+1}^{\infty} \frac{\theta^j}{j} = -\ln(1-\theta) - \sum_{j=1}^{x} j^{-1} + \int_0^{1-\theta} \phi^{-1}\{1 - (1-\phi)^x\}d\phi$$

$$\approx -\ln(1-\theta) - (\gamma + x^{-1}/2 - x^{-2}/12 + \log x)$$

$$+ [-Ei(-x(1-\theta)) + \gamma + \ln\{x(1-\theta)\}]$$

$$\approx -Ei[-x(1-\theta)] - x^{-1}/2 + x^{-2}/12 \qquad (7.24)$$

with an error less than $x(1 - \theta)^2$. (The approximation

$$\sum_{j=1}^{x} j^{-1} = \gamma + x^{-1}/2 - x^{-2}/12 + \ln x$$

has an error less than 10^{-6} for $x \geq 10$, and less than 10^{-10} for $x \geq 100$; the function $Ei(u) = \int_{-\infty}^{u} z^{-1} e^z dz$.)

If $x(1 - \theta)$ is small, then the approximation

$$\sum_{j=x+1}^{\infty} \frac{\theta^j}{j} \doteq - \ln[x(1 - \theta)] + x(1 - \theta) - \gamma \qquad (7.25)$$

has an error less than $[x(1 - \theta)]^2/4$. Also

$$\sum_{j=x+1}^{\infty} \frac{\theta^j}{j} \doteq Ei(x \ln \theta) + \frac{\theta^x}{2x}. \qquad (7.26)$$

These formulas are given in Gower (1961), and they lead to an approximation for the median of the distribution; see Section 4 of this chapter.

Owen (1965) gave the following formula, which is suitable when $x(1 - \theta)$ is large:

$$\sum_{j=x+1}^{\infty} \frac{\theta^j}{j} \doteq \theta^{x+1}(1 - \theta)^{-1} \left[x^{-1} - 1!\{x(x - 1)\}^{-1}(1 - \theta)^{-1} \right.$$

$$+ 2!\{x(x - 1)(x - 2)\}^{-1}(1 - \theta)^{-2} - \cdots$$

$$\left. + (-1)^r r!\{x(x - 1) \cdots (x - r)\}^{-1}(1 - \theta)^{-r} \right]. \qquad (7.27)$$

The inequalities

$$\theta^{x+1}(x + 1)^{-1} \left[1 - \frac{(x + 1)\theta}{(x + 2)} \right] < \sum_{j=x+1}^{\infty} \frac{\theta^j}{j} < \theta^{x+1}(x + 1)^{-1}(1 - \theta)^{-1} \qquad (7.28)$$

provide simple bounds for the tail probability and are useful for large x.

6 COMPUTATION AND TABLES

The one-term recurrence relationship

$$\Pr[X = x + 1] = \frac{\theta x \Pr[X = x]}{x + 1}, \qquad x = 1, 2, \ldots,$$

together with $\Pr[X = 1] = -\theta/\ln(1 - \theta)$, facilitates the computation of the probabilities and the cumulative probabilities. When $\theta > 0.99$, the tail length is great, and it may be advisable to use high precision arithmetic on a computer.

Tables of the individual probabilities (7.1) and cumulative probabilities have been provided by Williamson and Bretherton (1964). The argument of the tables is the expected value ($\mu = -\theta[(1 - \theta)\ln(1 - \theta)]^{-1}$) and not θ. However, for $\mu = 1.0(0.1)10.0(1)50$, values of θ corresponding to given μ (to five decimal places) are provided. Values of second and fourth differences are also given. Individual probabilities and cumulative probabilities are given (also to five decimal places) for $\mu = 1.1(0.1)2.0(0.5)5.0(1)10$, the tabulation in each case being continued until the cumulative probability exceeds 0.999.

Extensive tables of these probabilities are available in Patil et al. (1964). These give values to six decimal places for $\theta = 0.01(0.01)0.70(0.005)0.900$ $(0.001)0.999$. Patil and Wani (1965a) gave θ to four decimal places for $\mu = 1.02(0.02)2.00(0.05)4.00(0.1)8.0(0.2)16.0(0.5)30.0(2)40(5)60(10)140(20)200$. A table for μ as a function of θ for $\theta = 0.01(0.01)0.99$ appeared in Patil (1962d). The availability of tables for MVU estimation was described by Patil (1985).

Fisher et al. (1943) gave a table which enables the "index of diversity" α to be estimated from observed values of N and S, using $e^{E[S]/\alpha} = 1 + E[N]/\alpha$; see Section 1 of this chapter. Alternatively, the tables of Barton et al. (1963) may be employed (see Section 7.2 of this chapter).

7 ESTIMATION

7.1 Model Selection

Before considering numerical estimation of the parameter θ of the logarithmic distribution, we first give details of graphical methods of model selection.

Ord (1967a, 1972) has shown that plotting $u_x = xf_x/f_{x-1}$ against x can be expected to give a straight line with intercept $-\theta$ and slope θ, where f_x is the observed frequency of an observation x (see Chapter 2, Section 3.2) . Clearly successive u_x are dependent; to smooth the data, Ord suggested the use of the statistic $v_x = (u_x + u_{x-1})/2$. In Ord (1967a) he applied the method to Corbet's butterfly data [Fisher et al. (1943)] and in Ord (1972) to product-purchasing data.

Gart (1970) noted that $x \Pr[X = x]/\{(x - 1) \Pr[X = x - 1]\} = \theta$ and proposed the use of $xf_x/\{(x - 1)f_{x-1}\}$; when plotted against x, this can be expected to give a horizontal line with intercept θ.

A third method, due to Hoaglin and Tukey [Hoaglin et al. (1985)] is to plot $\log n_x^* + \log x - \log N$ against x, where N is the total frequency and n_x^* is obtained by smoothing f_x. This can be expected to give a straight line with intercept $-\ln\{-\ln(1 - \theta)\}$ and slope θ. Hoaglin and Tukey considered their method to be superior to Ord's. They too used Corbet's butterfly data for illustration.

These graphical methods would seem to have much to commend them with regard to distribution selection, but their usefulness for parameter estimation may be dubious.

7.2 Point Estimation and Confidence Intervals

Consider now the problem of the numerical estimation of θ, given values of n independent random variables $x_1, x_2, \ldots, x_n$, each having distribution (7.1). The maximum-likelihood estimator $\hat{\theta}$ is given by

$$\bar{x} = n^{-1} \sum_{j=1}^{n} x_j = \frac{\hat{\theta}}{-(1 - \hat{\theta}) \ln(1 - \hat{\theta})}; \tag{7.29}$$

the asymptotic variance is $n^{-1}(\theta^2/\mu_2)$, where μ_2 is the variance of the logarithmic distribution [Patil (1962d)]. To estimate the variance of $\hat{\theta}$, the usual practice is to replace θ by $\hat{\theta}$. Because (7.1) is a generalized power series distribution, maximum-likelihood and moment estimation are equivalent.

Patil (1962d) gave a table of μ as a function of θ, $\theta = 0.01(0.01)0.99$, whereas Williamson and Bretherton (1964) provided values of θ corresponding to μ for $\mu = 1.0(0.1)10.0(1)50$. (Williamson and Bretherton also tabulated the bias and standard error of $\hat{\theta}$ in a few cases.) Patil and Wani (1965a) gave a more extensive table of θ for values of μ (see Section 6). Barton et al. (1963) gave solutions of b (to seven decimal places) for the equation $e^b = 1 + b\bar{x}$, where $1 - (\bar{x})^{-1} = 0(0.001)0.999$. The maximum-likelihood estimator $\hat{\theta}$ can then be calculated as

$$\hat{\theta} = \frac{b\bar{x}}{1 + b\bar{x}}. \tag{7.30}$$

Chatfield (1969) noted that θ changes very slowly for large values of μ, and advocated re-parameterization, with $\psi = \theta/(1 - \theta)$; he tabulated ψ against μ for $\mu = 0.0(0.1)20(1)69$.

A detailed study of the maximum-likelihood estimator $\hat{\theta}$ was carried out by Bowman and Shenton (1970). They gave values for the mean, variance, skewness $(\sqrt{\beta_1})$, and kurtosis (β_2) of $\hat{\theta}$ for various combinations of $\hat{\theta}$ and the sample size n, where $0.1 \leq \hat{\theta} \leq 0.9$ and $n \geq 8$; they commented on the computational difficulties when $0.9 < \hat{\theta} < 1.0$. They also made the general statement that "the distribution of $\hat{\theta}$ is not too far removed from the normal distribution; however, departures from normality become serious when $\hat{\theta}$ exceeds about 0.9 [or] when the sample size is less than about nine." Bowman and Shenton remarked that the convergence of the Taylor expansion for $\hat{\theta}$ poses difficult problems.

Earlier, in a note unreferenced by Bowman and Shenton, Birch (1963) had provided a computer algorithm for $\hat{\theta}$. This uses $\xi = (1 - \hat{\theta})^{-1}$ and solves

$$\xi - 1 - \bar{x} \ln \xi = 0$$

by Newton-Raphson iteration, taking

$$\xi_{i+1} = \frac{1 - \bar{x}(\log \xi_i - 1)}{1 - \bar{x}/\xi_i}, \qquad i = 1, 2, \ldots.$$

The recommended first approximation is

$$\xi_1 = 1 + \{k(\bar{x} - 1) + 2\} \ln \bar{x},$$

with k some value between 1 and 5/3. When $\bar{x}$ is close to unity, Birch recommended, instead, the use of the first few terms of the Taylor series

$$\xi = 1 + 2(\bar{x} - 1) + \frac{2(\bar{x} - 1)^2}{3} - \frac{2(\bar{x} - 1)^3}{9} + \frac{14(\bar{x} - 1)^4}{135} - \cdots.$$

Böhning (1983a, b) treated maximum-likelihood estimation for the logarithmic distribution as a problem in numerical analysis. He referred to Birch (1963), but not to Bowman and Shenton (1970). Bowman and Shenton stated, without proof, the uniqueness of $\hat{\theta}$. Böhning has proved this. He has also shown that $\hat{\theta}$ is the maximum-likelihood estimator of θ iff $\hat{\theta}$ is a fixed point of

$$\Phi(\hat{\theta}) = \frac{\bar{x} \ln(1 - \hat{\theta})}{\bar{x} \log(1 - \hat{\theta}) - 1}.$$

He found that $\hat{\theta}_{i+1} = \Phi(\hat{\theta}_i)$ converges monotonically to $\hat{\theta}$ and that for this method, unlike Newton-Raphson, the choice of $\hat{\theta}_1$ is not critical. His computer timings, for his fixed-point method, for Newton-Raphson, and for a secant method, indicate that his fixed-point method sometimes converges much more slowly. Nethertheless, it converges more surely.

Patil (1962d) also suggested the following explicit estimators of θ (based on relationships that are exactly true in the population):

1. $1 - f_1/\bar{x}$.
2. $1 - \sum_{j \geq 1}\{j f_j\}/\sum_{j \geq 1}\{j^2 f_j\}$.
3. $\theta^\dagger = \sum_{j \geq 2}\{j/(j - 1)\}f_j$, where f_j, $j = 1, 2, \ldots$, is the proportion of observations that are equal to j.

Table 7.3 Asymptotic Efficiencies (%) of Estimators of θ for the Logarithmic Distribution

	Value of θ		
Estimator	0.1	0.5	0.9
1	98.3	89.7	73.9
2	22.8	44.9	48.8
3	89.5	44.7	5.7

Estimators 1 and 2 are asymptotically unbiased. Estimator 3, $\theta^\dagger$, is unbiased for all sizes of sample and has variance

$$\mathrm{Var}(\theta^\dagger) = n^{-1} \left[\sum_{j=2}^{\infty} \{j/(j-1)\}^2 \Pr[X=j] - \theta^2 \right].$$

The asymptotic efficiencies of these estimators (relative to $\hat{\theta}$) for certain values of θ are shown in Table 7.3.

It appears that estimation method 1 is preferable to estimation methods 2 or 3 on grounds of both accuracy and simplicity. Kemp (1986) showed that the equation for estimator 1 is an approximation to the maximum-likelihood equation.

An alternative approach is to use a computer optimization package to solve the ML equation. Such packages often require bounds for the parameter estimates to be specified. Kemp and Kemp (1988) gave the following bounds for $\hat{\theta}$:

$$\frac{(9\bar{x}-6) - (9\bar{x}^2 - 12\bar{x} + 12)^{1/2}}{6\bar{x}-2} < \hat{\theta} < \frac{(6\bar{x}-3) - (24\bar{x}^2 - 24\bar{x} + 9)^{1/2}}{\bar{x}}.$$

The minimum variance unbiased estimator, θ^*, of θ is given by the following equation:

$$\theta^* = \begin{cases} b\left(\sum_{j=1}^{n} x_j - 1\right) \Big/ b\left(\sum_{j=1}^{n} x_j\right) & \text{if } \sum_{j=1}^{n} x_j > n, \\ = 0 & \text{if } \sum_{j=1}^{n} x_j = n, \end{cases} \tag{7.31}$$

where $b(m)$ is the coefficient of θ^m in $[-\ln(1-\theta)]^n$. Tables of θ^*, given $\sum_{j=1}^{n} x_j$, for small n, to four decimal places, are given in Patil et al. (1964).

Wani and Lo (1975a, b) provided tables and charts for obtaining confidence intervals for θ, given the values of X in a random sample of size n. They used the observed value of $\sum_{j=1}^{n} x_j$ and the fact that the sum of independent logarithmic random variables has a Stirling distribution of the first kind; see

Patil and Wani (1965b) and Section 11 of this chapter. These confidence intervals were compared with ones obtained by other methods in Wani and Lo (1977).

Estimation for the log-series distribution has been studied by Anscombe (1950), Engen (1974), Lo and Wani (1983), and Rao (1971). Rao also discussed the special problems of testing goodness of fit for the log-series distribution.

8 CHARACTERIZATIONS

The following characterization of the logarithmic distribution has been given by Patil and Wani (1965b): "Let X and Y be two independent discrete random variables each taking the value 1 with nonzero probability; then if

$$\Pr[X = x | X + Y = z] = \frac{[x^{-1} + (z-x)^{-1}]\beta^x}{\sum_{j=1}^{z-1}[j^{-1} + (z-j)^{-1}]\beta^j}, \tag{7.32}$$

for $0 < \beta < \infty$, $z = 2, 3, \ldots$, then X and Y each has the logarithmic distribution with parameters in the ratio β."

Wani (1967) has also shown that a distribution is a logarithmic distribution with parameter θ iff its moments are $\mu_1' = a\theta/(1-\theta)$, $\mu_s = a\theta c_s(\theta)/(1-\theta)^{s+2}$, $s = 0, 1, 2, \ldots$, with $a = -[\ln(1-\theta)]^{-1}$ and $c_s(\theta) = \sum_{i=0}^{s} c(s,i)\theta^i$, where

$$c(s,i) = (i+1)c(s-1,i) + (s-i+1)c(s-1,i-1)$$

and $c(0,0) = 1$, $c(s,i) = 0$ for $i < 0$.

Kyriakoussis and Papageorgiou (1991b) have obtained the following two characterizations: First, suppose that X is a rv with nonzero pmf for every nonnegative integer value of X and finite mean, and also that Y is another nonnegative integer valued rv such that the conditional distribution of $Y|X = x$ is binomial. Then the regression function of X on Y has the form

$$E[X|Y = y] = \begin{cases} b & \text{if } y = 0, \\ ay & \text{if } y = 1, 2, \ldots, \end{cases} \tag{7.33}$$

where a and b are constants, iff X has a logarithmic distribution with added zeroes (a modified logarithmic distribution; see Section 10 of this chapter).

Their second characterization is a characterization of the logarithmic distribution itself, and it involves the assumption that $Y|X = x$ is a positive integer valued rv with pmf

$$\Pr[Y = y | X = x] = \binom{y-1}{x-1} p^x (1-p)^{y-x},$$

where $y = x, x + 1, \ldots, x = 1, 2, \ldots, 0 < p < 1$. Then the regression function $E[X|Y = y]$ has a specified form (more complicated than above) iff X has a logarithmic distribution.

9 APPLICATIONS

In Fisher, Corbet and Williams (1943), the log-series distribution was applied to the results of sampling butterflies (Corbet's data) and also to data obtained in connection with the collection of moths by means of a light-trap (Williams' data). In these experiments it was found that if the number of species represented by exactly one individual is n_1, then the numbers of species represented by two, three, etc., individuals are approximately $(n_1/\theta)\theta^2/2$, $(n_1/\theta)\theta^3/3$, etc., respectively, where θ is a positive number less than unity; see Section 1 of this chapter. The quantity $\alpha = n_1/\theta$ is sometimes called an "index of diversity" and is thought to be independent of the size of catch.

Table 7.4 gives the numbers (N) of mosquitoes caught in light traps in various cities in Iowa [Rowe (1942)], together with the numbers of species (S) observed. The last column gives the estimated value of α obtained by eliminating $\hat{\theta}$ from $S = -\hat{\alpha} \ln(1 - \hat{\theta})$, $N = \hat{\alpha}\hat{\theta}/(1 - \hat{\theta})$, that is, by solving $\exp(S/\hat{\alpha}) = 1 + N/\hat{\alpha}$. There seems sound evidence for a constant value of α (of about 2) for mosquito data.

Williams (1964) gave numerous other sets of ecological data on species abundance where the log-series distribution fits very well, such as Blackman's (1935) data on the average number of plant species found on quadrats of various sizes in a grassland formation, and also the distribution of species of British nesting birds by numbers of individuals [Witherby et al. (1941)]. Williams (1964) also fitted the logarithmic distribution to many instances of data concerning the distribution of parasites per host, such as the number of head lice per host [Buxton (1940)]. In Williams (1944), he had applied it to numbers of publications by entomologists. An application to an inventory control problem in the steel industry appears in Williamson and Bretherton (1964).

In a long series of papers, including Chatfield et al. (1966) and Chatfield (1970, 1986), the authors have described the use of the logarithmic distribution to represent the distribution of numbers of items of a product purchased by a buyer in a given period of time. They remarked that the logarithmic distribution is likely to be a useful approximation to a negative binomial distribution with a low value of the parameter k (e.g., less than 0.1). The logarithmic distribution has the advantage of depending on only one parameter, θ, instead of the two, k and P, that are needed for the negative binomial distribution.

Table 7.4 Numbers of Mosquitoes Caught, Numbers of Species Observed, and Estimated Index of Diversity

City	Number of individuals	Number of species	α
Ruthven	20,239	18	1.95
Des Moines	17,077	20	2.24
Davenport	15,280	18	2.02
Ames	12,504	16	1.81
Muscatine	6,426	16	1.98
Dubuque	6,128	14	1.71
Lansing	5,564	16	2.03
Bluffs	1,756	13	1.90
Sioux City	661	12	2.08
Burlington	595	12	2.13

Adapted from Rowe (1942).

10 TRUNCATED AND MODIFIED LOGARITHMIC DISTRIBUTIONS

The most common form of truncation is by exclusion of values greater than a specified value r. This kind of truncation is likely to be encountered with distributions of logarithmic type, as they have long positive tails and it is not always practicable to evaluate each large observed value. Paloheimo (1963) describes circumstances under which a truncated logarithmic distribution might be appropriate.

If $X_1, X_2, \ldots, X_n$ are independent random variables each with the distribution

$$\Pr[X = x] = \frac{\theta^x}{x} \left[\sum_{j=1}^{r} \frac{\theta^j}{j} \right]^{-1}, \qquad x = 1, 2, \ldots, r, \qquad (7.34)$$

then the maximum-likelihood estimator, $\hat{\theta}$, of θ satisfies the equation

$$\bar{x} = n^{-1} \sum_{j=1}^{n} x_j = \frac{\hat{\theta}(1 - \hat{\theta}^r)}{(1 - \hat{\theta}) \left\{ \sum_{j=1}^{r} (\hat{\theta}^j / j) \right\}}. \qquad (7.35)$$

Patil and Wani (1965a) have provided tables giving the value of $\hat{\theta}$ corresponding to selected values of $\bar{x}$ for $r = 4(1)8(2)12, 15(5)40(10)60(20)100, 200, 500, 1000$. They also gave some numerical values of the bias and standard error of $\hat{\theta}$.

Alternatively, a method based on equating sample and population moments may be used. This leads to

$$\theta^* = \frac{m_3' - (r+2)m_2' + (r+1)m_1'}{m_3' - rm_2'},$$ (7.36)

where $m_s' = n^{-1}\sum_{j=1}^n x_j^s$.

A modified logarithmic distribution, called the *logarithmic-with-zeroes distribution*, or *log-zero distribution*, with pgf

$$G(z) = c + (1-c)\frac{\ln(1-\theta z)}{\ln(1-\theta)}$$ (7.37)

was introduced by Williams (1947). Derivations of the distribution via models of population growth were given by Feller (1957, p. 276), Kendall (1948), and Bartlett (1960, p. 9). Its use as a model for stationary purchasing behavior was suggested by Chatfield et al. (1966). Khatri (1961), also Patil (1964b), obtained it by mixing binomials, the binomial index having a logarithmic distribution; we have

$$G(z) = \frac{\sum_{n=1}^\infty (q+pz)^n \theta^n/n}{-\log(1-\theta)} = c + (1-c)\frac{\ln(1-\phi z)}{\log(1-\phi)},$$ (7.38)

with $c = \ln(1-q\theta)/\ln(1-\theta)$ and $\phi = p\theta/(1-q\theta)$. The distribution is discussed in greater detail in Chapter 8, Section 2.2.

11 OTHER RELATED DISTRIBUTIONS

The distribution of the sum (X_n) of n independent random variables, each having the logarithmic distribution, has the pgf $[-a\ln(1-\theta z)]^n$. From this it follows that $\Pr[X_n = x]$ is proportional to the coefficient of z^x in $[-\ln(1-\theta z)]^n$:

$$\Pr[X_n = x] = \frac{n!|S_x^{(n)}|\theta^x}{x![-\ln(1-\theta)]^n}, \qquad x = n, n+1, \ldots,$$ (7.39)

where $S_x^{(n)}$ is the Stirling number of the first kind, with arguments n and x; see Chapter 1, Section A3. The distribution of X_n is called the *Stirling distribution of the first kind*.

Patil and Wani (1965b) state the following properties of the distribution:

1. If $\theta < 2n^{-1}$, the distribution has a unique mode at $x = n$ (the smallest possible value), and the values of $\Pr[X_n = x]$ decrease as x increases.
2. If $\theta = 2n^{-1}$, there are two equal modal values at $x = n$ and $x = n+1$.

3. If $\theta > 2n^{-1}$, the value of $\Pr[X_n = x]$ increases with x to a maximum (or pair of equal maxima) and thereafter decreases as x increases; see also Sibuya (1988).

From the definition of this distribution, it is clear that as n tends to infinity the standardized distribution, corresponding to (7.39), tends to the unit normal distribution.

Douglas (1971) and Shanmugam and Singh (1981) have studied properties and estimation for the distribution. A good general discussion is in Berg (1988a). Cacoullos (1975) has introduced a multiparameter extension, obtained from the convolution of a number of logarithmic distributions that have been left-truncated at different points.

A distribution obtained by assigning a lognormal distribution to the parameter of a Poisson distribution is termed a *discrete lognormal distribution* [Anscombe (1950)]. This mixed Poisson distribution is mentioned here because a zero-truncated form is an important competitor of the log-series distribution as a model for the distribution of observed abundancies of species and similar phenomena. Bliss (1965) made comparisons of the fidelity with which the two distributions represented five sets of data from moth-trap experiments – actually four distinct sets and the combined data [see also Preston (1948) and Cassie (1962)]. He found that in each case the truncated discrete lognormal distribution gave the better fit (as judged by χ^2-probabilities), though both distributions gave acceptable representations (again judged by a χ^2-test). If we suppose that the Poisson parameter λ is distributed lognormally, with expected value ξ and standard deviation η, then the compound distribution has

$$\mu = \exp\left(\xi + \frac{\eta^2}{2}\right),$$

$$\sigma^2 = \mu + \mu^2 \left\{1 - \exp(\eta^2)\right\}. \tag{7.40}$$

A summary of its properties appears in Shaban (1988).

Darwin (1960) has obtained an ecological distribution (with finite support) from the beta-binomial distribution by a process analogous to that used by Fisher when obtaining the logarithmic distribution from the negative binomial. Katti (1966) included a model with clusters distributed according to a Pascal distribution, and individuals within a cluster distributed logarithmically, in his investigation of interrelations among certain discrete distributions. Panaretos (1983a) has studied a model with a logarithmic distribution of clusters, where the number of individuals within a cluster has a Pascal distribution; he has given a species-abundance interpretation of the model.

Shenton and Skees (1970) found that the logarithmic distribution, depending as it does on only one parameter, was inadequate as a descriptor of storm duration measured in discrete units of time. They suggested (on an empirical basis) two J-shaped variants:

1. $\Pr[X = x] = a(1 - p)p^{x-1} + \dfrac{(1 - a)\theta^x}{-x \ln(1 - \theta)},$

$$0 \le p < 1,\ 0 \le \theta < 1,\qquad x = 1, 2, \ldots \qquad (7.41)$$

2. $\Pr[X = 1] = 1 - \dfrac{a\theta}{b + 1},$

$$\Pr[X = x] = a\theta^{x-1}\left\{(x + b - 1)^{-1} - \theta(x + b)^{-1}\right\},$$

$$0 < a,\ -1 < b,\ 0 < \theta < 1,\ x = 2, 3, \ldots . \qquad (7.42)$$

They noted that the latter distribution involves negative mixing, and considered that it fitted the data "rather well."

Kempton's (1975) generalization of the log-series distribution also involves two parameters, β and γ. He took

$$\Pr[X = x] = C \int_0^\infty \frac{t^{x-1}e^{-t}\,dt}{x!(1 + \beta t)^\gamma},\qquad 0 < \beta,\ 0 < \gamma, \qquad (7.43)$$

where C is a normalizing constant. The distribution is J-shaped, like the log-series distribution; however, it has an even longer tail. If $\beta \to 0$ and $q \to \infty$ in such a way that $\lambda = \beta q$ remains finite and positive, then it tends to a log-series distribution with

$$\Pr[X = x] = \frac{x - 1}{x(1 + \lambda)}\Pr[X = x - 1].$$

The *"generalized" logarithmic distribution* of Jain and Gupta (1973) is a limiting form as $n \to \infty$ of a zero-truncated "generalized" negative binomial distribution. It has pmf

$$\Pr[X = x] = \frac{-c^x(1 - c)^{x(\beta-1)}\Gamma(\beta x)}{\Gamma(x + 1)\Gamma(\beta x - x + 1)\ln(1 - c)},\qquad x = 1, 2, \ldots, \qquad (7.44)$$

where $0 < c \le c\beta < 1$. The ordinary logarithmic distribution is the special case $\beta = 1$. Jain and Gupta showed that

$$\mu_1' = \frac{-c}{(1 - \beta c)\ln(1 - c)},$$

$$\mu_2' = \frac{-c(1 - c)}{(1 - \beta c)^3 \ln(1 - c)},\qquad \text{etc.} \qquad (7.45)$$

They used the method of moments to fit Williams' data on numbers of papers by entomologists.

Gupta (1977) has examined MVU estimation for the parameters of the distribution. Rao (1981) has applied it to the study of correlation between two types of children in a family. The incomplete moments were studied by Tripathi, Gupta and Gupta (1986). The log-convexity of the probabilities, and hence the infinite divisibility of the distribution when shifted to the support $0, 1, 2, \ldots$, were shown by Hansen and Willekens (1990); these authors also discussed the use of the distribution in risk theory in a problem related to the total claim size up to time t [see also Hogg and Klugman (1984)]. A modified form of Jain and Gupta's "generalized" logarithmic distribution, with added zeroes, has been studied by Jani (1986). Gupta (1976) examined the distribution of the sum of a number of independent "generalized" logarithmic distributions; a special case is the Stirling distribution of the first kind.

A different "generalized" logarithmic distribution was developed by Tripathi and Gupta (1988). It is a limiting form of their "generalized" negative binomial distribution, and has a complicated pmf. The authors did not give moment properties. They refitted the data on entomologists' papers using a form of maximum likelihood.

The *logarithmic distribution of order k*, with pgf

$$G(z) = (k \ln p)^{-1} \ln \left\{ \frac{1 - z + qp^k z^{k+1}}{1 - pz} \right\}, \tag{7.46}$$

where $0 < p < 1$, $q = 1 - p$, k an integer, was introduced by Hirano et al. (1984); see Chapter 10, Section 6.2.

A second logarithmic distribution of order k with pgf

$$G(z) = \frac{-\ln \left\{ 1 - \theta(z + z^2 + \cdots + z^k)/(1 + \theta k) \right\}}{\ln(1 + \theta k)} \tag{7.47}$$

has been studied by Panaretos and Xekalaki (1986a); see Chapter 10, Section 6.3.

There are three negative binomial distributions of order k (see Chapter 10, Sections 6.2 and 6.3); the two logarithmic distributions of order k are related pairwise to two of them in the same manner that the ordinary logarithmic distribution is related to the ordinary negative binomial distribution. For both logarithmic distributions of order k taking $k = 1$ gives the ordinary logarithmic distribution.

A quite different generalization of the logarithmic distribution has been obtained by Tripathi and Gupta (1985) as a limiting form of their generalization of the negative binomial distribution. We have

$$G(z) = \lim_{\substack{a \to 0 \\ c=1}} \frac{c \, _2F_1[a/\beta + c, 1; \lambda + c; \beta z]}{_2F_1[a/\beta + c, 1; \lambda + c; \beta]} = \frac{_2F_1[1, 1; \lambda + 1; \beta z]}{_2F_1[1, 1; \lambda + 1; \beta]}, \tag{7.48}$$

with $-1 < \lambda$, $0 < \beta < 1$. It is a shifted generalized hypergeometric distribution, with the properties of such distributions. It tends to the ordinary logarithmic distribution as $\lambda \to 1$. Tripathi and Gupta gave recurrence formulas for the probabilities and factorial moments, some simple estimators of the parameters, minimum-χ^2 estimators, and maximum-likelihood estimators. Also they suggested a graphical selection procedure.

Recently Kemp (1992a,b) has investigated the Euler distribution used by Benkherouf and Bather (1988) in their oil-exploration problem. It is an infinitely divisible q-series analogue of the Poisson distribution; see Chapter 4, Section 12.6. Given a Poisson distribution for the number of clusters, an Euler distribution is the outcome when the cluster-size distribution has pgf

$$G(z) = \frac{\sum_{i \geq 1} \left[p^i z^i / \{ i(1 - q^i) \} \right]}{\sum_{i \geq 1} \left[p^i / \{ i(1 - q^i) \} \right]}, \tag{7.49}$$

$0 < p < 1$, $0 < q < 1$, p and q unrelated. This cluster-size distribution is a q-series analogue of the logarithmic distribution; it tends to the logarithmic distribution when $q \to 1$.

CHAPTER 8

Mixture Distributions

1 INTRODUCTION

The important class of distributions to be discussed in this chapter consists of *mixtures* of discrete distributions. The notion of *mixing* often has a simple and direct interpretation in terms of the physical situation under investigation. For instance, the random variable concerned may be the result of actual mixing of a number of different populations, such as the number of car insurance claims per driver, where the expected number of claims varies with category of driver. Alternatively, the random variable may come from one of a number of different sources, but the source is unknown; a mixture rv is then the outcome of ascribing a probability distribution to the possible sources. Sometimes, however, "mixing" is just a mechanism for constructing new distributions for which empirical justification must later be sought.

The term *compounding* has often been used in place of "mixing," as in the first edition of this book. There is, however, an alternative usage of the term "compounding;" see Feller (1957) and also Chapter 4, Section 11. To avoid confusion we now use the term "mixing."

The process of mixing has already been described in Chapter 1, Section B12. The two important categories of mixtures of discrete distributions are as follows:

1. A *k-component finite mixture distribution* is formed from k different component distributions with cdf's $F_1(x), F_2(x), \ldots, F_k(x)$ with mixing weights

$$\omega_1, \omega_2, \ldots, \omega_k \qquad \text{where } \omega_j > 0, \ \sum_{j=1}^{k} \omega_j = 1,$$

by taking the weighted average

$$F(x) = \sum_{j=1}^{k} \omega_j F_j(x) \tag{8.1}$$

as the cdf of a new (mixture) distribution. This corresponds to the *actual mixing* of a number of different distributions. In his book Medgyessy (1977) calls this a *superposition* of distributions. In the theory of insurance ω_j, $j = 1, \ldots$, is called the *risk function*. It follows from (8.1) that if the component distributions are defined on the nonnegative integers with

$$P_j(x) = F_j(x) - F_j(x - 1),$$

then the mixture distribution is a discrete distribution with pmf

$$\Pr[X = x] = \sum_{j=1}^{k} \omega_j P_j(x). \tag{8.2}$$

The concept of a finite mixture of discrete distributions was introduced into the literature by Pearson (1915). The support of the outcome for this type mixture is the union of the supports for the individual components of the mixture.

The logarithmic-with-zeros distribution (Section 2.2 below) is an example of a mixture distribution formed from two components. Here the support for the two components is disjoint, and the support for the outcome is $0,1,\ldots$.

An important method for the computer generation of pseudo-random variates involves the *decomposition* of a distribution into components, most of which are much simpler to generate than the target distribution (Chapter 1, Section C2). This involves the representation of the target cdf as a finite mixture; see Peterson and Kronmal (1980, 1982) and also Devroye (1986, pp. 66-75).

Discrete distributions that are formed as mixtures, using a finite number of components, are studied in detail in Section 2 of this chapter.

2. A mixture distribution also arises when the cumulative distribution function of a rv depends on the parameters $\theta_1, \theta_2, \ldots, \theta_m$ (i.e., has the form $F(x|\theta_1, \ldots, \theta_m)$) and some (or all) of those parameters *vary*. The new distribution then has the cumulative distribution function

$$E[F(X|\theta_1, \ldots, \theta_m)],$$

where the expectation is with respect to the joint distribution of the k parameters that vary. This includes situations where the source of a random variable is unknowable.

Suppose now that only one parameter varies (this will be the case in most instances in this volume). It is convenient to denote a mixture distribution of this type by the symbolic form

$$\mathcal{F}_A \bigwedge_{\Theta} \mathcal{F}_B,$$

where $\mathcal{F}_A$ represents the original distribution and $\mathcal{F}_B$ the mixing distribution (i.e., the distribution of Θ).

When Θ has a discrete distribution with probabilities $p_i, i = 0, 1, \ldots$, we will call the outcome a *countable mixture*; the cdf is

$$F(x) = \sum_{i \ge 0} p_i F_i(x), \tag{8.3}$$

where the probabilities p_i replace the weights ω_i in (8.1). The pmf of the mixture is

$$\Pr[X = x] = \sum_{i \ge 0} p_i P_i(x), \tag{8.4}$$

where $P_j(x) = F_j(x) - F_j(x-1)$.

As an example consider a mixture of Poisson distributions, where the Poisson parameter $\Theta = \alpha V$, α is constant, and V has a logarithmic distribution with parameter λ. The outcome is represented by the symbolic form

$$\text{Poisson}(\Theta) \bigwedge_{\Theta/\alpha} \text{Logarithmic}(\lambda);$$

the corresponding pmf is

$$\Pr[X = x] = \sum_{j=1}^{\infty} e^{-\alpha j} \frac{(\alpha j)^x}{x!} \times \frac{\lambda^j}{j[-\ln(1-\lambda)]}, \qquad x = 0, 1, \ldots. \tag{8.5}$$

Consider, on the other hand, a mixture of logarithmic distributions with parameter $\Phi = \beta Y$, where β is constant and Y now has a Poisson distribution with parameter θ. This gives rise to a mixture distribution of the form

$$\text{Logarithmic}(\Phi) \bigwedge_{\Phi/\beta} \text{Poisson}(\theta)$$

with pmf

$$\Pr[X^* = x] = \sum_{j=0}^{\infty} \frac{\beta^x j^x}{[-\ln(1-\beta j)]x} \times \frac{e^{-\theta}\theta^j}{j!}, \qquad x = 1, 2, \ldots. \tag{8.6}$$

When the points of increase of the mixing distribution are continuous, we will call the outcome a *continuous mixture*. The cdf is obtained by integration over the mixing parameter Θ; if $H(\Theta)$ is the cdf of Θ, then the mixture distribution has cdf

$$F(x) = \int F(x|\theta)dH(\theta), \tag{8.7}$$

where integration is over all values taken by Θ.

Thus, for example,

$$\text{Poisson}(\Theta) \bigwedge_{\Theta} \text{Gamma } (\alpha, \beta)$$

means a mixture of Poisson distributions formed by ascribing the gamma distribution with probability density

$$p_\Theta(t) = \frac{t^{\alpha-1}\exp(-t/\beta)}{\beta^\alpha \Gamma(\alpha)}, \qquad 0 \le t, \ \alpha, \beta > 0,$$

to the parameter Θ of a Poisson distribution. This mixed Poisson distribution is of course a negative binomial distribution (see Chapter 5, Section 3). Notice that the support of the outcome of mixing processes such as these is the same as the support of the initial distribution.

Discrete mixed distributions that have been obtained using countable and continuous mixing distributions are studied in Section 3 of this chapter.

There is an interesting interpretation of mixtures via Bayes' theorem. From (8.4)

$$\sum_i \frac{p_i P_i(x)}{\Pr[X = x]} = 1;$$

hence if p_i is the pmf for a discrete prior distribution, then $p_i P_i(x)/Pr[X = x]$ can be regarded as the pmf for a posterior distribution. Furthermore from (8.7) the pmf for a mixture of discrete distributions formed using a continuous mixing distribution is

$$\Pr[X = x] = \int \Pr[X = x|\theta]h(\theta)d\theta,$$

where integration is over all values of θ; the probability density function

$$\frac{h(\theta)\Pr[X = x|\theta]}{\Pr[X = x]}$$

can be looked upon as a posterior density function for the prior density function $h(\theta)$.

The negative binomial distribution was obtained above as a gamma mixture of Poisson distributions. It is also a Poisson-stopped-sum distribution. Many other discrete distributions are similarly both Poisson mixtures and also Poisson-stopped-sum distributions; see Chapter 9, Sections 3 et seq. Three important theorems that relate to this class of distributions are Lévy's theorem, Maceda's theorem and Gurland's theorem; see Section 3.1 of this chapter.

2 FINITE MIXTURES OF DISCRETE DISTRIBUTIONS

2.1 Parameters of Finite Mixtures

Finite mixture distributions arise in many probabilistic situations. Smith (1985), in a general overview of finite mixtures of continuous and discrete distributions, has listed the following applications:

Fisheries research, where the k components (categories) are different ages.

Sedimentology, where the categories are mineral types.

Medicine, where the categories are disease states.

Economics, where the categories are discontinuous forms of behavior.

Many other direct applications of finite mixtures are referenced in Table 2.1.3 of Titterington et al. (1985). Titterington (1990) has commented on their use in speech recognition and in image analysis. Indirect contexts that involve mixture distributions include the use of contaminated models for outliers, cluster analysis, latent structure models, empirical Bayes methods, kernel-based density estimation, and the classification problem of identifying population membership.

Let us consider a k-component (finite) mixture of discrete distributions with pmf

$$\Pr[X = x] = \sum_{j=1}^{k} \omega_j P_j(x), \qquad (8.8)$$

and let $g(X)$ be some function of X. It follows that

$$E[g(X)] = \sum_{j=1}^{k} \omega_j E[g(X|F_j(x))]; \qquad (8.9)$$

in particular

$$\mu_r'(X) = E[X^r] = \sum_{j=1}^{k} \omega_j \mu_r'(X|F_j(x)), \qquad (8.10)$$

$$\mu'_{[r]}(X) = E\left[\frac{X!}{(X-r)!}\right] = \sum_{j=1}^{k} \omega_j \mu'_{[r]}(X|F_j(x)), \qquad (8.11)$$

and
$$G(z) = E[z^X]$$

$$= \sum_{j=1}^{k} \omega_j E[z^X|F_j(x)]$$

$$= \sum_{j=1}^{k} \omega_j G_j(z), \qquad (8.12)$$

where $G_j(z)$ is the pgf of the jth component.

In most applications of finite mixtures the number of components k is quite small — from two to five, say. A special case with $k = 2$ is the formation of distributions with "added zeros" (see Section 2.2 below). Most of the work on finite mixtures of discrete distributions has concerned mixtures of binomials and mixtures of Poissons; these researches are reviewed in Sections 2.3 and 2.4 below.

In Titterington (1990) mixture data are regarded as realizations of X, or alternatively as realizations of (X, Z) but with Z missing, where X is the mixture variable and Z the mixing variable. The author points out that Z can be treated as an indicator vector of dimension k, with $\omega_j = \Pr[Z = c_j]$ and c_j a k-vector with unity in position j and zeros elsewhere. The set of all parameters is then $\psi = (\omega, \theta)$, where ω is the set $\{\omega_j\}$ and θ is the set $\{\theta_j\}$ of all the parameters of the $F_j(x)$, $j = 1, \ldots, k$.

The parameters of a finite mixture fall therefore into three categories. First, there is k, the number of components. If k is not known the problem of estimating it is usually quite difficult. Often the number of components is decided on an ad hoc basis because of the inherent problems of doing otherwise. Second, there are the mixing weights ω_j. Note that there are only $k - 1$ weight parameters to be estimated since their sum is necessarily unity. Third, there are the parameters of the component distributions. The components need not all have the same form, though they usually do. Suppose that each has s parameters unrelated to those of any other component. The total number of unknown parameters possessed by the mixture distribution is therefore $k - 1 + ks$, assuming that k itself is known.

It may not be possible to estimate all of these parameters, however. Consider, for instance, a mixture of two binomial distributions with parameters $n_1 = n_2 = 2$. We have

$$\Pr[X = 0] = \omega(1 - \pi_1)^2 + (1 - \omega)(1 - \pi_2)^2,$$
$$\Pr[X = 1] = 2\omega\pi_1(1 - \pi_1) + 2(1 - \omega)\pi_2(1 - \pi_2),$$
$$\Pr[X = 2] = \omega\pi_1^2 + (1 - \omega)\pi_2^2. \qquad (8.13)$$

Because $\sum_i \Pr[X = i] = 1$, there are here only two independent equations in three unknowns, ω, π_1, π_2. Clearly the solution is not unique — the mixture is said to be unidentifiable.

Seminal in-depth studies of the problem of identifiability are Teicher (1960, 1961, 1963), Barndorff-Nielson (1965), and Yakowitz and Spragins (1968). Chandra (1977) has reexamined the problem. Important early references concerning the general theory of finite mixtures of discrete distributions are Blischke (1965) and Behboodian (1975).

Titterington (1990) has listed the main objectives when analyzing mixture data. These are: (1) the estimation of the total set of parameters; (2) estimation directed specifically towards the $\{\omega_j\}$; (3) assessment of the number of components; and (4) imputation of the missing Z-data.

Let us suppose that k is known, that there are $k(s+1) - 1$ parameters to be estimated, and that the mixture is identifiable. Then, if the first $k(s+1) - 1$ sample moments (or sample factorial moments) of the mixture distribution have been obtained, the parameters can be estimated by setting the sample moments equal to their expected values and solving the resultant equations. Note that either the central moments or the uncorrected moments can be used (or indeed the factorial moments, since the factorial moments are linear functions of the uncorrected moments).

Titterington et al. (1985) have drawn attention to the problems inherent in moment estimation for finite mixtures, though they have also pointed out that there is a long history of its use. These authors, like Everitt and Hand (1981), have discussed a number of other estimation methods. In the last two decades much of the research on finite mixtures has been concerned with mixtures of members of the linear exponential family. Titterington et al. have described in detail the use of the EM algorithm for maximum likelihood estimation in the general case. They leave as an exercise for the student its application to simple special cases — such special cases include mixtures of (1) Poisson, (2) binomial, and (3) geometric distributions. They have also discussed, in the general case, Bayesian methods, minimum distance estimation based on the cdf, and minimum distance estimators based on transforms.

The mathematical level of the book by Everitt and Hand (1981) is aimed at the research worker with some statistical knowledge. One of their five chapters is devoted to finite mixtures of discrete distributions. Their emphasis is on moment estimation and on ML estimation using the EM algorithm; they demonstrate and discuss these methods using simulated data.

Smith (1985) has tried to explain why there should be so many approaches. In his discussion of maximum likelihood methods he has compared the merits of Newton-Raphson iteration, Fisher's scoring method, and the EM algorithm.

Titterington (1990) has stressed that the usual dependence structure associated with the random variables $Y = (X, Z)$ is not the only possibility. He uses the name "hidden multinomial" for the usual dependence structure; here the term "hidden multinomial" is motivated by regarding the "hidden"

$Z_1, \ldots Z_N$ (where N is the sample size) as a sample from a k-cell multinomial population. Alternatively, the $Z_1, \ldots Z_N$ may form a stationary Markov chain on the state space $\{c_1, \ldots, c_k\}$; this is Titterington's "hidden Markov chain" dependence structure, with particular relevance in speech recognition where Z is the articulatory configuration of the vocal tract and X the resultant signal. Another possibility is for the $Z_1, \ldots Z_N$ to form a Markov random field; this is Titterington's "MRF" dependence structure, for example for model-based image analysis. The difficulties of obtaining ML estimates of the parameters appear to increase dramatically with increasing complexity of the dependence structure.

Medgyessy (1977) has developed an approach that is quite different from the usual estimation approaches. It uses numerical analysis methods to *decompose* a mixture into its components — a prior estimate of the number of components is not required. Medgyessy has also suggested that Bellman's (1960) method for unscrambling exponential distributions can be applied in a similar manner to discrete distributions.

2.2 Zero-Modified Distributions

Empirical distributions obtained in the course of experimental investigations often have an excess of zeros compared with a Poisson distribution with the same mean. This has been a major motivating force behind the development of many distributions that have been used as models in applied statistics. The phenomenon can arise as the result of clustering; distributions with clustering interpretations (see Chapter 9) often do indeed exhibit the feature that the proportion of observations in the zero class is greater than $e^{-\bar{x}}$, where $\bar{x}$ is the observed mean.

A very simple alternative to the use of a cluster model is just to add an arbitrary proportion of zeros, decreasing the remaining frequencies in an appropriate manner. Thus a combination of the original distribution with pmf P_j, $j = 0, 1, 2, \ldots$, together with the degenerate distribution with all probability concentrated at the origin, gives a finite mixture distribution with

$$\Pr[X = 0] = \omega + (1 - \omega)P_0,$$
$$\Pr[X = j] = (1 - \omega)P_j, \qquad j \geq 1. \tag{8.14}$$

A mixture of this kind is referred to as a *zero-modified distribution*, or as a *distribution with added zeros*. Another epithet is *inflated distribution* [Singh (1963); Pandey (1965)].

It is also possible to take ω less than zero (*decreasing* the proportion of zeros), provided that

$$\omega + (1 - \omega)P_0 \geq 0,$$

that is,

$$\omega \geq \frac{-P_0}{1 - P_0}. \tag{8.15}$$

The probability generating function, moments, etc., of a zero-modified distribution are easily derived from those of the original distribution. For example, if $G(z)$ is the original pgf, then that of the modified distribution is

$$H(z) = \omega + (1 - \omega)G(z). \tag{8.16}$$

Similarly, if the original distribution has uncorrected moments μ'_r, then the rth moment about zero of the modified distribution is $(1 - \omega)\mu'_r$.

We note that when the original distribution $\mathcal{F}_A$ has a parameter θ with distribution $\mathcal{F}_B$, then the outcome can be modeled either as

$$(\mathcal{F}_A \text{ with added zeros}) \bigwedge_\Theta \mathcal{F}_B$$

or as

$$(\mathcal{F}_A \bigwedge_\Theta \mathcal{F}_B) \text{ with added zeros.}$$

In fitting a distribution with added zeros, the estimation of parameters other than ω can be carried out by ignoring the observed frequency in the zero class, and then using a technique appropriate to the original distribution truncated by omission of the zero class. After the other parameters have been estimated, the value of ω can then be estimated by equating the observed and expected frequencies in the zero class. (For the modified distribution an arbitrary probability has in effect been assigned to the zero class.)

An important modified distribution is the *Poisson-with-added-zeros*; this is defined by

$$\Pr[X = 0] = \omega + (1 - \omega)e^{-\lambda}$$

$$\Pr[X = j] = \frac{(1 - \omega)e^{-\lambda}\lambda^j}{j!}, \qquad j = 1, 2, \ldots; \tag{8.17}$$

the pgf is therefore

$$H(z) = \omega + (1 - \omega)e^{\lambda(z-1)}. \tag{8.18}$$

Goralski (1977) has called it a *z-Poisson distribution*.

Since this is a zero-modified distribution, one of the ML equations is

$$\hat{\omega} + (1 - \hat{\omega})e^{-\hat{\lambda}} = \frac{f_0}{N} \tag{8.19}$$

where f_0/N is the observed proportion of zeros. It is also a power series distribution, and so the other ML equation is

$$\bar{x} = \hat{\mu} = \hat{\lambda}(1 - \hat{\omega}). \tag{8.20}$$

Eliminating $\hat{\omega}$ gives

$$\bar{x}(1 - e^{-\hat{\lambda}}) = \hat{\lambda}\left(1 - \frac{f_0}{N}\right), \tag{8.21}$$

and hence $\hat{\lambda}$ (and $\hat{\omega}$) can be obtained by iteration. Singh (1963) obtained the approximate formulas

$$\text{Var}(\hat{\lambda}) \doteq (1 - \omega)^{-1}\lambda(1 - e^{-\lambda})(1 - e^{-\lambda} - \lambda e^{-\lambda})^{-1}, \tag{8.22}$$

$$\text{Var}(\hat{\omega}) \doteq (1 - \omega)[\omega(1 - \lambda e^{-\lambda}) + (1 - \omega)e^{-\lambda}](1 - e^{-\lambda} - \lambda e^{-\lambda})^{-1}. \tag{8.23}$$

Cohen (1960c) gave some examples of fitting the distribution to empirical data. Martin and Katti (1965) also fitted the distribution to a number of data sets, using ML with $\lambda^{(0)} = \bar{x}/(1 - f_0/N)$ for the initial estimate of λ. Kemp's (1986) approximation to the ML equation for $\hat{\lambda}$ gave $\lambda_{(0)} = \ln(N\bar{x}/f_1)$ as an initial estimate. She found that usually

$$\lambda_{(0)} < \hat{\lambda} < \lambda^{(0)}. \tag{8.24}$$

Kemp and Kemp's (1988) bounds for the maximum likelihood estimator $\hat{\lambda}$ are

$$\frac{6(\bar{x}^* - 1)}{\bar{x}^* + 2} < \hat{\lambda} < \bar{x}^*\{1 - (e^{\bar{x}^*} - \bar{x}^*)^{-1}\}, \tag{8.25}$$

where $\bar{x}^* = \bar{x}/(1 - f_0/N)$; they appear to lie closer to $\hat{\lambda}$ than $\lambda_{(0)}$ and $\lambda^{(0)}$. Umbach (1981) has also investigated inference for a mixture of a Poisson and a degenerate distribution.

Yoneda (1962) has considered an extended modification of the Poisson distribution. This allows the frequencies of the values $0, 1, 2, \ldots, K$ to take arbitrary values, the remaining frequencies (for $K + 1, K + 2, \ldots$) being proportional to Poisson probabilities. The distributions depend on $(K + 1)$ parameters $\alpha_0, \alpha_1, \ldots, \alpha_K$ as well as on the parameter λ of the Poisson "upper tail" and are defined by

$$\Pr[X = x] = \frac{(1 - \alpha_x)e^{-\lambda}\lambda^x/x!}{1 - e^{-\lambda}\sum_{j=0}^{K}\alpha_j\lambda^j/j!}, \quad x = 0, 1, 2, \ldots, K,$$

$$\Pr[X = x] = \frac{e^{-\lambda}\lambda^x/x!}{1 - e^{-\lambda}\sum_{j=0}^{K}\alpha_j\lambda^j/j!}, \quad x = K + 1, K + 2, \ldots. \tag{8.26}$$

Yoneda gave the equations satisfied by the ML estimators $\hat{\lambda}$, $\hat{\alpha}_0$, $\hat{\alpha}_1$, ..., and also tables whereby the ML equation for $\hat{\lambda}$ can be solved by inverse interpolation with accuracy sufficient for most purposes.

The *binomial-with-added-zeros* distribution can be formed similarly, giving

$$\Pr[X = 0] = \omega + (1 - \omega)q^n,$$

$$\Pr[X = x] = (1 - \omega)\binom{n}{x}p^x(1 - p)^{n-x}, \qquad x \geq 1, \qquad (8.27)$$

and the pgf

$$H(z) = \omega + (1 - \omega)(1 - p + pz)^n. \qquad (8.28)$$

The zero-modified binomial is the special case $m = 1$ of Khatri and Patel's (1961) binomial-binomial distribution with pgf $\{\omega + (1 - \omega)(1 - p + pz)^n\}^m$. The mean and variance for (8.27) are

$$\mu = (1 - \omega)np \quad \text{and} \quad \mu_2 = (1 - \omega)np\{1 - p + \omega np\}. \qquad (8.29)$$

In their study of the estimation of the parameters of this distribution, Kemp and Kemp (1988) showed that the maximum likelihood equations are

$$\frac{f_0}{N} = \hat{\omega} + (1 - \hat{\omega})(\hat{q})^n,$$

$$\bar{x} = n(1 - \hat{\omega})\hat{p}, \qquad (8.30)$$

where f_0/N is the observed relative frequency of zero. These equations do not have an explicit solution, so Kemp and Kemp also studied simple, explicit, estimation methods suitable for a rapid assessment of data or to provide initial estimates for ML estimation:

1. Method of moments,

$$\tilde{p} = \frac{(s^2/\bar{x}) + \bar{x} - 1}{n - 1},$$

$$\tilde{\omega} = 1 - \bar{x}/(n\tilde{p}). \qquad (8.31)$$

2. Method of mean and first-frequency,

$$1 - p^* = \left\{\frac{f_1}{N\bar{x}}\right\}^{1/(n-1)},$$

$$\omega^* = 1 - \bar{x}/(np^*), \qquad (8.32)$$

where f_1/N is the observed relative frequency of an observation equal to unity.

3-4. Two methods based on the empirical pgf.

The four methods were applied to a small ($N = 60$) and a large ($N = 53,680$) data set. The investigation demonstrated that estimates that are close to one another may not, however, be close to a maximum likelihood estimate, that a method that is satisfactory for one data set may be useless for another with parameters in another part of the parameter space, and that inefficient methods may lead to impossible estimates.

The binomial distribution modified in this way should not be confused with Dandekar's modified binomial distributions; see Dandekar (1955), Basu (1955), and Chapter 11, Section 12.

A zero-inflated geometric distribution has been studied by Holgate (1964) as a model for the length of residence of animals in a specified habitat. Holgate (1966) subsequently studied a more complicated model for such data. This led to a mixed zero-inflated geometric distribution with pgf involving an Appell function of the first kind; for the theory of Appell functions see Appell and Kampé de Fériet (1926).

Khatri (1961) [see also Patil (1964b)] has studied the *logarithmic-with-zeros distribution* (Chapter 7, Section 10). The probabilities are

$$\Pr[X = 0] = \omega,$$

$$\Pr[X = x] = \frac{(1 - \omega)\theta^x}{-x \ln(1 - \theta)}, \qquad x \geq 1. \tag{8.33}$$

The mean and variance are

$$\mu = \frac{(1 - \omega)\alpha\theta}{1 - \theta},$$

$$\mu_2 = \frac{(1 - \omega)\alpha\theta[1 - \alpha\theta(1 - \omega)]}{(1 - \theta)^2}, \tag{8.34}$$

where $\alpha = -1/\ln(1 - \theta)$, and the pgf is

$$H(z) = \omega + (1 - \omega)\frac{\ln(1 - \theta z)}{\ln(1 - \theta)}. \tag{8.35}$$

The distribution arises as a mixed binomial distribution with parameters (n, p), where n has a logarithmic distribution with parameter λ [see Khatri (1961)]; the connection between the parameters is

$$\omega = \frac{\ln(1 - \lambda + \lambda p)}{\ln(1 - \lambda)},$$

$$\theta = \frac{\lambda p}{1 - \lambda + \lambda p}. \tag{8.36}$$

Khatri has shown that the maximum likelihood estimators of ω and θ are given by

$$\sum_{j\geq 1} \frac{j f_j}{N} = \frac{\hat{\theta}}{(1 - \hat{\theta})\{-\ln(1 - \hat{\theta})\}},$$

$$\hat{\omega} = 1 - \frac{f_0}{N}, \tag{8.37}$$

where f_j/N is the observed relative frequency of an observation equal to j. The asymptotic variances and covariance of the ML estimators are

$$\text{Var}(\hat{\omega}) = \frac{\omega(1 - \omega)}{N}, \tag{8.38}$$

$$\text{Var}(\hat{\theta}) \doteq \frac{\theta(1 - \theta)^2 \{\ln(1 - \theta)\}^2}{N(1 - \omega)\{-\ln(1 - \theta) - \theta\}}, \tag{8.39}$$

$$\text{Cov}(\hat{\omega}, \hat{\theta}) = 0. \tag{8.40}$$

The *log-zero-Poisson (LZP) distribution* was constructed by Katti and A. V. Rao (1970). It can be obtained by modifying a Poisson$\bigwedge$Logarithmic distribution by "adding zeros." This gives

$$\Pr[X = 0] = \omega + (1 - \omega) \frac{\ln(1 - \lambda e^{-\phi})}{\ln(1 - \lambda)},$$

$$\Pr[X = x] = (1 - \omega) \frac{\phi^x}{x!} \sum_{j=1}^{\infty} \frac{j^{x-1}(\lambda e^{-\phi})^j}{\{-\ln(1 - \lambda)\}}, \qquad x \geq 1, \tag{8.41}$$

using the expression for the pmf of the Poisson$\bigwedge$Logarithmic distribution in Section 3.2. The pgf is

$$H(z) = \omega + (1 - \omega) \frac{\ln(1 - \lambda e^{\phi(z-1)})}{\ln(1 - \lambda)}. \tag{8.42}$$

Katti and Rao have prepared tables of the Poisson$\bigwedge$Logarithmic probabilities from which the LZP probabilities can be calculated straightforwardly. These authors claim that a wide variety of distributions can be reproduced, given an appropriate choice of the parameters ω, λ and θ. They used ML estimation to fit the distribution to the 35 data sets in Martin and Katti (1965), and they compared the fits with Martin and Katti's fits for a number of two-parameter distributions. The log-zero-Poisson (with three parameters) emerged quite well from these comparisons. Katti and Rao provided a model for the distribution, obtained a recurrence formula for the probabilities, made a thorough study of the distribution's properties, and explained in detail their method of ML estimation. In particular they showed that the variance is greater or less than the mean according to the values taken by the parameters. For most of the other distributions that have been used widely in biometry, the variance is necessarily greater than the mean.

One of the drawbacks to the use of the log-zero-Poisson distribution has been the complexity of (8.41). Willmot (1987a) has derived the following finite-sum expression for the pmf

$$\Pr[X = x] = \frac{(1 - \omega)\phi^x}{x![-\ln(1 - \lambda)]}$$

$$\times \sum_{j=1}^{x} \left[\left(\frac{\lambda e^{-\phi}}{1 - \lambda e^{-\phi}} \right)^j \sum_{m=1}^{j} \binom{j-1}{m-1} (-1)^{j-m} m^{x-1} \right] \quad (8.43)$$

for $x \geq 1$, and he has shown that asymptotically

$$\Pr[X = x] \approx \frac{(1 - \omega)\phi^x(\phi - \ln \lambda)^{-x}}{-x \ln(1 - \lambda)}. \quad (8.44)$$

2.3 Finite Poisson Mixtures

A *k-component Poisson mixture* arises when

$$\Pr[X = x] = \sum_{j=1}^{k} \omega_j \frac{e^{-\theta_j}(\theta_j)^x}{x!}. \quad (8.45)$$

The identifiability of mixtures of Poisson distributions with k components, $\omega_j \neq 0$, $j = 1, 2, \ldots, k$, has been established by Feller (1943) and Teicher (1960).

Given a sample of N observations, $x_1, \ldots, x_N$, from such a mixture, we can calculate the first $2k - 1$ sample factorial moments

$$T_r = \sum_{i=1}^{n} x_i(x_i - 1) \ldots (x_i - r + 1)/N, \qquad r = 1, 2, \ldots, 2k - 1;$$

estimates of the ω_j and θ_j, $j = 1, 2, \ldots, k$, can then be obtained by setting these equal to their expectations

$$\mu'_{[r]} = \sum_{j=1}^{k} \omega_j \theta_j^r.$$

If $k = 2$, then $\omega_1 + \omega_2 = 1$, $\omega_1, \omega_2 \neq 0$; estimates of θ_1 and θ_2 can be obtained as the roots of the quadratic equation in $\tilde{\theta}$,

$$(T_2 - T_1^2)\tilde{\theta}^2 - (T_3 - T_1 T_2)\tilde{\theta} + (T_1 T_3 - T_2^2) = 0, \quad (8.46)$$

see Jones (1933) and Rider (1962a). Rider explored the asymptotic variances of $\tilde{\theta}_1$ and $\tilde{\theta}_2$ given known values of ω_1 and ω_2. Everitt and Hand (1981) have illustrated the moments method of estimation using Hasselblad's (1969) data on death notices of elderly women (two components), and computer generated data with known values of the parameters (three components).

Tiago de Oliveira (1965) has devised a one-sided test of the hypothesis that there are two Poisson components against a null hypothesis of a pure Poisson distribution, based on $s^2 - \bar{x}$, where $\bar{x}$ and s^2 are the sample mean and variance.

ML estimation of the parameters of (8.45) is straightforward, but it does require the use of iteration. Everitt and Hand (1981) illustrated its use with a four-component Poisson mixture, using generated data with known parameters; as initial estimates they used (1) the known values and (2) guesses (the moments method failed to give a solution with real roots). The discrepancy between the two sets of final values is very noticeable.

Hasselblad (1969) used the EM algorithm to obtain ML estimates for a mixture of two Poissons for the death-notice data. Titterington et al. (1985) reexamined ML estimation for these data, using both the EM algorithm and also a Newton-Raphson technique, with various sets of initial values. The Newton-Raphson algorithm either failed to converge or converged to four decimal places in fewer than a dozen iterations; the EM algorithm, on the other hand, always converged but took between one and two thousand iterations to achieve the same accuracy.

A modified version of the minimum χ^2 method of estimation that had been suggested by Blischke (1964) was applied to Poisson mixtures by Saleh (1981).

For more complicated mixtures involving Poisson distributions, see Simar (1976) and Godambe (1977). Medgyessy (1977, pp. 200–224) has studied the decomposition of finite mixtures of an unknown number of Poisson components, (1) with means that are not too similar (i.e., have distinct, separated, values) and (2) with means less than unity. Medgyessy gives references to his earlier papers (many not in English).

2.4 Finite Binomial Mixtures

The most common mixture of binomial distributions is one in which the component distributions have a common (known) value of the exponent parameter n but differing values of p. The mixture distribution then has probabilities of the form

$$\Pr[X = x] = \binom{n}{x} \sum_{j=1}^{k} \omega_j \, p_j^x (1 - p_j)^{n-x}, \qquad (8.47)$$

where $\omega_j > 0$, $\sum_{j=1}^{k} \omega_j = 1$. A binomial mixture of this kind is identifiable iff $(2k - 1) \le n$. Teicher (1961) established the necessity of this condition, and Blischke (1964) proved its sufficiency.

A binomial mixture of the type

$$\Pr[X = x] = \sum_{j=1}^{k} \omega_j \binom{n_j}{x} p_j^x (1 - p_j)^{n_j - x},$$

where $\omega_j > 0$, $\sum_{j=1}^{k} \omega_j = 1$, $0 < p_j < 1$, and the n_j are different integers, is, nevertheless, identifiable [Teicher (1963)].

If $n_1 = n_2 = \cdots = n_k$, if k is small, and particularly if k is known, then it may be possible to obtain useful estimates of ω_j and p_j, $j = 1, \ldots, k$. Blischke (1964, 1965) has shown that estimation is possible, provided that there are available at least $(2k - 1)$ observations on the random variable with the binomial mixture distribution. Let $x_1, x_2, \ldots, x_N$ be a sample of $N \geq (2k - 1)$ observations, with pmf (8.47). Then the first $2k - 1$ *reduced factorial moments*

$$F_s = \frac{1}{N} \sum_{j=1}^{N} \frac{x_j(x_j - 1) \ldots (x_j - s + 1)}{n(n-1) \ldots (n - s + 1)}, \qquad s = 1, 2, \ldots, (2k - 1) \qquad (8.48)$$

are calculable from $x_1, x_2, \ldots, x_N$, and

$$E[F_s] = \sum_{j=1}^{k} \omega_j p_j^s. \qquad (8.49)$$

The method is then, essentially, to equate the observed values of $F_1, F_2 \ldots$, F_{2k-1} to their expected values, and then to solve the resulting equations for $\omega_1, \omega_2 \ldots, \omega_k, p_1, p_2, \ldots, p_k$ (subject to the condition $\sum_{j=1}^{k} \omega_j = 1$). This is done by first eliminating the ω_j's and then reducing the resultant set of equations to a single polynomial of degree k, the roots of which are the required estimators of $p_1, p_2, \ldots, p_k$. Approximations to the variances and covariances of the estimators of the ω_j's and p_j's can be obtained straight-forwardly. Blischke (1964) found that estimators based on moments have an asymptotic efficiency of 100% when $N = (2k - 1)$, and that the efficiency is high for n and N large enough. Everitt and Hand (1981) have provided a worked example using data generated from a mixture of four binomials with known parameters, including a common exponent parameter n. They have also discussed Blischke's (1962, 1964) work on the asymptotic properties of the moment estimators.

Rider (1962a) and Blischke (1962) considered the case of a mixture of two binomials in detail. The mixture pmf is

$$\Pr[X = x] = \omega \binom{n_1}{x} p_1^x q_1^{n_1 - x} + (1 - \omega) \binom{n_2}{x} p_2^x q_2^{n_2 - x}, \qquad (8.50)$$

with $0 < \omega < 1$, $0 < q_j = 1 - p_j < 1$, $j = 1, 2$. When $n_1 = n_2 = n$ (known), simple explicit formulas in terms of the first three moments can be obtained for the parameters ω, p_1 and p_2; see, e.g., Everitt and Hand (1981) for details.

The special cases $k = 2$, $n_1 = n_2 = 12$, with $p_1 = p_2 = p$ and with $p_1 \neq p_2$ occur in Gelfand and Solomon's (1975) analysis of jury decisions. They estimated the two models using the method of moments, maximum likelihood, and minimum χ^2, having first grouped the data into five classes. Reasonably good agreement between methods was observed.

Blischke (1962, 1964) also examined maximum likelihood estimation for a mixture of k binomials; this requires iteration. An advantage of ML estimation is that it is impossible to get estimates that lie outside the admissible limits if the initial estimates are within those limits [Hasselblad (1969)]. Everitt and Hand (1981) applied ML estimation to their mixture of four binomials, using (1) the population values and (2) the moment estimates as initial estimates. Their two sets of final estimates were very close; this would not necessarily occur with other data sets, especially if the likelihood surface were uneven.

Other estimation methods investigated by Blischke (1964) include a single cycle of maximum likelihood estimation, using the method of scoring with the moment estimates as initial estimates; he also explored a minimum χ^2 approach in which the χ^2 statistic is expanded as a Taylor series about the estimates given by the method of moments. With present-day computing power such methods are unlikely to receive much attention; maximum likelihood estimates can now be achieved iteratively without undue effort, using the moment estimates as initial values.

Medgyessy (1977, pp. 195–198) has considered in detail the decomposition of a mixture of an unknown number of binomial components that are not too similar (i.e., have means with distinct, separated, values).

Bondesson (1988) has described the application of a mixed binomial model (with seven components) to data on the success of pine seeds sown in batches of 20 seeds at 7 adjacent spots with 6 replicate locations. He used his analysis to recommend the optimal number of seeds per spot (for a fixed total number of seeds per location) needed in order to minimize the probability of no success per location.

2.5 Other Finite Mixtures of Discrete Distributions

A mixture of a Poisson and a binomial distribution was considered by Cohen (1965). The pmf is

$$\Pr[X = x] = \omega \frac{e^{-\theta} \theta^x}{x!} + (1 - \omega)\binom{n}{x} p^x (1 - p)^{n-x}, \qquad x = 0, 1, \ldots, \quad (8.51)$$

where n is assumed to be known and the second component is zero if $x > n$. For this distribution the pgf is

$$H(z) = \omega \, e^{\theta(z-1)} + (1 - \omega)(1 - p + pz)^n, \tag{8.52}$$

and the factorial moments are

$$\mu'_{[r]} = \omega \theta^r + (1 - \omega)n(n - 1)\ldots(n - r + 1)p^r. \tag{8.53}$$

Cohen obtained moment estimators for the parameters ω, θ and p of the mixture using the method of moments.

Cohen (1965) has also described how to fit a mixture of two positive Poissons. This distribution is defined by

$$\Pr[X = x] = \omega \left(\frac{\theta_1^x / x!}{e^{\theta_1} - 1} \right) + (1 - \omega) \left(\frac{\theta_2^x / x!}{e^{\theta_2} - 1} \right), \tag{8.54}$$

where $x = 1, 2, \ldots$, and now

$$H(z) = \omega \left(\frac{e^{\theta_1 z} - 1}{e^{\theta_1} - 1} \right) + (1 - \omega) \left(\frac{e^{\theta_2 z} - 1}{e^{\theta_2} - 1} \right) \tag{8.55}$$

and

$$\mu'_{[r]} = \omega \left(\frac{\theta_1^r}{1 - e^{-\theta_1}} \right) + (1 - \omega) \left(\frac{\theta_2^r}{1 - e^{-\theta_2}} \right). \tag{8.56}$$

Daniels' (1961) study of the busy-time distribution in a queueing process involved mixtures of geometric distributions (with components corresponding to priority levels). Mixtures of negative binomials have been studied by Rider (1962a) (who used moments to estimate the parameters), by Harris (1983) (who used a gradient method), and by Medgyessy (1977, pp. 203–222).

John (1970) has described the use of the method of moments, and also a maximum likelihood method, to identify the population of origin of individual observations drawn from a mixture of two distributions. He discussed the cases where both components in the mixture are (1) binomial, (2) Poisson, (3) negative binomial, and (4) hypergeometric.

3 CONTINUOUS AND COUNTABLE MIXTURES OF DISCRETE DISTRIBUTIONS

3.1 Three Important Theorems

The development of the theory of continuous and countable mixtures of discrete distributions (particularly mixtures of Poisson distributions) is linked closely to accident proneness theory and to actuarial risk theory. Greenwood and Yule (1920) and Lundberg (1940) are classical works that are particularly interesting concerning the concept of conditionality on the realization of the distribution of a parameter.

For example, in Greenwood and Yule's "accident proneness" model an individual is assumed to have accidents at random with intensity θ, where θ is assumed to have a gamma distribution over the population of individuals. The number of accidents per individual is therefore a Poisson distribution with the value of its parameter θ conditional on a realization of a gamma variable; this leads to a negative binomial distribution (see Section 1 of this chapter and Chapter 5, Section 3).

The identifiability of countable mixtures has been studied by Patil and Bildikar (1966), and Tallis (1969). For identifiability in the general case; see Teicher (1961), Tallis (1969), Blum and Susarla (1977), and Tallis and Chesson (1982). Blischke's (1965) paper on infinite as well as finite mixtures was at the time of writing a very thorough overview; Blischke included applications, implications, identifiability, estimation and hypothesis testing. Not all of his "areas requiring further investigation" have been resolved. Haight (1967) is another useful reference.

In this section we introduce three important theorems that assist in our understanding of relationships between distributions.

Lévy's Theorem [see Feller (1957)]. If and only if a discrete probability distribution on the nonnegative integers is infinitely divisible, then its pgf can be written in the form

$$G(z) = e^{\lambda\{g(z)-1\}}, \tag{8.57}$$

where $\lambda > 0$ and $g(z)$ is another pgf.

The implication of Lévy's theorem is that an infinitely divisible distribution with nonnegative support can be interpreted:

1. as a stopped sum of Poisson distributions, that is, as the sum of Y random variables with pgf $g(z)$, where Y has a Poisson distribution, and

2. as a convolution (sum) of a Poisson singlet, Poisson doublet, triplet, etc., where the successive parameters are proportional to the probabilities given by $g(z)$. This is a consequence of the reexpression of (8.57) as

$$G(z) = \exp[\lambda(p_0 + p_1 z + p_2 z^2 + \cdots - 1)]$$
$$= \exp[\lambda p_1 (z - 1)] \times \exp[\lambda p_2 (z^2 - 1)] \times \exp[\lambda p_3 (z^3 - 1)] \cdots \tag{8.58}$$

where $g(z) = \sum_{i \geq 0} p_i z^i$.

Maceda's Theorem [see Maceda (1948); also Godambe and Patil (1969)]. Consider a mixture of Poisson distributions where the mixing distribution has nonnegative support. Then the resultant distribution is infinitely divisible iff the mixing distribution is infinitely divisible.

Because infinitely divisible discrete distributions can be interpreted as Poisson-stopped-sum distributions, the implication from Lévy's and Maceda's theorems is that mixing Poisson distributions using an infinitely divisible distribution yields a Poisson-stopped-sum distribution. Furthermore, mixing a Poisson-stopped-sum distribution using an infinitely divisible distribution gives rise to another Poisson-stopped-sum distribution.

Gurland's Theorem [see Gurland (1957)]. Consider two distributions $\mathcal{F}_1$ and $\mathcal{F}_2$ with pgf's

$$G_1(z) = \sum_{k \geq 0} p_k z^k \tag{8.59}$$

and $G_2(z)$, respectively, where $G_2(z)$ depends on a parameter ϕ in such a way that

$$G_2(z|k\phi) = [G_2(z|\phi)]^k. \tag{8.60}$$

Then the mixed distribution represented by

$$\mathcal{F}_2(K\phi) \bigwedge_K \mathcal{F}_1 \tag{8.61}$$

has the pgf

$$\sum_{k \geq 0} p_k G_2(z|k\phi) = \sum_{k \geq 0} p_k [G_2(z|\phi)]^k$$
$$= G_1(G_2(z|\phi)). \tag{8.62}$$

Since Feller (1943), the adjective "generalized" has very often been applied in a restrictive sense to describe distributions with pgf's of the form $G_1(G_2(z))$. Confusingly the word "generalized" is now also used in other senses; see, for example, Feller's (1957) book, and the three different "generalized Poisson" distributions of Morlat (1952), Gerstenkorn (1962), and Consul (1989).

A distribution with pgf of the form $G_1(G_2(z))$ will here be called a *generalized $\mathcal{F}_1$ distribution*, more precisely an *$\mathcal{F}_1$ distribution generalized* by the *generalizing $\mathcal{F}_2$ distribution*. It will be represented by the symbolic form $\mathcal{F}_1 \bigvee \mathcal{F}_2$. Chapter 9 is devoted to this very important class of distributions.

Gurland's theorem can now be stated symbolically as

$$\mathcal{F}_2 \bigwedge \mathcal{F}_1 \sim \mathcal{F}_1 \bigvee \mathcal{F}_2, \tag{8.63}$$

provided that

$$G_2(z|k\phi) = [G_2(z|\phi)]^k. \tag{8.64}$$

Because the Poisson, binomial, and negative binomial distributions all have pgf's of the form (8.64), it follows that discrete mixtures of Poisson, binomial, and negative binomial distributions are also generalized distributions in the above sense.

Consider, for example, a mixture of binomial distributions with parameters nK and p, where $0 < p < 1$, and K is a random variable taking nonnegative integer values according to a Poisson distribution with parameter θ, $0 < \theta$. Symbolically this mixture is represented by

$$\text{Binomial}(nK, p) \underset{K}{\bigwedge} \text{Poisson}(\theta). \tag{8.65}$$

The probabilities are

$$\Pr[X = x] = \sum_{k=0}^{\infty} \frac{(nk)! p^x (1-p)^{nk-x}}{x!(nk-x)!} \times \frac{e^{-\theta}\theta^k}{k!}, \tag{8.66}$$

and the pgf is

$$G(z) = \sum_{x=0}^{\infty} \sum_{k=0}^{\infty} \frac{(nk)! p^x (1-p)^{nk-x} e^{-\theta}\theta^k z^x}{x!(nk-x)! k!}$$

$$= \sum_{k=0}^{\infty} (1-p+pz)^{nk} \frac{e^{-\theta}\theta^k}{k!}$$

$$= e^{\theta(1-p+pz)^n - \theta}. \tag{8.67}$$

The essential step in this argument is the *binomial* property

$$(1-p+pz)^{nk} = \{(1-p+pz)^n\}^k.$$

Returning to (8.67), we see that this has the form of a Poisson distribution (with parameter θ) generalized using a binomial distribution with parameters n and p, that is, (8.67) has the second symbolic representation

$$\text{Poisson}(\theta) \bigvee \text{Binomial}(n, p). \tag{8.68}$$

Notice that parameter K in (8.65) does not appear in (8.68).

A detailed examination of the implications of Lévy's, Maceda's and Gurland's theorems has been made by Molenaar (1965).

The most important of the mixed Poisson, mixed binomial and mixed negative binomial distributions will merely be listed in the following sections in this chapter; they will be dealt with in greater depth elsewhere in the book.

3.2 Mixtures of Poisson Distributions

The Poisson is a single-parameter distribution with variance equal to the mean. A common practical problem when analyzing sets of data thought to be Poissonian is a breakdown in this variance-mean relationship due to overdispersion (rarely underdispersion). The effects of overdispersion are twofold: first, the summary statistics have larger variances than expected; and second, there may be a loss of efficiency if an inadequate model is adopted. Cox (1983) has studied in detail the effect on analysis if a Poisson model is adopted in the presence of overdispersion. He showed that the efficiency of ML estimation remains high provided that the amount of overdispersion is "modest."

The other line of approach is to represent the overdispersion by a specific model. This leads to the theory of mixtures of Poisson distributions. The development of the theory of such mixtures is linked closely to accident-proneness theory and to actuarial-risk theory; seminal works are Greenwood and Yule (1920) and Lundberg (1940). These authors introduced the idea of conditionality on a realization of a variable parameter.

Ottestad (1944) and Consael (1952) obtained certain moment properties for mixed Poisson distributions. If

$$\Pr[X = x] = \int_0^\infty e^{-\theta}\theta^x(x!)^{-1}dF(\theta), \qquad (8.69)$$

then

$$G(z) = \sum_{x \geq 0}\Pr[X = x]z^x = \sum_{x \geq 0}\int_0^\infty e^{-\theta}(\theta z)^x(x!)^{-1}dF(\theta)$$

$$= \int_0^\infty e^{\theta(z-1)}dF(\theta). \qquad (8.70)$$

The factorial moment generating function is therefore

$$G(1+t) = \int_0^\infty e^{\theta t}dF(\theta), \qquad (8.71)$$

and so the factorial moment of order r for a mixture of Poisson distributions is equal to the rth moment about the origin for the mixing distribution. Ottestad and Consael also discovered a relationship between the cumulants of the two distributions. Because the factorial moment generating function of the mixture is equal to the uncorrected moment generating function of the mixing distribution, it follows that the factorial cumulant generating function of the mixture is equal to the cumulant generating function of the mixing distribution; see Table 1.1 in Chapter 1, Section B9. Let $\kappa_{[r]}$ and κ_r denote the factorial cumulants and the cumulants, respectively, for the mixture; let c_r denote the cumulants of the mixing distribution. Then

$$\kappa_{[1]} = \kappa_1 = c_1,$$

$$\kappa_{[2]} = \kappa_2 - \kappa_1 = c_2,$$

$$\kappa_{[3]} = \kappa_3 - 3\kappa_2 + 2\kappa_1 = c_3, \qquad \text{etc.,} \tag{8.72}$$

and

$$\kappa_1 = c_1,$$

$$\kappa_2 = c_2 + c_1,$$

$$\kappa_3 = c_3 + 3c_2 + c_1, \qquad \text{etc.} \tag{8.73}$$

The representation (8.70) for the pgf has the form of a Laplace transform. The uniqueness of the Laplace transform implies that for any $G(z)$ there is a corresponding $F(\theta)$; if $F(\theta)$ is a cdf then $G(z)$ is a mixture of Poisson distributions. The uniqueness of the Laplace transform also provides a proof of the identifiability of infinite mixtures of Poisson distributions; see Teicher (1960) and also Douglas (1980, pp. 59, 60).

Teicher's (1960) paper contains a number of other important results. For instance, no mixed Poisson with more than one component can itself be a Poisson distribution; also the variance of a mixed Poisson cannot be less than its mean. Teicher also showed that the convolution of two mixed Poissons is itself mixed Poisson, with a mixing distribution that is the convolution of the mixing distributions for the two component mixed Poissons.

Feller (1943) had previously proved that

$$\Pr[X = 0] > \Pr[Y = 0] \tag{8.74}$$

and that

$$\frac{\Pr[X = 1]}{\Pr[X = 0]} < \frac{\Pr[Y = 1]}{\Pr[Y = 0]}, \tag{8.75}$$

where $\Pr[X = x]$, $x = 0, 1, \ldots$ is the pmf of a mixed Poisson distribution with more than one component and Y is a Poisson random variable with the *same mean as* X.

Some mixed Poisson distributions are multimodal; many are unimodal. Holgate (1970) has proved that if the mixing distribution is a nonnegative continuous unimodal distribution, then the resultant mixture of Poisson distributions is unimodal. Al–Zaid (1989) has obtained Holgate's condition for the *unimodality of a Poisson mixture* as a corollary of a deeper theorem, and he has applied his theorem to mixtures of binomial distributions; see the next section. Further general results concerning mixtures of Poisson distributions are given in Haight (1967, pp. 35–43); see also Douglas (1980, pp. 52–74), Stam (1973), and Willmot (1989, 1990).

Blischke's (1965) overview of applications, implications, identifiability, estimation and hypothesis testing for Poisson mixtures is still worthy of attention. Not all of his "areas requiring further investigation" have been resolved. A recent review is by Willmot (1986).

The asymptotic tail behaviour of mixtures of Poisson distributions is discussed in Willmot (1989, 1990).

Lundberg (1940) initiated the study of mixed Poisson processes. For bibliographic notes concerning these processes and the more general doubly stochastic processes, see Cox and Isham (1980).

We now list a number of mixtures of Poisson distributions where the mixing distributions are continuous or countable, beginning with the mixtures that are infinitely divisible (Poisson-stopped-sum) distributions.

1. A gamma mixture of Poisson distributions is denoted by

$$\text{Poisson}(\Theta) \bigwedge_{\Theta} \text{Gamma } (\alpha, \beta).$$

The gamma distribution is infinitely divisible, and the outcome is the (infinitely divisible) negative binomial distribution; see Chapter 5, Section 3. A three-parameter gamma distribution used as the mixing distribution gives rise to the Delaporte distribution; see Chapter 5, Section 12.5. Chukwu and Gupta (1989) have studied a generalized gamma mixture of a generalized Poisson distribution.

2. An inverse Gaussian mixture of Poisson distributions,

$$\text{Poisson}(\Theta) \bigwedge_{\Theta} \text{Inverse-Gaussian } (m, \sigma^2),$$

was postulated by Holla (1967). An extension, with the symbolic representation Poisson$\bigwedge$Generalized inverse-Gaussian, is known in the literature as Sichel's distribution; it is a long-tailed distribution that is suitable for highly skewed data. These distributions have been the subject of a long series of papers and are discussed more fully in Chapter 11, Section 15.

3. A Poisson mixture of Poisson distributions,

$$\text{Poisson}(\Theta) \bigwedge_{\Theta/\phi} \text{Poisson}(\lambda),$$

is an important distribution where the mixing distribution is infinitely divisible and also countably infinite (its cdf is a step-function). The resultant distribution is known as the *Neyman Type A distribution*. By Gurland's theorem,

$$\text{Poisson}(\Theta) \bigwedge_{\Theta/\phi} \text{Poisson } (\lambda) \sim \text{Poisson}(\lambda) \bigvee \text{Poisson}(\phi).$$

The distribution is therefore a Poisson-stopped-sum distribution, with pgf

$$G(z) = \exp[\lambda\{e^{\phi(z-1)} - 1\}]; \qquad (8.76)$$

it is examined in detail in Chapter 9, section 6.

4. A negative binomial mixture of Poisson distributions,

$$\text{Poisson}(\Theta) \bigwedge_{\Theta/\phi} \text{Negative binomial}(k, P),$$

is another mixture with an infinitely divisible, countable mixing distribution. Gurland's theorem implies that

$$\text{Poisson}(\Theta) \bigwedge_{\Theta/\phi} \text{Negative binomial}(k, P)$$

$$\sim \text{Negative binomial } (k, P) \bigvee \text{Poisson}(\phi);$$

this distribution is again a Poisson-stopped-sum distribution. The pgf is

$$G(z) = [1 + P - P \exp\{\phi(z - 1)\}]^{-k}, \qquad (8.77)$$

where $P > 0$, the probabilities are

$$\Pr[X = x] = \frac{\phi^x}{x!} \sum_{j=0}^{\infty} e^{-j\phi} j^x \binom{k+j-1}{k-1} P^j Q^{-k-j}, \quad k = 0, 1, 2, \dots, \quad (8.78)$$

where $Q = 1 + P$, and the moments are

$$\mu = kP\phi, \qquad \mu_2 = kP\phi + kP(1+P)\phi^2, \qquad \text{etc.}$$

5. A Poisson mixture of negative binomial distributions has the structure

$$\text{Negative binomial}(Y, P) \bigwedge_{Y/k} \text{Poisson}(\lambda),$$

that is,

$$\text{Poisson}(\Theta) \bigwedge_{\Theta} \text{Gamma}(Y, P) \bigwedge_{Y/k} \text{Poisson}(\lambda).$$

This is called the Poisson-Pascal distribution; see Chapter 9, Section 8. It should not be confused with the previous distribution. By Gurland's theorem the pgf is

$$G(z) = \exp[\lambda\{(1 + P - Pz)^{-k} - 1\}]. \qquad (8.79)$$

The distribution has had a long history of use as a model for biological (especially entomological) data. Taking $k = 1$ gives a Poisson mixture of geometric distributions, known as the *Pólya-Aeppli distribution*. A Poisson mixture of shifted negative binomial (Pascal) distributions is called the *generalized Pólya-Aeppli distribution*. For further details concerning these distributions, see Chapter 9, Sections 7 and 8.

6. A negative binomial mixture of negative binomial distributions can be formed similarly.

The mixing distributions for the remaining distributions in this section are not infinitely divisible, and so do not lead to Poisson-stopped-sum distributions.

7. A (continuous) rectangular mixture of Poisson distributions,

$$\text{Poisson}(\Theta) \underset{\Theta/\phi}{\bigwedge} \text{Rectangular}(a,b),$$

was studied by Feller (1943). Bhattacharya and Holla (1965) studied it as a possible alternative to the negative binomial in the theory of accident proneness. It is a special case of a beta mixture of Poisson distributions.

8. A beta mixture of Poisson distributions,

$$\text{Poisson}(\Theta) \underset{\Theta/\phi}{\bigwedge} \text{Beta}(a,b),$$

was derived by Gurland (1958) by supposing that the number of insect larvae per egg mass has a Poisson distribution with parameter $\Theta = \phi P$, where P (the probability that an egg hatches into a larva) is a random variable having a beta distribution. The distribution was subsequently studied by Katti (1966) who called it a type H_1 distribution. Its pgf is

$$G(z) = {}_1F_1[a; a + b; \phi(z - 1)]. \tag{8.80}$$

The probabilities are

$$\Pr[X = x] = \frac{a \ldots (a + x - 1)\phi^x}{(a + b) \ldots (a + b + x - 1)x!} {}_1F_1[a + x; a + b + x; -\phi],$$

$$x = 0, 1, \ldots. \tag{8.81}$$

More conveniently, these may be obtained from $\Pr[X = 0]$ and $\Pr[X = 1]$ by using the recurrence relation

$$(x + 2)(x + 1)\Pr[X = x + 2] = (x + a + b + \phi)(x + 1)\Pr[X = x + 1]$$
$$- \phi(x + a)\Pr[X = x]; \tag{8.82}$$

see Chapter 2, Section 4.2, for the method for obtaining the recurrence relation. Because this is a generalized hypergeometric factorial moment distribution, the factorial moments are

$$\mu'_{[r]} = \frac{a(a+1)\ldots(a+r-1)\phi^r}{(a+b)(a+b+1)\ldots(a+b+r-1)}, \tag{8.83}$$

whence

$$\mu = \frac{a\phi}{a+b}$$

$$\mu_2 = \frac{a\phi}{a+b} + \frac{ab\phi^2}{(a+b)^2(a+b+1)}, \qquad \text{etc.} \tag{8.84}$$

9. A truncated-gamma mixture of Poisson distributions,

$$\text{Poisson}(tY) \bigwedge_Y \text{Truncated gamma}(a,b,p),$$

was studied by Kemp (1968c) in the context of limited collective risk theory. The pgf is

$$G(z) = \int_0^p \frac{e^{ty(z-1)} e^{-ay} a^b y^{b-1} dy}{\Gamma(b)} \bigg/ \int_0^p \frac{e^{-ay} a^b y^{b-1} dy}{\Gamma(b)}$$

$$= \frac{{}_1F_1[b; b+1; ptz - pt - ap]}{{}_1F_1[b; b+1; -ap]}. \tag{8.85}$$

This is a mixed (compound) Poisson process with time parameter t. Such processes have been studied widely; see, e.g., Lundberg (1940) and Cox and Isham (1980). Expansion of (8.85), followed by the use of Kummer's first transformation gives

$$\Pr[X = x] = \frac{b(pt)^x e^{-pt}}{(b+x)x!} \times \frac{{}_1F_1[1; b+x+1; pt+ap]}{{}_1F_1[1; b+1; ap]},$$

$$= \frac{(pt)^x (ap)^b}{(pt+ap)^{x+b} x!} \times \frac{\gamma(b+x; pt+ap)}{\gamma(b; ap)}, \tag{8.86}$$

where $\gamma(c,d)$ is an incomplete gamma function (Chapter 1, Section A5). The factorial moments are

$$\mu = \mu'_{[1]} = \frac{bpt}{(b+1)} \times \frac{{}_1F_1[b+1; b+2; -ap]}{{}_1F_1[b; b+1; -ap]},$$

$$\mu'_{[2]} = \frac{b(pt)^2}{(b+2)} \times \frac{{}_1F_1[b+2; b+3; -ap]}{{}_1F_1[b; b+1; -ap]}, \qquad \text{etc.} \tag{8.87}$$

Haight's (1965) insurance claims process is an unconditional risk process with removals, and can also be regarded as a mixture of Poisson distributions. The pgf is

$$G(z) = \int_0^\infty \frac{e^{Ty(z-1)}\,\Gamma(N,yt)e^{-ay}a^b y^{b-1}dy}{\Gamma(N)\Gamma(b)} \bigg/ \int_0^\infty \frac{\Gamma(N,yt)e^{-ay}a^b y^{b-1}dy}{\Gamma(N)\Gamma(b)}$$

$$= \frac{{}_2F_1[N+b,b;b+1;(Tz-T-a)/t]}{{}_2F_1[N+b,b;b+1;-a/t]} \tag{8.88}$$

[see also Kemp (1968c)].

10. A truncated Pearson Type III mixture of Poisson distributions,

$$\text{Poisson}(\lambda Y) \bigwedge_Y \text{Truncated Pearson Type III}(\lambda, \beta),$$

was proposed by Barton (1966); the pgf is

$$G(z) = \int_0^1 e^{\lambda y(z-1)} \times \frac{e^{\lambda y}(1-y)^{\beta-2}dy}{{}_1F_1[1;\beta;\lambda]}$$

$$= \frac{{}_1F_1[1;\beta;\lambda z]}{{}_1F_1[1;\beta;\lambda]}. \tag{8.89}$$

As Barton pointed out, this is a hyper-Poisson distribution; see Chapter 4, Section 12.4. Philipson (1960b) has made a systematic study of Poisson mixtures with mixing distributions that are members of the Pearson family of continuous distributions.

11. A lognormal mixture of Poisson distributions,

$$\text{Poisson}(\Theta) \bigwedge_\Theta \text{Lognormal } (\xi, \sigma, a),$$

has been regarded as a competitor to the logarithmic distribution for certain kinds of ecological data; see Chapter 7, Section 11.

12. A truncated-normal mixture of Poisson distributions,

$$\text{Poisson}(\Theta) \bigwedge_\Theta \text{Truncated-normal}(\xi, \sigma),$$

is closely related to the Hermite distribution; see Chapter 9, Section 4.

13. A Lindley (1958) mixture of Poisson distributions,

$$\text{Poisson}(\Theta) \bigwedge_\Theta \text{Lindley } (\phi),$$

has the pgf

$$G(z) = \int_0^\infty e^{\theta(z-1)} \frac{\phi^2(\theta+1)e^{-\theta\phi}d\theta}{(\phi+1)}$$

$$= \frac{\phi^2(\phi+2-z)}{(\phi+1)(\phi+1-z)^2}; \tag{8.90}$$

the pmf is

$$\Pr[X = x] = \frac{\phi^2(\phi + 2 + x)}{(\phi + 1)^{x+3}}$$

$$= \frac{(\phi + 2 + x)}{(\phi + 1)(\phi + 1 + x)} \Pr[X = x - 1]. \tag{8.91}$$

The fmgf is

$$G(1 + t) = \frac{1 - t/(1 + \phi)}{(1 - t/\phi)^2}, \tag{8.92}$$

whence

$$\mu = \frac{\phi + 2}{\phi(\phi + 1)} \quad \text{and} \quad \mu_2 = \frac{\phi^3 + 4\phi^2 + 6\phi + 2}{\phi^2(\phi + 1)^2}. \tag{8.93}$$

This distribution was studied by Sankaran (1970), with applications to errors and accidents. Sankaran called it the *Poisson-Lindley distribution*. It is a special case of Bhattacharya's (1966) more complicated mixed Poisson distribution.

14. A binomial mixture of Poisson distributions is represented by

$$\text{Poisson}(\Theta) \underset{\Theta/\phi}{\bigwedge} \text{Binomial}(n, p).$$

For this distribution

$$\Pr[X = x] = \sum_{j=0}^{n} \binom{n}{j} p^j q^{n-j} e^{-j\phi} (j\phi)^x / x!, \quad x = 0, 1, 2, \dots, \tag{8.94}$$

$$= \frac{\phi^x}{x!} \sum_{j=0}^{x} \binom{n}{j} \Delta^j 0^x (pe^{-\phi})^j (q + pe^{-\phi})^{n-j},$$

where $q = 1 - p$. From Gurland's theorem the pgf is

$$G(z) = [q + p \exp\{\phi(z - 1)\}]^n. \tag{8.95}$$

The moment properties can be obtained quite readily from those of the binomial distribution, since the fmgf is

$$(q + pe^{\phi t})^n = \sum_{i \geq 0} \frac{a_i \phi^i t^i}{i!}, \tag{8.96}$$

where a_i is the ith uncorrected moment of the binomial distribution. So

$$\mu = np\phi,$$

$$\mu'_{[2]} = \{np + n(n-1)p^2\}\phi^2,$$

that is, $\qquad \mu_2 = np\phi + npq\phi^2,$

and $\qquad \mu_3 = np\phi + 3npq\phi^2 + npq(q-p)\phi^3, \qquad$ etc. $\qquad$ (8.97)

This distribution should not be confused with the more commonly used Poisson-binomial distribution of Chapter 9, Section 5.

15. A logarithmic mixture of Poisson distributions is denoted by

$$\text{Poisson}(\Theta) \bigwedge_{\Theta/\phi} \text{Logarithmic } (\lambda).$$

The pgf is

$$G(z) = \frac{\log\{1 - \lambda e^{\phi(t-1)}\}}{\log(1-\lambda)}. \qquad (8.98)$$

The probabilities are given by

$$\Pr[X=0] = [\log(1-\lambda)]^{-1}[\log(1-\lambda e^{-\phi})],$$
$$\Pr[X=1] = [-\log(1-\lambda)]^{-1}\lambda\phi e^{-\phi}(1-\lambda e^{-\phi})^{-1},$$
$$\Pr[X=x] = [-\log(1-\lambda)]^{-1}\left(\frac{\phi^x}{x!}\right)\sum_{j=1}^{\infty} j^{x-1}(\lambda e^{-\phi})^j, \; x = 0,1,2,\ldots; \quad (8.99)$$

the restrictions on the parameters are $0 < \theta$ and $0 < \lambda < 1$. The mean and variance are

$$\mu = [-\log(1-\lambda)]^{-1}\frac{\lambda\phi}{(1-\lambda)},$$

$$\mu_2 = [-\log(1-\lambda)]^{-1}\frac{\lambda\phi}{(1-\lambda)}$$
$$+[-\log(1-\lambda)]^{-1}\frac{\lambda\phi^2}{(1-\lambda)^2}[1 - \lambda\{-\log(1-\lambda)\}^{-1}]. \qquad (8.100)$$

The zero-modified Poisson$\bigwedge$Logarithmic distribution [Katti and Rao (1970)] is known as the *log-zero-Poisson distribution*; see Section 2.2 of this chapter.

16. A hypergeometric mixture of Poisson distributions is

$$\text{Poisson}(\Theta) \bigwedge_{\Theta/\phi} \text{Hypergeometric}(n, Np, N).$$

For this distribution

$$\Pr[X=x] = \binom{N}{n}^{-1}\left(\frac{\phi^x}{x!}\right)\sum_j e^{-j\phi}(j\phi)^x\binom{Np}{j}\binom{N-Np}{n-j}; \qquad (8.101)$$

the summation is taken over the values of j for which $0 \le j \le Np$ and $0 \le n - j \le N - Np$. The pgf is

$$G(z) = \frac{(N - n)!(N - Np)!}{N!(N - n - Np)!} {}_2F_1[-n, -Np; N - Np - n + 1; e^{\phi(z-1)}]. \quad (8.102)$$

The case $n = 1$ is of course the Poisson$\bigwedge$Bernoulli distribution.

3.3 Mixtures of Binomial Distributions

We now consider binomial distributions. The binomial distribution has two parameters, n and p, and either or both of these may be supposed to have a probability distribution. We will not discuss cases in which both n and p vary, though it is easy to construct such examples.

In most cases discussed in the statistical literature, p has a continuous distribution, while n is discrete. The latter restriction is necessary, but the former is not. The reader will recall that for the Poisson parameter θ, both continuous and discrete distributions have been used as mixing distributions. However, discrete distributions for p have not been found to be useful, nor have they attracted much attention from a theoretical point of view.

Mixtures of binomial distributions have finite support and so cannot be infinitely divisible (and cannot be Poisson-stopped-sum distributions). Nevertheless, since

$$(1 - p + pz)^{mj} = \{(1 - p + pz)^m\}^j, \quad (8.103)$$

Gurland's theorem applies when the parameter n is allowed to vary, just as it applied for Poisson distributions with θ varying.

Hald (1968) has derived a general approximation for a binomial mixture where the parameter p has a continuous mixing distribution with probability density function $w(p)$. Hald found that

$$\Pr[X = x] = \binom{n}{x} \int_0^1 p^x (1 - p)^{n-x} w(p) dp$$

$$= \frac{w(x/n)}{n} \left[1 + \frac{b_1(x/n)}{n} + \frac{b_2(x/n)}{n^2} + O(n^{-3}) \right], \quad (8.104)$$

where $b_1(h) = [w(h)]^{-1}[-w(h) + (1 - 2h)w'(h) + \frac{1}{2}h(1 - h)w''(h)]$ and where $b_2(h) = [w(h)]^{-1}[w(h) - 3(1-2h)w'(h) + (1 - 6h + 6h^2)w''(h) + \frac{5}{6}h(1-h)(1-2h)w'''(h)] + \frac{1}{8}h^2(1-h)^2 w^{iv}(h)]$.

Al–Zaid (1989) has shown that a mixed binomial distribution is unimodal if the mixing distribution has support $(0,1)$ and is unimodal.

We now list a number of mixtures of binomials, beginning with p varying. Because p is restricted to the range $(0, 1)$, a Gamma mixture of binomials is impossible.

1. A beta mixture of binomial distributions,

$$\text{Binomial}(n, P) \bigwedge_{P} \text{Beta}(\alpha, \beta)$$

has already been discussed in Chapter 6 under the name *beta-binomial* (Ishii and Hayakawa (1960) called it the *binomial-beta*). It has been used to model variation in the number of defective items per lot in routine sampling inspection. The continuous rectangular-binomial is a special case. Horsnell (1957) has proposed the use of a continuous rectangular mixing distribution with range not extending over the entire interval $(0, 1)$. Horsnell has also used

$$\text{Binomial}(n, P) \bigwedge_{P} \text{Triangular}$$

in the same connection.

We come now to mixtures of binomials formed by n varying.

2. A Poisson mixture of binomial distributions is denoted by

$$\text{Binomial}(N, p) \bigwedge_{N/n} \text{Poisson } (\lambda).$$

This is known as the *Poisson-binomial distribution*. The *Hermite distribution* is the particular case $n = 2$. These important distributions are discussed in Chapter 9, Sections 4 and 5.

3. A binomial mixture of binomial distributions,

$$\text{Binomial}(N, p) \bigwedge_{N/n} \text{Binomial}(N', p'),$$

has the pgf

$$G(z) = [1 - p' + p'(1 - p + pz)^n]^{N'}, \tag{8.105}$$

and the probabilities are

$$\Pr[X = x] = \sum_{j \geq x/n} \left[\binom{N'}{j} (p')^j (1 - p')^{N'-j} \binom{nj}{x} p^x (1 - p)^{nj-x} \right], \tag{8.106}$$

where $x = 0, 1, 2, \ldots, N'n$. If $n = 1$, the pgf is $[1 - p' + p'(1-p) + pp'z]^{N'}$, and the distribution is binomial. This result is also evident on realizing that the model corresponds to N' repetitions of a two-stage experiment with probabilities of success p, p' at the two stages independently. The distribution has not been used very much. The fact that there are four parameters to be estimated (three, if n is known) makes fitting the distribution a discouraging task.

4. A negative binomial mixture of binomial distributions,

$$\text{Binomial}(N, p) \bigwedge_{N/n} \text{Negative binomial } (k, P'),$$

has similarly found little application in statistical work. The pgf is

$$G(z) = [1 + P' - P'(1 - p + pz)^n]^{-k}. \tag{8.107}$$

5. A logarithmic mixture of binomial distributions,

$$\text{Binomial}(N, p) \bigwedge_{N/n} \text{Logarithmic}(\theta),$$

has the pgf

$$G(z) = \frac{\ln\{1 - \theta(1 - p + pz)^n\}}{\ln(1 - \theta)}, \tag{8.108}$$

and the probabilities are

$$\Pr[X = x] = [-\ln(1 - \theta)]^{-1} \sum_{j \geq x/n} \frac{\theta^j}{j} \binom{nj}{x} p^x (1 - p)^{nj-x}, \tag{8.109}$$

where $x = 0, 1, 2, \ldots$, and $0 < \theta < 1$. The special case $n = 1$ yields the zero-modified logarithmic distribution; see Section 2.2 of this chapter.

Another kind of mixture of binomials is as follows.

6. A hypergeometric mixture of binomial distributions has the form

$$\text{Binomial}(m, Y/n) \bigwedge_{Y} \text{Hypergeometric}(n, Np, N).$$

This model corresponds to sampling without replacement from a population of size N containing Np "defective" items, followed by sampling (with sample size m) with replacement from the resultant set of n individuals. The probabilities are

$$\Pr[X = x] = \left[\binom{m}{x} \Big/ \binom{N}{n}\right] \sum_y \binom{Np}{y} \binom{N - Np}{n - y} \left(\frac{y}{n}\right)^x \left(1 - \frac{y}{n}\right)^{m-x}. \tag{8.110}$$

The range of summation for y is $\max(0, n - N + Np)$ to $\min(Np, n)$.

Very recent further work on mixtures of binomial distributions is that of Bowman, Kastenbaum and Shenton (1992).

3.4   Other Continuous and Countable Mixtures of Discrete Distributions

We have seen that a *continuous mixture* of discrete distributions arises when a parameter corresponding to some feature of a model for a discrete distribution can be regarded as a random variable taking a continuum of values.

If, as is common, the variable parameter can take any nonnegative value, a frequent assumption is that it has a gamma distribution. The unimodality of the gamma distribution makes it a realistic choice in many situations. The resultant mixture has pgf

$$G(z) = \int_0^\infty g(z|u)e^{-u/\beta}u^{\alpha-1}\{\beta^\alpha\Gamma(\alpha)\}^{-1}du, \qquad (8.111)$$

where $g(z|u)$ is the pgf of a distribution with parameter u. The relationship between this integral and the Laplace transform leads to mathematical tractability in many instances.

Suppose that the parameter can take only a restricted range of values, from a to b. It is then often scaled to take values from 0 to 1 and assumed to have a beta distribution on $(0,1)$. The variety of shapes that the beta density can take leads to its popularity for this purpose. The resultant distribution has the pgf

$$G(z) = \int_0^1 g(z|u)\frac{u^{c-1}(1-u)^{d-c-1}du}{B(c,d-c)}. \qquad (8.112)$$

This integral is related to the Mellin transform, and often the integration is reasonably tractable.

Kemp GHF distributions have pgf's of the form

$$_pF_q[a_1,\ldots,a_p;b_1,\ldots,b_q;\theta(z-1)];$$

see Chapter 2, Section 4.2. Gamma and beta mixtures of these distributions have the special property that the mixture distribution is also GHF. For gamma mixtures

$$G(z) = \int_0^\infty {}_pF_q[a_1,\ldots,a_p;b_1,\ldots,b_q;\theta u(z-1)] \times \frac{e^{-u}u^{c-1}du}{\Gamma(c)}$$

$$= {}_{p+1}F_q[a_1,\ldots,a_p,c;b_1,\ldots,b_q;\theta(z-1)]. \qquad (8.113)$$

The well-known model for the negative binomial as a gamma mixture of Poisson distributions is an example.

When the mixing distribution is a beta distribution, we have

$$G(z) = \int_0^1 {}_pF_q[a_1,\ldots,a_p;b_1,\ldots,b_q;\theta u(z-1)] \times \frac{u^{c-1}(1-u)^{d-c-1}du}{B(c,d-c)}$$

$$= {}_{p+1}F_{q+1}[a_1,\ldots,a_p,c;b_1,\ldots,b_q,d;\theta(z-1)], \qquad (8.114)$$

where $d > c > 0$. An example is the beta-Poisson (i.e., Poisson$\bigwedge$Beta) distribution of Gurland (1958); see also Holla and Bhattacharya (1965).

Another example is the beta mixed negative binomial distribution of Kemp and Kemp (1956a) — this is not the usual beta-negative binomial distribution of Chapter 6, Section 2.3, but the distribution mentioned there in passing; it is a special case of Katti's (1966) Type H_2 distribution (Chapter 2, Section 4.2). Here

$$G(z) = \int_0^1 {}_1F_0[k; \ ; u(z-1)] \times \frac{u^{c-1}(1-u)^{d-c-1}du}{B(c,d-c)}$$
$$= {}_2F_1[k,c;d;(z-1)].\tag{8.115}$$

Kemp GHP distributions possess analogous properties [Kemp (1968a)]. We recall from Chapter 2, Section 4.1 that their pgf's have the form

$$\frac{{}_pF_q[a_1,\ldots,a_p;b_1,\ldots,b_q;\theta yz]}{{}_pF_q[a_1,\ldots,a_p;b_1,\ldots,b_q;\theta y]}.\tag{8.116}$$

Mixtures of GHP distributions using gamma-type mixing distributions with pdf's

$$\frac{e^{-y}y^{c-1}{}_pF_q[a_1,\ldots,a_p,c;b_1,\ldots,b_q;\theta y]}{\Gamma(c)_{p+1}F_q[a_1,\ldots,a_p,c;b_1,\ldots,b_q;\theta]}\tag{8.117}$$

produce mixture distributions that are again GHP. For example, Bhattacharya's (1966) derivation of the negative binomial distribution as a mixed confluent hypergeometric distribution can be rewritten as

$$G(z) = \int_0^\infty \frac{{}_1F_1[v;c;uz]}{{}_1F_1[v;c;u]} \times \frac{e^{-(\alpha+1)u}u^{c-1}\alpha^v(\alpha+1)^{c-v}{}_1F_1[v;c;u]\,du}{\Gamma(c)}$$
$$= \int_0^\infty \frac{{}_1F_1[v;c;yz/(\alpha+1)]}{{}_1F_1[v;c;y/(\alpha+1)]} \times \frac{e^{-y}y^{c-1}{}_1F_1[v;c;y/(\alpha+1)]dy}{\Gamma(c)_1F_0[v; \ ;1/(\alpha+1)]}$$
$$= \frac{{}_1F_0[v; \ ;z/(\alpha+1)]}{{}_1F_0[v; \ ;1/(\alpha+1)]} = \left(\frac{\alpha}{\alpha+1-z}\right)^v.\tag{8.118}$$

Mixtures of GHP distributions using beta-type mixing distributions with pdf's

$$\frac{y^{c-1}(1-y)^{d-c-1}{}_pF_q[a_1,\ldots,a_p;b_1,\ldots,b_q;\theta y]}{B(c,d-c)_{p+1}F_{q+1}[a_1,\ldots,a_p,c;b_1,\ldots,b_q,d;\theta]}\tag{8.119}$$

are similarly again GHP. Barton's (1966) derivation of the hyper-Poisson distribution as a mixture of Poisson distributions can be rewritten as

$$G(z) = \int_0^1 \frac{{}_0F_0[\ ; \ ;\theta yz]}{{}_0F_0[\ ; \ ;\theta y]} \times \frac{(1-y)^{d-2}{}_0F_0[\ ; \ ;\theta y]dy}{{}_1F_1[1;d;\theta]}$$
$$= \frac{{}_1F_1[1;d;\theta z]}{{}_1F_1[1;d;\theta]}\tag{8.120}$$

(where ${}_0F_0[\ ; \ ;\theta y] = e^{\theta y}$).

We note that these mixing processes are only meaningful when the initial and the resultant distributions have pgf's that either converge or terminate. Only positive numerator and denominator parameters can be added by these processes.

A number of other discrete mixtures will now be cataloged. We will not discuss mixed logarithmic distributions; they do not lend themselves to analysis and are not at present used in statistical modeling.

Gurland's theorem applies to mixtures of negative binomial distributions as well as to mixtures of Poisson and binomial ones, since

$$(1 + P - Pz)^{-mj} = \{(1 + P - Pz)^{-m}\}^j,$$

where $k = mj$. The power parameter of the negative binomial distribution can take all nonnegative values, however, unlike the power parameter of the binomial distribution.

1. The usual beta mixture of negative binomial distributions is

$$\text{Negative binomial}(k, P) \bigwedge_{Q^{-1}} \text{Beta}(\alpha, \beta),$$

where $Q = 1 + P$ (i.e., using the parameterization $(1 + P - Pz)^{-k}$ for the negative binomial pgf). Here $p = Q^{-1}$ has the pdf

$$\frac{p^{\alpha-1}(1 - p)^{\beta-1}}{B(\alpha, \beta)},$$

and the resultant distribution is the beta-negative binomial distribution of Chapter 6, Section 2.3.

2. A gamma mixture of negative binomial distributions,

$$\text{Negative binomial}(K, P) \bigwedge_{K} \text{Gamma}(\alpha, \beta),$$

has the pgf

$$G(z) = (\beta^\alpha \Gamma(\alpha))^{-1} \int_0^\infty (1 + P - Pz)^{-k} k^{\alpha-1} e^{-k/\beta} dk$$

$$= \int_0^\infty \frac{\exp[-y\{1 + \beta \ln(1 + P - Pz)\}] y^{\alpha-1} dy}{\Gamma(\alpha)}$$

$$= [1 + \beta \ln(1 + P - Pz)]^{-\alpha}. \tag{8.121}$$

This can also be regarded as a negative binomial stopped sum of logarithmic distributions,

$$\text{Negative binomial}(\alpha, \rho) \bigvee \text{Logarithmic}(\lambda)$$

with pgf

$$G(z) = \left[1 + \rho - \rho \frac{\ln(1 - \lambda z)}{\ln(1 - \lambda)}\right]^{-\alpha}, \tag{8.122}$$

by taking $\rho = \beta \ln(1 + P)$ and $\lambda = P/(1 + P)$. The mean and variance are

$$\mu = \alpha \beta P,$$
$$\mu_2 = \alpha \beta P(1 + P + \beta P). \tag{8.123}$$

3. A binomial mixture of negative binomial distributions,

$$\text{Negative binomial}(kY, P) \bigwedge_Y \text{Binomial}(n, p),$$

is not often used. The pgf is

$$G(z) = [1 - p + p(1 + P - Pz)^{-k}]^n. \tag{8.124}$$

4. A hypergeometric mixture of negative binomial distributions, i.e.

$$\text{Negative binomial}(kY, P) \bigwedge_Y \text{Hypergeometric}(n', N'p', N'),$$

is similarly rarely encountered.

5. A logarithmic mixture of negative binomial distributions,

$$\text{Negative binomial}(kY, P) \bigwedge_Y \text{Logarithmic}(\theta),$$

has the pgf

$$G(z) = \frac{\ln\{1 - \theta(1 + P - Pz)^{-k}\}}{\ln(1 - \theta)}. \tag{8.125}$$

When $k = 1$, the pgf can be written as

$$G(z) = \frac{\ln(1 + P - Pz - \theta) - \ln(1 + P - Pz)}{\ln(1 - \theta)}.$$

In the following mixtures of hypergeometric distributions the parameter Y is assumed to vary. Hald (1960) has pointed out that if the rth factorial moment of the distribution of Y is $\xi_{[r]}'$, then the rth factorial moment of the mixture distribution is

$$\mu_{[r]}' = \frac{n!(N - r)!}{(n - r)!N!} \xi_{[r]}'.$$

6. A binomial mixture of hypergeometric distributions,

$$\text{Hypergeometric}(n, Y, N) \bigwedge_Y \text{Binomial}(N, p),$$

gives rise to another binomial distribution with parameters n, p. This is self-evident on realizing that it represents the results of choosing a random subset of a random sample of fixed size.

7. A hypergeometric mixture of hypergeometric distributions is represented by

$$\text{Hypergeometric}(n, Y, N) \bigwedge_Y \text{Hypergeometric}(N, N'p', N').$$

By an argument similar to that for a binomial mixture of hypergeometric distributions, it can be seen that this is a hypergeometric distribution with parameters $n, N'p', N'$. A special case is

$$\text{Hypergeometric}(n, Y, N) \bigwedge_Y \text{Discrete rectangular},$$

where the probabilities for the discrete rectangular distribution are

$$\Pr[X = x] = \frac{1}{N+1}, \qquad x = 0, 1, 2, \ldots, N. \tag{8.126}$$

More details concerning mixed hypergeometric distributions are given in Hald (1960). Horsnell (1957) has examined them in relation to economical acceptance sampling schemes.

CHAPTER 9

Generalized (Stopped-Sum) Distributions

The distributions considered in this chapter result from the combination of two independent distributions in a particular way. This process was called "generalization" by Feller (1943). The use of the term "generalization" was reinforced by Gurland (1957), who introduced the symbolic notation that is customarily employed to represent the process. However, "generalized" is a term that is greatly overused in statistics. Some authors (for example Douglas, 1971, 1980) have chosen to use the term "stopped-sum" instead for this type of distribution, because the principal model for the process can be interpreted as the summation of observations from the distribution $\mathcal{F}_2$, where the number of observations to be summed is determined by an observation from the distribution $\mathcal{F}_1$ (that is, summation of $\mathcal{F}_2$ observations is stopped by the value of the $\mathcal{F}_1$ observation).

A number of distributions that were merely listed in the previous chapter are dealt with more fully here; this is because they arise from two distinct models, a mixture model and a stopped-sum model. The relationship between the two kinds of model is a consequence of Gurland's theorem. Readers are advised to refer back to the previous chapter when reading the present chapter, and to pay particular attention to the three important theorems in Section 3.1 of that chapter.

Much research has been devoted to stopped-sum distributions. Within the space available we have tried to give the main results for the more important members. Certain of the distributions, including the negative binomial, the order k, and the lost-games distributions, have other, perhaps more important, modes of genesis, however; these distributions are discussed elsewhere in the book.

1 INTRODUCTION

Neyman (1939) constructed a statistical model of the distribution of larvae in a unit area of a field (in a unit of habitat) by assuming that the variation in the number of clusters of eggs per unit area (per unit of habitat) could be represented by a Poisson distribution with parameter λ, while the numbers of larvae developing from the clusters of eggs are assumed to have independent Poisson distributions all with the same parameter ϕ. This is a model of heterogeneity. Neyman described the model as "contagious," but before long there arose a substantial body of opinion distinguishing *heterogeneity* from *true contagion*, that is to say, from situations in which the events under observation depend on the pattern of previous occurrences of the events.

Consider the initial (zero) and first generations of a branching process. Let the pgf for the size N of the initial (parent) generation be $G_1(z)$, and suppose that each individual i of this initial generation independently gives rise to a random number Y_i of first generation individuals, where $Y_1, Y_2, \ldots$ have a common distribution, that of Y with pgf $G_2(z)$. The random variable for the total number of first generation individuals is then

$$S_N = Y_1 + Y_2 + \ldots + Y_N, \tag{9.1}$$

where N and Y_i, $i = 1, 2 \ldots, N$ are all random variables. The pgf of the distribution of S_N is

$$
\begin{aligned}
E[z^{S_N}] &= E_N[E[z^{S_N}|N]] \\
&= E_N[G_2(z)] \\
&= G_1(G_2(z)). \tag{9.2}
\end{aligned}
$$

This result seems to have been formulated for the first time by Watson and Galton (1874), and to have been rediscovered repeatedly. See, however, an earlier attribution to Bienaymé by Heyde and Seneta (1972).

Some authors have termed distributions with pgf's of the form $G_1(G_2(z))$ "clustered" or "compound" (the term "compound" is very ambiguous as it is frequently used for mixtures distributions such as those in Chapter 8). Douglas (1980) has discussed terminology for these distributions at some length; see also Chapter 4, Section 11. In Chapter 8, Section 3.1, a distribution with pgf of the form (9.2) was called a *generalized $\mathcal{F}_1$ distribution*, or more precisely an $\mathcal{F}_1$ *distribution generalized* by the *generalizing $\mathcal{F}_2$ distribution*. Another common terminology is to say that S_n has a randomly $\mathcal{F}_1$-stopped summed-$\mathcal{F}_2$ distribution. ($G_1(z)$ is the pgf for $\mathcal{F}_1$; $G_2(z)$ is the pgf for $\mathcal{F}_2$).

One of the authors who has consistently used the term "generalized" to denote these distributions is Gurland (1957). Gurland's notation is especially useful. It enables us to refer to an $\mathcal{F}_1$-stopped summed-$\mathcal{F}_2$ distribution, that is, an $\mathcal{F}_1$ distribution generalized by an $\mathcal{F}_2$ distribution, by means of the symbolic representation

$$S_N \sim \mathcal{F}_1 \bigvee \mathcal{F}_2. \tag{9.3}$$

For example, the negative binomial distribution can be represented as

$$\text{Negative binomial} \sim \text{Poisson} \bigvee \text{Logarithmic;}$$

see equation (5.15) in Chapter 5, Section 3.

Consider now the characteristic function of S_N. Let $\varphi_1(t) = G_1(e^{it})$ and $\varphi_2(t) = G_2(e^{it})$ be the cf's for N and X, respectively. Then the cf of S_N is

$$\begin{aligned} \varphi(t) &= G_1(G_2(e^{it})) \\ &= G_1(\varphi_2(t)) \\ &= \varphi_1[-i \ln\{\varphi_2(t)\}]. \end{aligned} \tag{9.4}$$

Clearly discrete distributions can arise provided Y has a discrete distribution, even when $\varphi_1(t)$ is the cf of a continuous distribution; see Douglas (1980). However, under these circumstances the model loses its physical meaning.

If the pgf of Y (i.e., $G_2(z)$) depends on a parameter ϕ in such a way that

$$G_2(z|j\phi) = [G_2(z|\phi)]^j, \tag{9.5}$$

then Gurland's theorem holds, and

$$\mathcal{F}_1 \bigvee \mathcal{F}_2 \sim \mathcal{F}_2 \bigwedge_j \mathcal{F}_1; \tag{9.6}$$

in such circumstances the generalized distribution has an alternative genesis as a mixed distribution. (It is customary in the case of a mixture distribution to write under the inverted vee sign the name of the parameter that is summed or integrated out when the mixture is formed.) To clarify this duality in modes of genesis, readers should refer back to Chapter 8, Section 3.1 for the general theory and for an example.

There are special relationships between stopped sum distributions and their components. Let

$$G_1(z) = \sum_{i \geq 0} a_i z^i, \quad G_2(z) = \sum_{i \geq 0} b_i z^i, \quad \text{and} \quad G_1(G_2(z)) = \sum_{i \geq 0} c_i z^i. \tag{9.7}$$

Then by direct expansion

$$\Pr[X = 0] = c_0 = \sum_{j=0}^{\infty} a_j b_0^j,$$

$$\Pr[X = 1] = c_1 = \sum_{j=0}^{\infty} j a_j b_0^{j-1} b_1,$$

$$\Pr[X = 2] = c_2 = \sum_{j=0}^{\infty} a_j \left\{ j b_0^{j-1} b_2 + \frac{j(j-1)}{2} b_0^{j-2} b_1^2 \right\}, \qquad \text{etc.} \quad (9.8)$$

A general expression for $\Pr[X = x]$, $x \geq 1$, can be obtained by the use of Faà di Bruno's formula [Jordan (1950)]:

$$\frac{d^x}{dz^x} G_1(G_2(z)) = \sum_{\pi(x)} \left\{ \frac{x!}{n_1! n_2! \dots n_x!} \left[\frac{d^n G_1(u)}{du^n} \right]_{u = G_2(z)} \right.$$

$$\left. \times \left(\frac{G_2^{(1)}}{1!} \right)^{n_1} \left(\frac{G_2^{(2)}}{2!} \right)^{n_2} \dots \left(\frac{G_2^{(x)}}{x!} \right)^{n_x} \right\}, \quad (9.9)$$

where $G_2^{(r)} = d^r G_2(z)/dz^r$ and summation is over all partitions $\pi(x)$ of x with $n_i \geq 0$, $i = 1, \dots, x$, such that

$$x = n_1 + 2n_2 + \dots + xn_x$$

and

$$n = n_1 + n_2 + \dots + n_x \qquad (9.10)$$

(for $x > 1$ at least one of the n_i will be zero).

An alternative notation for the derivatives of a composite function involves the use of the Bell polynomials; see, e.g., Riordan (1958, pp. 34-37), but note that Riordan uses k instead of our n, n instead of our x, and t instead of z. Continuing to use n, x and z as in (9.7), (9.9) and (9.10), let

$$f_r = \left[\frac{d^r G_1(u)}{du^r} \right]_{u = G_2(z)},$$

$$g_r = \frac{d^r G_2(z)}{dz^r},$$

$$A_r = \frac{d^r G_1(G_2(z))}{dz^r}. \qquad (9.11)$$

Then

$$A_x = \sum_{\pi(x)} \frac{x! f_n}{n_1! n_2! \dots n_x!} \left(\frac{g_1}{1!} \right)^{n_1} \left(\frac{g_2}{2!} \right)^{n_2} \dots \left(\frac{g_x}{x!} \right)^{n_x}$$

$$= \sum_{j=1}^{x} f_j A_{x,j}(g_1, g_2, \dots, g_x), \qquad (9.12)$$

where the A_x are known as Bell's polynomials. For an exposition of their use in the present context, see Charalambides (1977a).

We have

$$A_1 = f_1 g_1,$$
$$A_2 = f_1 g_2 + f_2 g_1^2,$$
$$A_3 = f_1 g_3 + f_2(3g_2 g_1) + f_3 g_1^3,$$
$$A_4 = f_1 g_4 + f_2(4g_3 g_1 + 3g_2^2) + f_3(6g_2 g_1^2) + f_4 g_1^4, \qquad \text{etc.} \qquad (9.13)$$

Expressions for A_5, A_6, A_7 and A_8 are given in Riordan (1958, p. 49). Interested readers may care to compare these polynomials with the expressions for uncorrected moments in terms of cumulants, given for instance in Stuart and Ord (1987, pp. 86–87), of which the first four are

$$\mu_1' = \kappa_1$$
$$\mu_2' = \kappa_2 + \kappa_1^2$$
$$\mu_3' = \kappa_3 + 3\kappa_2 \kappa_1 + \kappa_1^3$$
$$\mu_4' = \kappa_4 + 4\kappa_3 \kappa_1 + 3\kappa_2^2 + 6\kappa_2 \kappa_1^2 + \kappa_1^4 \qquad (9.14)$$

Let $u = G_2(z)$. Then returning to (9.9), we have

$$[u]_{z=0} = [G_2(z)]_{z=0} = b_0,$$
$$[G_2^{(x)}(z)]_{z=0} = x! b_x,$$
$$\left[\frac{d^n G_1(u)}{du^n} \right]_{z=0} = \sum_{j \geq 0} \frac{(n+j)!}{j!} a_{n+j} b_0^j, \qquad (9.15)$$

whence

$$\Pr[X = x] = c_x = \frac{1}{x!} \left[\frac{d^x}{dz^x} G_1(G_2(z)) \right]_{z=0}$$
$$= \sum_{\pi(x)} \left[\left(\sum_{j \geq 0} \frac{(n+j)! a_{n+j} b_0^j}{j!} \right) \frac{b_1^{n_1} b_2^{n_2} \dots b_x^{n_x}}{n_1! n_2! \dots n_x!} \right], \qquad x \geq 1. \quad (9.16)$$

If $\sum_i a_i z^i$ is an infinite series and $b_0 \neq 0$, then the inner summation over j leads to an infinite series expression for $\Pr[X = x]$. However, whenever $b_0 = 0$ the only nonzero term in the summation over j is $n! a_n$ and we obtain the finite series expansion

$$\Pr[X = x] = \sum_{\pi(x)} n! a_n \left(\frac{b_1^{n_1} b_2^{n_2} \dots b_x^{n_x}}{n_1! n_2! \dots n_x!} \right), \qquad x \geq 1. \qquad (9.17)$$

Rearrangement of $G_1(G_2(x))$ to give

$$G_1(G_2(x)) = G_1^*(G_2^*(x)), \tag{9.18}$$

where $G_2^*(z)$ is the pgf of a distribution with positive support (i.e., where $b_0^* = 0$) is sometimes possible; it is advantageous in that it leads to simpler expressions for the probabilities of the generalized distribution. Such rearrangement is always possible for generalized Poisson (Poisson-stopped sum) distributions.

Consider now moment properties. These have been studied by Katti (1966), by Douglas (1980), and in a series of papers by Charalambides [e.g., Charalambides (1977a, 1986a, c)]. Let the factorial moment generating functions of N and Y be

$$G_1(1+t) = \sum_r \frac{{}_1\mu'_{[r]} t^r}{r!} \quad \text{and} \quad G_2(1+t) = \sum_r \frac{{}_2\mu'_{[r]} t^r}{r!}, \tag{9.19}$$

respectively. Then the factorial moments of $X = S_N$ are

$$\mu'_{[r]} = \sum_{n=1}^r \left[{}_1\mu'_{[n]} \sum \frac{r!}{n_1! n_2! \dots n_r!} \left(\frac{{}_2\mu'_{[1]}}{1!} \right)^{n_1} \left(\frac{{}_2\mu'_{[2]}}{2!} \right)^{n_2} \cdots \left(\frac{{}_2\mu'_{[r]}}{r!} \right)^{n_r} \right], \tag{9.20}$$

with the inner summation over all partitions of r with $n_i \geq 0$, $i = 1, \dots, r$, such that

$$r = n_1 + 2n_2 + \cdots + rn_r$$

and

$$n = n_1 + n_2 + \cdots + n_r. \tag{9.21}$$

The factorial cumulants of $X = S_N$ are similarly

$$\kappa_{[r]} = \sum_{n=1}^r \left[{}_1\kappa_{[n]} \sum \frac{r!}{n_1! n_2! \dots n_r!} \left(\frac{{}_2\mu'_{[1]}}{1!} \right)^{n_1} \left(\frac{{}_2\mu'_{[2]}}{2!} \right)^{n_2} \cdots \left(\frac{{}_2\mu'_{[r]}}{r!} \right)^{n_r} \right] \tag{9.22}$$

with the same inner summation range as before.

Setting $r = 1, 2$ in these formulas gives

$$\mu_X = \mu_N \mu_Y$$

and

$$\sigma_X^2 = \mu_N \sigma_Y^2 + \mu_Y^2 \sigma_N^2, \tag{9.23}$$

where (μ_X, σ_X^2), (μ_N, σ_N^2) and (μ_Y, σ_Y^2) are the means and variances of X, N and Y, respectively. In terms of the index of dispersion, $I = \sigma^2/\mu$, (9.23) becomes

$$I_X = I_Y + \mu_Y I_N; \tag{9.24}$$

hence I_X is always greater than I_Y.

Katti (1966) has demonstrated how to obtain expressions for the probabilities for a generalized distribution using (9.20) and

$$\Pr[X = x] = \sum_{j \geq x} \frac{(-1)^{j-x} \mu'_{[j]}}{x!(j-x)!} \tag{9.25}$$

(inversion of the factorial moment generating function).

2 DAMAGE PROCESSES

The concept of a damage process is inherent in the work of Catcheside (1948). For a given dosage and length of exposure to radiation, the number N of chromosome breakages in individual cells can be assumed to have a Poisson distribution with pgf $e^{\theta(z-1)}$. If each breakage has a fixed and independent probability p of persisting and probability $q = 1 - p$ of healing, then the number of observed breakages has pgf $e^{\theta\{(q+pz)-1\}} = e^{\theta p(z-1)}$ (i.e., has a Poisson distribution with parameter θp; see Chapter 4, Section 8).

More generally [see, e.g., Kendall (1948)], if $G(z)$ is the pgf for the distribution of the number of animals of a particular species per unit of habitat in animal-trapping experiments, and if all the animals have an independent and equal probability p of being trapped, then the number of animals trapped per unit of habitat will have pgf $G(q+pz)$. Similarly, if each child in a family has the same independent probability of having a certain genetic defect, and this probability is the same for all families under consideration, then the number of children per family who have the defect has the pgf $G(q+pz)$, where $G(z)$ is the pgf for family size.

The pgf of X is

$$G_X(z) = G_Y(1 - p + pz)$$
$$= \sum_{y \geq 0} \sum_{j=0}^{y} \Pr[Y = y]\binom{y}{j} q^{y-j} p^j z^j, \tag{9.26}$$

whence

$$\Pr[X = x] = \sum_{y \geq x} \Pr[Y = y]\binom{y}{x} p^x q^{y-x}. \tag{9.27}$$

(The distribution of $X|Y = y$ is therefore binomial with parameters y and p.) The fmgf of X is

$$G_X(1 + t) = G_Y(1 + pt), \tag{9.28}$$

and so

$$x\mu'_{[r]} = p^r {}_Y\mu'_{[r]}. \tag{9.29}$$

The first two moments of X are

$$\mu_X = p\mu_Y$$

and
$$\sigma_X^2 = p^2\sigma_Y^2 + p(1 - p)\mu_Y. \qquad (9.30)$$

The Rao-Rubin characterization theorem for the Poisson distribution [Rao and Rubin (1964); Rao (1965)] has initiated much work on characterizations for damage processes. This has concentrated on conditions characterizing the distribution of Y from that of X; see Chapter 3, Section 9, and Chapter 4, Section 8, and, for example, C. R. Rao et al. (1980), Johnson and Kotz (1982), M. B. Rao and Shanbhag (1982), Panaretos (1982, 1987a), B. R. Rao and Janardan (1985), and Talwalker (1986). Rao and Rubin showed that the Bernoulli survival pattern described above is the only one that preserves a Poisson model.

Consider now the general class of distributions for which the underlying model is preserved under a Bernoulli damage process. Clearly their pgf's must be of the form

$$G(\alpha(z - 1)), \qquad (9.31)$$

with maybe other parameters beside α. This class of distributions includes the Poisson, binomial, negative binomial, and other GHF distributions (Chapter 2, Section 4.2), also the Neyman Type A, Hermite, Poisson-binomial, Pólya-Aeppli, Poisson-Pascal, and Thomas distributions; see, e.g., Boswell, Ord, and Patil (1979). Sprott (1965) showed that this class of distributions has an important maximum likelihood feature. Since

$$\alpha\frac{\partial G}{\partial \alpha} = (z - 1)\frac{\partial G}{\partial z}, \qquad (9.32)$$

that is, since

$$\alpha\frac{\partial \Pr[X = x]}{\partial \alpha} = x \Pr[X = x] - (x + 1) \Pr[X = x + 1], \qquad (9.33)$$

it follows that

$$\frac{\partial}{\partial \alpha} \sum_x f_x \ln \Pr[X = x] = \sum_x f_x \left\{ x - \frac{(x + 1) \Pr[X = x + 1]}{\Pr[X = x]} \right\}, \qquad (9.34)$$

where f_x is the frequency of the observation x in a sample of size N. Hence for this class of distributions the ML estimate of α is obtained from the ML equation

$$\bar{x} = \sum_x \frac{(x + 1)f_x\hat{p}_{x+1}}{N\hat{p}_x}, \qquad (9.35)$$

where $\hat{p}_x$ is the ML estimate of $\Pr[X = x]$; see, for example, Section 6.4 below concerning ML estimation for the Neyman Type A distribution.

3 POISSON-STOPPED-SUM DISTRIBUTIONS: GENERALIZED POISSON DISTRIBUTIONS

A random variable is said to be *infinitely divisible* iff it has a cf $\varphi(t)$ that can be represented for every positive integer n as the nth power of some cf $\varphi_n(t)$:

$$\varphi(t) = \{\varphi_n(t)\}^n. \tag{9.36}$$

Poisson-stopped-sum distributions (generalized Poisson distributions) have pgf's of the form

$$G(z) = \sum_{x \geq 0} \Pr[X = x] z^x = e^{\lambda\{g(z)-1\}}, \tag{9.37}$$

where $g(z)$ is the pgf of the generalizing distribution. In terms of branching processes, $g(z)$ is the pgf for the number of subsequent generation individuals per initial generation individual. Since $\varphi(t) = G(e^{it})$ and

$$e^{\lambda\{g(z)-1\}} = \left[e^{\lambda n^{-1}\{g(z)-1\}} \right]^n, \tag{9.38}$$

it follows that $G(z)$ is infinitely divisible; therefore Poisson-stopped-sum (generalized Poisson) distributions belong to the important class of infinitely divisible distributions.

The converse of this result is also true. If $G(z)$ is the pgf of an infinitely divisible nonnegative integer-valued rv, then it has the form

$$G(z) = e^{\lambda\{g(z)-1\}}, \tag{9.39}$$

where $g(z)$ is a pgf; $G(z)$ is therefore the pgf of a Poisson-stopped-sum distribution [see Feller (1957) for an elementary proof]. De Finetti (1931) has proved furthermore [see also Lukacs (1970)] that *all* infinitely divisible distributions are limiting forms of generalized Poisson distributions.

The importance of the property of infinite divisibility in modeling was stressed by Steutel (1983); see also the monograph by Steutel (1970). Infinitely divisible discrete distributions with pgf's of the form

$$G(z) = \frac{(1-q)g(z)}{1-qg(z)} \tag{9.40}$$

were studied by Steutel (1990) under the name *geometrically infinitely divisible distributions*. Conditions for a discrete distribution to be infinitely divisible are discussed in Katti (1967), Warde and Katti (1971), and Chang (1989).

Suppose now that

$$G(z) = \sum_{j \geq 0} Pr[X = x] z^x$$

is the pgf of a distribution with a finite mean; suppose also that it is a generalized Poisson distribution where the Poisson parameter is λ and the generalizing pgf is $g(z) = \sum_{j \geq 0} b_j z^j$. Then the mean of $G(z)$ is $\lambda \sum_{j \geq 0} j b_j$. This is finite; also $b_j \geq 0$. Consequently $\Pr[X = 0] = \exp(-\sum_{j \geq 0} b_j) > 0$.

Conversely, if $\Pr[X = 0] = 0$ for a distribution with a finite mean, then the distribution cannot be a generalized Poisson distribution. Similarly a distribution with finite support, $x = 0, 1, \ldots, n$, cannot be a generalized Poisson distribution.

It is straightforward to show that the representation of an infinitely divisible distribution on the nonnegative integers as a Poisson-stopped-sum distribution is unique only if we treat it as a Poisson-stopped sum of a distribution with *positive* support. This can be achieved by writing

$$G(z) = \exp\left[\lambda(1 - b_0)\left\{\frac{g(z) - b_0}{1 - b_0} - 1\right\}\right], \tag{9.41}$$

that is, by use of the representation

$$X \sim \text{Poisson}(\lambda(1 - b_0)) \bigvee \mathcal{F}^*, \tag{9.42}$$

where the pgf for $\mathcal{F}^*$ is

$$\frac{g(z) - b_0}{1 - b_0} = \frac{b_1 z + b_2 z^2 + \cdots}{b_1 + b_2 + \cdots}. \tag{9.43}$$

Such rearrangement is always possible for an infinitely divisible distribution with nonnegative integer support and a finite mean.

This rearrangement, together with the use of (9.17), enables the probabilities to be expressed as

$$\Pr[X = x] = \sum_{\pi(x)} e^{\lambda(b_0 - 1)} \frac{\lambda^n (1 - b_0)^n (b_1^*)^{n_1} \ldots (b_x^*)^{n_x}}{n_1! \ldots n_x!}, \tag{9.44}$$

where the summation is over all partitions $\pi(x)$ of x with $n_i \geq 0$, $i = 1, \ldots, x$, $x = n_1 + 2n_2 + \cdots + xn_x$ and $n = n_1 + n_2 + \cdots + n_x$, also $b_i^* = b_i/(1 - b_0)$, $i \geq 1$.

The following recurrence relationship

$$(x + 1)\Pr[X = x + 1] = \lambda \sum_{j=0}^{x} (x + 1 - j) b_{x+1-j} \Pr[X = j] \tag{9.45}$$

was derived by Kemp (1967a) by the simple method of differentiating the pgf once to give

$$\frac{dG(z)}{dz} = \sum_x (x + 1)\Pr[X = x + 1]z^x$$

$$= G(z)\frac{\lambda dg(z)}{dz}$$

$$= \lambda \left[\sum_{j\geq 0} \Pr[X=j]z^j\right]\left[\sum_{k\geq 0}(k+1)b_{k+1}z^k\right]; \qquad (9.46)$$

the recurrence relationship is then obtained by equating coefficients of z^x. It can also be obtained by repeated differentiation of $G(z)$; see Khatri and Patel (1961) and Gurland (1965).

We find that

$$\Pr[X=1]=\lambda b_1\Pr[X=0],$$

$$\Pr[X=2]=\lambda\left(b_2\Pr[X=0]+\frac{b_1}{2}\Pr[X=1]\right)$$

$$=\left(\frac{\lambda^2 b_1^2}{2!}+\lambda b_2\right)\Pr[X=0],$$

$$\Pr[X=3]=\lambda\left(b_3\Pr[X=0]+\frac{2b_2}{3}\Pr[X=1]+\frac{b_1}{3}\Pr[X=2]\right)$$

$$=\left(\frac{\lambda^3 b_1^3}{3!}+\lambda^2 b_1 b_2+\lambda b_3\right)\Pr[X=0], \qquad (9.47)$$

and so on.

Consider now the moment properties of X. Let the moment and the factorial moment generating functions of the generalizing distribution be

$$g(e^t)=\sum_{r\geq 0}\frac{{}_g\mu_r' t^r}{r!}$$

and

$$g(1+t)=\sum_{r\geq 0}\frac{{}_g\mu_{[r]}' t^r}{r!}, \qquad (9.48)$$

where ${}_g\mu_r'$ and ${}_g\mu_{[r]}'$ denote the rth uncorrected moment and the rth factorial moment of the generalizing distribution. Satterthwaite (1942) showed that the first four moments of X are

$$\mu_1'=\lambda\ {}_g\mu_1',$$

$$\mu_2=\lambda\ {}_g\mu_2',$$

$$\mu_3=\lambda\ {}_g\mu_3',$$

$$\mu_4=\lambda\ {}_g\mu_4'+3\lambda^2({}_g\mu_2')^2. \qquad (9.49)$$

Since $g(z) = \sum_x b_x z^x$, we have

$$_g\mu_1' = \sum_x x b_x \leq \sum_x x^2 b_x = {_g\mu_2'}.$$

Consequently $\mu_1' \leq \mu_2$ (i.e., the mean is less than or equal to the variance) for all generalized Poisson distributions. Equality is achieved only for the Poisson distribution itself.

More generally, from Feller (1943),

$$\ln\, G(e^t) = \lambda\, g(e^t) - \lambda, \tag{9.50}$$

and hence the cumulants of X are

$$\kappa_r = \lambda\, {_g\mu_r'}. \tag{9.51}$$

Similarly, $\ln\{G(1+t)\} = \lambda\, g(1+t) - \lambda$, whence the factorial cumulants are

$$\kappa_{[r]} = \lambda\, {_g\mu_{[r]}'}. \tag{9.52}$$

For relationships between the factorial moments of X and $_g\mu_{[r]}'$, see Katti (1966).

Generalized Poisson distributions can of course be interpreted as mixture distributions (Chapter 8, Section 3.1). We have

$$G(z) = \sum_{j=0}^{\infty} \frac{e^{-\lambda}\lambda^j}{j!}\{g(z)\}^j. \tag{9.53}$$

Let $\mathcal{F}(n)$ denote the distribution with pgf $\{g(z)\}^n$. Then this interpretation can be stated symbolically as

$$X \sim \text{Poisson}(\lambda) \bigvee \mathcal{F}(1) \sim \mathcal{F}(N) \bigwedge_N \text{Poisson}(\lambda). \tag{9.54}$$

Generalized Poisson distributions can also be regarded as convolutions of Poisson singlet, Poisson doublet, Poisson triplet, ..., etc., distributions, since

$$G(z) = e^{\lambda\{g(x)-1\}} = \exp\left\{\sum_{i \geq 0} \lambda(b_i z^i - b_i)\right\} = \prod_{i \geq 1} e^{\lambda b_i(z^i - 1)}. \tag{9.55}$$

Kemp and Kemp (1965) numerically decomposed several generalized Poisson distributions into Poisson singlet, Poisson doublet, etc., components in an investigation into the similarity of fits to data by different generalized Poisson distributions. Maritz (1952) showed that (9.55) can also arise as the pgf for the sum of k correlated Poisson singlet distributions ($b_i = 0$ for $i > k$).

Poisson-stopped-sum distributions with pgf's of the form

$$G(z) = e^{\lambda\{g(x)-1\}} = \exp\left\{\sum_{i=0}^{k} \lambda(b_i z^i - b_i)\right\} \qquad (9.56)$$

(that is, with a finite number of components) have further special properties. From (9.56) their factorial cumulants are

$$\kappa_{[r]} = \sum_{i=1}^{k} \frac{\lambda b_i \, i!}{(i-r)!} \qquad (9.57)$$

(cf. 9.52), and their moments can be derived therefrom. Also their maximum likelihood equations have a particularly simple form [Kemp (1967a)]. Let $p_j = \Pr[X = j]$ and $a_i = \lambda b_i$. Then differentiation of the pgf with respect to a_i, $i = 1, 2, \ldots, k$, gives

$$\sum_{j=0}^{\infty} \frac{\partial p_j z^j}{\partial a_i} = \frac{\partial G(z)}{\partial a_i}$$

$$= (z^i - 1)G(z)$$

$$= (z^i - 1)\sum_{j=0}^{\infty} p_j z^j, \qquad (9.58)$$

and equating coefficients of z yields

$$\frac{\partial p_j}{\partial a_i} = p_{j-i} - p_j, \qquad (9.59)$$

where $p_{j-i} = 0$ for $j < i$. The ML equations are therefore

$$\frac{\partial L}{\partial a_i} \equiv \sum_{j=0}^{\infty} f_j \frac{1}{p_j} \frac{\partial p_j}{\partial a_i} \equiv \sum_{j=0}^{\infty} f_j \left(\frac{p_{j-i}}{p_j} - 1\right) = 0, \qquad (9.60)$$

where $L \equiv \sum_j f_j \ln p_j$ is the log-likelihood and f_x is the observed frequency of x. Patel (1976a) has used the method to estimate the parameters of the two distributions for which $k = 3$ and $k = 4$ (he called these the *triple stuttering Poisson* and the *quadruple stuttering Poisson* distributions).

All Poisson-stopped sum distributions have the useful property of reproducibility, for if X_i, $i = 1, \ldots, s$, are independent rv's with generalized Poisson distributions, then $\sum_{i=1}^{s} X_i$ has the pgf

$$\prod_{i=1}^{s} \left\{e^{\lambda_i \, g_i(z) - \lambda_i}\right\} = \exp\left[\sum_{i=1}^{s} \{\lambda_i \, g_i(z) - \lambda_i\}\right], \qquad (9.61)$$

and so is also a generalized Poisson distribution.

Consider now a generalization of Fisher's derivation of the logarithmic distribution as a limiting form of the negative binomial distribution (Chapter 7, Section 2). Let $h(z) = \lambda g(z)$, where $g(z)$ is the pgf of the generalizing distribution. Then $G(z) = \exp\{h(z) - h(1)\}$. Using l'Hôpital's rule it can be proved that for all generalized Poisson distributions

$$\lim_{\gamma \to 0} \left[\frac{\{G(z)\}^\gamma - \{G(0)\}^\gamma}{1 - \{G(0)\}^\gamma} \right] = \frac{h(z)}{h(1)} = g(z) \qquad (9.62)$$

[Kemp (1978a)].

Parzen's axioms for the Poisson distribution were discussed in Chapter 4, Section 2. If Parzen's axiom 3 is modified to allow a random number of events (possibly more than one) to occur at a given instant of time, then the outcome is a generalized Poisson distribution for the number of events in a given interval of time. A stochastic process of this kind is often called a *compound Poisson process*; however, see Chapter 4, Section 11, and Chapter 8, Section 1, for an alternative usage of the term "compound."

The many models that give rise to Poisson-stopped sum (generalized Poisson) distributions have led to their study under a variety of names in the older literature. Besides "compound" Poisson distributions [Feller (1957)], the following terms have been used: Pollaczek-Geiringer distributions of multiple occurrences of rare events [Lüders (1934), Haight (1961b)]; composed Poisson distributions [Jánossy et al. (1950)]; multiple Poisson distributions [Feller (1957)]; stuttering-Poisson distributions [Galliher et al. (1959), Kemp (1967a)]; distributions "par grappes" [Thyrion (1960)]; Poisson power series distributions [Khatri and Patel (1961)]; and Poisson distributions with events in clusters [Castoldi (1963)]. Thompson (1954) has described a spatial mode of genesis (Darwin's model) that can lead to generalized Poisson distributions.

From the definition of a generalized Poisson distribution it follows that none of the b_j, $j = 0, 1, \ldots$, in $g(z) = \sum_j b_j z^j$ can be negative. Nevertheless, it is interesting to ask whether

$$\exp \sum_{j=1}^\infty \lambda b_j (z^j - 1), \qquad \lambda > 0,$$

can be a pgf, albeit not for a generalized Poisson distribution, if any of the b_j, $j \geq 1$, is negative. Lévy (1937a), quoted by Lukacs (1970, p. 252), has answered the question by proving that it cannot be a pgf unless a term with a negative coefficient is preceeded by one term and followed by at least two terms with positive coefficients.

Most of the distributions in the following sections of this chapter were developed using one or other of the models described in this section and are generalized Poisson distributions.

4 HERMITE DISTRIBUTION

The simplest generalized Poisson distribution arises when the generalizing distribution is a Bernoulli distribution with pgf $(1-p+pz)$, giving

$$X \sim \text{Poisson}(\lambda) \bigvee \text{Bernoulli}(p),\qquad(9.63)$$

$0 < p \leq 1$. This is equivalent to a damage process, and it leads to another Poisson distribution with parameter λp; see Section 2. Taking $p = 1$ gives a degenerate mixing distribution with pgf $g(z) = 1$ that leaves the initial Poisson distribution unaltered.

Consider now the generalization of a Poisson distribution with a binomial distribution with parameter $n = 2$:

$$X \sim \text{Poisson}(\lambda) \bigvee \text{Binomial}(2,p).\qquad(9.64)$$

The resultant distribution has pgf

$$G(z) = \exp[\lambda\{2pq(z - 1) + p^2(z^2 - 1)\}]\qquad(9.65)$$
$$= \exp[\alpha\beta z + \alpha^2 z^2/2 - \alpha\beta - \alpha^2/2]\qquad(9.66)$$

in the notation of Kemp and Kemp (1965). These authors showed that the pgf can be expanded in terms of Hermite polynomials, giving

$$\Pr[X = 0] = e^{-\alpha\beta - \alpha^2/2}\qquad(9.67)$$

and
$$\Pr[X = x] = \frac{\alpha^x H_x^*(\beta)}{x!}\,\Pr[X = 0],\qquad(9.68)$$

where $H_x^*(\beta)$ is the modified Hermite polynomial of Fisher (1951):

$$H_x^*(\beta) = \sum_{j=0}^{[n/2]} \frac{n!\,x^{n-2j}}{(n - 2j)!\,j!\,2^j},$$

where $[n/2]$ denotes the integer part of $(n/2)$. Hence

$$\Pr[X = 1] = \alpha\beta\,\Pr[X = 0],$$

$$\Pr[X = 2] = \frac{\alpha^2(\beta^2 + 1)}{2!}\,\Pr[X = 0],$$

$$\Pr[X = 3] = \frac{\alpha^3(\beta^3 + 3\beta)}{3!}\,\Pr[X = 0],$$

$$\Pr[X = 4] = \frac{\alpha^4(\beta^4 + 6\beta^2 + 3)}{4!}\,\Pr[X = 0],$$

$$\Pr[X = 5] = \frac{\alpha^5(\beta^5 + 10\beta^3 + 15\beta)}{5!} \Pr[X = 0], \qquad \text{etc.} \qquad (9.69)$$

The probabilities can also be expressed in terms of confluent hypergeometric functions. Setting $a_1 = \alpha\beta$, $a_2 = \alpha^2/2$ gives

$$\Pr[X = 0] = e^{-a_1 - a_2},$$

$$\Pr[X = x] = e^{-a_1 - a_2} \sum_{j=0}^{[x/2]} \frac{a_1^{x-2j} a_2^j}{(x - 2j)! j!}, \qquad (9.70)$$

and hence

$$\Pr[X = 2r] = e^{-a_1 - a_2} \left(\frac{a_2^r}{r!}\right) {}_1F_1[-r; 1/2; -a_1^2/(4a_2)],$$

$$\Pr[X = 2r + 1] = e^{-a_1 - a_2} \left(\frac{a_2^r}{r!}\right) a_1 \, {}_1F_1[-r; 3/2; -a_1^2/(4a_2)]. \qquad (9.71)$$

Kummer's transformation for the ${}_1F_1[\cdot]$ series,

$$ {}_1F_1[\alpha; \beta; \xi] = e^\xi {}_1F_1[\beta - \alpha; \beta; -\xi] \qquad (9.72)$$

(Chapter 1, Section A7) yields the alternative infinite series formula

$$\Pr[X = x] = \exp\{-a_1 - a_2 - a_1^2/(4a_2)\} \sum_{j=[(x+1)/2]}^{\infty} \frac{(2j)!(a_1/2)^{2j-x} a_2^{x-j}}{x!(2j - x)! j!}. \qquad (9.73)$$

The recursion relationship

$$(x + 1)\Pr[X = x + 1] = \alpha\beta \Pr[X = x] + \alpha^2 \Pr[X = x - 1],$$

that is, $$(x + 1)\Pr[X = x + 1] = a_1 \Pr[X = x] + 2a_2 \Pr[X = x - 1], \qquad (9.74)$$

is useful. It holds for $x \geq 0$, with $\Pr[X = -1] = 0$.

Y. C. Patel (1985) has obtained an asymptotic formula for the cumulative probabilities.

The cumulant generating function is

$$\ln \, G(e^t) = a_1(e^t - 1) + a_2(e^{2t} - 1), \qquad (9.75)$$

whence $\kappa_r = a_1 + 2^r a_2$.

Thus
$$\mu = \kappa_1 = a_1 + 2a_2 = \alpha(\alpha + \beta),$$
$$\mu_2 = \kappa_2 = a_1 + 4a_2 = \alpha(2\alpha + \beta),$$

$$\mu_3 = \kappa_3 = a_1 + 8a_2 = \alpha(4\alpha + \beta),$$

$$\mu_4 = \kappa_4 + 3\kappa_2^2 = a_1 + 16a_2 + 3(a_1 + 4a_2)^2,$$

$$= \alpha(8\alpha + \beta) + 3\alpha^2(2\alpha + \beta)^2, \qquad \text{etc.} \qquad (9.76)$$

The factorial moments are given by

$$G(1 + t) = \exp\{a_1 t + a_2(t^2 + 2t)\} = \exp\{(\alpha + \beta)\alpha t + \alpha^2 t^2/2\}, \qquad (9.77)$$

and so

$$\mu'_{[r]} = \alpha^r H_r^*(\alpha + \beta), \qquad (9.78)$$

and

$$\mu'_{[r+1]} = \alpha(\alpha + \beta)\mu'_{[r]} + \alpha^2 r \mu'_{[r-1]}. \qquad (9.79)$$

The indices of skewness and kurtosis are

$$\beta_1 = \alpha_3^2 = \frac{\mu_3^2}{\mu_2^3} = \frac{(4\alpha + \beta)^2}{\alpha(2\alpha + \beta)^3},$$

$$\beta_2 = \alpha_4 = \frac{\mu_4}{\mu_2^2} = 3 + \frac{(8\alpha + \beta)}{\alpha(2\alpha + \beta)^2}. \qquad (9.80)$$

A number of models give rise to the Hermite distribution. Clearly it is the special case $n = 2$ of the Poisson-binomial distribution, the widespread use of which for biometrical data was stimulated by Skellam (1952) and McGuire et al. (1957); see Section 5 below.

By Gurland's theorem, the Hermite distribution can also be regarded as a Poisson mixture of binomial distributions; we have

$$X \sim \text{Binomial}(N, p) \bigwedge_{N/2} \text{Poisson}(\lambda) \qquad (9.81)$$

(cf. (9.64)), giving

$$G(z) = \sum_{j=0}^{\infty} \frac{e^{-\lambda}\lambda^j}{j!} (1 - p + pz)^{2j}. \qquad (9.82)$$

The Hermite distribution appears in the highly innovative paper by Mc-Kendrick (1926) that was quoted extensively by Irwin (1963). McKendrick derived the distribution from the sum of two correlated Poisson random variables. The bivariate Poisson distribution may be defined as $(X_1, X_2) = (U + V, U + W)$, where U, V and W are independent Poisson variables with parameters a_2, c_1 and c_2, respectively [Ahmed (1961)]. The joint pgf of X_1 and X_2 is

$$G(z_1, z_2) = E[z_1^{X_1} z_2^{X_2}] = E[z_1^V z_2^W (z_1 z_2)^U]$$

$$= \exp\{c_1 z_1 + c_2 z_2 + a_2(z_1 z_2) - c_1 - c_2 - a_2\}. \qquad (9.83)$$

The pgf of $X = X_1 + X_2$ is

$$g(z) = E[z^{X_1+X_2}] = \exp\{(c_1 + c_2)z + a_2 z^2 - c_1 - c_2 - a_2\}, \qquad (9.84)$$

and therefore the distribution of X is Hermite with parameters $a_1 = c_1 + c_2$ and a_2. McKendrick fitted the distribution to counts of bacteria in leucocytes and obtained a very much better fit than with a Poisson distribution. Like certain other Poisson-stopped-sum (generalized Poisson) distributions, the Hermite distribution can have any number of modes. As an example, the fitted distribution for McKendrick's data has $a_1 = 0.0135$, $a_2 = 0.0932$, and the first five calculated probabilities are 0.899, 0.012, 0.084, 0.001, 0.004.

The Hermite distribution can also be derived as the sum $X = Y_1 + Y_2$, where Y_1 is a Poisson rv with parameter a_1 and support $0, 1, 2, \ldots$, and Y_2 is a Poisson rv with parameter a_2 and support $0, 2, 4, \ldots$, that is to say, as the convolution

$$X \sim \text{Poisson singlet}(a_1) \, * \, \text{Poisson doublet}(a_2). \qquad (9.85)$$

The good fit to McKendrick's data could also be explained by supposing that some of the bacteria occurred as singletons and many others occurred in pairs.

Kemp and Kemp also examined the Hermite distribution as a penultimate limiting form for other generalized Poisson distributions. We have

$$\lim_{a_i \to 0, i > 2} \exp[a_1 z + a_2 z^2 + a_3 z^3 + \ldots - a_1 - a_2 - a_3 - \ldots]$$

$$= \exp[a_1 z + a_2 z^2 - a_1 - a_2]. \qquad (9.86)$$

When a_2 also tends to zero the limiting form is of course Poissonian. The negative binomial, Neyman Type A, and Pólya-Aeppli are generalized Poisson distributions with generalizing distributions that are logarithmic, Poisson, and geometric, respectively. When the generalizing distributions are reversed J-shaped the corresponding generalized Poisson distributions are often approximated well by Hermite distributions.

McKendrick (1926) fitted the distribution using estimation by the method of moments. Sprott (1958) derived a maximum likelihood method for the more general Poisson-binomial distribution (see Section 5), and applied it in the particular (Hermite) case $n = 2$. Kemp and Kemp (1965) re-examined and discussed ML estimation for the Hermite distribution using both a "false position" and a Newton-Raphson iterative procedure. Because the distribution is a power series distribution and also has a pgf of the form $G(z) = h(\alpha(z - 1))$, the ML equations are

$$\bar{x} = \hat{\mu} = \hat{\alpha}(\hat{\alpha} + \hat{\beta}) \qquad (9.87)$$

and
$$\bar{x} = \hat{\mu} = \sum_j (j+1)\frac{f_j}{N}\frac{\hat{p}_{j+1}}{\hat{p}_j} \qquad (9.88)$$

in the notation of Section 2. Use of the recurrence formula (9.74) for the probabilities together with (9.88) gives alternatively,

$$\bar{x} = \hat{\alpha}(\hat{\alpha} + \hat{\beta}) \qquad (9.89)$$

and
$$1 = \sum_j \frac{f_j}{N}\frac{\hat{p}_{j-1}}{\hat{p}_j}. \qquad (9.90)$$

In both cases the solution requires iteration.

Y. C. Patel (1971), in his doctoral thesis, made a very thorough study of various estimation procedures for the Hermite distribution. Much of this work has appeared subsequently in journal papers. In Patel, Shenton and Bowman (1974) the properties of the maximum likelihood estimators are discussed, and numerical assessments of the biases, variances and covariances of the ML estimators are made. Moment estimators and mean-and-zero-frequency estimators are investigated in Patel (1976b). In this paper he introduced a new type of estimation procedure for discrete distributions, the method of even-points; see, e.g., Kemp and Kemp (1989). He compared the asymptotic efficiencies for the Hermite distribution using this method, the method of moments, and the mean-and-zero-frequency method. In Patel (1977) he obtained the higher moments of both the moment estimators and the even-point estimators for the Hermite distribution; he showed that for $a_1 < 1/2$ and large values of a_2 the even-point estimators are very close to the ML estimators.

Kemp and Kemp (1966) showed that formally the Hermite distribution can be obtained by mixing a Poisson distribution with parameter θ using a normal mixing distribution for θ; see also Greenwood and Yule (1920). This model assumes, however, that θ can take all real values, not merely nonnegative ones, and so lacks physical interpretation. For this model the recursion formula for the probabilities becomes

$$(x+1)Pr[X = x+1] = (\mu - \sigma^2)\Pr[X = x] + \sigma^2\Pr[X = x-1], \qquad x \ge 0,$$
$$(9.91)$$

with $\Pr[X = -1] = 0$ and $\Pr[X = 0] = e^{-\mu + \sigma^2/2}$. Here μ and σ^2 are the mean and variance of the normal distribution on $(-\infty, \infty)$. A necessary restriction on the parameters of the normal mixing distribution is $\mu \ge \sigma^2$.

A closely related distribution, obtained by letting θ have a normal distribution truncated to the left at the origin (the *Poisson truncated normal distribution*), was found by Kemp and Kemp (1967) to be a special case of Fisher's (1931) *modified Poisson distribution*. The Poisson truncated normal distribution has also been studied by Berljand, Nazarov and Pressman (1962) and Patil (1964b). Its pmf is

$$\Pr[X = x] = \exp(\sigma^2/2 - \mu)\sigma^x \frac{I_x(\sigma - \mu/\sigma)}{I_0(-\mu/\sigma)}, \qquad x \geq 0, \qquad (9.92)$$

where

$$I_0(x) = \frac{1}{\sqrt{2\pi}} \int_x^\infty e^{-t^2/2} dt,$$

$$I_r(x) = \frac{1}{\sqrt{2\pi}} \int_x^\infty \frac{(t-x)^r}{r!} e^{-t^2/2} dt, \qquad (9.93)$$

and μ and σ^2 are again the mean and variance of the complete normal distribution. Kemp and Kemp found that a more convenient method of handling the probabilities is via

$$\Pr[X = 0] = \exp(\sigma^2/2 - \mu) \frac{I_0(\sigma - \mu/\sigma)}{I_0(-\mu/\sigma)},$$

$$\Pr[X = 1] = \Pr[X = 0]\sigma \frac{I_1(\sigma - \mu/\sigma)}{I_0(\sigma - \mu/\sigma)},$$

$$(x+1)\Pr[X = x+1] = (\mu - \sigma^2)\Pr[X = x] + \sigma^2 \Pr[X = x - 1], \qquad x \geq 1$$
$$(9.94)$$

(cf. (9.91)). Alternatively, the pgf can be expressed as

$$G(z) = \left\{ \sum_{j=0}^\infty \left[\frac{\mu + \sigma^2(z-1)}{\sigma\sqrt{2}} \right]^j \Big/ \left(\frac{j}{2}\right)! \right\} \div \left\{ \sum_{j=0}^\infty \left[\frac{\mu}{\sigma\sqrt{2}} \right]^j \Big/ \left(\frac{j}{2}\right)! \right\},$$
$$(9.95)$$

and the probabilities derived therefrom by expansion. The various formulas simplify considerably when $\mu = 0$ and also when $\mu = \sigma^2$. The closeness of this distribution to the Hermite distribution increases as μ/σ^2 increases.

Gupta and Jain (1974) have investigated an *extended form* of the Hermite distribution with $X = X_1 + mX_2$ where X_1 and X_2 are independent Poisson rv's. The pgf is

$$G(z) = \exp[a_1 z + a_m z^m - a_1 - a_m], \qquad (9.96)$$

and the pmf is

$$\Pr[X = 0|m] = e^{-a_1 - a_m},$$

$$\Pr[X = r|m] = \Pr[X = 0|m] \sum_{j=0}^{[x/m]} \frac{(a_m)^j a_1^{x-mj}}{j!(x - mj)!}. \tag{9.97}$$

The cumulants are $\kappa_r = a_1 + m^r a_m$, and hence

$$\mu = a_1 + m a_m \quad \text{and} \quad \mu_2 = a_1 + m^2 a_m. \tag{9.98}$$

The *Gegenbauer distribution* was derived by Plunkett and Jain (1975) by mixing a Hermite distribution with pgf

$$G(z) = \exp\{\gamma(z - 1) + \gamma\rho(z^2 - 1)\},$$

using a gamma distribution for γ; see Chapter 11, Section 7.

Patil and Raghunandanan (1990) allowed all the parameters a_i, $i = 1, \ldots, k$ in

$$G(z) = \exp[a_1(z - 1) + a_2(z^2 - 1) + \ldots + a_k(z^k - 1)]$$

to have gamma distributions, and thereby obtained a distribution with pgf of the form

$$G(z) = \prod_{i=1}^{k} \left(\frac{1 - q_i}{1 - q_i z^i} \right)^{n_i}, \tag{9.99}$$

which they called a *stuttering negative binomial distribution*. When $a_i = a$, $i = 1, 2, \ldots, k$, this is the *negative binomial distribution of order k* discussed by Panaretos and Xekalaki (1986a); see Chapter 10, Section 6.2. Patil and Raghunandanan (1990) have also studied the distribution that arises when $a_1 = a_2 = \ldots = a_k = a$, and a has a reciprocal gamma distribution.

The *Borel-Hermite* distribution was developed by Jain and Plunkett (1977) by replacing the Poisson singlet and doublet distributions in (9.85) with singlet and doublet forms of Consul's "generalized Poisson" distribution; see Section 11 of this chapter. The resultant distribution has pmf

$$\Pr[X = x]$$

$$= \sum_{j=0}^{[x/2]} \left[\frac{m_1(m_1 + j)^{j-1} e^{-\lambda_1(m_1+j)} m(m + x - 2j)^{x-2j-1} \lambda^{x-2j} e^{-\lambda(m+x-2j)}}{j!(x - 2j)!} \right],$$

$$\tag{9.100}$$

and mean and variance

$$\mu = \frac{m\lambda}{1 - \lambda} + \frac{2m_1\lambda_1}{1 - \lambda_1} \quad \text{and} \quad \mu_2 = \frac{m\lambda}{(1 - \lambda)^3} + \frac{4m_1\lambda_1}{(1 - \lambda_1)^3}. \tag{9.101}$$

Here there are four parameters. Jain and Plunkett defined four cases of the distribution (A, B, C and D) by imposing restrictions on the parameters, thus reducing their number to either two or three. When fitting the distribution to McKendrick's leucocyte data, Jain and Plunkett used estimation by the method of moments.

5 POISSON-BINOMIAL DISTRIBUTION

The Poisson binomial distribution has the representation

$$X \sim \text{Poisson}(\lambda) \bigvee \text{Binomial}(n,p) \sim \text{Binomial } (N,p) \bigwedge_{N/n} \text{Poisson}(\lambda), \quad (9.102)$$

where N/n takes integer values; see Chapter 8, Section 3.1. The distribution was discussed by Skellam (1952) and fitted by him to quadrat data on the sedge *Carex flacca*. In McGuire et al. (1957) it was used to represent variation in the numbers of corn-borer *Pyrausta nubilalis* larvae in randomly chosen areas of a field. Since then it has found an increasing number of applications. Computational difficulties at first impeded its wider adoption, but these have been overcome with improvements in computational facilities and the development of formulas better suited to computational requirements.

The pgf is

$$G(z) = \exp[\lambda\{(q + pz)^n - 1\}]. \quad (9.103)$$

Expansion as an infinite series gives

$$\Pr[X = x] = e^{-\lambda} \sum_{j \geq x/n} \frac{\lambda^j}{j!} \binom{nj}{x} p^x q^{nj-x}$$

$$= \frac{e^{-\lambda}(p/q)^x}{x!} \sum_{j \geq x/n} \frac{(nj)!(\lambda q^n)^j}{(nj - x)!/j!}. \quad (9.104)$$

An alternative expression for the probabilities can be obtained in terms of $\mu^*_{[x]}$, the xth factorial moment of nY, where Y has a Poisson distribution with expected value λ:

$$\Pr[X = x] = \frac{e^{\lambda(q^n-1)}(p/q)^x \mu^*_{[x]}}{x!} \quad (9.105)$$

[Shumway and Gurland (1960b)]. This is equivalent to a finite series expansion for $\Pr[X = x]$. It follows from (9.105) that

$$\frac{\Pr[X = x + 1]}{\Pr[X = x]} = \frac{p \mu^*_{[x+1]}}{(x + 1)q \mu^*_{[k]}} = \frac{p}{(x + 1)q} R_{[x]}; \quad (9.106)$$

Shumway and Gurland (1960b) provided tables of $R_{[x]}$. Shumway and Gurland (1960a) suggested yet another way of calculating the probabilities based on the direct calculation of the $\mu^*_{[k]}$ as

$$\mu^*_{[k]} = \sum_{j=1}^{k} A_{kj}(\lambda q^n)^j; \tag{9.107}$$

they gave formulas and tables for A_{kj}.

The recurrence relation

$$\Pr[X = x + 1] = \frac{np\lambda}{(x + 1)} \sum_{j=0}^{x} \binom{n-1}{j} p^j q^{n-1-j} \Pr[X = x - j], \tag{9.108}$$

$x = 0, 1, \ldots$, can be used to calculate successive probabilities starting from

$$\Pr[X = 0] = \exp[\lambda(q^n - 1)]. \tag{9.109}$$

In particular

$$\Pr[X = 1] = \exp[\lambda(q^n - 1)]np\lambda q^{n-1}, \tag{9.110}$$

$$\Pr[X = 2] = \exp[\lambda(q^n - 1)]np^2\lambda q^{n-2}(n - 1 + nq^n\lambda)/2. \tag{9.111}$$

If $npq^{n-1}\lambda < 1$, then there is a mode at the origin. Note that if $npq^{n-1}\lambda < 1$ and $(n - 1 + nq^n\lambda) > 2q/p$ as well, then $\Pr[X = 0] > \Pr[X = 1] < \Pr[X = 2]$ and the distribution is at least bimodal. (Multimodality is a typical property of many Poisson-stopped-sum distributions.) Douglas (1980) has suggested that if there are many modes, then, when n is large, they will be roughly at $n, 2n, 3n, \ldots$ because the distribution will then tend to a Neyman Type A distribution (see below). He found that this will also happen as $p \to 1$, when the limiting form is an n-let Poisson distribution with support $0, n, 2n, \ldots$.

The cumulants of the Poisson-binomial distribution can be obtained from the uncorrected moments of the binomial distribution. Simpler still are the factorial cumulants. The fcgf is

$$\ln G(1 + t) = \lambda(1 + pt)^n - \lambda, \tag{9.112}$$

giving $\kappa_{[r]} = \lambda p^r n!/(n - r)!$. The first four moments are

$\mu = \lambda np,$

$\mu_2 = \lambda n^2 p^2 + \lambda npq,$

$\mu_3 = \lambda n^3 p^3 + 3\lambda n^2 p^2 q + \lambda npq(1 - 2p),$

$$\mu_4 = \lambda n^4 p^4 + 6\lambda n^3 p^3 q + \lambda n^2 p^2 q(7 - 11p) + \lambda npq(1 - 6pq) + 3\lambda^2(n^2 p^2 + npq)^2.$$
$$(9.113)$$

The factorial cumulant generating function has been used by Douglas (1980) to explore limiting forms of the distribution. If the generalizing binomial distribution tends to a Poisson distribution as $n \to \infty$, then the limiting form is Neyman Type A (see Section 6). When $p \to 1$, Douglas found that the limiting form is Poissonian with support $0, n, 2n, 3n, \ldots$.

Estimation for the three parameters n, λ, and p is made difficult by the integer restriction on n. A common estimation practice has been to estimate λ and p for several fixed values of n (say, $n = 2, 3, 4$) and to choose the set of three values for n, λ and p that appear to give the best fit judged by some criterion such as a χ^2 goodness-of-fit test. See also Section 9, where estimation for the distribution as a member of the Poisson-Katz family is discussed.

Consider now estimation for λ and p, given a sample of N observed values $x_1, x_2, \ldots, x_N$ from the distribution (9.103), n being supposed known. The equations satisfied by the maximum likelihood estimators $\hat{\lambda}$ and $\hat{p}$ of λ and p were obtained in the following form by Sprott (1958):

$$\bar{x} = \sum_{j=1}^{N} \frac{x_j}{N} = n\hat{\lambda}\hat{p}, \qquad (9.114)$$

$$\sum_{j=1}^{N} (x_j + 1)\hat{p}_{x_{j+1}}/\hat{p}_{x_j} = Nn\hat{\lambda}\hat{p}, \qquad (9.115)$$

where $\hat{p}_{x_j}$ is $\Pr[X = x_j]$ with λ and p replaced by $\hat{\lambda}$ and $\hat{p}$, respectively. Shumway and Gurland (1960b) introduced an alternative formulation of these equations.

The variances and covariances of the ML estimates are

$$\mathrm{Var}(\hat{p}) \doteq \frac{q^2 A + [np + (n-1)pq](n\lambda)^{-1}}{[n\lambda(n + q/p)A - (n-1)^2]N},$$

$$\mathrm{Var}(\hat{\lambda}) \doteq \frac{n^2\lambda^2 A - n\lambda(n - 1 + p^{-1})}{[\lambda n(n + q/p)A - (n-1)^2]N},$$

$$\mathrm{Cov}(\hat{\lambda}, \hat{p}) \doteq \frac{n\lambda q A - nq - p}{[\lambda n(n + q/p)A - (n-1)^2]N}, \qquad (9.116)$$

where

$$A = -1 + \left[\sum_{j=0}^{\infty} \frac{(j+1)^2(\Pr[X = j + 1])^2}{\Pr[X = j]} \right] (n\lambda p)^{-2}. \qquad (9.117)$$

The ML equations have to be solved iteratively. Martin and Katti (1965) have remarked on the practical difficulty of obtaining convergence using Newton-Raphson iteration. It is important therefore to begin with good initial estimates.

There are several ways of obtaining initial estimates. The method of moments involves equating the first and second sample moments. This gives

$$\tilde{p} = \frac{s^2 - \bar{x}}{(n-1)\bar{x}},$$

$$\tilde{\lambda} = \frac{\bar{x}}{n\tilde{p}}, \tag{9.118}$$

where
$$s^2 = \sum_{i=1}^{N} \frac{(x_i - \bar{x})^2}{N-1};$$

see Sprott (1958).

An alternative method of estimation uses the equations

$$n\lambda^* p^* = \bar{x},$$

$$n\lambda^* p^* (q^*)^{n-1} = \frac{f_1}{f_0}, \tag{9.119}$$

where f_j/N denotes the proportion of j's among the N observations. These equations give

$$p^* = 1 - q^* = 1 - \left(\frac{f_1}{f_0\bar{x}}\right)^{1/(n-1)},$$

$$\lambda^* = \frac{\bar{x}}{np^*}. \tag{9.120}$$

Instead of the ratio f_1/f_0, the proportion f_0/N alone might be used. Distinguishing this case by primes, we have

$$n\lambda' p' = \bar{x},$$

$$\exp[-\lambda'(1 - (q')^n)] = \frac{f_0}{N}, \tag{9.121}$$

whence
$$\frac{\bar{x}}{\ln(f_0/N)} = \frac{np'}{(q')^n - 1} \tag{9.122}$$

which can be solved by iteration. Katti and Gurland (1962b) found that this last method is markedly superior to that using the first and second moments.

Table 9.1 Efficiency of the Method of Moments for the Poisson-Binomial Distribution

		λ				
n	p	0.1	0.3	0.5	1.0	2.0
2	0.1	0.928	0.865	0.843	0.840	0.870
2	0.3	0.732	0.569	0.525	0.533	0.635
2	0.5	0.494	0.307	0.264	0.267	0.392
3	0.1	0.896	0.823	0.793	0.779	0.810
3	0.3	0.658	0.501	0.452	0.446	0.542
3	0.5	0.426	0.268	0.231	0.231	0.333
5	0.1	0.816	0.726	0.688	0.671	0.715
5	0.3	0.527	0.379	0.337	0.332	0.435
5	0.5	0.345	0.210	0.178	0.176	0.277

Adapted from Katti and Gurland (1962b).

Tables 9.1 and 9.2 show the approximate efficiencies (ratios of generalized variances) relative to ML estimation.

Linear minimum chi-squared estimation for the Poisson-binomial distribution has been studied by Katti and Gurland (1962b) and Hinz and Gurland (1967). Fitting a truncated Poisson-binomial distribution, with the zero class missing, has been described by Shumway and Gurland (1960b).

6 NEYMAN TYPE A DISTRIBUTION

6.1 Definition

This is the Poisson-stopped-summed-Poisson distribution that was discussed at the beginning of Section 1 of this chapter; its symbolic representation is

$$X \sim \text{Poisson}(\lambda) \bigvee \text{Poisson}(\phi). \qquad (9.123)$$

The pgf is

$$G(z) = \exp[\lambda\{e^{\phi(z-1)} - 1\}]. \qquad (9.124)$$

(David and Moore (1954) called ϕ the *index of clumping*.) By Gurland's theorem (Chapter 8, Section 3.1),

$$X \sim \text{Poisson}(\lambda) \bigvee \text{Poisson}(\phi) \sim \text{Poisson}(\Theta) \underset{\Theta/\phi}{\bigwedge} \text{Poisson }(\lambda). \qquad (9.125)$$

The distribution is therefore both a Poisson-stopped sum of Poisson distributions, and also a Poisson mixture of Poisson distributions (Chapter 8, Section 3.2).

Table 9.2　Efficiency of the Method of the First Moment and First Frequency for the Poisson-Binomial Distribution

n	p	λ				
		0.1	0.3	0.5	1.0	2.0
2	0.1	0.984	0.974	0.977	0.991	0.947
2	0.3	0.937	0.888	0.883	0.923	0.981
2	0.5	0.862	0.740	0.700	0.717	0.851
3	0.1	0.994	0.986	0.984	0.944	0.994
3	0.3	0.968	0.930	0.918	0.935	0.974
3	0.5	0.896	0.798	0.763	0.765	0.850
5	0.1	0.995	0.987	0.985	0.993	0.989
5	0.3	0.969	0.924	0.905	0.911	0.950
5	0.5	0.889	0.769	0.716	0.690	0.793

Adapted from Katti and Gurland (1962b).

There are two standard expressions for the probabilities, corresponding to the two different ways of expanding the pgf. First,

$$G(z) = e^{-\lambda} \sum_{j=0}^{\infty} \frac{\lambda^j e^{-j\phi} e^{j\phi z}}{j!}$$

$$= e^{-\lambda} \sum_{j=0}^{\infty} \left[\frac{(\lambda e^{-\phi})^j}{j!} \sum_{x=0}^{\infty} \frac{(j\phi z)^x}{x!} \right], \qquad (9.126)$$

whence

$$\Pr[X = x] = \frac{e^{-\lambda} \phi^x}{x!} \sum_{j=0}^{\infty} \frac{(\lambda e^{-\phi})^j j^x}{j!}, \qquad x = 0, 1, \dots. \qquad (9.127)$$

Alternatively,

$$G(z) = e^{-\lambda + \lambda e^{-\phi}} \sum_{j=0}^{\infty} \frac{\lambda^j e^{-j\phi} (e^{\phi z} - 1)^j}{j!}$$

$$= e^{-\lambda + \lambda e^{-\phi}} \sum_{j=0}^{\infty} \left[(\lambda e^{-\phi})^j \sum_{x=j}^{\infty} \frac{S(x, j)\phi^x z^x}{x!} \right], \qquad (9.128)$$

using the generating expression for Stirling numbers of the second kind in

Chapter 1, Section A3. Hence

$$\Pr[X = x] = \frac{e^{-\lambda + \lambda e^{-\phi}} \phi^x}{x!} \sum_{j=1}^{x} S(x, j) \lambda^j e^{-j\phi}. \tag{9.129}$$

This expression was given by Cernuschi and Castagnetto (1946); it can also be obtained from (9.44). Whereas (9.127) involves an infinite series, (9.129) involves only a polynomial of degree x in $\lambda e^{-\phi}$; (9.129) is therefore more useful for computation of the probabilities for low values of x. In particular

$$\Pr[X = 0] = e^{-\lambda + \lambda e^{-\phi}},$$

$$\Pr[X = 1] = \lambda \phi e^{-\phi} e^{-\lambda + \lambda e^{-\phi}},$$

$$\Pr[X = 2] = \lambda \phi^2 e^{-\phi} (1 + \lambda e^{-\phi}) e^{-\lambda + \lambda e^{-\phi}} / 2. \tag{9.130}$$

It will be noticed that $\Pr[X = 1] < \Pr[X = 0]$ if $\lambda \phi e^{-\phi} < 1$; also $\Pr[X = 1] < \Pr[X = 2]$ if $1 < \phi(\lambda e^{-\phi} + 1)/2$. It is easy to see that for every value of ϕ greater than 2, there are values of λ such that $\Pr[X = 0] > \Pr[X = 1] < \Pr[X = 2]$, giving a local minimum at $x = 1$. The distribution can moreover be *multimodal*. Barton (1957) made a systematic study of the modality of the Type A distribution and has given an interesting diagram of the regions of the parameter space having a given number of modes; whenever the distribution has more than one mode, one of the modes is at the origin. The degree of multimodality increases as ϕ increases; this is to be expected since for larger ϕ, the accretion of each additional "group" makes a larger contribution to the overall distribution.

The modal values are *approximately* multiples of ϕ. Shenton and Bowman (1967) gave modal values of X for several Type A distributions, illustrating this point. For example, with $\lambda = 7$ and $\phi = 25$, the modal values are 0, 25, 50, 76, 103, 129, 153, 173.

Beall (1940) derived the recurrence relation

$$\Pr[X = x] = \frac{\lambda \phi e^{-\phi}}{x} \sum_{j=0}^{x-1} \frac{\phi^j}{j!} \Pr[X = x - 1 - j]. \tag{9.131}$$

Douglas (1955) pointed out that

$$\Pr[X = x] = \frac{\phi^x}{x!} \mu_x'^* \Pr[X = 0], \tag{9.132}$$

where $\mu_x'^*$ is the xth moment about zero for a Poisson distribution with expected value $\lambda e^{-\phi}$.

6.2 Moment Properties

The cumulants for the Neyman Type A distribution are proportional to the uncorrected moments of a Poisson distribution with parameter ϕ (the constant of proportionality is λ). Shenton (1949) established the following recurrence relations among the cumulants:

$$\kappa_{r+1} = \phi \left\{ \sum_{j=0}^{r-1} \binom{r}{j} \kappa_{r-j} + \lambda \right\}, \qquad (9.133)$$

$$\kappa_{r+1} = \phi \left(\kappa_r + \frac{\partial \kappa_r}{\partial \lambda} \right). \qquad (9.134)$$

These are similar to the recurrence relations for the uncorrected moments of the Poisson distribution.

The factorial cumulants can be obtained from $\ln G(1+t) = \lambda(e^{\phi t} - 1)$ and hence are

$$\kappa_{[r]} = \lambda \phi^r, \qquad r = 1, 2, \ldots. \qquad (9.135)$$

From the factorial cumulants, or otherwise, the rth factorial moment is found to be

$$\mu'_{[r]} = \phi^r \sum_{j=1}^{r} \frac{\Delta^j 0^r}{j!} \lambda^r = \phi^r \sum_{j=1}^{r} S(r, j) \lambda^j; \qquad (9.136)$$

see Chapter 1, Section A3, for the relationship between differences of zero and Stirling numbers of the second kind. It follows that

$$\mu = \lambda \phi,$$
$$\mu_2 = \lambda \phi (1 + \phi),$$
$$\mu_3 = \lambda \phi (1 + 3\phi + \phi^2),$$
$$\mu_4 = \lambda \phi (1 + 7\phi + 6\phi^2 + \phi^3) + 3\lambda^2 \phi^2 (1 + \phi)^2. \qquad (9.137)$$

(Comparison with the corresponding formulas for the Thomas distribution is interesting; see Section 10 of this chapter.)

The moment ratios are

$$\beta_1 = \alpha_3^2 = \frac{\mu_3^2}{\mu_2^3} = \frac{(1 + 3\phi + \phi^2)^2}{\lambda \phi (1 + \phi)^3},$$

$$\beta_2 = \alpha_4 = \frac{\mu_4}{\mu_2^2} = 3 + \frac{(1 + 7\phi + 6\phi^2 + \phi^3)}{\lambda \phi (1 + \phi)^2}. \qquad (9.138)$$

We recall that $\beta_2 - \beta_1 - 3 = 0$ for the Poisson distribution. For the Neyman Type A distribution we find that

$$\frac{\beta_2 - 3}{\beta_1} = \frac{(1 + 7\phi + 6\phi^2 + \phi^3)(1 + \phi)}{(1 + 3\phi + \phi^2)^2}. \qquad (9.139)$$

This ratio is independent of λ. When $\phi = 0$, it is equal to unity. It rises to a maximum value of approximately 1.215 near $\phi = 0.5$, falling slowly thereafter to 1 as ϕ tends to infinity:

ϕ	0	1	2	3	4	5	6	7	10	20
$(\beta_2 - 3)/\beta_1$	1	1.20	1.16	1.14	1.12	1.11	1.10	1.09	1.07	1.04

The narrow limits of this ratio restrict the field of applicability of the distribution. The Neyman Type B and Type C and the more general contagious distributions described later in this chapter in Section 9 extend the flexibility of this class of distributions.

6.3   Tables and Approximations

Grimm (1964) gave values of $\Pr[X = x]$ for

$$E[X] = \lambda\phi = 0.1(0.1)1.0(0.2)4, 6, 10,$$

$$\phi = 0.2, 0.5, 1.0, 2, 3, 4, 5,$$

to five decimal places, up to $\Pr[X = x] = 0.99900$. Douglas (1955) has provided tables of μ'^*_{k+1}/μ'^*_k to assist in using equation (9.132) to calculate the probabilities to from two to four decimal places, for $k=0(1)19$ and

$$\lambda e^{-\phi} = 0.000(0.001)0.030(0.01)0.30(0.1)3.0.$$

Martin and Katti (1962) have considered three approximations to the Neyman Type A distribution, applicable when the parameters λ and ϕ take "extreme" values. These approximations are in fact limiting forms of the distributions.

Limiting Form I

The distribution of the standardized variable

$$Y = (X - \lambda\phi)[\lambda\phi(1 + \phi)]^{-1/2}$$

is approximately unit normal. This approximation is useful when λ is large and the mean $(= \lambda\phi)$ does not approach zero.

Limiting Form II

If λ *is small*, then X is approximately distributed as a *modified* Poisson variable ("Poisson with zeroes" distribution, in Chapter 8, Section 2.2), with

$$\Pr[X = 0] \doteq (1 - \lambda) + \lambda e^{-\phi},$$

$$\Pr[X = x] \doteq \frac{\lambda e^{-\phi} \phi^x}{x!}, \qquad x = 1, 2, \ldots.$$

Limiting Form III

If ϕ *is small*, then X is approximately distributed as a Poisson variable with expected value $\lambda\phi$.

Martin and Katti (1962) gave diagrams that provide a general picture of the regions in which these three limiting forms give useful practical approximations. Their diagrams are based on contours of ξ^2 and ζ^2 where ξ^2 and ζ^2 are measures of the goodness of the approximations. Let P_i and P_i^* be the probabilities for the Neyman Type A and the approximating distribution, respectively. Then

$$\xi^2 = \sum_i (P_i - P_i^*)^2$$

and

$$\zeta^2 = \sum_i \frac{(P_i - P_i^*)^2}{P_i}.$$

For the diagrams, more details of these approximation methods, and some examples of their fit, see Martin and Katti's paper. Douglas (1965) has suggested the approximate formula

$$\Pr[X = x] \doteq \left(\frac{e^{-\lambda}}{\sqrt{2\pi}} \right) \frac{\phi^x \exp[x/g(x)]}{[g(x)]^x [x(1 + g(x))]^{1/2}}, \tag{9.140}$$

where

$$g(x) \exp[g(x)] = x(\lambda e^{-\phi})^{-1}.$$

Bowman and Shenton (1967) quoted the following formula due to Philpot (1964):

$$\Pr[X = x] \doteq \Pr[X = 0] \frac{\phi^x x_0 \exp[f(x_0) - \lambda e^{-\phi}]}{x!(x_0 + x - 0.5)^{1/2}}, \tag{9.141}$$

where

$$x_0 \ln(x_0 e^{\phi} / \lambda) = x - 0.5$$

and

$$f(x_0) = x_0 + (x - 0.5)[\ln(\lambda e^{-\phi}) + x_0^{-1}(x - x_0 - 0.5)].$$

These authors also considered approximations of the form

$$\Pr[X = x] \doteq \sum_{j=1}^{s} A_{j,s} \frac{e^{-\theta_j} \theta_j^x}{x!},$$

with the A's and θ_j's chosen to give the correct values for the first $2s - 1$ moments.

6.4 Estimation

The two parameters of the Type A distribution are λ and ϕ. The distribution is a power series distribution, and also the pgf is a function of $\phi(z-1)$. Given observations on N independent random variables $X_1, X_2, \ldots, X_N$, each having the same Neyman Type A distribution, the maximum likelihood equations are therefore

$$\sum_{j=1}^{N} \frac{x_j}{N} = \bar{x} = \hat{\lambda}\hat{\phi}, \tag{9.142}$$

$$\sum_{j=1}^{N} \frac{(x_j + 1)\Pr[X = x_j + 1|\hat{\lambda}, \hat{\phi}]}{\Pr[X = x_j|\hat{\lambda}, \hat{\phi}]} = N\bar{x}, \tag{9.143}$$

where $\hat{\lambda}$ and $\hat{\phi}$ are the ML estimates of λ and ϕ.

These equations do not have an explicit solution. Shenton (1949) solved them using a Newton-Raphson iterative procedure. For hand computation the tables given by Douglas (1955) can be used to facilitate the calculations. Computer methods for solving the equations are discussed in Douglas (1980). Comments on difficulties associated with the ML estimators have been made by Shenton and Bowman (1967, 1977).

Initial values can be obtained by the method of moments. This gives

$$\hat{\lambda} = \frac{\bar{x}}{\hat{\phi}}, \tag{9.144}$$

$$\hat{\phi} = \frac{s^2 - \bar{x}}{\bar{x}}, \tag{9.145}$$

where

$$s^2 = \sum_{j-1}^{N} \frac{(x_j - \bar{x})^2}{N - 1}.$$

(Equations (9.142) and (9.144) are the same because the Neyman Type A is a power series distribution.) If the sample size is large, then the variances and covariance of these estimators are approximately

$$\text{Var}(\hat{\lambda}) \doteq \frac{\lambda[2 + \phi^2 + 2\lambda(1 + \phi)^2]}{\phi^2 N}, \tag{9.146}$$

$$\text{Var}(\hat{\phi}) \doteq \frac{2 + \phi + 2\lambda(1 + \phi)^2}{\lambda N}, \tag{9.147}$$

$$\text{Cov}(\hat{\lambda}, \hat{\phi}) \doteq -\frac{2[1 + \lambda(1 + \phi)^2]}{\phi N}. \tag{9.148}$$

For $\phi < 0.2$ the asymptotic efficiency of the moment estimators is at least 85% (whatever the value of λ), while for $0.2 < \phi < 1.0$ the efficiency is generally between 75% and 90%. A table of efficiencies is in Shenton (1949).

Katti and Gurland (1962a) and Bowman and Shenton (1967) have provided diagrams of the contours of efficiency of the moment estimators with respect to the ML estimators; see also Shenton and Bowman (1977). Shenton and Bowman (1967) showed that even in samples of size 100, there is substantial bias in both the maximum likelihood and moment estimators. For λ the bias is positive; for ϕ it is negative. Generally the biases in the two kinds of estimators are of comparable magnitude, the bias in the moment estimators usually being slightly larger. The biases in the estimators of ϕ do not greatly depend on λ and are of order -1% to -2%. The biases in the estimators of λ decrease as ϕ increases; for $\phi = 1$ they are about 10%.

A third method of estimation uses the sample mean ($\bar{x}$) and the observed proportion of zeros (f_0/N). The equations for these estimators, λ^*, ϕ^* , are

$$\bar{x} = \lambda^* \phi^*, \tag{9.149}$$

$$\frac{f_0}{N} = \exp[\lambda^*(e^{-\phi^*} - 1)]. \tag{9.150}$$

Elimination of λ between these equations gives the following equation for ϕ^*:

$$\frac{\bar{x}}{\ln(f_0/N)} = \frac{\phi^*}{e^{-\phi^*} - 1}. \tag{9.151}$$

A fourth method uses the ratio of the frequencies of ones and zeros, f_1/f_0, and the sample mean. This leads to the equations

$$\bar{x} = \lambda^{**} \phi^{**}, \tag{9.152}$$

$$\frac{f_1}{f_0} = \lambda^{**} \phi^{**} e^{-\phi^{**}}, \tag{9.153}$$

whence

$$\phi^{**} = \ln(\bar{x} f_0/f_1). \tag{9.154}$$

Katti and Gurland (1962a) and Bowman and Shenton (1967) tabled the efficiencies (relative to the moment estimators $\hat{\lambda}$, $\hat{\phi}$) of the two latter pairs of estimators. They also obtained contours of their efficiency with respect to the moment estimators; see also Shenton and Bowman (1977). The method of moments would seem to be the best of the three if λ is greater than about 5, for most values of ϕ. For $\lambda < 4.5$, the estimators λ^*, ϕ^*, appear to be preferable.

Katti (1965) has suggested that the different methods of estimation might be combined, with some gain in efficiency. In particular he proposed using statistics that are functions of the sample mean and variance, together with the observed proportion f_0/N of zeroes. His equations have to be solved by some sort of iterative process (Katti suggested a graphical method). The process could well be tedious, but Katti obtained approximate values of efficiency relative to ML estimation that are very high.

This idea has been taken further by Gurland (1965) and Hinz and Gurland (1967), who developed a generalized minimum chi-squared method of estimation based on the first few moments and a function of the zero-frequency; this too has very high asymptotic efficiency. Its implementation is described in detail in Douglas (1980). Hinz and Gurland (1967) also studied goodness-of-fit tests for the distribution. The modified test of Bhalerao, Gurland, and Tripathi (1980) has high power.

If a series of sample values are available, and it is only desired to estimate ϕ (the mean number per cluster), the following method of estimation has found favor among practical workers. This is simply to plot $\ln(f_0/N)$ against $\bar{x}$. Since

$$\frac{\ln \Pr[X = 0 | \lambda, \phi]}{E[X]} = \frac{e^{-\phi} - 1}{\phi}, \qquad (9.155)$$

a value of ϕ can be estimated from the slope of the graph. (A similar method has been suggested for the Thomas distribution; see Section 10 below.) Pielou (1957) has discussed this method of estimation.

Maximum likelihood, moment, and minimum chi-squared estimation for a common value of ϕ (or alternatively for a common value of λ), given a series of samples, has been investigated at length by Douglas (1980). Hinz and Gurland (1968, 1970) examined tests of linear hypotheses for series of samples from different Neyman Type A distributions. Gurland (1965), Hinz and Gurland (1967), and Grimm (1970) have investigated graphical methods for assessing the suitability of a Neyman Type A model.

6.5 Applications

The Neyman Type A distribution has often been used to describe plant distributions, especially when reproduction of the species produces clusters. This frequently happens when the species is generated by offshoots from parent plants, or by seeds falling near the parent plant or carried by living creatures, as is the case with many fruits. However, Archibald (1948) found that there is not enough evidence to make an induction from the type of fitted distribution to the type of reproduction. Evans (1953) found that while the Neyman Type A gave good results for plant distributions, negative binomial distributions (Chapter 5) were better for insect distributions; see also Wadley (1950). Martin and Katti (1965) fitted 35 data sets with a number of standard distibutions; they found that both the negative binomial and the Neyman Type A have wide applicability.

Pielou (1957) also has investigated the use of Neyman Type A distributions in ecology. She compared them with the Thomas distributions discussed in Section 10, but she found that neither type of distribution is likely to be applicable to describe plant populations unless the clusters of plants are so compact as not to lie across the edge of the quadrat used to select sample areas. Skellam (1958) did indeed point out that the compactness of the clusters is a hidden assumption in Neyman's original derivation of the distribution. The choice of quadrat size was found to greatly affect the results. Pielou decided that the Neyman Type A distributions fitted a wider variety of plant data than did the Thomas distributions.

Cresswell and Froggatt (1963) [see also Kemp (1967b)] derived the Neyman Type A in the context of bus-driver accidents, on the basis of the following assumptions:

1. Every driver is liable to "spells," where the number of spells per driver in a given period of time is Poissonian, with the same parameter λ for all drivers.
2. The performance of a driver during a spell is substandard; he is liable to have a Poissonian number of accidents during a spell with the same parameter ϕ for all drivers.
3. Each driver behaves independently.
4. No accidents can occur outside of a spell.

These assumptions lead to a Neyman Type A distribution via the Poisson(λ) $\bigvee$ Poisson(ϕ) model. Cresswell and Froggatt named this the "long distribution," because of its long tail, in contradistinction to their "short distribution" (for which see Section 9 below). Irwin (1964) remarked that a Type A distribution would also be obtained by assuming that different drivers have differing levels K of proneness, taking values $0, \phi, 2\phi, 3\phi, \ldots,$ with probability $\Pr[K = k\phi] = \exp(-\lambda)\lambda^k/k!$, and that a driver with proneness $k\phi$ has X accidents where $\Pr[X = x|k\phi] = \exp(-k\phi)(k\phi)^x/x!$. This is the Poisson($K\phi$) $\bigwedge$ Poisson(λ) model with mixing over the values taken by K.

Rogers (1965, 1969) has applied the Neyman Type A distribution to the clustering of retail food stores.

It has been suggested by David and Moore (1954) that complete distributions need not be fitted to data if one only wants to estimate indices of clustering (i.e., contagiousness) or the mean number of entities per cluster. Among the methods they discussed is a regression method of estimating ϕ.

7 PÓLYA-AEPPLI DISTRIBUTION

The *Pólya-Aeppli distribution* arises in a model formed by supposing that objects (which are to be counted) occur in clusters, the number of clusters having a Poisson distribution, while the number of objects per cluster has the geometric distribution with pmf

$$\Pr[Y = y] = qp^{k-1}, \qquad x = 1, 2, \ldots, \quad q = 1 - p. \qquad (9.156)$$

The Pólya-Aeppli distribution can therefore be represented as

$$X \sim \text{Poisson}(\theta) \bigvee \text{Shifted geometric}(p). \qquad (9.157)$$

It also arises as a generalization of the Poisson distribution with parameter θ/p, using an unshifted geometric distribution (with parameter p) as the generalizing distribution; hence it can be represented as

$$X \sim \text{Poisson}(\theta/p) \bigvee \text{Geometric}(p). \qquad (9.158)$$

The Pólya-Aeppli distribution was described by Pólya (1930); he ascribed the derivation of the distribution to his student Aeppli in a Zurich thesis in 1924. It is a special case of the Poisson-Pascal distribution; the latter is sometimes called the generalized Pólya-Aeppli distribution (see Section 8 below).

The distribution is defined by the pgf

$$G(z) = \exp\left[\theta\left(\frac{(1-p)z}{1-pz} - 1\right)\right] \qquad (9.159)$$

$$= \exp\left[\theta\left(\frac{z-1}{1-pz}\right)\right]$$

$$= \exp\left[\frac{\theta}{p}\left(\frac{1-p}{1-pz} - 1\right)\right]. \qquad (9.160)$$

Expression (9.159) corresponds to model (9.157), while expression (9.160) corresponds to model (9.158). In the literature, formulas for the Pólya-Aeppli distribution are given either in terms of θ and p, or alternatively in terms of ζ and p, where $\zeta = \theta/p$. This led to a confusion of notations in the first edition of this book.

By direct expansion of the pgf

$$\Pr[X = 0] = e^{-\theta} \tag{9.161}$$

and

$$\Pr[X = x] = e^{-\theta} p^x \sum_{j=1}^{x} \binom{x-1}{j-1} \frac{(\theta q/p)^j}{j!}, \qquad x = 1, 2, \ldots, \tag{9.162}$$

$$= e^{-\theta} \left(\frac{\theta q}{p} \right) p^x {}_1F_1[1-x; 2; -\theta q/p], \tag{9.163}$$

where ${}_1F_1[\cdot]$ is the confluent hypergeometric function of Chapter 1, Section A7. Algebraic manipulation that is equivalent to the use of Kummer's transformation,

$${}_1F_1[a; b; y] = e^y {}_1F_1[b-a; b; -y],$$

gives

$$\Pr[X = x] = e^{-\theta/p} \left(\frac{\theta q}{p} \right) p^x {}_1F_1[x+1; 2; \theta q/p], \qquad x = 1, 2, \ldots \tag{9.164}$$

[Evans (1953), Philipson (1960a)]; see also Galliher et al. (1959) for a Laguerre polynomial expression (Chapter 1, Section A11).

Evans (1953), using the notation $m = \theta/q$, $a = 2p/q$, derived a recurrence formula for the probabilities; in our notation this becomes

$$(x+1)\Pr[X = x+1] = (\theta q + 2px)\Pr[X = x] - p^2(x-1)\Pr[X = x-1], \tag{9.165}$$

for $x = 0, 1, 2, \ldots$, with $\Pr[X = 0] = e^{-\theta}$, $\Pr[X = -1] = 0$. An alternative recurrence formula is

$$(x+1)\Pr[X = x+1] = \theta q \sum_{j=0}^{x} (x+1-j) p^{x-j} \Pr[X = j] \tag{9.166}$$

Kemp (1967a). Douglas (1986) has reported that use of the three-term recurrence relation (9.165) may run into numerical difficulties because of the differencing of small terms.

Taking $\alpha = \theta q/p$ we obtain

$$\Pr[X = 1] = e^{-\theta} \theta q = e^{-\theta} \alpha p,$$

$$\Pr[X = 2] = e^{-\theta} \theta q \left(p + \frac{\theta q}{2} \right) = e^{-\theta} \alpha p^2 \left(1 + \frac{\alpha}{2} \right),$$

$$\Pr[X = 3] = e^{-\theta} \alpha p^3 \left(1 + \alpha + \frac{\alpha^2}{6} \right),$$

$$\Pr[X=4] = e^{-\theta}\alpha p^4 \left(1 + \frac{3\alpha}{2} + \frac{\alpha^2}{2} + \frac{\alpha^3}{24}\right). \tag{9.167}$$

There is a mode at the origin if $\theta q < 1$. When $2 < \theta < 1/q$ there is a local minimum probability at $X = 1$; that is, the distribution has at least two modes. Douglas (1965, 1980) has obtained approximate formulas for the probabilities; these are useful for probabilities in the tail of the distribution. Evans (1953) gave inequalities that can be used to assess the cumulative probability $\Pr[X \leq x - 1]$.

The Pólya-Aeppli distribution was called the *geometric Poisson* by Sherbrooke (1968), who gave tables of individual and cumulative probabilities to four decimal places for

$$\mu = \theta/q = 0.10, 0.25(0.25)1.00(0.5)3.0(1)10,$$
$$(1+p)(1-p)^{-1} = 1.5(0.5)5.0(1)7.$$

The expected value and first three central moments of the distribution are most easily found by using the factorial cumulant generating function. The factorial moment generating function is

$$G(1+t) = \exp[\theta t(q - pt)^{-1}]. \tag{9.168}$$

Therefore the factorial cumulant generating function is

$$\ln G(1+t) = \theta t(q - pt)^{-1} = \frac{\theta t}{q}\left(1 - \frac{pt}{q}\right)^{-1}, \tag{9.169}$$

and the rth factorial cumulant is

$$\kappa_{[r]} = \frac{r!\theta}{q}\left(\frac{p}{q}\right)^{r-1}, \qquad r = 1, 2, \ldots. \tag{9.170}$$

Hence the central moments are

$$\mu = \kappa_1 = \frac{\theta}{q},$$

$$\mu_2 = \kappa_2 = \frac{\theta(1+p)}{q^2},$$

$$\mu_3 = \kappa_3 = \frac{\theta(1+4p+p^2)}{q^3},$$

$$\mu_4 = \kappa_4 + 3\kappa_2^2 = \frac{\theta(1+11p+11p^2+p^3)}{q^4} + \frac{3\theta^2(1+p)^2}{q^4}. \tag{9.171}$$

From these equations we find that

$$\alpha_3^2 = \beta_1 = \frac{(1 + 4p + p^2)^2}{(1 + p)^3 \theta}, \qquad (9.172)$$

$$\alpha_4 = \beta_2 = 3 + \frac{(1 + 11p + 11p^2 + p^3)}{(1 + p)^2 \theta}. \qquad (9.173)$$

The distribution is the limiting form as $\beta \to \infty$, $\phi \to \infty$, $\beta/(\phi + \beta) \to p$ of Beall and Rescia's generalization of the Neyman Type A, B, and C distributions (see Section 9). The limiting form of the distribution as $p \to 0$ is a Poisson distribution with parameter θ.

We now consider the problem of estimation, given a sample of size N from the distribution (9.159), with observed values $x_1, x_2, \ldots, x_N$. The maximum likelihood equations are obtained from

$$\frac{\partial P_x}{\partial \theta} = (x\theta^{-1} - 1)P_x - (x - 1)\theta^{-1}pP_{x-1}$$

$$\frac{\partial P_x}{\partial p} = q^{-1}[(x - 1)P_{x-1} - xP_x],$$

where $P_x = \Pr[X = x]$, and hence they are

$$\bar{x} = \frac{\hat{\theta}}{1 - \hat{p}}, \qquad (9.174)$$

$$\bar{x} = \sum_{j=1}^{N} \frac{f_j}{N} \frac{(j - 1)\widehat{P_{j-1}}}{\widehat{P_j}}, \qquad (9.175)$$

or, equivalently,

$$\bar{x} = \frac{\hat{\theta}}{1 - \hat{p}}, \qquad (9.176)$$

$$\bar{x} = \sum_{j=1}^{N} \frac{f_j}{N} \frac{(j + 1)\widehat{P_{j+1}}}{\widehat{P_j}} \qquad (9.177)$$

(because of the three-term recurrence relationship for the probabilities), where f_j/N is the observed proportion of the observations that are equal to j, and $\hat{P}_j$ denotes P_j with θ replaced by $\hat{\theta}$ and p by $\hat{p}$. These equations require iteration.

The moment estimators $\tilde{\theta}$ and $\tilde{p}$ are

$$\tilde{\theta} = \frac{2\bar{x}^2}{s^2 + \bar{x}}, \qquad (9.178)$$

$$\tilde{p} = \frac{s^2 - \bar{x}}{s^2 + \bar{x}}, \tag{9.179}$$

where s^2 is the sample variance.

The biases, variances and covariances of the maximum likelihood and moment estimators are tabulated for certain parameter combinations in Shenton and Bowman (1977).

Use of the mean and zero frequency leads to the estimators θ^* and p^* given by

$$\theta^* = -\ln(f_0/N), \tag{9.180}$$

$$p^* = 1 - \frac{\theta^*}{\bar{x}}, \tag{9.181}$$

where f_j/N is as above [Evans (1953)].

Another simple method of estimation is based on the first two frequencies. From $\Pr[X = 0] = e^{-\theta}$ and $\Pr[X = 1] = e^{-\theta}\theta q$, we have

$$\theta = -\ln(\Pr[X = 0]) \quad \text{and} \quad q = \frac{-\Pr[X = 1]}{\Pr[X = 0]\ln\Pr[X = 0]}.$$

This suggests using as estimators

$$\theta^{**} = -\ln(f_0/N) \tag{9.182}$$

$$q^{**} = -\frac{f_1}{f_0 \ln(f_0/N)}. \tag{9.183}$$

For minimum chi-squared estimation see Hinz and Gurland (1967), where the method is described in detail, and Douglas (1980), where its application to the Pólya-Aeppli distribution is examined. Tripathi, Gurland and Bhalerao (1986) have discussed the use of the method for the Poisson-Katz family of distributions to which the Pólya-Aeppli belongs (see Section 9 below).

8 POISSON-PASCAL DISTRIBUTION: POISSON-NEGATIVE BINOMIAL DISTRIBUTION, GENERALIZED PÓLYA-AEPPLI DISTRIBUTION

The *Poisson-Pascal distribution* was introduced in the context of the spatial distribution of plants by Skellam (1952) who regarded it as a *generalized Pólya-Aeppli distribution*. Katti and Gurland (1961) studied its properties and estimation, and derived it from an entomological model. Consider the number of surviving larvae when the number of egg masses per plot has a Poisson distribution, and the number of survivors per egg mass has a more heterogeneous distribution than a binomial or a Poisson. The assumption that the number of survivors in an egg mass is Poissonian, with an expected value that varies according to a gamma distribution, leads to a negative binomial distribution for the number per egg mass. The number of survivors per plot will then have a Poisson-stopped sum of negative binomial distributions. This is the Poisson-Pascal distribution:

$$X \sim \text{Poisson}(\theta) \bigvee \text{Negative binomial}(k, P), \qquad (9.184)$$

$0 < \theta, 0 < k, 0 < P$. The pgf is

$$G(z) = \exp\{\theta[(Q - Pz)^{-k} - 1]\}, \qquad (9.185)$$

where $Q = 1 + P$. The special case with $k = 1$, $P = p/(1 - p)$, $\theta = \theta_1/p$ is the Pólya-Aeppli distribution.

Note that the generalizing distribution in (9.185) is negative binomial with support $0, 1, 2, \ldots$, and <u>not</u> a binomial waiting-time distribution with support $k, k + 1, \ldots$, as the name Poisson-Pascal might suggest. A distribution with pgf $\exp\{\theta[z^k(Q - Pz)^{-k} - 1]\}$ can of course be constructed, but it lacks biological realism in the contexts in which the Poisson-Pascal distribution has been used.

Katti and Gurland pointed out that the distribution can also be obtained as

$$X \sim \text{Negative binomial}(\Psi, P) \bigwedge_{\Psi/k} \text{Poisson}(\theta), \qquad (9.186)$$

with pgf

$$G(z) = E[(Q - Pz)^{\Psi}] = \exp\{\theta[(Q - Pz)^{-k} - 1]\}, \qquad (9.187)$$

where Ψ/k has a Poisson distribution with parameter θ, and k is a positive constant.

The individual probabilities are

$$\Pr[X = 0] = e^{\theta(Q^{-k} - 1)}, \qquad (9.188)$$

$$\Pr[X = x] = \frac{e^{-\theta}(P/Q)^x}{x!} \sum_{j=1}^{\infty} \frac{(kj + x - 1)!}{(kj - 1)!j!}(\theta Q^{-k})^j; \qquad (9.189)$$

that is,

$$\Pr[X = 1] = \theta k P Q^{-k-1} \Pr[X = 0],$$

$$\Pr[X = 2] = \theta k P^2 Q^{-2k-2} \left[\frac{\theta k}{2} + \frac{(k + 1)Q^k}{2} \right] \Pr[X = 0],$$

$$\Pr[X = 3] = \theta k P^3 Q^{-3k-3} \left[\frac{\theta^2 k^2}{6} + \frac{\theta k (k+1) Q^k}{2} \right.$$

$$\left. + \frac{(k+1)(k+2) Q^{2k}}{6} \right] \Pr[X = 0], \qquad \text{etc.} \qquad (9.190)$$

Shumway and Gurland (1960a) have provided tables to assist in the direct calculation of $\Pr[X = x]$ from this formula. Their method is very similar to that for the Poisson-binomial distribution (Section 5), except that instead of (descending) factorial moments it uses *ascending* factorial moments. Katti and Gurland (1961) obtained a recurrence relation for the probabilities that is particularly useful for use on a computer. Restated in the above notation it becomes

$$\Pr[X = x+1] = \frac{\theta k P}{(x+1) Q^{k+1}} \sum_{j=0}^{x} \binom{k+x-j}{k} \left(\frac{P}{Q} \right)^{x-j} \Pr[X = j]. \quad (9.191)$$

The factorial cumulant generating function is

$$\ln G(1 + t) = \theta [(1 - Pt)^{-k} - 1], \qquad (9.192)$$

and hence

$$\kappa_{[r]} = \frac{\theta (k + r - 1)! P^r}{(k-1)!}. \qquad (9.193)$$

The mean and variance are therefore

$$\mu = \kappa_1 = \theta k P, \qquad (9.194)$$

$$\mu_2 = \kappa_2 = \theta k P (Q + k P); \qquad (9.195)$$

see Douglas (1980) for further moment properties. The flexibility of the Poisson-Pascal compared with certain other widely used stopped-sum distributions was assessed quantitively by Katti and Gurland, by evaluating their relative skewness and kurtosis for fixed mean and variance.

As $k \to \infty$, $P \to 0$ such that $kP \to \lambda$, the generalizing negative binomial tends to a Poisson distribution, and the Poisson-Pascal tends to a Neyman Type A distribution. As $\theta \to \infty$, $k \to 0$ such that $k\theta \to \lambda$, the limiting form is a negative binomial distribution, and as $\theta \to \infty$, $p \to 0$ so that $\theta P \to \lambda$, a Poisson distribution is the outcome; see Katti and Gurland (1961).

Katti and Gurland described three methods of fitting the distribution and gave examples using plant quadrat data. Their first method uses the first three sample moments:

$$\tilde{k} = \frac{\kappa_{[3]}^{\dagger} \kappa_{[1]}^{\dagger}}{\kappa_{[3]}^{\dagger} \kappa_{[1]}^{\dagger} - (\kappa_{[2]}^{\dagger})^2} - 2,$$

$$\tilde{P} = \frac{\kappa^\dagger_{[2]}}{\kappa^\dagger_{[1]}(\tilde{k} + 1)},$$

$$\tilde{\theta} = \frac{\kappa^\dagger_{[1]}}{\tilde{k}\tilde{P}}, \tag{9.196}$$

where $\kappa^\dagger_{[j]}$ is the jth sample factorial cumulant.

Their second method obtains the estimators P^*, k^* and θ^* via the first two sample moments and the proportion of zeros (f_0/N) in the sample. Again their formulas are given in terms of the sample factorial cumulants:

$$P^* \ln\left[1 + \left(\frac{\kappa^\dagger_{[2]}}{\kappa^\dagger_{[1]}} - P^*\right)\frac{\ln (f_0/N)}{\kappa^\dagger_{[1]}}\right] = \left(\frac{\kappa^\dagger_{[2]}}{\kappa^\dagger_{[1]}} - P^*\right)\ln(1 + P^*),$$

$$k^* = \frac{\kappa^\dagger_{[2]}}{\kappa^\dagger_{[1]}P^*} - 1,$$

$$\theta^* = \frac{\kappa^\dagger_{[1]}}{k^*P^*}. \tag{9.197}$$

Iteration is required for the solution of these equations.

Katti and Gurland's third method is based on the first two sample moments and the ratio f_1/f_0. Their estimator P^{**} is the solution of

$$\frac{\ln(1 + P^{**})}{P^{**}} = \frac{\kappa^\dagger_{[1]}}{\kappa^\dagger_{[2]}} \ln(\kappa^\dagger_{[1]}f_0/f_1). \tag{9.198}$$

Iteration is again necessary. k^{**} and θ^{**} are then calculated from the last two equations of (9.196) with $\tilde{P}$, $\tilde{k}$, and $\tilde{\theta}$ replaced by P^{**}, k^{**}, and θ^{**}.

Katti and Gurland (1961) calculated the asymptotic efficiency (the ratio of the generalized variances) of each method relative to the method of maximum likelihood, for

$$\theta = 0.1, 0.5, 1.0, 5.0 \quad \text{and} \quad P = 0.1, 0.3, 0.5, 1.0, 2.0.$$

For small θ (≤ 1) they found the third method to be generally the best of the three, with an efficiency greater than 90% for $k \leq 1$, $P \leq 0.5$. For larger θ, the second method appeared to be better. In general the first method was worst.

Derivation of the ML estimates via the ML equations is described in detail in Douglas (1980). The procedure parallels that for the Poisson-binomial distribution [Shumway and Gurland (1960b)]. Alternatively, the ML estimates may be obtained by direct search of the likelihood surface; with modern computing facilities this method is straightforward. Good initial estimates are advantageous. Minimum chi-squared estimation is examined in Hinz and Gurland (1967).

9 GENERALIZATIONS OF THE NEYMAN TYPE A DISTRIBUTION

Various generalizations of the Neyman Type A distribution have been sought by modifying the assumptions described in Section 6.1 above. Suppose in addition that while the number of larvae produced have a Neyman Type A distribution, the number surviving to be observed is subjected to a Rao damage process; that is, given that m are produced, the number observed has a binomial distribution with parameters m, p. Then [Feller (1943)], the overall distribution is still a Neyman Type A distribution but with parameters $\lambda, p\phi$. We do not get a new distribution this way.

Gurland (1958) made the further assumptions that the parameter p varies from cluster to cluster (egg mass to egg mass) and that this variation can be represented by a beta distribution. The distribution of the number of larvae per cluster now has the form

$$\text{Poisson}(\phi) \bigvee \text{Binomial}(1, P) \bigwedge_{P} \text{Beta}\,(\alpha, \beta), \qquad (9.199)$$

that is, it is a Beta-Poisson distribution with the probability generating function

$$\int_{0}^{1} e^{\phi p(z-1)} \frac{p^{\alpha-1}(1-p)^{\beta-1}dp}{B(\alpha, \beta)} = {}_1F_1[\alpha; \alpha + \beta; \phi(z-1)]; \qquad (9.200)$$

see Chapter 8, Section 3.2.

Gurland assumed also that the number of egg masses per plot is Poissonian. The symbolic representation for the resultant distribution of the number of surviving larvae per plot is

$$X \sim \text{Poisson}(\lambda) \bigvee [\{\text{Poisson}(\phi) \bigvee \text{Binomial}(1, P)\} \bigwedge_{P} \text{Beta}(\alpha, \beta)] \qquad (9.201)$$

$$\sim \text{Poisson}(\lambda) \bigvee [\{\text{Binomial}(M, P) \bigwedge_{M} \text{Poisson}(\phi)\} \bigwedge_{P} \text{Beta}(\alpha, \beta)], \qquad (9.202)$$

and the pgf is

$$G(z) = \exp\left\{ \lambda \left[\int_{0}^{1} e^{\phi(1-p+pz)} \frac{p^{\alpha-1}(1-p)^{\beta-1}dp}{B(\alpha, \beta)} \right] - \lambda \right\},$$

$$= \exp\{\lambda\, {}_1F_1[\alpha; \alpha + \beta; \phi(z-1)] - \lambda\}. \qquad (9.203)$$

This corresponds to a family of distributions with four parameters. There are the original two parameters λ (expected number of clusters) and ϕ (expected number of larvae per cluster), and α and β, defining the distribution of p (the probability of survival or, more generally, the probability of being observed).

Neyman's (1939) own generalizations of the Type A distribution are obtained by putting $\alpha = 1$ and $\beta = 1$, $\beta = 2$. He called these Type B and Type C respectively. (Type A corresponds to $\alpha = 1$, $\beta = 0$.) Feller (1943) gave an alternative derivation of the Type B and Type C distributions. For an early paper on applications of the distributions, see Beall (1940).

The subfamily which is obtained when $\alpha = 1$, $\beta \geq 0$, was studied by Beall and Rescia (1953). Members of this subfamily have pgf's that can be rewritten as

$$G(z) = e^{-\lambda} \exp\left[\lambda\Gamma(\beta + 1) \sum_{j=0}^{\infty} \frac{\phi^j(z - 1)^j}{\Gamma(\beta + j + 1)}\right]. \qquad (9.204)$$

Formulas for the probabilities and moments of the Type B and Type C distributions (with two parameters) and the Beall and Rescia distributions (three parameters) are obtainable as special cases of the formulas that Gurland derived for his four parameter family. His expressions for the probabilities are

$$\Pr[X = 0] = \exp\{\lambda \,_1F_1[\alpha; \alpha + \beta; -\phi] - \lambda\}, \qquad (9.205)$$

$$\Pr[X = x + 1] = \frac{\lambda}{x + 1} \sum_{j=0}^{x} F_j \Pr[X = x - j], \qquad x = 0, 1, 2, \ldots, \qquad (9.206)$$

where $_1F_1[\cdot]$ is a confluent hypergeometric function (Chapter 1, Section A7). F_j is defined as

$$F_j = \frac{\phi^{j+1}(\alpha + j)!(\alpha + \beta - 1)!\,_1F_1[\alpha + j + 1; \alpha + \beta + j + 1; -\phi]}{j!(\alpha - 1)!(\alpha + \beta + j)!} \qquad (9.207)$$

when $j = 0, 1, \ldots$, and zero otherwise; F_j satisfies

$$F_j = \left(\frac{\phi + \alpha + \beta + j - 1}{j}\right) F_{j-1} - \frac{\phi(\alpha + j - 1)}{j(j - 1)} F_{j-2}, \qquad j \geq 2. \qquad (9.208)$$

(Care should be taken not to confuse the notations $_1F_1$ and F_j.) Gurland pointed out that the use of the recurrence relation (9.208) may lead to numerical difficulties. He recommended the use of (9.207), with (9.208) kept as a check on the calculations.

Gurland found from the factorial cumulant generating function that

$$\kappa_{[r]} = \frac{(\alpha + r - 1)!(\alpha + \beta - 1)!}{(\alpha - 1)!(\alpha + \beta + r - 1)!}\lambda\phi^r, \qquad (9.209)$$

whence

$$\mu = \lambda\phi\,\alpha(\alpha + \beta)^{-1},$$

$$\mu_2 = \lambda\phi\,\alpha(\alpha + \beta)^{-1}[1 + \phi(\alpha + 1)(\alpha + \beta + 1)^{-1}],$$

$$\mu_3 = \lambda\phi\,\alpha(\alpha + \beta)^{-1}[1 + 3\phi(\alpha + 1)(\alpha + \beta + 1)^{-1}$$

$$+\phi^2(\alpha + 1)(\alpha + 2)(\alpha + \beta + 1)^{-1}(\alpha + \beta + 2)^{-1}]. \tag{9.210}$$

For the estimation of λ and ϕ, given α and β, equating the first and second sample and population moments would appear to suffice. Maximum likelihood estimation by direct search of the likelihood surface would, nevertheless, be straightforward given modern computing facilities, but little is known about the properties of the estimators.

When Beall and Rescia fitted their subfamily, with $\alpha = 1$ and β, λ, and ϕ all unknown, they admitted only integer values for β, though fractional values are possible. They suggested fitting their three-parameter distribution by a method that essentially consists of first fixing β, then estimating λ and ϕ by equating first and second sample and population moments. Their equations are

$$\phi^* = \frac{(\beta + 2)(s^2 - \bar{x})}{2\bar{x}},$$

$$\lambda^* = \frac{(\beta + 1)\bar{x}}{\phi^*}, \tag{9.211}$$

where $\bar{x}$ and s^2 are the sample mean and variance. The estimated and observed distributions are then compared by means of a χ^2 test. The process is repeated for a succession of integer values of β, and the value of β that gives the best fit to the data is selected as the estimated value of β. The corresponding values of λ^* and ϕ^* from (9.211) are taken to be the estimated values of λ and ϕ. Fortunately, the optimal solution does not seem to be very sensitive to the exact value chosen for β.

If all four parameters are to be estimated, Gurland's suggestion is that the first two observed relative frequencies and the first two sample moments should each be equated to their theoretical values.

Gurland pointed out that as $\beta \to \infty$ and ϕ varies so that the first two moments remain fixed, then the limiting form of his family is a Pólya-Aeppli distribution (see Section 7 above). He also commented that for a fixed general value of α, the limiting distribution is a generalized Pólya-Aeppli distribution (see Section 8 above).

Kocherlakota Subrahmaniam(1966) and Kathleen Subrahmaniam (1978) have made a special study of the Negative binomial⋁Poisson distribution, which they called the Pascal-Poisson distribution (Chapter 5, Section 12.6). They obtained it as a limiting case of a more general contagious distribution.

For their more general model the number of *egg masses per plot* is assumed to be Poissonian, though subject to a Rao damage process with parameter p having a beta distribution; the number of surviving larvae per egg mass is also assumed to be Poissonian. The symbolic representation for this model is

$$X \sim [\{\text{Poisson}(\lambda) \bigvee \text{Binomial}(1, P)\} \bigvee \text{Poisson}(\phi)] \underset{P}{\bigwedge} \text{Beta}(a, b) \quad (9.212)$$

$$\sim [\{\text{Binomial}(N, P) \bigvee \text{Poisson}(\phi)\} \underset{N}{\bigwedge} \text{Poisson}(\lambda)] \underset{P}{\bigwedge} \text{Beta}(a, b). \quad (9.213)$$

This is <u>not</u> the same model as (9.201) or (9.202). The corresponding pgf is

$$G(z) = \int_0^1 \exp\{\lambda[(1 - p + pe^{\phi(z-1)}) - 1]\} \frac{p^{a-1}(1 - p)^{b-1} dp}{B(a, b)}$$

$$= {}_1F_1[a; a + b; \lambda\{e^{\phi(z-1)} - 1\}]. \quad (9.214)$$

Subrahmaniam (1966) showed how to obtain the probabilities recursively and found that

$$\mu = \frac{\lambda \phi a}{a + b}$$

$$\mu_2 = \frac{\lambda \phi a}{a + b} \left[1 + \lambda + \frac{\lambda \phi b}{(a + b)(a + b + 1)}\right]. \quad (9.215)$$

He commented that $a = 1$, $b = 0$ gives a Neyman Type A distribution and that the limiting forms of (9.213) are nearly always the same as those of (9.202). There is one exception, however. When $\beta \to \infty$ for fixed α, μ and μ_2, Gurland's family tends to a generalized Pólya-Aeppli (Poisson$\bigvee$Pascal) distribution. Contrariwise, when $b \to \infty$ for fixed a, μ and μ_2, the limiting form for the Subrahmaniam family has the pgf

$$G(z) = \lim_{b \to \infty} {}_1F_1[a; a + b; \frac{(a + b)\mu}{a\phi}\{e^{\phi(z-1)} - 1\}]$$

$$= {}_1F_0[a; \ ; \frac{\mu}{a\phi}\{e^{\phi(z-1)} - 1\}]$$

$$= \left[1 + \frac{\mu}{a\phi} - \frac{\mu}{a\phi} e^{\phi(z-1)}\right]^{-a}, \quad (9.216)$$

that is, it has a Negative binomial$\bigvee$Poisson ($\equiv$ Pascal-Poisson) distribution; further details concerning this distribution are in Chapter 5, Section 12.6.

A different extension of the Type A distribution was made by Cresswell and Froggatt (1963). They supposed, in addition to their assumptions in Section 6.5 for the Type A distribution, that accidents can also occur outside a spell according to a Poisson distribution with parameter ρ. This yields a convolution of a Neyman Type A with a Poisson distribution that can be represented symbolically as

$$X \sim [\text{Poisson}(\lambda) \bigvee \text{Poisson}(\phi)]^* \text{Poisson}(\rho).$$

They used the term "short" distribution to describe this distribution because the tail length is reduced compared with that for the Type A component distribution. The pgf is

$$G(z) = \exp\{\lambda[e^{\phi(z-1)} - 1] + \rho(z - 1)\}, \tag{9.217}$$

with $\lambda, \phi, \rho > 0$. These authors fitted the distribution by the method of moments to many data sets on accidents to public transport bus-drivers. Froggatt (1966) has also examined its usefulness for fitting data on industrial absenteeism.

The probabilities can be obtained by convoluting the Type A probabilities and the probabilities for a Poisson distribution with parameter (ρ); this gives

$$\Pr[X = 0] = \exp\{\lambda[e^{-\phi} - 1] - \rho\}, \tag{9.218}$$

$$\Pr[X = x] = \frac{\exp[-(\lambda + \phi)]}{x!} \sum_{j=0}^{\infty} (j\phi + \rho)^x \frac{(\lambda e^{-\phi})^j}{j!} \tag{9.219}$$

$$= \exp[\lambda(e^{-\phi} - 1) - \rho] \sum_{j=0}^{x} \sum_{k=0}^{j} S(j,k) \frac{\rho^{x-j}\phi^j}{(x-j)!j!}(\lambda e^{-\phi})^k, \tag{9.220}$$

$x = 1, 2, \ldots$. The polynomial form (9.220) is advantageous for computing the probabilities.

Kemp (1967b) derived the following useful recurrence relationship:

$$\Pr[X = x + 1] = \frac{\rho}{x+1} \Pr[X = x] + \frac{\lambda\phi e^{-\phi}}{x+1} \sum_{j=0}^{x} \frac{\phi^j}{j!} \Pr[X = x - j]. \tag{9.221}$$

Cresswell and Froggatt derived formulas for the moments from expressions for the cumulants that involve Stirling numbers. Kemp showed that they can be obtained more simply from the factorial cumulants, $\kappa_{[1]} = \rho + \lambda\phi$, $\kappa_{[r]} = \lambda\phi^r$, $r \geq 2$. We have

$$\mu = \rho + \lambda\phi,$$
$$\mu_2 = \rho + \lambda\phi(1 + \phi),$$
$$\mu_3 = \rho + \lambda\phi(1 + 3\phi + \phi^2). \tag{9.222}$$

Kemp also studied maximum likelihood estimation of the (three) parameters using Newton-Raphson iteration and showed how to obtain their variances and covariances; he found that the relative efficiency of estimation by the method of moments is generally poor, particularly for $\phi \geq 1$.

The Poisson$\bigvee$Katz family is yet another generalization of the Neyman Type A distribution; see Bhalerao and Gurland (1977) and Tripathi, Gurland, and Bhalerao (1986). Its pgf is

$$G(z) = \exp\left\{\lambda[\{1 - \frac{\beta}{(1-\beta)}(z-1)\}^{-\alpha/\beta} - 1]\right\}, \qquad (9.223)$$

$\lambda > 0$, $\alpha > 0$, $\beta < 1$. It was so named because it can be obtained via a model where the number of clusters has a Poisson distribution, and the number of entities within a cluster has a distribution belonging to the Katz family of distributions (see Chapter 2, Section 3.1, for the Katz family). The Katz family includes the binomial, Poisson, and negative binomial distributions, and hence the Poisson$\bigvee$Katz family includes the Hermite ($-\alpha/\beta = 2$), Poisson-binomial ($-\alpha/\beta > 0$, an integer), Neyman Type A ($\beta \to 0$), Pólya-Aeppli ($\alpha/\beta = 1$), and Poisson-Pascal ($0 < \beta < 1$). The binomial, Poisson and negative binomial distributions are obtainable as limiting forms.

The probabilities for this family are given by

$$\Pr[X = 0] = \exp\{\lambda[(1-\beta)^{\alpha/\beta} - 1]\}, \qquad (9.224)$$

$$\Pr[X = x+1] = \frac{\lambda}{x+1} \sum_{j=0}^{x} \frac{(\alpha/\beta + j)!\beta^{j+1}(1-\beta)^{\alpha/\beta}}{(\alpha/\beta - 1)!j!} \Pr[X = x - j], \quad (9.225)$$

$x = 0, 1, 2, \ldots$. The moments can be obtained from the factorial cumulants,

$$\kappa_{[r]} = \lambda \frac{(\alpha/\beta + r - 1)!}{(\alpha/\beta - 1)!} \left(\frac{\beta}{1-\beta}\right)^r; \qquad (9.226)$$

in particular

$$\mu = \frac{\lambda\alpha}{1-\beta} \quad \text{and} \quad \mu_2 = \frac{\lambda\alpha(\alpha+\beta)}{(1-\beta)^2}. \qquad (9.227)$$

Tripathi, Gurland and Bhalerao developed a generalized minimum chi-squared method for the estimation of the (three) parameters; they noted, in passing, that maximum likelihood estimation could also be used.

The rationale underlying the study of the Poisson$\bigvee$Katz family is the desire to provide a method for selecting and fitting one out of a number of possibly suitable generalized Poisson distributions, by applying an overall estimation procedure to the data and then observing the consequences of the values of the parameter estimates (particularly their signs).

10 THOMAS DISTRIBUTION

This is similar to the Neyman Type A distribution in Section 6 above, except that the generalizing Poisson distribution is replaced by the distribution of a Poisson variable increased by unity. Symbolically the distribution can be represented by

$$X \sim \text{Poisson}(\lambda) \bigvee \text{Shifted Poisson}(\phi) \tag{9.228}$$

where the shifted Poisson has support $1, 2, \ldots$.

This distribution was used by Thomas (1949) in constructing a model for the distribution of plants of a given species in randomly placed quadrats. It is well suited to situations in which *the parent as well as the offspring* is included in the count for each cluster arising from an initial individual, assuming that the initial distribution is Poissonian with parameter λ and that the number of offspring per initial individual is also Poissonian but with parameter ϕ. The pgf is therefore

$$G(z) = \exp[\lambda\{ze^{\phi(z-1)} - 1\}]. \tag{9.229}$$

Thomas called this distribution a "double Poisson" distribution, though Douglas (1980) has pointed out that the term applies more appropriately to a Neyman type A distribution than to a Thomas distribution. (The name double Poisson has also been used for the bivariate Poisson distribution.)

By Gurland's thorem the distribution can also be regarded as a mixture distribution with the symbolic representation

$$X \sim \text{Shifted Poisson}(\Theta) \bigwedge_{\Theta/\phi} \text{Poisson}(\lambda), \tag{9.230}$$

where the shifted Poisson is on the positive integers $1, 2, \ldots$.

It is easy to verify by differentiation of the pgf that

$$\Pr[X = 0] = e^{-\lambda},$$

$$\Pr[X = 1] = \lambda e^{-(\lambda+\phi)},$$

$$\Pr[X = 2] = \lambda e^{-(\lambda+\phi)}(\phi + \lambda e^{-\phi})/2, \qquad \text{etc.} \tag{9.231}$$

Thomas obtained the following general expression for the probabilities:

$$\Pr[X = x] = \frac{e^{-\lambda}}{x!} \sum_{j=1}^{x} \binom{x}{j} (\lambda e^{-\phi})^j (j\phi)^{x-j}. \tag{9.232}$$

Like the Neyman type A, the distribution can have more than one mode; this can be seen by considering $\phi > \max(2, \ln \lambda)$, in which case $\Pr[X = 0] > \Pr[X = 1] < \Pr[X = 2]$.

The factorial cumulant generating function is

$$\ln G(1+t) = \lambda(1+t)e^{\phi t} - \lambda, \tag{9.233}$$

whence

$$\kappa_{[r]} = \lambda\phi^{r-1}(r+\phi) \tag{9.234}$$

[Ord (1972)]. From the relationship between cumulants and factorial cumulants

$$\mu = \kappa_{[1]} = \lambda(1+\phi),$$

$$\mu_2 = \kappa_2 = \lambda(1+3\phi+\phi^2),$$

$$\mu_3 = \kappa_3 = \lambda(1+7\phi+6\phi^2+\phi^3),$$

$$\mu_4 = \kappa_4 + 3\kappa_2^2 = \lambda(1+15\phi+25\phi^2+10\phi^3+\phi^4)$$
$$+ 3\lambda^2(1+3\phi+\phi^2)^2 \tag{9.235}$$

(comparison with the corresponding formulas for the Neyman Type A distribution in Section 6.2 above is interesting). The index of dispersion is

$$1 + \frac{\phi(2+\phi)}{1+\phi};$$

for constant λ this increases with ϕ, as one would expect from the model. The moments can also be obtained directly from the central moment generating function

$$e^{-\mu}G(e^t) = \exp\{\lambda[\exp(t+\phi e^t - \phi) - (1+t+\phi t)]\}. \tag{9.236}$$

The distribution is positively skew, with

$$\alpha_3^2 = \beta_1 = \frac{\mu_3^2}{\mu_2^3} = \frac{(1+7\phi+6\phi^2+\phi^3)^2}{\lambda(1+3\phi+\phi^2)^3},$$

$$\alpha_4 = \beta_2 = \frac{\mu_4}{\mu_2^2} = 3 + \frac{(1+15\phi+25\phi^2+10\phi^3+\phi^4)}{\lambda(1+3\phi+\phi^2)^2}. \tag{9.237}$$

The values of the ratio $(\beta_2-3)/\beta_1$ are remarkably stable; the ratio tends to unity as $\phi \to 0$ and $\phi \to \infty$, and it has a maximum of about 1.19 near $\phi = 0.25$:

ϕ	0	0.1	0.5	1	2	3	4	5	10
$(\beta_2-3)/\beta_1$	1	1.17	1.18	1.16	1.13	1.11	1.10	1.09	1.06

These results are very similar to those for the Neyman Type A distribution (Section 6 above); they suggest that the distribution is no more flexible than the Neyman Type A.

Thomas applied the distribution to observed distributions of plants per quadrat (*Armeria maritima* and *Plantago maritima*) and obtained marked improvements over fits with Poisson distributions. In particular, Thomas distributions can be bimodal, and bimodality was a feature in both observed distributions. The data were fitted by the method of moments; this is discussed in detail in Gleeson and Douglas (1975). They were also fitted by setting the observed proportions of zero and unit values (f_0/N and f_1/N where N is the sample size) equal to their expectations. The latter method gives

$$\lambda^* = -\ln(f_0/N)$$
$$\phi^* = \ln(f_0\lambda^*/f_1). \tag{9.238}$$

Thomas described this as a maximum likelihood method — it is in fact the ML method for the censored distribution with support $0, 1, 2$, and probabilities

$$\Pr[X = 0] = e^{-\lambda},$$
$$\Pr[X = 1] = \lambda e^{-(\lambda+\phi)},$$
$$\Pr[X \geq 2] = 1 - e^{-\lambda} - \lambda e^{-(\lambda+\phi)}. \tag{9.239}$$

The use of only the observed proportions of zero and unit counts is advantageous when an exhaustive count of actual numbers per quadrat for $x \geq 2$ is not required and would be tedious to obtain. Thomas gave approximate values of the ratios of the standard errors of estimation of the expected value $\mu = \lambda(1 + \phi)$ using these two methods.

Methods of solution of the ML equations for a complete sample of data from a Thomas distribution have been investigated by Russell (1978).

Pielou (1957), by hand, and Gleeson and Douglas (1975), by computer, have carried out simulation studies comparing the usefulness of the Neyman type A and Thomas distributions for clustered data.

11   LAGRANGIAN POISSON DISTRIBUTION: SHIFTED BOREL-TANNER DISTRIBUTION

The *Borel-Tanner distribution* (Tanner-Borel distribution) describes the distribution of the total number of customers served before a queue vanishes given a single queue with random arrival times of customers (at constant rate ℓ), and a constant time (β) occupied in serving each customer. We suppose that the probability of arrival of a customer during the period $(t, t + (\Delta t))$ is $\ell(\Delta t) + o(\Delta t)$ and that the probability of arrival of two or more customers in this period is $o(\Delta t)$. If there are initially n customers in the queue, then the probability that the total number (Y) of customers served before the queue vanishes is equal to y is

$$\Pr[Y = y] = \frac{n}{(y-n)!} y^{y-n-1} (\ell\beta)^{y-n} e^{-\ell\beta y}, \qquad y = n, n+1, \dots . \qquad (9.240)$$

The distribution was obtained by Borel (1942) for the case $n = 1$, and for general values of n by Tanner (1953). The parameters ℓ and β appear only in the form of their product $\ell\beta$. It is convenient to use a single symbol for this product and to put $\ell\beta = a$, say. For (9.240) to represent a proper distribution, it is necessary to have $0 < a < 1$. If $a < 0$, the "probabilities" change sign, while if $a > 1$, $\sum_{x=n}^{\infty} \Pr[Y = y] < 1$.

Let $a(b)$ be the solution of the equation $b = ae^{-a}$. Using this inverse function Haight and Breuer (1960) were able to show that the pgf can be expressed as

$$H(z) = \left[\frac{a(bz)}{a(b)} \right]^n$$

$$= z^n e^{na(bz) - na(b)}$$

$$= z^n e^{n\ell\beta[\{H(z)\}^{1/n} - 1]}. \qquad (9.241)$$

Clearly $H(z)/z^n$ is a Poisson-stopped-sum (generalized Poisson) distribution. The pgf of the generalizing distribution is $\{H(z)\}^{1/n}$, that is, the generalizing distribution is a Borel distribution.

Haight and Breuer found from (9.241) that

$$H'(1) = n + aH'(1)$$

$$H''(1) = n(n-1) + naH'(1) + \frac{a\{H'(1)\}^2}{n} + aH''(1),$$

and so
$$\mu = \frac{n}{1-a}, \qquad (9.242)$$

$$\mu'_{[2]} = \frac{n(n-1)}{(1-a)} + \frac{n^2 a}{(1-a)^2} + \frac{na}{(1-a)^3}, \qquad (9.243)$$

and
$$\mu_2 = \frac{na}{(1-a)^3}. \qquad (9.244)$$

Haight and Breuer remarked that the moment properties can also be obtained by successive differentiation of

$$K(t) = \ln G(e^t) = n(t-a) + na[G(e^t)]^{1/n}, \qquad (9.245)$$

where $K(t)$ is is the cumulant generating function and $G(e^t)$ is the uncorrected moment generating function.

The modal value lies approximately between $k - 1$ and k, where

$$ae^{1-a} = (k - n)k^{n+1/2}(k - 1)^{-n-3/2} \qquad (9.246)$$

and $n + 1 \leq k \leq 2n^2/3 + 3n$.

Tables of the cdf $\Pr[X \leq x]$ to five decimal places are given in Haight and Breuer (1960), for $n = 1$ and $a = 0.01(0.01)0.62$. Owen (1962) gave values of $\Pr[X \leq x]$ to five decimal places for $n = 1(1)5$ and $a = 0.01(0.01)0.25$. One of the problems of tabulating the distribution is that it has a very long tail except when a is small.

The distribution is a power series distribution (Chapter 2, Section 2), and hence the ML equation for a, assuming n is known, is the first-moment equation, giving

$$\hat{a} = \frac{\bar{x} - n}{\bar{x}}. \qquad (9.247)$$

A Lagrangian Poisson distribution can be obtained by shifting the Tanner-Borel distribution so that it has support $0, 1, 2, \ldots$, that is, by transforming to the random variable $X = Y - n$. It is a member of the Lagrangian family of distributions — the probabilities can be obtained by Lagrangian expansion of the pgf. This distribution has been studied in considerable detail by Consul and his co-workers (it is sometimes called *Consul's generalized Poisson distribution*). For the very extensive literature on the distribution, see Consul (1989). The usual notation sets $\theta = an$ and $\lambda = a$. Once the distribution has been shifted to the origin, it is no longer necessary to have $n = \theta/\lambda$ an integer. The pmf becomes

$$\Pr[X = x] = \frac{\theta(\theta + x\lambda)^{x-1}e^{-\theta-x\lambda}}{x!}, \qquad x = 0, 1, 2, \ldots. \qquad (9.248)$$

When the parameter λ is considered to be linearly proportional to the parameter θ, the model is referred to as "restricted" by Consul, and the parameterization $\theta = an$, $\alpha = a/\theta$ (i.e., $\alpha = \lambda/\theta$) is used.

There has been controversy in the literature regarding the parameter space. The distribution with full properties undoubtedly exists for $\theta > 0$, $0 < \lambda < 1$ (i.e., $0 < \alpha < \theta^{-1}$). Nelson (1975) commented that for $\lambda < 0$ (i.e., $\alpha < 0$), the probabilities sooner or later become negative. The response of Consul and Shoukri (1985) to this problem has been to recommend the use of truncation when λ is negative, where $\max(-1, -\theta/m) < \lambda < 0$, and the support of the distribution is restricted to $0, 1, \ldots, m$, where m is the largest integer for which $\theta + m\lambda > 0$; that is, $m = [-\theta/\lambda]$, where $[\cdot]$ denotes the integer part. They imposed the arbitrary condition $m \geq 4$ in order to ensure that there are at least five classes in the resultant distribution. In their 1985 paper they made a detailed anlysis of the numerical effect of such truncation; see also Sect. 9.1.1, of Consul's (1989) book.

When the distribution is obtained by truncation in this manner, the probabilities no longer sum to unity. Many of its properties cease to hold exactly, even when the probabilities are normalized. For instance, the formulas for the moments become approximations, and the generalized Poisson model no longer holds (no distribution with finite support can be infinitely divisible).

Consider now the properties of the distribution when $0 < \theta$, $0 < \lambda < 1$ (i.e. $0 < \alpha < \theta^{-1}$). Let $t = z\, e^{\lambda(t-1)}$. Then the pgf has the form

$$G(z) = e^{\theta(t-1)}$$
$$= \exp\{\theta[z\, e^{\lambda(t-1)} - 1]\}$$
$$= \exp\{\theta z[G(z)]^{\lambda/\theta} - \theta\}; \qquad (9.249)$$

from this the distribution is clearly seen to be a Poisson-stopped-sum (generalized Poisson) distribution; hence it is infinitely divisible.

It is straightforward to show that the convolution of two such distributions with parameters (θ_1, λ) and (θ_2, λ) is another such distribution with parameters $(\theta_1 + \theta_2, \lambda)$. Consul (1975, 1989) has characterized the distribution by the property that if the sum of two independent random variables has a Lagrangian Poisson distribution, then each must have a Lagrangian Poisson distribution; Letac (1991) has, however, put forward a counterexample. A second characterization theorem in Consul (1975, 1989) concerns a Lagrangian Poisson random variate Z that is split into two components X and Y such that the conditional distribution $\Pr[X = x,\ Y = c - x | Z = c]$ is quasibinomial with parameters (c, p, θ). The proof of this characterization would seem to depend on the previous characterization.

The pmf (9.240) can be derived directly from the pgf (9.249) by Lagrangian expansion (Chapter 1, Section A4). From (9.248)

$$\Pr[X = 0] = e^{-\theta},$$
$$\Pr[X = 1] = \theta e^{-\lambda - \theta},$$

and generally

$$\Pr[X = x] = \frac{(\theta + x\lambda)^{x-1} e^{-\lambda}}{(\theta + x\lambda - \lambda)^{x-2} x} \Pr[X = x - 1],\ x = 1, 2, \ldots. \qquad (9.250)$$

The distribution is unimodal. Recurrence-type formulas for the cumulative probabilities are given in Sect. 1.7 of Consul's (1989) book.

Consul and Jain (1973b) gave the following formulas for the first four moments:

$$\mu = \frac{\theta}{1 - \lambda},$$

$$\mu_2 = \frac{\theta}{(1-\lambda)^3},$$

$$\mu_3 = \frac{\theta(1+2\lambda)}{(1-\lambda)^5},$$

$$\mu_4 = \frac{3\theta^2}{(1-\lambda)^6} + \frac{\theta(1+8\lambda+6\lambda^2)}{(1-\lambda)^7}. \tag{9.251}$$

The indices of skewness and kurtosis are

$$\beta_1 = \alpha_3^2 = \frac{\mu_3^2}{\mu_2^3} = \frac{(1+2\lambda)^2}{\theta(1-\lambda)}, $$

$$\beta_2 = \alpha_4 = \frac{\mu_4}{\mu_2^2} = 3 + \frac{(1+8\lambda+6\lambda^2)}{\theta(1-\lambda)}. \tag{9.252}$$

Consul and Jain also gave a method for calculating the higher moments using Stirling numbers. Consul and Shenton (1975) showed that the uncorrected moments satisfy

$$(1-\lambda)\mu_{r+1}' = \theta\mu_r' + \theta\frac{\partial\mu_r'}{\partial\theta} + \lambda\frac{\partial\mu_r'}{\partial\lambda}, \qquad r = 0,1,2,\dots. \tag{9.253}$$

The central moments and the cumulants satisfy

$$\mu_{r+1} = \frac{r\theta}{(1-\lambda)^3}\mu_{r-1} + \frac{1}{(1-\lambda)}\left[\frac{\partial\mu_r(t)}{\partial t}\right]_{t=1}, \qquad r = 1,2,3,\dots, \tag{9.254}$$

$$(1-\lambda)L_{r+1} = \lambda\frac{\partial L_r}{\partial\lambda} + \theta\frac{\partial L_r}{\partial\theta}, \qquad r = 1,2,3,\dots, \tag{9.255}$$

where $\mu_r(t)$ is μ_r with λ and θ replaced by λt and θt respectively, and $L_1 = \theta(1-\lambda)^{-1}$.

Gupta (1974) and Gupta and Singh (1981) found that

$$\mu_{r+1} = \frac{\theta}{(1-\alpha\theta)}\frac{\partial\mu_r}{\partial\theta} + r\mu_2\mu_{r-1} \tag{9.256}$$

and

$$\mu_{[r]}' = \frac{\theta}{(1-\alpha\theta)}\frac{\partial\mu_{[r]}'}{\partial\theta} + \mu_{[r]}'\{\mu_{[1]}' - r\} \tag{9.257}$$

for the central and factorial moments of the restricted model. Further formulas for the uncorrected moments and for the factorial moments appear in Janardan (1984).

Negative moments for (9.248) have been studied by Consul and Shoukri (1986) (see also Sect. 3.8 in Consul (1989)); formulas for the incomplete moments are in Consul (1989) Sect. 3.10.

Limiting forms of the distribution include the normal and the inverse Gaussian [Consul (1989)]. Consul (1989) has also studied the distribution of the difference $X_1 - X_2$ of two Lagrangian Poisson variables with parameters (θ_1, λ) and (θ_2, λ). The left-truncated distribution has been researched by Medhi (1975). Consul (1984) and Gupta and Gupta (1984) have investigated the distribution of the order statistics of a sample whose *size* has a Lagrangian Poisson distribution.

Estimation for the Lagrangian Poisson distribution is described in Chapter 4 of Consul P (1989). The distribution is a modified power series distribution, and hence when $\alpha = \lambda/\theta$ is known, the maximum likelihood estimator $\hat{\theta}$ and the moment estimator $\tilde{\theta}$ of θ are

$$\hat{\theta} = \tilde{\theta} = \frac{\bar{x}}{1 + \alpha\bar{x}}; \tag{9.258}$$

this is known to be negatively biased.

When estimation is for both parameters, the moment estimators are given by

$$\tilde{\theta} = \sqrt{\frac{m_1^3}{m_2}} \tag{9.259}$$

$$\tilde{\alpha} = \sqrt{\frac{m_2}{m_1^3}} - \frac{1}{m_1}, \tag{9.260}$$

where m_1 and m_2 are the first two sample moments. Kumar and Consul (1980) have derived expressions for the asymptotic biases, variances, and covariance of $\tilde{\theta}$ and $\tilde{\alpha}$.

The mean-and-zero-frequency estimators are

$$\theta^* = \ln(f_0/N),$$

$$\lambda^* = 1 - \frac{\theta^*}{\bar{x}}, \tag{9.261}$$

where f_0/N is the observed relative frequency of zero.

Maximum likelihood by direct search of the likelihood surface is not straightforward when the sample variance is less than the sample mean. Certain properties and existence theorems for the ML estimators were obtained by Consul and Shoukri (1984). For generalized minimum chi-square estimation, see Sect. 4.7 in Consul (1989).

Shoukri and Consul (1987) have given a helpful account of modes of genesis leading to the distribution. These include the distribution as a limit of the "generalized negative binomial" distribution, as a Lagrangian probability distribution, as the distribution of the total progeny in some Galton-Watson-type branching processes, and as the outcome of a birth-and-death process. These authors drew attention to the following uses of the distribution: modeling the spread of ideas, rumours, fashions and sales; the distribution of salespersons in 'pyramid' dealership; modeling population counts of biological organisms; thermodynamic processes; and the spread of burnt trees in forest fires. Further applications are cited in Chapter 5 of Consul (1989), where more than fifty data sets are fitted; in more than twenty of these the estimated value of λ is negative. It should be noted that many of the data sets are very short-tailed, in some cases consisting entirely of counts of $0, 1$, or 2.

12   OTHER FAMILIES OF STOPPED-SUM DISTRIBUTIONS

Sections 3–11 of this chapter have concentrated on Poisson-stopped-sum distributions. These are the Type A distributions of Khatri and Patel (1961). The Type B and Type C families in Khatri and Patel's somewhat neglected paper have pgf's of the form

Type B, $G_B(z) = \{h(z)\}^n$,   (9.262)

Type C, $G_C(z) = c \ln\{h(z)\}$,   (9.263)

where $h(z) = a + b\, g(z)$ and $g(z)$ is a pgf. These Type A, B, and C distributions are not the same as the Neyman Type A, B, and C distributions.

Let $f^{(r)}(z)$ denote the rth derivative of $f(z)$. Then

$$G(z) = \sum_x \Pr[X = x]z^x = \sum_x \left[\frac{G^{(x)}(z)}{x!}\right]_{z=0} z^x, \qquad (9.264)$$

$$g(z) = \sum_x \pi_x z^x = \sum_x \left[\frac{g^{(x)}(z)}{x!}\right]_{z=0} z^x. \qquad (9.265)$$

For the type B family $a = 1 - b$; successive differentiation of

$$\{a + b\,g(z)\}G_B^{(1)}(z) = nb\, g^{(1)}(z)\, G_B(z)$$

yields

$$aG^{(x)}(z) + b\sum_{j=1}^{x} \binom{x-1}{j-1} g(z)^{(j-1)}(z) G_B^{(x-j+1)}(z)$$

$$= nb \sum_{j=1}^{x} \binom{x-1}{j-1} g^{(j)}(z)\, G_B^{(x-j)}(z), \quad (9.266)$$

and setting $z = 0$ gives

$$\Pr[X = x] = \sum_{j=1}^{x} \frac{(nj + j - x)\pi_j}{x(\pi_0 + a/b)} \Pr[X = x - j] \qquad (9.267)$$

with $\qquad \Pr[X = 0] = (a + b\pi_0)^n. \qquad (9.268)$

For the type C family

$$G_C(z) = c \ln\{a + b\,g(z)\},$$

where $c = [\ln(a + b)]^{-1}$. Successive differentiation of

$$[a + b\,g(z)]G_C^{(1)}(z) = bc\,g^{(1)}(z)$$

gives

$$[a + b\,g(z)]G_C^{(x)}(z) = bc\,g^{(x)}(z) - b\sum_{j=1}^{x-1} \binom{x-1}{j} g^{(j)}(z)\,G_C^{(x-j)}(z), \qquad (9.269)$$

whence setting $z = 0$ yields

$$\Pr[X = 0] = c\ln(a + b\pi_0),$$

$$\Pr[X = x] = (a + b\pi_0)^{-1}\{bc\pi_x - \sum_{j=1}^{x-1} \frac{b(x-j)}{x}\pi_j\,\Pr[X = x - j]\}, \qquad x \geq 1;$$

$$(9.270)$$

Note the slight alterations to the formulas in Khatri and Patel (1961). When $a = 1$ and $-1 < b < 0$ the Type C family comprises logarithmic-stopped-sum distributions.

Khatri and Patel developed not only these probability properties but also moment properties.

Let $\qquad \mu'_{[r]} = [h^{(r)}(z)]_{z=1},$

$$M'_{[r]} = [G^{(r)}(z)]_{z=1},$$

$$K_{[r]} = \left[\frac{d^r}{dz^r} \ln G(z)\right]_{z=1}.$$

Then $\mu'_{[r]}$ is the rth factorial moment for $h(z)$ if $h(z)$ is a pgf. Also $M'_{[r]}$ and

$K_{[r]}$ are, respectively, the rth factorial moment and rth factorial cumulant for $G(z)$.

For the Type B family $\ln G_B(z) = n \ln h(z)$, so the rth factorial cumulant of $G_B(z)$ is n times the rth factorial cumulant for $h(z)$ if $h(z)$ is a pgf. Also

$$K_{[r]} = n\mu'_{[r]} - \sum_{j=1}^{r-1} \binom{r-1}{j} \mu'_{[j]} K_{[r-j]}. \tag{9.271}$$

For the Type C family Khatri and Patel found that

$$M'_{[r]} = \left\{ c\mu'_{[r]} - \sum_{j=1}^{r-1} \binom{r-1}{j} \mu'_{[j]} M'_{[r-j]} \right\} / \mu'_0, \tag{9.272}$$

where $\mu_0^1 = h(1)$.

For Khatri and Patel's Type B family the pgf must have the form

$$G_B(z) = \{1 - p + p g(z)\}^n;$$

consequently, when n is a positive integer and $0 < p < 1$, this family consists of generalized binomial distributions. Furthermore, for n a positive integer and $0 < p < \{1 - g(0)\}^{-1}$, it can be interpreted as an n-fold convolution of zero-modified distributions. When n and p are both negative (n not necessarily an integer), the family consists of generalized negative binomial distributions.

Khatri and Patel (1961) gave the recurrence formulas for the probabilities for binomial and negative binomial distributions with the following generalizing distributions: hypergeometric, binomial, negative binomial, and Poisson. Special attention was paid to the Binomial√Negative binomial distribution; Khatri and Patel developed a model for it based on the distribution of organisms that occur in colonies; they studied maximum likelihood and other forms of estimation, and they fitted the distribution to insect data. See also Khatri (1962).

The Negative binomial√Hypergeometric appears in Gurland's (1958) paper. The Negative binomial√Poisson has been investigated in depth by Subrahmaniam (1966) and Subrahmaniam (1978); see Chapter 5, Section 12.6, and also Section 9 of the present chapter.

The modeling of insurance claims is studied in Bowers et al. (1986). Verrall (1989) has taken their ideas further by putting forward a binomial-stopped-sum model for individual risk claims. He showed that if the rv X has the pgf

$$G(z) = \{1 - q + q g(z)\}^n,$$

where the mean and variance for $g(z)$ are μ and σ^2, then the mean and variance of X are

$$E[X] = nq\mu \quad \text{and} \quad \text{Var}(X) = n\{q\sigma^2 + q(1-q)\mu^2\}. \tag{9.273}$$

Given heterogeneous claims, the pgf becomes

$$G(z) = \prod_{k=1}^{K}\{1 - q_k + q_k g_k(z)\}^{n_k},$$

and the corresponding mean and variance are

$$E[X] = \sum_{k=1}^{K} n_k q_k \mu_k,$$

$$\text{Var}(X) = \sum_{k=1}^{K} n_k \{q_k \sigma_k^2 + q_k(1-q_k)\mu_k^2\}, \tag{9.274}$$

where μ_k and σ_k^2 are the mean and variance of the kth individual claim distribution. Verrall also studied Poisson-stopped-sum approximations to his binomial-stopped-sum models.

The aim of Katti's (1966) paper was to examine interrelations between generalized distributions with pgf's of the more general form $G(z) = g_1(g_2(z))$ and their component distributions $g_1(z)$ and $g_2(z)$. He showed that

$$\mu'_{[1]} = {}_1\mu'_{[1]2}\mu'_{[1]},$$

$$\mu'_{[2]} = {}_1\mu'_{[2]}({}_2\mu'_{[1]})^2 + {}_1\mu'_{[1]2}\mu'_{[2]},$$

$$\mu'_{[3]} = {}_1\mu'_{[3]}({}_2\mu'_{[1]})^3 + 3\,{}_1\mu'_{[2]2}\mu'_{[2]2}\mu'_{[1]} + {}_1\mu'_{[1]2}\mu'_{[3]} \quad \text{etc.,} \tag{9.275}$$

and commented on the similarity between these formulas and those for moments in terms of cumulants. He also showed how to obtain corresponding formulas for the factorial cumulants for $G(z)$ in terms of the factorial cumulants for $g_1(z)$ and $g_2(z)$. His formulas simplify not only for Poisson-stopped-sum distributions but also when $g_2(z) = \exp[\phi(z-1)]$.

In addition Katti gave a table containing skewness and kurtosis comparisons for some twenty-one stopped sum distributions and appended detailed comments. In making these comparisons he adopted Anscombe's (1950) method of fixing the first two factorial cumulants as kp and kp^2 and then evaluating $\kappa_{[3]}/kp^3$ and $\kappa_{[4]}/kp^4$.

CHAPTER 10

Matching, Occupancy, and Runs Distributions

1 INTRODUCTION

Problems concerning *coincidences* (i.e., *matching problems*) arise when sequences of characteristics are compared, for example when two sequences of n items are compared a pair at a time. The occurrence of the same characteristic at the jth comparison is called a *match*.

Occupancy distributions relate to the number of occupied categories when r objects are assigned in a random manner to n categories. Many variants of the classical occupancy problem (involving restraints on the method of assignment) have been studied, some for their practical import and others for their theoretical interest.

A *run* is the occurrence of an uninterrupted sequence of a particular attribute in an observed series of attributes. In a series of variate values, an uninterrupted sequence that is monotonically nondecreasing or nonincreasing is termed a run *up* or *down*, respectively. Results concerning runs are important in the theory of distribution free tests; see, e.g., Weiss (1988).

The theory of occupancy and matching distributions developed in the context of gambling problems. Twentieth century uses include the following: Maxwell-Boltzmann, Bose-Einstein and Fermi-Dirac systems in statistical physics [e.g. Desloge (1966), Feller (1957)]; personality assessment in psychology [Vernon (1936)]; applications in genetics [Stevens (1937, 1939)], in the estimation of insect populations [Craig (1953)], and in computer storage analysis [Meilijson et al. (1982)]; and matches between two DNA sequences [Goldstein (1990)].

Matching, occupancy, and runs problems are of special interest in discrete distribution theory for the impetus that they have given to methodology. The solution of such problems often depends on the inclusion-exclusion principle. Because of its importance in this area of discrete distribution theory, the next section is devoted to the principle and to associated theorems for the probabilities of combined events.

2 PROBABILITIES OF COMBINED EVENTS

In the (1718) edition of the *Doctrine of Chances*, de Moivre was able to generalize results that had previously been used by Montmort in the (1713) edition of his *Essai d'Analyse sur les Jeux de Hasard*.

Consider k events $E_1, E_2, \ldots, E_k$, and suppose that the probabilities of the simultaneous occurrence of any number of them are known. Suppose also, like de Moivre, that the events are exchangeable, that is, that their probabilities satisfy

$$P[E_{i_1} E_{i_2} \ldots E_{i_j}] = P[E_1 E_2 \ldots E_j], \qquad (10.1)$$

where $\{i_1, i_2, \ldots, i_j\}$ is any reordering of $\{1, 2, \ldots, j\}$. Also let

$$S_j = \sum P[E_{i_1} E_{i_2} \ldots E_{i_j}], \qquad (10.2)$$

where summation is over $\{i_1, i_2, \ldots, i_j\}$ such that $1 \le i_1 < i_2 < \ldots < i_j \le k$, and $j = 1, 2, \ldots, k$; also let $S_0 = 1$.

Then, in modern notation rather than de Moivre's, the probability that at least m of the k events occur is

$$P_m = S_m - \binom{m}{1} S_{m+1} + \binom{m+1}{2} S_{m+2} - \ldots + (-1)^{k-m} \binom{k-1}{k-m} S_k$$

$$= \sum_{i=0}^{k-m} (-1)^i \binom{m+i-1}{i} S_{m+i} \qquad (10.3)$$

(with $P_0 = 1$), and the probability that exactly m of the k events occur is

$$P_{[m]} = S_m - \binom{m+1}{1} S_{m+1} + \binom{m+2}{2} S_{m+2} - \ldots + (-1)^{k-m} \binom{k}{k-m} S_k$$

$$= \sum_{i=0}^{k-m} (-1)^i \binom{m+i}{i} S_{m+i}. \qquad (10.4)$$

Clearly

$$P_m = P_{[m]} + P_{[m+1]} + \ldots + P_{[k]} \qquad (10.5)$$

and

$$P_{[m]} = P_m - P_{m+1}. \qquad (10.6)$$

The method of proof used by de Moivre involves the repeated use of

$$P[\overline{E}_1 \cap E_2] = P[E_2] - P[E_1 \cap E_2];$$

see Hald's (1990) lucid historical discussion.

M. C. Jordan (1867) proved that these formulas are also valid when the events are not exchangeable (note the definition of S_j given above). This more general case can be proved by a method such as de Moivre's. It can also be proved by the use of the inclusion-exclusion principle. Given N objects, suppose that $N(a)$ have property a, $N(b)$ have property b, ..., $N(ab)$ have both a and b,..., $N(abc)$ have a, b and c, and so on. Then the inclusion-exclusion principle states that the number of objects with none of these properties is

$$N(\bar{a}\bar{b}\bar{c}\ldots) = N - N(a) - N(b) - \ldots$$
$$+N(ab) + N(ac) + \ldots$$
$$-N(abc) - \ldots$$
$$+\ldots. \tag{10.7}$$

The principle is fundamental to Boole's (1854) algebra of classes and is used extensively in McMahon's (1915, 1916) *Combinatory Analysis*.

The formulas can also be proved by Loève's (1963) indicator function method. Let X_i be an indicator variable that is equal to 1 when A_i occurs and zero otherwise. Then, for example,

$$1 - (1 - X_1)(1 - X_2)\ldots(1 - X_k)$$

is the indicator variable for the event that at least one of the A_i occur, and

$$P_1 = E[1 - (1 - X_1)(1 - X_2)\ldots(1 - X_k)]$$
$$= E[\sum_i X_i - \sum_{i \neq j} X_i X_j + \ldots + (-1)^{k+1} X_1 X_2 \ldots X_k]$$
$$= S_1 - S_2 + \ldots + (-1)^{k+1} S_k. \tag{10.8}$$

The probabilities P_m, $P_{[m]}$, $m = 2, \ldots, k$ can be obtained similarly; see, e.g., Parzen (1960) or Moran (1968). Besides a very clear account, Moran gives references to further work by Geiringer (1938) and to a long series of papers from 1941 to 1945 by Chung.

The formulas for P_m and $P_{[m]}$ have been used extensively in life assurance mathematics ever since de Moivre's (1725) work on the subject. In King's (1902) comprehensive text on life insurance mathematics, the author developed the symbolic representation

$$P_m \equiv S^m(1 + S)^{-m},$$
$$P_{[m]} \equiv S^m(1 + S)^{-m-1}, \tag{10.9}$$

where after expansion S^j is replaced by S_j, and $S_j = 0$ for $j > m$. This is the method adopted by K. Jordan (1972) and Riordan (1958). Takacs (1967) has generalized the formulas. Broderick (1937) and Fréchet (1940, 1943) have extended them to situations where the events are not independent.

Consider now the probability generating function

$$
G(z) = \sum_{m=0}^{k} P_{[m]} z^m = \sum_{m=0}^{k} S_m (z-1)^m \tag{10.10}
$$

for the rv X taking the values $0, 1, \ldots, k$ with probabilities $P_{[0]}, P_{[1]}, \ldots, P_{[k]}$. The corresponding factorial moment generating function is

$$
G(1+t) = \sum_{m=0}^{k} P_{[m]} (1+t)^m = \sum_{m=0}^{k} S_m t^m, \tag{10.11}
$$

and hence

$$
S_r = \frac{\mu'_{[r]}}{r!}, \tag{10.12}
$$

where $\mu'_{[r]}$ is the rth factorial moment. See Iyer (1949, 1958) for an interpretation of this result.

Reversing the series in the formulas gives

$$
S_r = \sum_{j=r}^{k} \binom{j}{r} P_{[j]} \tag{10.13}
$$

$$
= \sum_{j=r}^{k} \binom{j-1}{r-1} P_j. \tag{10.14}
$$

A consequence of (10.3) and (10.4) is

$$
S_m - (m+1)S_{m+1} \le P_{[m]} \le S_m \tag{10.15}
$$

and

$$
S_m - m S_{m+1} \le P_m \le S_m; \tag{10.16}
$$

these are widely known as Bonferroni's inequalities, see Galambos (1975, 1977) for a review of methods of proof of such inequalities.

Fréchet's (1940, 1943) more general inequalities include

$$
P_m \le S_m - \binom{m}{1} S_{m+1} + \ldots + \binom{m+\ell-1}{\ell} S_{m+\ell},
$$

$$
P_m \ge S_m - \binom{m}{1} S_{m+1} + \ldots - \binom{m+\ell}{\ell+1} S_{m+\ell+1},
$$

$$P_{[m]} \leq S_m - \binom{m+1}{1}S_{m+1} + \ldots + \binom{m+\ell}{\ell}S_{m+\ell},$$

$$P_{[m]} \geq S_m - \binom{m+1}{1}S_{m+1} + \ldots - \binom{m+\ell+1}{\ell+1}S_{m+\ell+1}, \qquad (10.17)$$

where ℓ is an even integer. Collectively such inequalities are termed Boole-Bonferroni-Fréchet inequalities, for example, by David and Barton (1962). An improved Bonferroni upper bound has been obtained by Worsley (1982); see also Hunter (1976), Seneta (1988), and Hoppe and Seneta (1990).

In Whitworth's (1878, 1948) *Choice and Chance*, the author proves and uses extensively a result sometimes called *Whitworth's theorem*; it is a re-statement of (10.4) with $m = 0$, giving

$$P_{[0]} = \sum_{i=0}^{k}(-1)^i S_i. \qquad (10.18)$$

Irwin (1967) drew the attention of his contempories to Whitworth's theorem; previously, in Irwin (1955), he had used it to derive three interesting distributional results.

The history of the compound probability theorems (10.3) and (10.4) has been discussed by Takacs (1967), and by Hald (1990) in his scholarly book on the history of probability and statistics. Fréchet's (1940, 1943) monograph (in French) gives not only his generalizations and applications but also delineates the work of certain earlier writers, some of which is not easy to obtain. Another useful and readily accessible reference is Feller (1957).

3 MATCHING DISTRIBUTIONS

The classical *problem of coincidences* (*problème de rencontre*) arises in the following way: Suppose that a set of k entities, numbered $1, 2, \ldots, k$, are arranged in a random order. Let X be the number of entities for which their position in the random order is the same as the number assigned to them. We seek to find the distribution of X.

For this problem $S_j = 1/j!$; see, e.g., Alt (1985) for an explanation. Applying this expression to (10.4) gives

$$\Pr[X = x] = P_{[x]} = \frac{1}{x!} - \binom{x+1}{1}\frac{1}{(x+1)!} + \cdots + (-1)^{k-x}\binom{k}{k-x}\frac{1}{k!}$$

$$= \frac{1}{x!}\sum_{i=1}^{k-x}\frac{(-1)^i}{i!}, \qquad (10.19)$$

where $x = 0, 1, \ldots, k$. Note that $\Pr[X = k - 1] = 0$.

From (10.19) it can be seen that, except when $(k - x)$ is small,

$$\Pr[X = x] \doteq \frac{e^{-1}}{x!} \tag{10.20}$$

to a close degree of approximation. The successive probabilities can therefore be approximated by the corresponding probabilities for a Poisson distribution with parameter 1. Irwin (1955) showed numerically that for $k = 10$ there is at most a discrepancy of 1 in the fifth decimal place for the two distributions, and stressed that for practical purposes the probabilities are the same for $k \geq 10$.

The close connection between the two distributions is brought out by considering their factorial moments [Olds (1938)]. For the matching distribution the fmgf is

$$G(1 + t) = \sum_{m=0}^{k} S_m t^m = \sum_{m=0}^{k} \frac{t^m}{m!}, \tag{10.21}$$

and so $\mu'_{[r]} = 1$, $r = 1, 2, \ldots, k$. For a Poisson distribution with parameter $\theta = 1$, we find that $\mu'_{[r]} = 1$, $r = 1, 2, \ldots$. The first k factorial moments, moments, cumulants, etc., for the two distributions are therefore identical. In particular

$$\mu = \mu_2 = 1. \tag{10.22}$$

The papers by Chapman (1934) and Vernon (1936) dealt with the use of the distribution in psychology, for example, for "coincidences" between character sketches of people and samples of their handwriting. Vernon (1936) included an extensive bibliography. Kendall (1968) has drawn attention to the work of Young (1819) on coincidences in linguistics concerning the numbers of word root forms in common in various languages.

There are many variants of this simple classical model. Suppose, for example, that there are N objects. Let na of these be labeled $1, 2, \ldots, n$, each label occurring a times, and let $N - na$ be unlabeled. Assume also that the labels are assigned at random. We will say that a match has occurred if an object with label j occurs at the jth position [Haag (1924)]. Fréchet (1940, 1943) called this the *Laplace-Haag matching problem*, and showed that

$$S_r = \binom{n}{r} \frac{(N - r)! a^r}{N!}.$$

It follows that

$$\mu'_{[r]} = \frac{n!(N - r)! a^r}{(n - r)! N!}, \tag{10.23}$$

and that the pgf is

$$G(z) = {}_1F_1[-n; -N; a(z - 1)] \tag{10.24}$$

[Kemp (1978b)]. This is a GHF distribution (see Chapter 2, Section 4.2) with a pgf that strongly resembles the pgf for the Poisson-beta distribution.

The distribution exists for all positive real a, not just when a is a positive integer. Fréchet recognized that a can be less than unity and interpreted the case $0 < a < 1$ via a visibility bias model. The probabilities for this distribution are

$$\Pr[X = x] = \sum_{i=0}^{n-x} \frac{(-1)^i n! (N - x - i)! a^{x+i}}{x! i! (n - x - i)! N!}. \tag{10.25}$$

Kemp showed from the theory of GHF distributions that the probabilities may more conveniently be calculated from

$$(x + 2)(x + 1) \Pr[X = x + 2] = (x + 1)(x + a - N) \Pr[X = x + 1]$$
$$+ a(n - x) \Pr[X = x], \tag{10.26}$$

where

$$\Pr[X = n] = a^n (N - n)! / N!, \tag{10.27}$$

$$\Pr[X = n - 1] = (1 + N - n - a) n a^{n-1} (N - n)! / N!. \tag{10.28}$$

Kemp also showed that as both N and a become large the pgf tends to

$$G(z) = {}_1F_0[-n; \ ; (1 - z)(a/N)], \tag{10.29}$$

that is, it becomes binomial. When both n and N become large, the limiting distribution is Poissonian with parameter an/N. However, Barton (1958) has argued that the Poisson limit may not be a very good approximation for even quite large N.

Fréchet (1940, 1943) considered several special cases of this distribution. For the Laplace matching distribution $N = an$. For the Gumbel distribution $N = n$, $a \neq 1$, and for the classical matching problem $N = n$, $a = 1$. Fréchet provided appropriate references.

Many authors, in particular McMahon (1894, 1898, 1902, 1915), have examined the problem of coincidences in the context of matching packs of cards. McMahon's method of specifying the composition of a pack is as follows: Suppose that there are k sorts of cards in a pack and that there are a_i cards of the ith sort, $i = 1, 2, \ldots, k$. Then the composition of the pack is said to be $(a_1, a_2, \ldots, a_k)$; for brevity a set of am cards of m kinds each replicated a times is represented by (a^m). The composition of a standard pack of playing cards with one joker is thereby represented as $(1, 13^4)$. It is usual to speak of one pack as the target pack and the other as the matching pack. For example, for the Laplace-Haag problem above, the target pack is $(N - n, 1^n)$, and the matching pack is $(N - an, a^n)$; the $(N - n)$ component in the target pack, and the $(N - an)$ component in the matching pack both consist of blank cards.

The equivalence of problems of coincidences and two-pack matching problems occurs because the order of the cards in the target pack is immaterial. What is important is the randomness of the matching pack.

Suppose, for example, that we have two identical packs of cards and that each pack of cards has n suits of a (distinct) cards per suit (so that each pack contains $N = an$ cards). The two packs are then dealt out simultaneously, in random order, and at the same rate. Let a match be said to occur if the cards dealt from the two packs at the same time have the same face value. (The case $n = 1$ is a special case of the problem of coincidences.) For general values of n

$$\Pr[X = x] = \sum_{j=x}^{N} (-1)^{j-x} \binom{j}{x} \frac{(n-j)! H_j}{n!}, \tag{10.30}$$

where H_j is the coefficient of s^j in the expansion of

$$\left[\sum_{i=0}^{n} \frac{(c!)^2 s^i}{i!(c-i)!(c-i)!} \right]^a.$$

Tables of $\Pr[X \geq x]$ to five decimal places for

$n = 1, 2$ and $a = 2(1)11$	also	$a = 2$ and $n = 6(1)12$
$n = 3$ and $a = 2(1)8$		$a = 3$ and $n = 6, 7$
$n = 4, 5$ and $a = 2(1)5$		

are given in Gilbert (1956) and are also included in Owen (1962). Silva (1941) has shown that for a large there are the following approximations for $\Pr[X = x]$:

$$\Pr[X = x] \doteq \frac{n^x e^{-n}}{x!} \left[1 - \left(\frac{n-1}{2na} \right) \left(\frac{(n-x)^2 - x}{c} \right) \right] \tag{10.31}$$

and

$$\Pr[X = x] \doteq \frac{n^x}{x!} \left(1 - \frac{1}{a} \right)^{na} \left[1 + \frac{1}{2a} + \frac{(n-1)(2n+1-x)x}{2n^2 a} \right]. \tag{10.32}$$

Greville (1941) gave an appropriate expression for the more general problem in which there are $n_{11}, \ldots, n_{1a}$ cards per face value in the first pack and $n_{21}, \ldots, n_{2a}$ cards per face value in the second pack. Joseph and Bizley (1960) gave formulas from which values of $\Pr[X = x]$ can be calculated for the case $n_{1i} = n_{2i}$ for all i.

McMahon (1915) has solved the problem of two packs with completely general composition; however, his formulas are not very tractable. A somewhat simpler operator technique was devised by Kaplansky and Riordan (1945). Barton (1958) has discussed their work and also that of Stevens (1939); a particular strength of Barton's paper is the derivation of Poissonian limiting forms.

Yet more matching problems arise when there is simultaneous matching among K packs of cards. Barton has defined the following types of matches:

1. All K cards in the same position have the same label.
2. The card in a given position in the target pack has the same label as at least one of the $K - 1$ corresponding cards in the $K - 1$ other packs.
3. At least two of the K cards in a given position have the same label.

McMahon managed to give some highly abstract formulas in certain of these more complex situations, but otherwise there has been little progress in this area.

A different kind of matching problem has arisen in genetics; see Levene (1949). It can be formulated as the random splitting of a pack with composition (2^N) into two equally sized packs. Barton (1958) showed that the factorial moments for the number of matches are given by

$$\mu'_{[r]} = \frac{N!(N - 1/2 - r)!}{(N - r)!(N - 1/2)!2^n} ,$$
(10.33)

where $\xi!$ is taken as $\Gamma(\xi+1)$ when ξ is positive but not an integer (Chapter 1, Section A2).

Barton (1958) next investigated the generalization to a pack of composition (K^N) randomly split into K equally sized packs; the factorial moments for the number of matches of type (1) are

$$\mu'_{[r]} = \frac{N!N!(K!)^r(KN - Kr)!}{(N - r)!(N - r)!(KN)!}.$$
(10.34)

Anderson's (1943) problem concerned a pack of composition (K^N) split into K packs each with composition (1^N). He obtained the following expression for the probabilities of the number of matches of type (1):

$$\Pr[X = x] = \sum_{j=0}^{N-x} \frac{(-1)^j}{x!j!} \left[\frac{(N - x - j)!}{N!} \right]^{K-2}.$$
(10.35)

Kemp (1978b) reexamined these three matching schemes. She showed that the pgf corresponding to (10.33) is

$$G(z) = {}_1F_1[-N; 1/2 - N; (z - 1)/2];$$
(10.36)

as N becomes large this tends to $\exp[N(z-1)/(2N-1)]$, and hence as $N \to \infty$ it tends to a Poisson pgf with parameter $\theta = 1/2$. She showed furthermore that the pgf's corresponding to (10.34) and (10.35) are

$$G(z) = {}_1F_{K-1}\left[-N; \frac{1}{K} - N, \frac{2}{K} - N, \ldots, \frac{K-1}{K} - N; \frac{K!(z-1)}{(-K)^K}\right] \quad (10.37)$$

and $\quad G(z) = {}_1F_{K-1}[-N; -N, -N, \ldots, -N; (-1)^K(z-1)],$ \hfill (10.38)

respectively. In both cases as N becomes large the pgf's tend to those of Poisson distributions (with parameters $N^2/\binom{NK}{K}$ and N^{2-K}, respectively) when N becomes infinite the distributions become the degenerate distribution with pgf $G(z) = 1$.

Historical aspects of matching problems have been reviewed by Takács (1980); see also Barton (1958) and David and Barton (1962).

4 OCCUPANCY DISTRIBUTIONS

4.1 Classical Occupancy and Coupon Collecting

Consider the placement of b balls into c cells (categories, classes). The number of ways in which this can be achieved is clearly c^b, since each of the b balls can be put into any one of the c cells. In the classical occupancy situation each of these ways is assumed to be equiprobable.

There has been considerable interest in the distribution of the number of empty cells. Because this distribution arises in many different contexts, there is no standard notation. We note that previous writers have used the following symbols:

Author	Number of balls (b)	Number of cells (c)	Number of empty cells (x)
Tukey (1949)	m	N	$m - b$
Feller (1957), Harkness (1970)	r	n	m or x
Riordan (1958)	n	m	x
Barton and David (1959a)	n	N	k
Parzen (1960)	n	M	m
Moran (1968)	N	n	m
Johnson and Kotz (1969)	N	k	$k - j$
Kemp (1978b)	N	k	x
Johnson and Kotz (1977)	n	m	k
Fang (1985)	n	m	x
Hald (1990)	n	f	$f - i$

Suppose that the cells are numbered $1, 2, \ldots, c$. Then the probability that there are no balls in any of a specified set of j cells is $(1 - j/c)^b$. Using the notation and methodology of Section 2 we have

$$S_j = \binom{c}{j}\left(\frac{c-j}{c}\right)^b ,$$ (10.39)

and the probability that <u>at least</u> x cells are empty is

$$\Pr[X \geq x] = P_x = \sum_{j=x}^{c} \frac{(-1)^{j-x}c!x}{x!(j-x)!(c-j)!j}\left(\frac{c-j}{c}\right)^b .$$ (10.40)

Moran (1968) has used the same method but not the same notation.

The probability that <u>exactly</u> x are empty can be written in a number of (equivalent) ways:

$$\Pr[X = x] = P_{[x]} = \sum_{i=0}^{c-x}(-1)^i\binom{x+i}{i}\binom{c}{x+i}\left(\frac{c-x-i}{c}\right)^b$$

$$= \sum_{j=x}^{c}(-1)^{j-x}\frac{c!}{x!(j-x)!(c-j)!}\left(\frac{c-j}{c}\right)^b$$

$$= \binom{c}{x}\sum_{v=0}^{c-x}(-1)^v\binom{c-x}{v}\left(1-\frac{x+v}{c}\right)^b$$

$$= \binom{c}{x}\frac{\Delta^{c-x}0^b}{c^b} = \frac{c!S(b,c-x)}{x!c^b} ,$$ (10.41)

where $S(b,c-x)$ is a Stirling number of the second kind (Chapter 1, Section A3).

A familiar textbook example of this occupancy distribution relates to birthdays. Consider a community of b people whose birthdays occur randomly and independently, and suppose that a year contains $c = 365$ days. Let X be the number of days of the year when nobody has a birthday. Then

$$\Pr[X = x] = \binom{365}{x}\sum_{v=0}^{365-x}(-1)^v\binom{365-x}{v}\left(\frac{365-x-v}{365}\right)^b .$$ (10.42)

Moran has pointed out that (10.41) still holds for $b < c$ and that alternative derivations are possible; see, for example, Domb's (1952) application of the result to a cosmic ray problem.

The factorial moments for the *distribution of the number of empty cells* are

$$\mu'_{[r]} = \frac{c!}{(c-r)!}\left(\frac{c-r}{c}\right)^b ;$$ (10.43)

in particular

$$\mu = (c - 1)^b c^{1-b},$$

$$\mu_2 = (c - 1)(c - 2)^b c^{1-b} + (c - 1)^b c^{1-b} - (c - 1)^{2b} c^{2-2b}. \qquad (10.44)$$

The pgf is

$$G(z) = \sum_{r=0}^{c} \binom{c}{r} \left(\frac{c - r}{c} \right)^b (z - 1)^r, \qquad (10.45)$$

which can be restated as

$$G(z) = {}_b F_{b-1}[1 - c, \ldots, 1 - c; -c, \ldots, -c; 1 - z], \qquad (10.46)$$

showing that the distribution is GHF [Kemp (1978b)]; see Chapter 2, Section 4.2.

Von Mises (1939) seems to have been the first to prove that the distribution tends to a Poisson distribution with parameter θ when b and c both tend to infinity such that $c \exp(-b/c) = \theta$ remains bounded. Feller (1957, pp. 93–95) has given an alternative proof and a short illustrative table. David (1950) gave tables of the probabilities for $b = c$ when $c = 3(1)20$; see Nicholson (1961) for certain probability levels in the tails of the distribution, and also see Owen (1962).

Many generalizations of the classical occupancy distribution have been studied. Parzen (1960) and David and Barton (1962) have examined, for example, the problem of specified occupancy. Suppose that precisely ℓ of the c cells are specified ($\ell \leq c$). Then the distribution of the number of empty cells among the ℓ specified cells has the factorial moments

$$\mu'_{[r]} = \frac{\ell!}{(\ell - r)!} \left(\frac{c - r}{c} \right)^b \qquad (10.47)$$

and the pmf

$$\Pr[X = x] = \sum_{j=x}^{\ell} \frac{(-1)^{j-x} \ell!}{(\ell - j)! x! (j - x)!} \left(\frac{c - j}{c} \right)^b. \qquad (10.48)$$

This distribution is also a GHF distribution; its pgf can be restated [Kemp (1978b)] as

$$G(z) = {}_{b+1} F_b[-\ell, 1 - c, \ldots, 1 - c; -c, \ldots, -c; 1 - z]. \qquad (10.49)$$

The moments of the numbers of cells with given numbers of balls in them have been investigated by von Mises (1939), Tukey (1949), and Johnson and Kotz (1977) inter alia; see also Fang (1985).

An interesting problem of a more general kind was formulated by Tukey (1949) in the contexts of breakages of chains in the oxidation of rubber, and irradition and mutation of bacteria. Suppose that a total of b balls are distributed at random among an *unknown* number of cells including ℓ specified cells. The total number of balls falling in the ℓ cells is b if the total number of cells is ℓ; otherwise it has a binomial distribution. Let the probability of a specified ball falling into any given one of the ℓ specified cells be p. Then the distribution of the number X_u of (specified) cells containing u balls each has the expected value

$$E[X_u] = \ell \binom{b}{u}(1-p)^b \left(\frac{p}{1-p}\right)^u, \tag{10.50}$$

variance

$$\mathrm{Var}(X_u) = E[X_u]\{1 - [1 - \omega(u,u)]E[X_u]\}, \tag{10.51}$$

and covariance

$$\mathrm{Cov}(X_u, X_v) = [\omega(u,v) - 1]E[X_u]E[X_v], \tag{10.52}$$

where

$$\omega(u,v) = \left(\frac{\ell-1}{\ell}\right)\frac{(b-u)!(b-v)!}{b!(b-u-v)!}\left[1 - \left(\frac{p}{1-p}\right)^2\right]^b \left(\frac{1-p}{1-2p}\right)^{u+v}. \tag{10.53}$$

Harkness (1970) [see also Sprott (1957), and Johnson and Kotz (1977)] has examined the problem of "leaking urns"; this is a randomized occupancy problem. Consider, like Sprott, the number of targets hit when b shots are fired randomly at c targets and the probability that any particular shot actually hits its target is p. This is the classical occupancy problem with a probability p that a ball randomly placed into a cell stays in the cell and a probability $1 - p$ that it immediately leaves the cell. The probability that x cells are unoccupied is now

$$\Pr[X = x] = P_{[x]} = \sum_{i=0}^{c-x} \frac{(-1)^i c!}{x!i!(c-x-i)!}\left(\frac{c-(x+i)p}{c}\right)^b. \tag{10.54}$$

Harkness has pointed out that these probabilities satisfy

$$\Pr[X = x|b+1, c, p] = (1-px/c)\Pr[X = x|b, c, p] + p(x+1)c^{-1}\Pr[X = x+1|b, c, p] \tag{10.55}$$

(the same recursion relationship holds with $p = 1$ for the classical occupancy distribution).

It can be shown that this leaking urn distribution tends to a Poisson distribution with parameter $\lambda = c(1 - p/c)^b$ as b and c become large. Harkness obtained certain moment properties, discussed applications and a Markov chain derivation, and suggested a binomial approximation to the distribution. His tables of probabilities demonstrate the superiority of his binomial approximation compared with a Poisson approximation.

For work on restricted occupancy models, see Fang (1985) and references therein. Yet other types of occupancy models, such as contagious occupancy models, have been studied by Barton and David (1959a) and David and Barton (1962).

Johnson and Kotz (1977) have written a very full account of occupancy distributions; besides the classical problem and the problem of leaking urns, they have discussed sequential occupancy and committee problems. A typical committee problem is as follows [Mantel and Pasternack (1968)]: A group consists of c individuals, any w of whom may be selected to serve on a committee. If b committees, each of size w, are chosen randomly, what is the probability that exactly x individuals will be on committees?

Occupancy problems can be reinterpreted as sampling problems. Parzen (1960) gives a very clear account. The classical occupancy problem can, for instance, be viewed as the withdrawal with replacement of a sample of size b from a collection containing c distinguishable items. The number $c - x$ of *occupied cells* now corresponds to the number of different items that have been sampled. The distribution of $Y = c - X$ is known by the following names:

Stevens-Craig	Stevens (1937), Craig (1953)
Arfwedson	Arfwedson (1951)
Coupon-collecting	David and Barton (1962)
Dixie-cup	Johnson and Kotz (1977).

The number of different items in the sample has the pmf

$$\Pr[Y = y] = \binom{c}{y} \frac{\Delta^y 0^b}{c^b} = \frac{c! S(b, y)}{(c - y)! c^b}. \qquad (10.56)$$

The mean and variance are

$$\mu = c \left[1 - \left(\frac{c - 1}{c} \right)^b \right],$$

$$\mu_2 = c(c - 1)(c - 2)^b c^{-b} + c(c - 1)^b c^{-b} - c^2 (c - 1)^{2b} c^{-2b}, \qquad (10.57)$$

and so on.

The joint distribution of the numbers of the c different types of item can be regarded as multinomial. A modified form of the Stevens-Craig distribution is obtained if it is supposed that there is an upper limit s to the value of each of the multinomial variables. This corresponds to a situation in which the c cells are each capable of containing up to s balls, and the b balls are placed at random in the available cells, with the proviso that, if any cells are full, then a choice must be made among the remaining cells. The probability that there are exactly x cells, each containing at least one ball, is

$$\Pr[X = x] = \frac{c!b!(cs - b)!}{x!(c - x)!(cs)!} \left[\binom{sx}{b} - \binom{x}{1}\binom{s(x - 1)}{b} + \binom{x}{2}\binom{s(x - 2)}{b} \right.$$
$$\left. - \ldots + (-1)^{\nu}\binom{x}{\nu}\binom{s(x - \nu)}{b} \right], \tag{10.58}$$

where ν is the greatest integer such that $s(x - \nu) \geq b$. The mean and variance of the distribution of X are

$$\mu = c(1 - q_1),$$
$$\mu_2 = c[q_1 - q_2 + c(q_2 - q_1^2)], \tag{10.59}$$

where

$$q_1 = \binom{s(c - 1)}{b} \bigg/ \binom{sc}{b},$$

$$q_2 = \binom{s(c - 2)}{b} \bigg/ \binom{sc}{b}.$$

A more general problem of this kind has been solved by Richards (1968) using an exponential-type generating function; the use of exponential-type generating functions is explained in Riordan (1958). Barton and David (1959a) have obtained limiting distributions for this and a number of similar distributions.

Feller (1957) has listed no fewer than sixteen applications of the classical occupancy distribution. These include the birthday example mentioned above, irradiation in biology, cosmic ray counts, gene distributions, polymer reactions, and the theory of photographic emulsions. Harkness (1970) has put on record further instances, including Mertz and Davies' (1968) predator-prey interpretation, Peto's (1953) dose-response model for invasion by microorganisms, and the epidemic spread model of Weiss (1965).

In Barton and David (1959b) the problem of placement of one-colored balls in cells is generalized to many colors and is interpreted in terms of dispersion of species. An application to computer storage is implicit in Denning and Schwartz (1972). In the usual birthday problem the question is the probability of at least one birthday match among a group of n people. Meilijson et al. (1982) have examined the distribution of the number of matches, also the distribution of the number of matched people; their work has implications regarding computer storage overflow.

A unified approach to a number of occupancy problems has been achieved by Holst (1986), by embedding in a Poisson process; this shows that many of these problems are closely related to the properties of order statistics from the gamma distribution. Holst gives a number of relevant references. Limit theorems for occupancy problems are studied in Johnson and Kotz (1977, Ch. 6), Kolchin et al. (1978), and also Barton and David (1959a).

4.2   Maxwell-Boltzmann, Bose-Einstein, and Fermi-Dirac Statistics

Consider a physical system comprising a very large number b of "particles" of some kind, for example, electrons, protons, and photons. Suppose that there are c states (energy levels) in which each particle can be. The overall state of the system is then $(b_1, b_2, \ldots, b_c)$, and equilibrium is defined as the overall state with the highest probability of occurrence.

If all c^b arrangements are equally likely the system is said to behave according to Maxwell-Boltzmann statistics (the term "statistics" is used here in a sense meaningful to physicists). The probability that there are x particles $(x \leq b)$ in a particular state is

$$\Pr[X = x] = \binom{b}{x} \left(\frac{1}{c}\right)^x \left(1 - \frac{1}{c}\right)^{b-x}, \tag{10.60}$$

and the probability that exactly k out of ℓ specified states ($\ell \leq c$) are unoccupied is given by the classical occupancy problem documented in the previous section. The classical theory of gases (at low densities and not very low temperatures) was based on Maxwell-Boltzmann statistics.

Modern theory and experimentation (particularly at very low temperatures) have, however, yielded two far more plausible hypotheses concerning the behavior of physical systems, namely Bose-Einstein and Fermi-Dirac statistics. In both cases particles are assumed to be indistinguishable (rather than distinguishable as for Maxwell-Boltzmann statistics). The difference between the two cases is the assumption of whether or not the Pauli exclusion principle holds. This postulates that there cannot be more than one particle in any particular state.

For Bose-Einstein statistics particles are assumed to be indistinguishable but not to obey the Pauli exclusion principle. The total number of arrangements of b particles in the c states can be derived by choosing b particles one at a time, with replacement, from the set of c states, and hence is equal to

$$\binom{b + c - 1}{b}.$$

Muthu (1982) reports that particles having integer spin (intrinsic momentum equal to $0, h/2\pi, 2h/2\pi, \ldots$), such as photons and pions, obey Bose-Einstein statistics and are called *bosons*. The probability that there are x particles $(x \leq b)$ in a particular state is now

$$\Pr[X = x] = \binom{b+c-x-2}{b-x} \bigg/ \binom{b+c-1}{b}, \qquad (10.61)$$

and the probability that exactly k out of ℓ specified states $(\ell \leq c)$ are unoccupied is

$$\Pr[K = k] = \binom{\ell}{\ell-k}\binom{b+c-1-\ell}{b+k-\ell} \bigg/ \binom{b+c-1}{b}; \qquad (10.62)$$

this is of course a hypergeometric distribution (Chapter 6, Section 2.1).

For Fermi-Dirac statistics particles are assumed not only to be indistinguishable but also to satisfy the Pauli exclusion principle. There can now be no more than one particle per state. The total number of arrangements of the b particles in the c states is now

$$\binom{c}{b}.$$

The probability that a specified state has x particles in it is

$$\Pr[X = x] = \binom{c-1}{b-x} \bigg/ \binom{c}{b}, \qquad x = 0, 1, \qquad (10.63)$$

and the probability that exactly k out of ℓ specified states $(\ell \leq c)$ are empty is

$$\Pr[K = k] = \binom{\ell}{\ell-k}\binom{c-\ell}{b+k-\ell} \bigg/ \binom{c}{b}. \qquad (10.64)$$

Again we have a hypergeometric distribution.

All known elementary particles have either integer spin or half-integer spin (this determines whether their wave form is symmetric or antisymmetric). Particles with half-integer spin, such as electrons and protons, are thought to obey Fermi-Dirac statistics; they have been called *fermions*.

Johnson and Kotz (1977) have given a full account of the properties of Maxwell-Boltzmann, Bose-Einstein, and Fermi-Dirac occupancy schemes. They pointed out, inter alia, that the expected number of particles X in a specified state is b/c under each of the three systems. The variance of X is not the same for the three schemes, however. For Maxwell-Boltzmann statistics $\text{Var}(X) = b(c-1)/c^2$, for Bose-Einstein statistics $\text{Var}(X) = b(c-1)(b+c)/\{c^2(c+1)\}$, and for Fermi-Dirac statistics $\text{Var}(X) = b(c-b)/c^2$.

Loève (1963) has studied these three occupancy models as particular cases of a more general model.

5 RUNS DISTRIBUTIONS

5.1 Runs of Like Elements

A considerable variety of forms of distributions have been discovered from the study of *runs* in sequences of observations. Among the simplest of such "runs" are sequences of identical values where the random variables giving rise to the observed values are independent and take the values 0 and 1 with probabilities p and $q = 1 - p$, respectively. A concise account of the early history of attempts to find the distribution of the total number of runs in such binomial (Bernoulli) sequences has been given by Mood (1940). Von Bortkiewicz (1917) obtained the mean and variance of this distribution. An asymptotic result by Wishart and Hirschfeld (1936) showed that the asymptotic standardized distribution of the total number of runs, as the length of the sequence increases, is normal.

Formulas for the mean and variance of the number of runs of a specified length were obtained by Bruns (1906); von Mises (1921) demonstrated that the distribution can be approximated by a Poisson distribution.

Results concerning the distribution of runs of like elements when the numbers n_0, n_1 of 0's and 1's in the sequence are fixed (i.e., the conditional distribution given n_0 and n_1) were published by Ising (1925) and Stevens (1939). Let the number of runs of i consecutive 1's be denoted by r_{1i}, and the number of runs of j consecutive 0's be denoted by r_{0j}. Then the probability of obtaining such sets of runs of various lengths, conditional on $\sum_i i r_{1i} = n_1$, $\sum_j j r_{0i} = n_0$, with $\sum_i r_{1i} = r_1$, $\sum_j r_{0j} = r_0$, $n_1 + n_0 = N$, is

$$F(r_1, r_0) \binom{r_1}{r_{11}, r_{12}, \ldots, r_{1N}} \binom{r_0}{r_{01}, r_{02}, \ldots, r_{0N}} \bigg/ \binom{N}{n_1}, \qquad (10.65)$$

where $F(r_1, r_0) = 0, 1, 2$ according as $|r_1 - r_0| > 1$, $|r_1 - r_0| = 1$, $r_1 = r_0$, respectively.

The joint distribution of the total numbers of runs R_1, R_0 is

$$\Pr[R_0 = r_0, R_1 = r_1] = F(r_1, r_0) \binom{n_1 - 1}{r_1 - 1} \binom{n_0 - 1}{r_0 - 1} \bigg/ \binom{N}{n_1}, \qquad (10.66)$$

where $r_0 + r_1 \geq 2$, $r_1 \leq n_1$, $r_0 \leq n_0$. From this result it is straightforward to deduce that the marginal distribution of R_1 is

$$
\begin{aligned}
\Pr[R_1 = r_1] &= \binom{n_1 - 1}{r_1 - 1}\binom{n_0 + 1}{n_0 + 1 - r_1} \Big/ \binom{n_0 + n_1}{n_1} \\
&= \binom{n_0}{r_1 - 1}\binom{n_1}{n_1 - r_1} \Big/ \binom{n_0 + n_1}{n_1 - 1},
\end{aligned}
\tag{10.67}
$$

where $r_1 = 1, 2, \ldots, \min(1 + n_0, n_1)$.

This is sometimes called the *Ising-Stevens distribution*. The pgf is

$$
G(z) = z \frac{{}_2F_1[-n_0, 1 - n_1; 2; z]}{{}_2F_1[-n_0, 1 - n_1; 2; 1]},
\tag{10.68}
$$

and hence the distribution is a shifted form of a special case of the classical hypergeometric distribution (see Chapter 6, Section 2.1). The mean and variance are

$$
E[R_1] = \frac{(n_0 + 1)n_1}{n_0 + n_1},
$$

$$
\mathrm{Var}(R_1) = \frac{n_0(n_0 + 1)n_1(n_1 - 1)}{(n_0 + n_1)^2(n_0 + n_1 - 1)}.
\tag{10.69}
$$

Finally the mean and variance of the *total* number of runs $R_0 + R_1$ are

$$
E[R_0 + R_1] = 1 + \left(\frac{2n_0n_1}{n_0 + n_1}\right),
$$

$$
\mathrm{Var}(R_0 + R_1) = \frac{2n_0n_1(2n_0n_1 - n_0 - n_1)}{(n_0 + n_1)^2(n_0 + n_1 - 1)}.
\tag{10.70}
$$

The asymptotic normality of the distribution of R_1 (or of $R_0 + R_1$) as $n_0 + n_1$ increases was established by Wald and Wolfowitz (1940).

Instead of assuming that n_0 and n_1 are known, one can suppose that n_0 and n_1 have arisen by some form of sampling. David and Barton (1962) have investigated the following scenarios:

1. There is a finite population of n elements with np of one kind and $n - np$ of the other. A sample of size N is drawn with replacement from this population; n_0 and n_1 are now the numbers of the two kinds of elements in a binomial sample ($n_0 + n_1 = N$).

2. Alternatively, inverse sampling with replacement for a fixed number n_0 of elements of one kind can be used; n_1 is now the number of elements of the second kind that have been sampled.

Consider now arrangements around a circle. When a red objects and b blue objects are arranged at random in a circle, the number of red runs X is identically equal to the number of blue runs; the pmf is

$$\Pr[X = x] = \binom{a-1}{x-1}\binom{b}{b-x} \bigg/ \binom{a+b-1}{b-1}$$

$$= \binom{b-1}{x-1}\binom{a}{a-x} \bigg/ \binom{a+b-1}{a-1}, \qquad x = 1, 2, \ldots, \min(a, b),$$

and the pgf is

$$G(z) = z \frac{{}_2F_1[1-a, 1-b; 2; z]}{{}_2F_1[1-a, 1-b; 2; 1]}. \tag{10.71}$$

This again is an Ising-Stevens distribution [Stevens (1939)]. The mean and variance are

$$\mu = \frac{ab}{a+b-1},$$

$$\mu_2 = \frac{a(a-1)b(b-1)}{(a+b-1)^2(a+b-2)}. \tag{10.72}$$

A natural generalization of runs of two kinds along a line is to consider arrangements of $k > 2$ kinds of elements, conditional on there being $n_1, n_2, \ldots, n_k$ elements of the 1st, 2nd, $\ldots$, kth kind, respectively (with $\sum_j n_j = N$). Let r_{ij} now denote the number of runs of elements of the ith kind that are exactly of length j, and put

$$r_i = \sum_{j=1}^{n_i} r_{ij}, \qquad i = 1, 2, \ldots, k. \tag{10.73}$$

Then the probability of obtaining the array $\{r_{ij}\}$ of runs is

$$F(r_1, r_2, \ldots, r_k) \left[\prod_{i=1}^{k} \binom{r_i}{r_{i1}, r_{i2}, \ldots, r_{in_i}} \right] \bigg/ \binom{N}{n_1, n_2, \ldots, n_k},$$

where $F(r_1, r_2, \ldots, r_k)$ is the coefficient of $\prod_{i=1}^{k} x_i^{r_i}$ in the expansion of

$$\left(\sum_{i=1}^{k} x_i \right)^k \prod_{j=1}^{k} \left(\sum_{i=1}^{k} x_i - x_j \right)^{r_i - 1}.$$

Various additional aspects of runs distributions are discussed in David and Barton (1962), in the references therein, and in Part IV of the Introduction to David, Kendall and Barton (1966).

An early example of the statistical use of distributions concerning runs is the two-sample test of Wald and Wolfowitz (1940). A more powerful test of this type is that of Weiss (1976). The theory of runs also provides tests of randomness; see Gibbons (1986) for a careful discussion and useful bibliography. Runs are also used for certain tests of adequacy of fit; for this and other applications, for example in quality control, see the helpful review article by Weiss (1988).

5.2 Runs Up and Down

Another kind of "run" that has received attention is a run of increasing or decreasing values in a sequence of N independent random variables each having the same distribution, usually assumed to be continuous (obviating the need to consider tied values). Evidently the numbers of such runs are unaltered if each value is replaced by its "rank order", that is, 1 for the smallest value, 2 for the next smallest, and so on, up to N for the largest value. Accordingly, it is only necessary to study the occurrence of runs by considering rearrangements of the integers $1, 2, \ldots, N$. The results will be valid, whatever the common continuous distribution of the random variables giving rise to the observations.

Wallis and Moore (1941) constructed a test of randomness of a sequence by considering the distribution of the number of *turning-points* in a sequence of N values, $x_1, x_2, \ldots, x_N$. The jth observation constitutes a turning-point if either $x_j = \min(x_{j-1}, x_j, x_{j+1})$ or $x_j = \max(x_{j-1}, x_j, x_{j+1})$. Neither x_1 nor x_N can be turning-points, but for each of the other $(N - 2)$ observations, the probability that it is a turning-point is $2/3$. Hence the expected value of the number of turning-points is $2(N - 2)/3$. The variance, and third and fourth moments are

$$\mu_2 = \frac{16N - 29}{90},$$

$$\mu_3 = -\frac{16(N + 1)}{945},$$

$$\mu_4 = \frac{448N^2 - 1976N + 2301}{4725}. \tag{10.74}$$

For $N > 12$, the distribution can be treated as normal. The number of runs (up or down) is *one more* than the number of turning-points.

Wallis and Moore (1941) have defined *phases* as runs excluding those beginning with x_1 or ending with x_N. There are no phases if there are no turning-points; otherwise the number of phases is <u>one less</u> than the number of turning-points. They defined the *duration* of a phase between turning-points x_j and $x_{j'}$ (where $j < j'$) to be $j' - j$; they showed that the expected number of phases of duration d is

$$\frac{2(d^2 + 3d + 1)(N - d - 2)}{(d + 3)!}.$$

Wallis and Moore (1941) traced the history of these distributions back to work by Bienaymé (1874). They ascribed credit for the formula for the distribution of phase duration to Besson. Further references concerning statistical tests based on runs up and down are given in Gibbons (1986).

6 DISTRIBUTIONS OF ORDER k

6.1 Success Runs Distributions

"The Probability of throwing a Chance assigned a given number of times without intermission, in any given number of Trials" [De Moivre (1738)] was interpreted by Todhunter (1865, pp. 184–186) to mean the probability that a run of r successes is completed at the nth trial in a sequence of Bernoulli trials each with probability of success p. Let the probability of this event be u_n, using Todhunter's notation. Todhunter argued that

$$u_{n+1} = u_n + (1 - u_{n-r})qp^r, \qquad \text{where } q = 1 - p,$$

and hence found the generating function of u_n; he obtained a formula substantially in agreement with the result given by De Moivre.

Feller (1957, pp. 299–303) treated this problem as an application of the theory of recurrent events, and hence showed that the probability that the *first* run of length r occurs at the xth trial has pgf

$$G(z) = \frac{p^r z^r (1 - pz)}{1 - z + qp^r z^{r+1}}, \tag{10.75}$$

where $x = r, r + 1, r + 2, \ldots$, $r = 1, 2, \ldots$, and $0 < p < 1$. The mean and variance are

$$\mu = \frac{1 - p^r}{qp^r}, \qquad \mu_2 = \frac{1}{(qp^r)^2} - \frac{(2r + 1)}{qp^r} - \frac{p}{q^2}. \tag{10.76}$$

Clearly $r = 1$ gives an ordinary geometric distribution with support $x = 1, 2, 3, \ldots$.

De Moivre gave without proof a formula for the number of trials needed in order to have a probability of approximately one-half of getting a run of r successes. Feller proved that the probability q_r that there is no success run of length r in n trials is

$$q_n \approx \frac{1 - p\zeta}{\zeta^{n+1} q(r + 1 - r\zeta)}, \tag{10.77}$$

where ζ is the smallest in absolute value of the roots of the denominator of (10.75), and demonstrated that the formula gives a remarkably good approximation, even for n as low as 4. Feller (1957, pp. 303–304) also considered success runs of either kind and success runs occurring before failure runs.

Shane (1973) gave the name *Fibonacci distribution* to the distribution with pgf (10.75) when $r = 2$ and $p = q = 1/2$. He showed that in this particular case

$$\Pr[X = x + 2] = 0.5 \Pr[X = x + 1] + 0.25 \Pr[X = x],$$

and hence that

$$\Pr[X = x] = 2^{-x} F_{x-1}, \qquad x = 2, 3, \ldots, \qquad (10.78)$$

where $\{F_1, F_2, F_3, F_4, F_5, \ldots\}$ are the Fibonacci numbers $\{1, 1, 2, 3, 5, \ldots\}$. Taillie and Patil (1986) related this distribution to a problem in disassortative mating. Patil et al. (1984) and Taillie and Patil (1986) named the distribution with pgf (10.75), where $r \geq 1$ and $p = q = 1/2$, a *poly-nacci distribution*, and expressed the probabilities in terms of rth order Fibonacci numbers. For general values of both r and p, $r \geq 1$, $0 < p < 1$, Uppuluri and Patil (1983) and Patil et al. (1984) called the distribution (10.75) a *generalized poly-nacci distribution*.

The papers by Philippou and Muwafi (1982) and Philippou, Georghiou and Philippou (1983) have created an upsurge of interest in these distributions under the name *geometric distributions of order k*. In the following sections we discuss a number of related distributions of order k that are extant in the literature.

6.2 Philippou, Aki, and Hirano's Distributions of Order k

From 1984 onward there has been a great number of papers on various *distributions of order k*. Many of these have been written by Philippou and his colleagues, and by Aki and Hirano; later these authors were joined by Panaretos and Xekalaki. Within the confines of this volume it has proved impossible to document all the work that is taking place. Philippou (1986), Charalambides (1986b), and Aki and Hirano (1988) included in their papers good overviews that were up-to-date at the time of writing.

The formulas for the probabilities for these distributions are not at first sight very illuminating; also in some cases there is more than one correct published expression for the probabilities. In general the pgf's give more insight into the models that lead to these the distributions. Also the properties of these distributions can be obtained more readily from their pgf's than from their pmf's. We will quote the pgf's where we can.

Consider (10.75) with $r = k$. This is Philippou and Muwafi's (1982) *geometric distribution of order k*. Immediately related distributions are the *negative binomial of order* k, the *Poisson of order* k, the *logarithmic of order* k, and the *compound Poisson of order* k; see Philippou, Georghiou and Philippou (1983), Philippou (1983, 1984), Aki, Kuboki and Hirano (1984), Charalambides (1986b), Hirano (1986), Philippou (1986).

Geometric Distribution of Order k
This distribution has the pgf (10.75) with $r = k$:

$$G(z) = \frac{p^k z^k (1 - pz)}{1 - z + qp^k z^{k+1}}$$

$$= \frac{p^k z^k}{1 - (1 - p^k)z \left(\frac{1-p^k z^k}{1-pz}\right) \Big/ \left(\frac{1-p^k}{1-p}\right)}, \qquad (10.79)$$

which has the form $G(z) = z^k g_1(g(z))$, where

$$g_1(z) = \frac{p^k}{1 - (1 - p^k)z} \qquad (10.80)$$

is the pgf of a geometric distribution,

and

$$g(z) = z \left(\frac{1 - p^k z^k}{1 - pz}\right) \Big/ \left(\frac{1 - p^k}{1 - p}\right) \qquad (10.81)$$

is the pgf of a truncated geometric distribution with a different parameter and support $1, 2, \ldots, k$. The mean and variance are given by (10.76) with $r = k$. Philippou and Muwafi's (1982) expression for the pmf is

$$\Pr[X = x] = p^x \sum_{x_1, \ldots, x_k} \binom{x_1 + \cdots + x_k}{x_1, \ldots, x_k} \left(\frac{q}{p}\right)^{x_1 + \cdots + x_k} \qquad (10.82)$$

for $x = k, k + 1, \ldots$, where summation takes place over $x_1, \ldots, x_k$ such that $x_1 + 2x_2 + \cdots + kx_k = x - k$.

Negative Binomial Distribution of Order k
This is the waiting-time distribution for b runs of successes of length k, and hence it is a b-fold convolution of geometric distributions of order k, that is, it has the pgf

$$G(z) = [z^k g_1(g(z))]^b = z^{bk} g_2(g(z)) \qquad (10.83)$$

with $g_1(z)$ and $g(z)$ as above; note that $g_2(z)$ is the pgf of a negative binomial distribution. The mean and variance are

$$\mu = \frac{b(1 - p^k)}{qp^k} \quad \text{and} \quad \mu_2 = \frac{b\{1 - (2k + 1)qp^k - p^{2k+1}\}}{q^2 p^{2k}}, \qquad (10.84)$$

and the pmf can be expressed as

$$\Pr[X = x] = p^x \sum_{x_1,\dots,x_k} \binom{x_1 + \cdots + x_k + b - 1}{x_1,\dots,x_k, b-1} \left(\frac{q}{p}\right)^{x_1 + \cdots + x_k} \tag{10.85}$$

for $x = kr, kr + 1, \dots$, where summation is over $x_1, \dots, x_k$ such that

$$x_1 + 2x_2 + \cdots + kx_k = x - kr.$$

Logarithmic Distribution of Order k
This distribution was obtained as a limiting form of a left-truncated negative binomial distribution of order k; the pgf is

$$G(z) = \ln\left(\frac{1 - pz}{1 - z + pq^k z^{k+1}}\right) \Big/ (-k \ln p); \tag{10.86}$$

this has the form

$$G(z) = g_3(g(z)), \tag{10.87}$$

with $g(z)$ as above and with $g_3(z)$ of the form $\ln(1 - \alpha z)/\ln(1 - \alpha)$, where $\alpha = 1 - p^k$. The mean and variance are

$$\mu = \frac{1 - p^k - kqp^k}{qp^k(-k \ln p)},$$

$$\mu_2 = \frac{1 - p^{2k+1} - (2k+1)qp^k}{qp^{2k}(-k \ln p)} - \mu^2. \tag{10.88}$$

The pmf can be expressed as

$$\Pr[X = x] = \frac{p^x}{(-k \ln p)} \sum_{x_1,\dots,x_k} \frac{(x_1 + \cdots + x_k - 1)!}{x_1! \cdots x_k!} \left(\frac{q}{p}\right)^{x_1 + \cdots + x_k} \tag{10.89}$$

for $x = 1, 2, \dots$, where summation is over $x_1, \dots, x_k$ such that $x_1 + 2x_2 + \cdots + kx_k = x$.

Poisson Distribution of Order k
This was derived as the limiting form as $b \to \infty$ of the negative binomial distribution of order k shifted to the support $0, 1, \dots$. Its pgf can be written as

$$G(z) = \exp\left\{-\lambda \left(k - \sum_{i=1}^{k} z^i\right)\right\} \tag{10.90}$$

$$= e^{\lambda k \{h(z) - 1\}}, \tag{10.91}$$

where $h(z)$ is the pgf of a *discrete rectangular distribution* (Chapter 6, Section 10.1) with support $1, 2, \ldots, k$. This representation of the pgf shows that the distribution is a particular stuttering Poisson distribution. The mean and variance are

$$\mu = \frac{k(k+1)\lambda}{2} \quad \text{and} \quad \mu_2 = \frac{k(k+1)(2k+1)\lambda}{6}, \tag{10.92}$$

and the pmf is

$$\Pr[X = x] = e^{-k\lambda} \sum_{x_1, \ldots, x_k} \frac{\lambda^{x_1 + \cdots + x_k}}{x_1! \cdots x_k!} \tag{10.93}$$

for $x = 0, 1, 2, \ldots$, where summation is over $x_1, \ldots, x_k$ such that $x_1 + 2x_2 + \cdots + kx_k = x$.

Compound Poisson Distribution of Order k

This is a gamma-mixed Poisson distribution of order k. The pgf is

$$G(z) = \left\{ 1 + \alpha^{-1} \left(k - \sum_{i=1}^{k} z^i \right) \right\}^{-c} \tag{10.94}$$

$$= \{ 1 - \alpha^{-1} k (h(z) - 1) \}^{-c}, \tag{10.95}$$

where $h(z)$ is again the pgf of a discrete rectangular distribution with support $1, 2, \ldots, k$. The mean and variance are

$$\mu = \frac{k(k+1)c}{2\alpha} \quad \text{and} \quad \mu_2 = \frac{k(k+1)(2k+1)c}{6\alpha} + \frac{k^2(k+1)^2 c}{4\alpha^2}, \tag{10.96}$$

and the pmf is

$$\Pr[X = x] = \left(\frac{\alpha}{k+\alpha} \right)^c \sum_{x_1, \ldots, x_k} \binom{x_1 + \cdots + x_k + c - 1}{x_1, \ldots x_k, c - 1} \left(\frac{1}{k+\alpha} \right)^{x_1 + \cdots + x_k} \tag{10.97}$$

for $x = 0, 1, 2, \ldots$, where summation is over $x_1, \ldots, x_k$ such that $x_1 + 2x_2 + \cdots + kx_k = x$. Panaretos and Xekalaki (1986a) referred to this distribution as a "negative binomial distribution of order k."

Other formulas have been given for the probability mass functions of these distributions; see, e.g., Godbole (1990a). Charalambides (1986b) helped to clarify relationships between order k distributions and stopped sum distributions (Chapter 9) by writing the pgf's of the geometric, negative binomial, logarithmic, and Poisson distributions of order k in the form $g_u(g_v(z))$ (such relationships are not readily seen from the form of the pmf's). Charalambides recognized the relevance of Bell polynomials and hence the relevance of Faà di Bruno's formula (Chapter 9, Section 1). He gave expressions for the probabilities and also the factorial moments of these order k distributions in terms of truncated Bell and partial Bell polynomials. Aki, Kuboki, and Hirano (1984) and Hirano (1986) also understood the connection with stopped sum distributions and gave symbolic representations as $\mathcal{F}_1 \vee \mathcal{F}_2$ distributions (i.e., using the notation of Chapter 9).

Consider now the number of occurrences X of the kth consecutive success in n independent Bernoulli trials with probability of success p. The distribution of X has been called the *binomial distribution of order k*. An important application of this distribution is in the theory of consecutive-k-out-of-n:F failure systems [see, e.g., Shanthikumar (1982)], and hence it is relevant to certain telecommunication and oil pipeline systems and in the design of integrated circuitry [Chiang and Niu (1981); Bollinger and Salvia (1982)]. A consecutive-k-out-of-n:F system is one comprising n ordered components, each with independent probabilities q and p of operating or failing; the system fails when k consecutive components fail. Papers on such systems often appear in the journal *IEEE Transactions on Reliability*, which sometimes devotes entire sections to them.

Binomial Distribution of Order k

The pmf of X has been given independently by both Hirano (1986) and Philippou and Makri (1986) as

$$\Pr[X = x] = p^n \sum_{i=0}^{k-1} \sum_{x_1,\ldots,x_k} \binom{x_1 + \cdots + x_k + x}{x_1, \ldots x_k, x} \left(\frac{q}{p}\right)^{x_1 + \cdots + x_k} \tag{10.98}$$

for $x = 0, 1, 2, \ldots, [n/k]$, where the inner summation is over $x_1, \ldots, x_k$ such that $x_1 + 2x_2 + \cdots + kx_k = n - i - kx$. Neither paper gave an elementary formula for the pgf. Recurrence relations for the probabilities were obtained by Aki and Hirano (1988). If the distribution of Y is negative binomial of order k, with parameters b and p as in the previous section, then

$$\Pr[X \geq b] = \Pr[Y \leq n], \qquad b = 0, 1, \ldots, [n/k]; \tag{10.99}$$

this result appears in Feller (1957, p. 297), and parallels the well-known relationship between the tails of ordinary binomial and negative binomial distributions (Chapter 5, Section 6). Assuming that $\sum_{x=0}^{[n/k]} \Pr[X = x] = 1$ and $k \leq n \leq 2n - 1$, Philippou (1986) found the mean and variance to be

$$\mu = p^k\{1 + (n - k)q\},$$
$$\mu_2 = p^k\{1 + (n - k)q\} - p^{2k}\{1 + (n - k)q\}^2. \tag{10.100}$$

With the notable exception of Aki and Hirano (1989) there has been little work on estimation of parameters for distributions of order k.

6.3 Further Distributions of Order k

Many other distributions of order k have been created; we do not have space to give formulas for all of these.

In the last section we saw that the negative binomial distribution of order k and the gamma-mixed Poisson (compound Poisson) distribution of order k are not the same, though both are negative binomial stopped sum distributions. A third type of negative binomial distribution of order k is that of Ling (1989).

Philippou (1986) derived a second type of binomial distribution of order k with pgf $g_4(h(z))$ analogous to the gamma-mixed Poisson distribution of order k. Here $g_4(z)$ is the pgf of a binomial distribution and $h(z)$ is the pgf of a discrete rectangular distribution with support $1, 2, \ldots, k$. Ling (1988) has also constructed an alternative binomial-type distribution of order k. The relationship between Ling's binomial and negative binomial distributions of order k has been investigated by Hirano et al. (1991).

The emphasis in Panaretos and Xekalaki (1986a) is on urn models for distributions of order k. They gave one such model for Philippou's (1983) gamma-mixed Poisson of order k, and they showed that a limiting form is a second logarithmic distribution of order k with pgf

$$G(z) = -\ln\left\{1 - \left(\frac{\theta k}{1 + \theta k}\right)h(z)\right\}\bigg/ \ln(1 + \theta k), \qquad (10.101)$$

where $h(z)$ is the pgf for a discrete rectangular distribution on $1, 2, \ldots, k$.

Hirano (1986) obtained a (shifted) Pólya-Aeppli distribution of order k and a Neyman Type A distribution of order k, using stopped-sum models (Chapter 9). Hypergeometric, inverse hypergeometric, Pólya, and inverse Pólya distributions of order k are called cluster hypergeometric, etc. distributions by Panaretos and Xekalaki (1986a, 1989). A cluster negative binomial, a cluster binomial, and a cluster generalized Waring distribution of order k appear in Panaretos, and Xekalaki (1986a), (1986b), and (1986c) respectively. Xekalaki, Panaretos and Philippou (1987) dealt with mixtures of distributions of order k. Certain distributions of order k arising in inverse sampling are the subject of Xekalaki and Panaretos (1989).

New Pólya and inverse Pólya distributions of order k were examined by Philippou, Tripsiannis and Antzoulakos (1989). Philippou (1989) investigated Poisson of order k mixtures of binomial, negative binomial, and Poisson distributions, and their interrelationships. Godbole (1990b) re-examined the hypergeometric and related distributions of order k of Panaretos and Xekalaki (1986a), and derived a waiting-time distribution of order k.

A number of papers in the literature deal with multiparameter distributions of order k (sometimes called *extended distributions of order k*). We mention in particular Aki (1985), Hirano and Aki (1987), Philippou (1988), Philippou and Antzoulakos (1990), and Ling (1990).

CHAPTER 11

Miscellaneous Discrete Distributions

This chapter contains descriptions of various distributions that have not fitted straightforwardly into the previous chapters. In some cases, e.g. the lost-games and the Sichel distributions, an earlier section would have been made unwieldly by their inclusion. However, most of the distributions in this chapter, such as the discrete Adès and the Gram-Charlier Type-B distributions, do not relate naturally to the distributions discussed earlier in the book. A few of the distributions that we include have lain dormant for many years; we hope that these have potential, either for modeling purposes or for the further development of theory of discrete distributions.

The arrangement within this chapter is alphabetical.

1 ABSORPTION DISTRIBUTION

The name *absorption distribution* was given to the distribution that represents the number of individuals that fail to cross a specified region containing hazards of a certain kind (such as a minefield) by Blomqvist (1952) and Zacks and Goldfarb (1966). More recently Dunkl (1981) has described the model in an alternative manner, in terms of a manuscript containing m errors. A proofreader goes through the manuscript line by line looking for errors. If the proofreader finds an error, he or she corrects it, goes back to the beginning, and repeats the process; if he or she reaches the end without finding an error, he or she likewise goes back to the beginning and repeats the process. The probability of finding any particular error is assumed to be constant and equal to $(1 - q)$. The random variable of interest is the number of errors X that are found in n attempts to read through the manuscript.

Blomqvist showed that

$$\Pr[X = 0] = q^{mn},$$

and

$$\Pr[X = x] = q^{(m-x)(n-x)}$$

$$\times \frac{(1-q^m)(1-q^{m-1})\cdots(1-q^{m-x+1})(1-q^n)(1-q^{n-1})\cdots(1-q^{n-x+1})}{(1-q)(1-q^2)\cdots(1-q^x)}, \quad (11.1)$$

where $x = 1, 2, \ldots, \min(m, n)$, and $0 < q < 1$. The symmetry of the pmf with respect to m and n is unexpected. Dunkl restated the pmf in terms of q-binomial coefficients:

$$\Pr[X = x] = \frac{q^{(m-x)(n-x)}(q^m; q^{-1})_x (q^n; q^{-1})_x}{(q; q)_x} \quad (11.2)$$

$$= \frac{q^{mn+x}(q^{-m}; q)_x (q^{-n}; q)_x}{(q; q)_x}, \quad x = 0, 1, \ldots, \min(m, n), \quad (11.3)$$

where

$$(a; q)_0 = 1,$$

$$(a; q)_x = (1 - a)(1 - aq)\cdots(1 - aq^{x-1}),$$

$$(a; q^{-1})_x = (1 - a)\left(1 - \frac{a}{q}\right)\cdots\left(1 - \frac{a}{q^{x-1}}\right)$$

(see Chapter 1, Section A12). It is implicit in Dunkl's paper that the pgf can be stated in terms of a basic hypergeometric series.

Earlier Borenius (1953) had said that the expected value of the distribution is

$$\mu = (1 - q^m)(1 - q^n) \quad (11.4)$$

and that the variance is approximately

$$\mu_2 \doteq \frac{q^{m+n}(1 - q^m)(1 - q^n)}{(1 - q)[1 - (1 - q^m)(1 - q^n)]^2}. \quad (11.5)$$

However, Dunkl has commented that "the mean and higher moments can not be found in closed form." The approximations that he gave for the mean and variance are

$$\mu = \frac{(1 - c)(1 - q^n)}{1 - q} + \frac{(1 - c)(1 - c/q)(1 - q^n)(1 - q^{n-1})}{1 - q^2} + o((1 - q)^2), \quad (11.6)$$

$$\mu_2 = \frac{c(1 - c)q^{n-1}(1 - q^n)}{1 - q} + \frac{2c(1 - c)(1 - c/q)q^{n-2}(1 - q^n)(1 - q^{n-1})}{1 - q} + o((1 - q)^2),$$

$$(11.7)$$

with $c \doteq q^m$, assuming that q is close to unity. While Dunkl's approximation for the mean is symmetric in m and n, his approximation for the variance is not.

Borenius standardized the distribution using his formulas for the mean and variance, and then he plotted the cumulative distribution function for his standardized distribution; his plots showed that when $q \geq 0.9$, his standardized distribution tends rapidly to normality as m and n increase.

Dunkl remarked that there is a q-analogue for the negative binomial distribution. The distribution of Y, the number of attempts to read through the manuscript that are required in order to detect exactly k of the m errors, has the pmf

$$\Pr[Y = y] = \frac{q^{(y-k)(m-k+1)}(q^m; q^{-1})_k(q^k; q)_{y-k}}{(q; q)_{y-k}}, \tag{11.8}$$

where $k = 0, 1, 2, \ldots$, and $y \geq k$. Dunkl noted that $\sum_{x=k}^{\infty} \Pr[X = x] = 1$ is a consequence of Heine's theorem for the sum of a $_1\Phi_0[\cdot]$ basic hypergeometric series (Chapter 1, Section A12).

2 DANDEKAR'S MODIFIED BINOMIAL AND POISSON DISTRIBUTIONS

Consider a sequence of n trials where the probability of a success in a single trial is p (constant), with the proviso that if a trial produces a success then the probability of success for the next $(m - 1)$ trials is zero. Dandekar's (1955) first modified binomial distribution is obtained by supposing in addition that the probability of success in the first trial is p. Let X be the number of successes in the n trials; then

$$\Pr[X \leq x] = q^{n-xm} \sum_{j=0}^{x} \binom{n - xm + j - 1}{j} p^j, \tag{11.9}$$

where $q = 1 - p$, and $x = 0, 1, 2, \ldots$ such that $x \leq [n/m]$, with $[\cdot]$ denoting the integer part.

As $n \to \infty$, $m \to \infty$, and $p \to 0$ so that $np \to \lambda$ and $m/n \to k$, λ and k constant, Dandekar's first modified binomial distribution tends to his first modified Poisson distribution, for which

$$\Pr[X \leq x] = e^{\lambda(kx-1)} \sum_{j=0}^{x} \frac{(1 - kx)^j \lambda^j}{j!}, \tag{11.10}$$

where $0 < \lambda$, $0 < k$, and $x = 0, 1, 2, \ldots$ such that $x \leq [1/k]$, where $[\cdot]$ denotes the integer part.

Dandekar's (1955) second modified binomial distribution arises when the first trial is assumed to occur at a random point in an infinite sequence of such trials; his second modified Poisson distribution is then the result of a limiting process analogous to that for his first modified Poisson distribution; see Dandekar (1955) and also Patil et al. (1984).

Dandekar's distributions have been re-examined by Basu (1955). Simple forms for the pgf's, means and variances of these four distributions appear to be unknown.

3 DIGAMMA AND TRIGAMMA DISTRIBUTIONS

The psi (digamma) function is defined as $\psi(z) = d \ln \Gamma(z)/dz$ (see Chapter 1, Section A2). Hence

$$\Delta \psi(z) = \psi(z+1) - \psi(z) = 1/z,$$

$$\Delta^n \psi(z) = \Delta^{n-1}(1/z) = \frac{(-1)^{n-1}(n-1)!(z-1)!}{(z+n-1)!},$$

and Newton's forward difference formula (Chapter 1, Section A3) gives

$$\psi(z+n) = \sum_{j=0}^{n} \binom{n}{j} \Delta^j \psi(z) = \psi(z) + \sum_{j=1}^{n} \frac{(-1)^{j-1} n!(z-1)!}{j(n-j)!(z+j-1)!}. \tag{11.11}$$

More generally, when $\mathrm{Re}(z+\nu) > 0$, z not a negative integer,

$$\psi(z+\nu) - \psi(z) = \sum_{j \geq 1} \frac{(-1)^{j-1} \nu(\nu-1) \dots (\nu-j+1)}{jz(z+1) \dots (z+j-1)} \tag{11.12}$$

is a convergent series; see Nörlund (1923). Taking $\nu = -\alpha$ and $z = \alpha + \gamma$, where α and γ are real and $\gamma > 0$, $\alpha > -1$, $\alpha + \gamma > 0$, the series becomes

$$\psi(\alpha+\gamma) - \psi(\gamma) = \sum_{x \geq 1} \frac{(\alpha+x-1)!(\alpha+\gamma-1)!}{x(\alpha-1)!(\alpha+\gamma+x-1)!}, \tag{11.13}$$

with all terms having the same sign. Moreover, for the trigamma function $\psi'(\gamma)$,

$$\psi'(\gamma) = \lim_{\alpha \to 0} \frac{\psi(\alpha+\gamma) - \psi(\gamma)}{\alpha} = \sum_{x \geq 1} \frac{(x-1)!(\gamma-1)!}{x(\gamma+x-1)!}. \tag{11.14}$$

Sibuya's (1979) *digamma and trigamma distributions* have probabilities that are proportional to the terms in (11.13) and (11.14), respectively. For the digamma distribution therefore

$$\Pr[X = x] = \frac{1}{\psi(\alpha+\gamma) - \psi(\gamma)} \times \frac{(\alpha+x-1)!(\alpha+\gamma-1)!}{x(\alpha-1)!(\alpha+\gamma+x-1)!}, \tag{11.15}$$

and for the trigamma distribution

$$\Pr[X = x] = \frac{1}{\psi'(\gamma)} \times \frac{(x-1)!(\gamma-1)!}{x(\gamma+x-1)!} \qquad (11.16)$$

(in both cases the support is $x = 1, 2, \ldots$). When $\gamma = 1$, the trigamma distribution becomes a zeta distribution (see Section 20 below).

Sibuya obtained these distributions as limiting cases of the zero-truncated hypergeometric Type IV (beta-negative binomial) distribution (Chapter 6, Section 2.3), with pgf

$$\frac{{}_2F_1[\alpha, \beta; \alpha+\beta+\gamma; z] - 1}{{}_2F_1[\alpha, \beta; \alpha+\beta+\gamma; 1] - 1} \qquad (11.17)$$

and probabilities

$$\Pr[X=x] = \frac{(\alpha+x-1)!(\beta+x-1)!(\alpha+\beta+\gamma-1)!}{x!(\alpha-1)!(\beta-1)!(\alpha+\beta+\gamma+x-1)!} \bigg/ \left\{\frac{(\alpha+\beta+\gamma-1)!(\gamma-1)!}{(\alpha+\gamma-1)!(\beta+\gamma-1)!} - 1\right\}, \qquad (11.18)$$

$x = 1, 2, \ldots$. The digamma distribution is the outcome as $\beta \to 0$ with $\alpha > 0$; the trigamma is obtained when $\alpha \to 0$ and $\beta \to 0$.

Bernardo (1976) has written a computer algorithm for the psi (digamma) function which can be used to facilitate calculation of the probabilities for the digamma distribution. Schneider (1978) and Francis (1991) have developed an algorithm for the trigamma function.

4 DISCRETE ADÈS DISTRIBUTION

Perry and Taylor (1985) put forward the *discrete Adès distribution* in order to model counts of individuals per unit of habitat in population ecology. There is much empirical evidence, especially in entomology, to support the view that, in a data set containing several samples of data, the theoretical mean and variance vary from sample to sample in such a way that

$$\mu_2 = \alpha \mu^\beta, \qquad (11.19)$$

that is, $\ln \mu_2 = \ln \alpha + \beta \ln \mu$; this is known as *Taylor's power law*.

Suppose that X has a gamma distribution with density function

$$f(x) = \frac{\lambda^r e^{-\lambda x} x^{r-1}}{\Gamma(r)}, \qquad x \geq 0, \qquad (11.20)$$

and that

$$
Y = \begin{cases} 0 & \text{if } 0 \le x \le 1, \\ (\ln x)^b & \text{if } x \ge 1. \end{cases} \tag{11.21}
$$

Then Y is deemed to have a (continuous) Adès distribution with parameters r, λ, b; sometimes $c = 1/b$ has been used instead of b. Perry and Taylor (1985) checked both theoretically and numerically that (11.19) holds to a very good degree of approximation given a family of Adès distributions with constant b and constant coefficient of variation.

The Adès distribution defined in this way is a continuous distribution with an atom of probability at the origin. It does not have closed forms for its mean and variance. Holgate (1989) has obtained asymptotic formulas for the central moments and hence has been able to re-examine the relationship between the variance and the mean. Holgate has also looked at estimation by the method of moments.

The discrete Adès distribution of Perry and Taylor (1985, 1988) is the distribution of W where

$$
\Pr[W = 0] = \Pr[0 \le Y < 0.5],
$$
$$
\Pr[W = x] = \Pr[x - 0.5 \le Y < x + 0.5], \qquad x = 1, 2, \dots. \tag{11.22}
$$

Perry and Taylor (1988) fitted this discrete Adès distribution to twenty-two entomological data sets made up altogether of 215 samples, with results that they found very encouraging. To do this, they used the Maximum Likelihood Program (MLP) of Ross (1980).

Kemp (1987b) pointed out that constrained (one-parameter) families of the negative binomial, Neyman Type A, Pólya-Aeppli, and Poisson-with-zeros distributions can be constructed so that they obey Taylor's power law. There has been controversy between Perry and Taylor (1988) and Kemp (1988a) as to whether these constrained distributions are able to exhibit the same flexibility of form as the discrete Adès distribution.

5   DISCRETE STUDENT'S *t*-DISTRIBUTION

Type IV of Ord's family of distributions (Chapter 2, Section 3.2) satisfies the recurrence relationship

$$
Pr[X = x] = P_x = \left\{ \frac{(x + a)^2 + d^2}{(x + k + a)^2 + b^2} \right\} P_{x-1}, \qquad k > 0, \tag{11.23}
$$

for all integer values of x (positive, negative or zero). A special case of this when $d = b$ is the Type VII (discrete Student's t-distribution) for which

$$P_x = \alpha_k \left[\prod_{j=1}^{k} \{(j + x + a)^2 + b^2\} \right]^{-1}, \qquad -\infty < x < \infty, \qquad (11.24)$$

where $0 \le a \le 1$, $0 < b < \infty$, k is a nonnegative integer, and α_k is a normalizing constant.

Ord (1967c, 1968b) has shown that

$$\alpha_k = b \prod_{j=1}^{k} (j^2 + 4b^2) \left\{ \binom{2k}{k} w(a,b) \right\}^{-1}, \qquad (11.25)$$

where $w(a,b)$ can be expressed in terms of a and b only, using imaginary parts of the digamma function $\psi(\cdot)$ with a complex argument. For $b = 1$, $w(a,b)$ varies from 3.12988 for $a = 0.5$ to 3.15334 for $a = 0$ or 1; for $b \ge 2$, Ord found $|w(a,b) - \pi| < 0.0001$.

Ord commented that the *discrete Student's t-distribution* has interest because of its curious property that all finite odd moments about the mean are zero, although the distribution is symmetric only when $a = 0, 0.5$ or 1. Also

$$\mu = \mu_1' = -(k/2 + a)$$

$$\mu_2 = \frac{k^2/4 + b^2}{2k - 1}$$

$$\mu_4 = \frac{k^3(k-4)/16 + (3k-2)kb^2/2 + 3b^4}{(2k-1)(2k-3)}. \qquad (11.26)$$

For every k, there exist moments up to order $2k$. The asymmetry has the unusual form

$$P_0 > P_1 > P_{-1} > P_2 > P_{-2} > \cdots.$$

No natural interpretation of the distribution has been found.

6 GEETA DISTRIBUTION

The *Geeta distribution* is both a Lagrangian-type distribution (Chapter 2, Section 5.2) and a modified power series distribution (Chapter 2, Section 2.2) on the positive integers $1, 2, 3, \ldots$, with pmf

$$\Pr[X = x] = \frac{1}{\beta x - 1} \binom{\beta x - 1}{x} \theta^{x-1} (1 - \theta)^{\beta x - x}, \qquad (11.27)$$

and zero otherwise, where $0 < \theta < 1$ and $1 < \beta < \theta^{-1}$; see Consul (1990a). The mean is infinite if $\beta\theta \geq 1$. As $\beta \to 1$, we find that $\Pr[X = 1] \to 1$. The mean and variance of the distribution are

$$\mu = \frac{1 - \theta}{1 - \beta\theta}$$

$$\mu_2 = \frac{(\beta - 1)\theta(1 - \theta)}{(1 - \beta\theta)^3}. \tag{11.28}$$

By reparameterizing with $\theta = (1 - \mu)/(1 - \mu\beta)$, the distribution can be shown to be a location-parameter distribution in the sense introduced into the literature by Consul (1990c). The pmf becomes

$$\Pr[X = x] = \frac{(\beta x - 2)!}{x!(\beta x - x - 1)!} \left(\frac{\mu - 1}{\mu\beta - 1}\right)^{x-1} \left(\frac{\mu\beta - \mu}{\mu\beta - 1}\right)^{\beta x - x},$$

$$x = 1, 2, 3, \ldots \tag{11.29}$$

where the mean is $\mu > 1$, and $\beta > 1$. The variance is $\mu_2 = \mu(\mu - 1)(\beta\mu - 1)/(\beta - 1)$; this clearly increases with μ for fixed β, and decreases to $\mu^2(\mu - 1)$ as β increases for fixed μ. The general recurrence formulas for the moments of Lagrangian distributions (Chapter 2, Section 5.2) simplify and enable the higher moments to be obtained recursively. A numerical approach has indicated that these distributions are reversed J-shaped with a mode at $x = 1$ and with the length and weight of the tail dependent on the values of β and θ.

Estimation using (1) moments, (2) sample mean and first frequency, (3) ML, and (4) MVUE were studied by Consul (1990a). Two modes of genesis (a two-urn model and a regenerative stochastic process) are given in Consul (1990b).

7 GEGENBAUER DISTRIBUTION: NEGATIVE BINOMIAL*PSEUDO-NEGATIVE BINOMIAL CONVOLUTION

The *Gegenbauer distribution* with pgf of the form

$$G(z) = (1 - \xi - \eta)^\ell (1 - \xi z - \eta z^2)^{-\ell} \tag{11.30}$$

was so named by Plunkett and Jain (1975), who derived it as a gamma-mixed Hermite distribution and expressed the probabilities in terms of Gegenbauer polynomials; for Gegenbauer polynomials, see Rainville (1960, Ch. 17).

The distribution has a long history in the theory of stochastic processes. Factorizing the quadratic expressions in the pgf gives

$$G(z) = (1 - \xi - \eta)^\ell (1 - \xi z - \eta z^2)^{-\ell}$$

$$= \left[\frac{(1-a)(1+b)}{(1-az)(1+bz)} \right]^\ell, \tag{11.31}$$

where $a - b = \xi$, $ab = \eta$ and $0 < b < a < 1$; this "factorized" nota-tion makes the distribution easy to handle, especially its moment properties. Kemp (1979), unaware of the paper by Plunkett and Jain (1975), pointed out that McKendrick (1926) had obtained the distribution with pgf (11.31) as the outcome of a non-homogeneous birth-and-death process with ℓ initial individuals; see also Irwin (1963). Kemp commented that the distribution is related to the Galton-Watson branching process and to the geometric-with-zeros distribution (Chapter 8, Section 2.2).

Plunkett and Jain's derivation of the distribution mixes a Hermite distri-bution with pgf

$$G(z) = \exp\{\gamma(z-1) + \gamma\rho(z^2 - 1)\}$$

(Chapter 9, Section 4), using a gamma distribution for γ. The outcome has the pgf

$$G(z) = \int_0^\infty e^{\gamma(z-1) + \gamma\rho(z^2-1)} \frac{\exp(-\gamma\delta)\gamma^{(\ell-1)}\delta^\ell \, d\gamma}{\Gamma(\ell)}$$

$$= (1 - \xi - \eta)^\ell (1 - \xi z - \eta z^2)^{-\ell}$$

$$= (1 - \xi - \eta)^\ell \sum_{x=0}^\infty \{G_x^\ell(\xi, \eta) z^x / x!\}, \tag{11.32}$$

where the support is $x = 0, 1, 2, \dots$, $\xi = (1 + \rho + \delta)^{-1}$, $\eta = \rho(1 + \rho + \delta)^{-1}$ and $G_x^\ell(\xi, \eta)$ is a modified Gegenbauer polynomial of order x. The pmf can therefore be stated as

$$\Pr[X = x] = (1 - \xi - \eta)^\ell G_x^\ell(\xi, \eta) / x!$$

$$= (1 - \xi - \eta)^\ell \sum_{j=0}^{[x/2]} \frac{\Gamma(x + \ell - j)\xi^{x-2j}\eta^j}{j!\Gamma(x + 1 - 2j)\Gamma(\ell)}. \tag{11.33}$$

This is a three-parameter distribution with parameters $\delta, \rho, \ell > 0$, that is, with $\xi, \eta > 0$ and $\xi + \eta < 1$ (since $\delta > 0$). Plunkett and Jain obtained certain of the distribution's probability and moment properties, and some of its limiting forms. They also fitted the distribution to a set of accident data using the method of moments.

Kemp (1979) studied the distribution with pgf in the form (11.31) as an instance of the distributions with pgf's of the form

$$G(z) = \frac{(1 - Q_1 z)^{U_1} (1 - Q_2 z)^{U_2}}{(1 - Q_1)^{U_1} (1 - Q_2)^{U_2}} \tag{11.34}$$

that can be obtained with various (specified) restrictions on Q_1, Q_2, U_1, U_2 by convoluting binomial and pseudo-binomial variables; see Chapter 3, Section 12.5, for probability and moment properties.

Borah (1984) has also studied the probability and moment properties of (11.30) and has used estimation via the first two sample moments and the ratio of the first two sample frequencies (i.e., $\bar{x}$, s^2, and f_1/f_0).

Medhi and Borah's (1984) four-parameter *generalized Gegenbauer distribution* has the pgf

$$G(z) = (1 - \xi - \eta)^{\ell} (1 - \xi z - \eta z^m)^{-\ell}; \tag{11.35}$$

they have studied moment estimation and estimation via $\bar{x}$, s^2, and f_1/f_0, assuming a known small integer value for m.

Other generalizations of (11.30) are those of Patil and Raghunandanan (1990), who examined, for example, a gamma mixture of stuttering Poisson distributions (Chapter 9, Section 3). This gives the pgf

$$G(z) = \prod_{j=1}^{k} \left(\frac{a_j}{1 + a_j - z^j} \right), \tag{11.36}$$

where $a_j > 0$, $j = 1, 2, \ldots, k$.

Kemp (1992c) has derived (11.31) by a quite different field observation model in which entities occur either as singlets or as doublets and become aware of an observer according to a nonhomogeneous stochastic process; visibility is then determined by Rao damage processes (see Chapter 9, Section 2, for Rao's damage process).

When $\ell = 1$, various formulas simplify very considerably; the case $\ell = 1$ corresponds to an exponential mixing distribution in Plunkett and Jain's mixed Hermite model, and to a nonhomogeneous stochastic awareness process leading to a geometric awareness distribution in Kemp's (1992c) model. Kemp studied various forms of estimation for this special case, including ML estimation.

8 GRAM-CHARLIER TYPE B DISTRIBUTIONS

Helpful approximations to discrete distributions can occasionally be obtained by using distributions formed from the first few terms of a Gram-Charlier Type B expansion. (The Type A expansions, applicable to continuous distributions, are of considerably wider use; see Chapter 12.) The basic idea is to express the pmf, $\Pr[X = x] = p_x$, as a linear series in the backward differences (with respect to x) of the Poisson probabilities

$$\omega_x = \frac{e^{-\theta}\theta^x}{x!}, \qquad x \geq 0,$$

$$= 0, \qquad\qquad x < 0. \tag{11.37}$$

Thus

$$p_x = \sum_{i=0}^{\infty} a_i \nabla^i \omega_x, \tag{11.38}$$

where $\nabla \omega_x = \omega_x - \omega_{x-1}$. Note that if (11.38) is summed from $x = c$ to $x = \infty$, then

$$\sum_{x=c}^{\infty} p_x = \sum_{i=0}^{\infty} a_i \nabla^i \left(\sum_{x=c}^{\infty} \omega_x \right) \tag{11.39}$$

(i.e., the cumulative sum of the probabilities p_x is expressed in terms of the corresponding cumulative sums of Poisson probabilities). With the definition (11.37), $\sum_{x=c}^{\infty} \omega_x = 1$ for all $c \leq 0$.

Since the factorial moments are determined from the probability generating function, and conversely (provided that the moments exist), we might expect values of p_x to be determined to an adequate degree of approximation by a sufficient number of sample values of $\mu'_{[1]}, \mu'_{[2]}, \ldots$. In fact if $p_x = 0$ for $x > m$, then

$$p_k = \sum_{x=0}^{m-k} \frac{(-1)^x \mu'_{[k+x]}}{(x!k!)}, \qquad k = 0, 1, 2, \ldots, m, \tag{11.40}$$

and the value of p_k lies between any two successive partial sums obtained by terminating summation at $x = s$ and $x = s + 1$.

Kendall (1943) demonstrated the use of Gram-Charlier Type B series with two, three, and four terms by quoting calculations from Aroian (1937) on the emission of a-particles. The representation of quasi-binomial, Lagrangian Poisson, and quasi-negative binomial distributions by means of modified Gram-Charlier expansions has been studied by Berg (1985).

9 "INTERRUPTED" DISTRIBUTIONS

Interrupted distributions arise under circumstances in which some values of a random variable are not observed because of their juxtaposition in time or space to some other value. Consider, for example, a Geiger-Muller counter recording the arrival of radioactive particles. Suppose that there is a constant resolving "dead" time D during which the counter is unable to record any arriving particles. The distribution of the number of arrivals *recorded* in a fixed period of time of length τ will not be the same as that of the actual number of arrivals. Assume that the actual arrivals occur randomly in time at a rate λ (so that the actual number of arrivals in time period τ has a Poisson distribution with mean $\lambda\tau$), that the times T between successive arrivals are independent, and that each interarrival time has the same density function

$$f(t) = \lambda e^{-\lambda t}, \quad t \geq 0. \tag{11.41}$$

This is the case examined in connection with telephone calls by Erlang (1918), and later by Giltay (1943), Feller (1948), and Feix (1955).

An especially simple case is that in which no particle has been recorded in time D preceding the beginning of a time period of length τ. A full analysis of this case appeared in the first edition of this volume. Given that there are $K \geq 1$ actual arrivals (the probability of this is $e^{-\lambda\tau}(\lambda\tau)^K/K!$), then the probability $p_{k,K}$ that exactly k will be recorded was shown to be

$$p_{k,K} = \binom{K}{k} \left[\left(\frac{kD}{\tau} \right)^{K-k} g^k + k \int_g^{1-(k-1)D/\tau} y^{k-1}(1-y)^{K-k} dy \right], \tag{11.42}$$

where

$$g = \begin{cases} 0 & \text{if } k \geq \tau/D, \\ 1 - kD/\tau & \text{if } k \leq \tau/D. \end{cases}$$

The overall probability of recording exactly k (≥ 1) arrivals is

$$p_k = e^{-\lambda\tau} \sum_{K=k}^{\infty} \frac{(\lambda\tau)^K}{K!} p_{k,K}. \tag{11.43}$$

Oliver (1961) has used a model leading to the same distribution to represent traffic counts over a fixed interval of time τ. In this model the distribution of times between successive arrivals of vehicles is assumed to have the density function

$$f(t) = \begin{cases} 0 & \text{if } t < D, \\ \lambda \exp[-\lambda(t-D)] & \text{if } t \geq D. \end{cases}$$

The number of arrivals in time τ then has the distribution (11.43) if the time period starts later than time D after a vehicle passes. Oliver also considered the situation in which the interval starts immediately after the arrival of a vehicle and so obtained a different distribution, which can, however, be obtained from (11.43) by replacing τ by $\tau - D$.

A model of a similar kind has been used by Singh (1964) to represent the distribution of numbers of conceptions, over a fixed interval of time τ, to married couples in a specified population. Singh assumed that for a time D following a conception no further conception is possible. Instead of representing the time between conceptions as a continuous random variable, however, he divided τ into m subintervals of length (τ/m). He supposed that, provided that at least $(mD\tau^{-1} - 1)$ subintervals have elapsed since a subinterval in which there was a conception, the probability of conception in a subinterval is p. ($mD\tau^{-1}$ is taken to be an integer.) The distribution that Singh obtained tends to distribution (11.43) as m increases and p decreases, with $mp = \lambda\tau$. For further work on this topic, see Singh (1968) and Singh, Bhattacharya and Yadava (1974).

10 LOST-GAMES DISTRIBUTIONS

The distribution of the number of games lost by the ruined gambler in the classical gambler's ruin problem has the pgf

$$G(z) = z^a \left\{ \frac{1 - (1 - 4pqz)^{1/2}}{2qz} \right\}^a$$

$$= z^a p^a {}_2F_1[a/2, (a+1)/2; a+1; 4pqz], \qquad (11.44)$$

$x = a, a+1, \ldots$, where a is the gambler's initial capital (a positive integer) and $q = 1 - p$ is the probability that he wins an individual game. This is also the pgf for the distribution of the number of customers served during a busy period of an M/M/1 queue (Poissonian arrivals, exponential service times, and one server), starting with a customers; see Takács (1955) and Haight (1961a). Moreover it is the pgf for the total size of an epidemic; see McKendrick (1926). The relationships between the random walks corresponding to these three stochastic processes were studied by Kemp and Kemp (1968).

A more general form of the *lost-games distribution* has the pgf

$$K(z|p, j, a) = z^j \left\{ \frac{1 - (1 - 4pqz)^{1/2}}{2qz} \right\}^a$$

$$= z^j p^a {}_2F_1[a/2, (a+1)/2; a+1; 4pqz], \qquad (11.45)$$

where $x = j, j+1, \ldots$, and the parameter constraints are $0.5 < p < 1$, $q = 1-p$, $0 < a$, and $j \geq 0$, a nonnegative integer. From (11.45) the pmf is

$$p_x = \Pr[X = x|p, j, a] = \frac{(2x + a - 2j - 1)! a p^{a+x-j} q^{x-j}}{(x + a - j)!(x - j)!}, \qquad x = j, j+1, \ldots,$$

$$(11.46)$$

where $y!$ is taken to mean $\Gamma(y+1)$ when y is not an integer. The probabilities can be computed very easily by means of the recurrence relation

$$(x-j)(x+a-j)p_x = (2x+a-2j-1)(2x+a-2j-2)pqp_{x-1}, \quad x>j, \quad (11.47)$$

with $p_x = 0$ for $x < j$, $p_j = p^a$. The unimodality of the distribution follows from (11.47).

The mean, variance, and third corrected moment are

$$\mu = j + \frac{aq}{(p-q)}, \quad \mu_2 = \frac{apq}{(p-q)^3}, \quad \mu_3 = \frac{apq(1+2pq)}{(p-q)^5}; \quad (11.48)$$

see Haight (1961a) and Kemp and Kemp (1968).

The distribution is a member of the class of modified power series distributions (Chapter 2, Section 2.2); see Gupta (1984). It is a (possibly shifted) generalized hypergeometric probability distribution (Chapter 2, Section 4.1) of the kind considered in Chapter 6, Section 11; see Kemp (1968b). Also it belongs to the Gould series family (Chapter 2, Section 5.3); see Charalambides (1986a). When $j = 0$, the distribution is infinitely divisible and hence is a Poisson-stopped sum distribution (Chapter 9, Section 3); the pgf for the distribution that is Poisson-stopped (the cluster-size distribution) is given in Kemp and Kemp (1969a).

Let $K(p,j,a)$ denote the lost-games distribution with the parameters p, j and a as in (11.45). Then if $X_1 \sim K(p,m,a_1)$ and $X_2 \sim K(p,n,a_2)$, and if X_1 and X_2 are independent, it follows that $X_1 + X_2 \sim K(p,m+n,a_1+a_2)$.

There are a number of models for the general lost-games distribution. Otter (1949) investigated a multiplicative process characterized by the equation $G(z,w) \equiv zf(w) - w = 0$ where $w = P(z)$ is the pgf of interest; this is the transformation that defines a Lagrangian distribution (see Chapter 2, Sections 5.1 and 5.2). Example 1 in Otter's paper, with $f(w) = (p+qz)^2$, yields the distribution with pgf $K(z|p,2,1)$. The relationships between Otter's branching model, the epidemiological model of Neyman and Scott, (1964) and the lost-games distribution were discussed by Kemp and Kemp (1969a), who also studied clustering models where the distribution of clusters is (1) binomial, (2) negative binomial, and (3) Poisson. Kemp and Kemp (1971) also investigated mixing processes; these included mixed negative binomial, mixed Poisson, and mixed confluent hypergeometric models.

The name *inverse binomial distribution* was used by Yanagimoto (1989) for the lost-games distribution with pgf $K(z|p,0,a)$ because its cumulant generating function can be regarded as the inverse function of the cgf of the binomial distribution. If the parameter space for this form of the distribution is extended to include $p < 0.5$, then there is a positive probability that the variate value is infinite. Yanagimoto used this probability to estimate the proportion of discharged patients who can be expected to stay completely free from some disease; in the context of the $M/M/1$ queue it is the probability that the busy period will never end.

Kemp and Kemp (1992) have developed a state-dependent equilibrium birth-and-death process for the distribution and related this to the size of naturally occurring groups of individuals; they used ML estimation to fit the distribution to data on group sizes.

11 NAOR'S DISTRIBUTION

Naor (1957) has studied the following urn model: Suppose that an urn contains n balls of which one is red and the remainder are white. Sampling with replacement of a white ball (if drawn) by a red ball continues until a red ball is drawn. Let Y be the requisite number of draws. Then

$$\Pr[Y = y] = \frac{(n-1)\dots(n-y+1)y}{n^y} = \frac{(n-1)!y}{(n-y)!n^y} \tag{11.49}$$

(after $n-1$ draws the urn contains only red balls and so no more than n draws are required). This problem is closely related to problem 6.24 in Wilks (1962, pp. 152–153). Consider a collection of n objects labeled $1, 2, \dots, n$. An object is chosen and returned, another is chosen and returned, and so on, until an object is chosen that has already been chosen before. Sampling then ceases. The number of draws that are required is the rv $Y + 1$, with Y as above; at least 2 draws are needed, and at most $n + 1$.

The pmf for the number of draws $X = n - Y$ not required is

$$\Pr[X = x] = \frac{(n-1)!(n-x)}{x!n^{n-x}}, \qquad x = 0, 1, \dots, n-1. \tag{11.50}$$

Naor (1957) showed that the pgf of X is

$$G(z) = \frac{n!}{n^n}\left\{(1-z)\sum_{j=0}^{n-1}\frac{(nz)^j}{j!} + \frac{(nz)^n}{n!}\right\}, \tag{11.51}$$

and found that when n is large

$$\mu \doteq n + \frac{1}{3} - \sqrt{\frac{n\pi}{2}}$$

$$\mu_2' \doteq n^2 + \frac{8n}{3} + \frac{1}{3} - (2n+1)\sqrt{\frac{n\pi}{2}}. \tag{11.52}$$

Kemp (1968b) showed that the pgf of X can be rewritten as

$$G(z) = \frac{{}_1F_1[-n+1; -n; nz]}{{}_1F_1[-n+1; -n; n]} \tag{11.53}$$

and that the corrected moments satisfy the recurrence relation

$$\mu_{r+2} + (2\mu - n - 1)\mu_{r+1} + (\mu - n - 1)\mu\mu_r$$

$$= r \sum_{j=0}^{r} \binom{r}{j} [\mu_{r+1-j} + (\mu - n + 1)\mu_{r-j}]; \qquad (11.54)$$

hence $$\mu_2 = n + 1 - \mu.$$

Naor's interest in the distribution arose from a machine-minding problem with n repairmen, Naor (1956, 1957); his solution of the machine-minding problem involved the rv W that is a convolution of X with a Poisson rv with mean $n\xi$. He showed that W has mean and variance

$$E[W] \approx n\xi + n + \frac{1}{3} - \sqrt{\frac{n\pi}{2}},$$

$$\mathrm{Var}(W) \approx n\xi + \frac{n(4 - \pi)}{2} + \frac{2}{9} - \frac{\sqrt{2n\pi}}{6}, \qquad (11.55)$$

for n sufficiently large, and hence derived a normal approximation.

12 PARTIAL-SUMS DISTRIBUTIONS

Consider a "parent" distribution with support $0, 1, 2, \ldots$, and probabilities $p_0, p_1, p_2, \ldots$, such that $\sum_{j=0}^{\infty} p_j = 1$. From this parent a new distribution can be derived with probabilities proportional to the complement of the parent cumulative distribution function. This distribution has pmf

$$\Pr[X = x] = (\mu_1'^*)^{-1} \sum_{j=x+1}^{\infty} p_j, \qquad x = 0, 1, \ldots. \qquad (11.56)$$

(The rth moment about zero and the rth factorial moment of the parent distribution are denoted by $\mu_r'^*$ and $\mu_{[r]}'^*$, respectively.)

The moment generating function is

$$(\mu_1'^*)^{-1}(1 - e^t)^{-1}(1 - \phi^*(t)), \qquad (11.57)$$

where

$$\phi^*(t) = \sum_{j=0}^{\infty} p_j e^{jt}. \qquad (11.58)$$

The r-th moment about zero is

$$\mu_r' = (\mu_1'^*)^{-1} \sum_{j=1}^{\infty} \left\{ p_j \sum_{k=0}^{j-1} k^r \right\}$$

$$= (\mu_1'^*)^{-1} \sum_{i=1}^{\infty} \frac{\Delta^i 0^r \mu_{[i+1]}'^*}{(i+1)!}, \qquad r = 1, 2, \ldots. \qquad (11.59)$$

The distribution that is obtained when the parent distribution is Poisson, with $p_j = e^{-\theta}(\theta^j/j!)$, has been called *Poisson's exponential binomial limit*. For this distribution $\mu_j'^* = \theta^j$ and

$$\mu = \frac{\theta}{2},$$

$$\mu_2 = \frac{\theta}{2} + \frac{\theta^2}{12},$$

$$\mu_3 = \frac{\theta}{2} + \frac{\theta^2}{4},$$

$$\mu_4 = \frac{\theta}{2} + \frac{4\theta^2}{3} + \frac{\theta^3}{4} + \frac{\theta^4}{80}. \qquad (11.60)$$

As $\theta \to \infty$, the moment ratios $\sqrt{\beta_1}$ and β_2 tend to 0 and 1.8, respectively.

Gold (1957) and Gerstenkorn (1962), however, defined *Poisson's exponential binomial limit distribution* as the special case $q = 1$ of the more general distribution with pmf

$$\Pr[X = x] = \frac{q^x e^{-q\omega}}{c} \sum_{j=x}^{\infty} \frac{\omega^j}{j!}, \qquad x = 0, 1, \ldots. \qquad (11.61)$$

where c is a normalizing constant (see Chapter 4, Section 12.5).

Bissinger (1965) has constructed systems of distributions derived from a parent distribution by another type of summation process. The *Bissinger system distributions* are defined by

$$\Pr[X = x] = (1 - p_0)^{-1} \sum_{j=x+1}^{\infty} j^{-1} p_j, \qquad x = 0, 1, \ldots. \qquad (11.62)$$

It is easily verified that $\sum_{x=0}^{\infty} \Pr[X = x] = 1$, so (11.62) defines a proper distribution. Bissinger gave this class of distributions the name *STER distributions*, from the phrase "Sums of Truncated forms of the Expected value of the Reciprocal" (of a variable having the parent distribution). These distributions arose in connection with an inventory decision problem, with the distribution of the number of demands as the parent distribution. From (11.62),

$$\Pr[X = x - 1] - \Pr[X = x] = (1 - p_0)^{-1} x^{-1} p_x \geq 0, \qquad x \geq 1. \qquad (11.63)$$

Hence the successive values of $\Pr[X = x]$ are nonincreasing as x increases.

Bissinger obtained the following relationship with the moments $(\mu_r'^*)$ about zero of the parent distribution:

$$\mu_r'^* = (1 - p_0) \sum_{j=0}^{r} \binom{r+1}{j} \mu_j'. \qquad (11.64)$$

In particular

$$\mu_1' = [(1 - p_0)^{-1} \mu_1'^* - 1]/2$$
$$\mu_2' = [(1 - p_0)^{-1} (2\mu_2'^* - 3\mu_1'^*) + 1]/6. \qquad (11.65)$$

Patil and Joshi (1968) gave the following results. Let $G^*(z)$, μ^* and μ_2^* be the pgf, mean, and variance, respectively, of the parent distribution. Then for the corresponding STER distribution,

$$G(z) = \frac{\theta}{(1 - z)} \int_{z}^{1} \frac{[G^*(u) - p_0] du}{u}, \qquad (11.66)$$

$$\mu = \frac{\theta \mu^* - 1}{2} \quad \text{and} \quad \mu_2 = \frac{\theta \{\mu_2^* + (\mu^*)^2\} - 1}{3} - \mu(1 + \mu), \quad (11.67)$$

where $\theta = (1 - p_0)^{-1}$. For work on STER distributions for hypergeometric-type demand distributions, see Kemp and Kemp (1969b) and Chapter 2, Section 4.2.

We further note that

$$\Pr[X = x] = \sum_{j=x}^{\infty} (j + 1)^{-1} p_j, \qquad x = 0, 1, 2, \ldots, \qquad (11.68)$$

defines a proper distribution, and that Haight (1961b) listed the distribution with pmf

$$\Pr[X = x] = \left\{ \sum_{j=x+k}^{\infty} p_j / (j - k + 1) \right\} \bigg/ \sum_{j=k}^{\infty} p_j, \qquad (11.69)$$

where $x = 0, 1, \ldots, k = 1, 2, \ldots$.

Another distribution that is defined in terms of partial sums is the *discrete Weibull distribution*; see Nakagawa and Osaki (1975). Here

$$\Pr[X \geq x] = (q)^{x^\beta}, \qquad x = 0, 1, 2, \ldots, \tag{11.70}$$

where $\beta > 0$ and $0 < q < 1$; this gives

$$\Pr[X = x] = (q)^{x^\beta} - (q)^{(x+1)^\beta}. \tag{11.71}$$

The failure rate is

$$r_x = 1 - (q)^{(x+1)^\beta}; \tag{11.72}$$

the distribution is therefore DFR for $0 < \beta < 1$ and IFR for $\beta > 1$. Nakagawa and Osaki noted that when $\beta = 1$ the distribution becomes a geometric distribution. They further commented that it is very difficult to obtain useful expressions for the mean and other characteristics of the distribution. Their calculations suggested that it can be J-shaped or unimodal. They considered the distribution to have useful potential as a failure-time distribution for failure data measured in discrete time (e.g., cycles, blows, shocks, and revolutions) and also as an approximation to the continuous Weibull distribution.

The distribution of the rv X with pmf

$$\Pr[X = x] = \sum_{y=x}^{\infty} p_y \Bigg/ \sum_{y=0}^{\infty} y p_y, \qquad x = 0, 1, \ldots, \tag{11.73}$$

is called the *renewal distribution* for the rv Y with pmf p_y. The mutual characterization of Y and X has been discussed by Nair and Hitha (1989), Hitha and Nair (1989), and Kotz and Johnson (1991).

13   QUEUEING THEORY DISTRIBUTIONS

We are not concerned here with the more general aspects of the theory of queues but only with certain distributions arising naturally from simple models of queueing situations. There is a very extensive literature concerning stochastic processes, with a great amount devoted to queueing theory. For lack of space we mention here only the very accessible introduction to applied stochastic processes by Bhat (1984) and the two substantial volumes by Kleinrock (1975, 1976). Chapter 11 of Bhat (1984) contains a useful short bibliography; Bhat (1978) gave a comprehensive bibliography.

Consider first a stochastic process in which customers arrive, wait in a single queue before being served, take a finite length of time to be served, and then depart. Let $p_n(t)$ denote the probability that at time t the queue size is n (including the customer being served). Suppose that the customers' arrival rate is α_n and that their departure rate is β_n, where α_n and β_n are functions of the queue size n. Then

$$\frac{dp_n(t)}{dt} = \alpha_{n-1}p_{n-1}(t) - (\alpha_n + \beta_n)p_n(t) + \beta_{n+1}p_{n+1}(t). \qquad (11.74)$$

When the time-dependent instantaneous probability distribution $\{p_0(t), p_1(t), \ldots\}$ tends to an equilibrium distribution as $t \to \infty$, with probabilities $\{p_0, p_1, \ldots\}$ that do not depend on the initial probability distribution $\{p_0(0), p_1(0), \ldots\}$, we have

$$0 = -\alpha_0 p_0 + \beta_1 p_1,$$
$$0 = \alpha_{n-1}p_{n-1} - (\alpha_n + \beta_n)p_n + \beta_{n+1}p_{n+1}, \qquad n \geq 1. \qquad (11.75)$$

These equations have the solution

$$p_n = \frac{\alpha_0 \alpha_1 \cdots \alpha_{n-1}}{\beta_1 \beta_2 \cdots \beta_n} p_0 = \frac{\alpha_0 \alpha_1 \cdots \alpha_{n-1}}{\beta_1 \beta_2 \cdots \beta_n} \left/ \left\{ 1 + \sum_{n=1}^{\infty} \frac{\alpha_0 \alpha_1 \cdots \alpha_{n-1}}{\beta_1 \beta_2 \cdots \beta_n} \right\} \right. . \qquad (11.76)$$

This queueing model is closely related to the generalized birth-and-death process with birth rate α_n and death rate β_n, where n is the current population size and the initial population size is unity. If

$$G(z) = \left\{ 1 + \sum_{n=1}^{\infty} \frac{\alpha_0 \alpha_1 \cdots \alpha_{n-1}}{\beta_1 \beta_2 \cdots \beta_n} z^n \right\} \left/ \left\{ 1 + \sum_{n=1}^{\infty} \frac{\alpha_0 \alpha_1 \cdots \alpha_{n-1}}{\beta_1 \beta_2 \cdots \beta_n} \right\} \right. \qquad (11.77)$$

is the pgf for the equilibrium population-size distribution given one initial individual at time $t = 0$, then $[G(z)]^M$ is the pgf for the corresponding equilibrium population-size distribution given M, instead of one, initial individuals. Many of the distributions considered elsewhere in this volume, such as the Poisson, negative binomial, Yule, and lost-games distributions, have modes of genesis of this kind.

Exact results in the nonequilibrium theory of queues and birth-and-death processes are remarkably intransigent to obtain, even for the simplest of situations. Consider, for example, the very simple M/M/1 queueing system with random arrivals (that is, constant arrival rate α), one server, first-come-first-served queueing discipline, and exponential service times (that is, constant service rate β). Then $\rho = \alpha/\beta$ is said to be the traffic intensity. We have

$$\frac{dp_0(t)}{dt} = -\alpha p_0(t) + \beta p_1(t), \qquad (11.78)$$

$$\frac{dp_n(t)}{dt} = \alpha p_{n-1}(t) - (\alpha + \beta)p_n(t) + \beta p_{n+1}(t), \qquad n \geq 1. \qquad (11.79)$$

The difficulty in solving these equations is that (11.78) is different from the general form (11.79). Nevertheless, much effort has been expended, and several methods of solution have been devised, notably by Bailey (1954), Champernowne (1956), Cox and Smith (1961), and Cohen (1982); for more recent work on the transient M/M/1 queue, see Parthasarathy (1987), Abate and Whitt (1988), and Syski (1988).

The pmf satisfying (11.78) and (11.79) can be stated in several ways; one of the simpler statements is as an infinite series of modified Bessel functions of the first kind (Chapter 1, Section A5):

$$p_n(t) = \rho^{(n-M)/2} e^{-(1+\rho)t} \{ I_{n-M}(2t\sqrt{\rho}) + \rho^{1/2} I_{M+n+1}(2t\sqrt{\rho})$$

$$+ (\rho - 1) \sum_{i=1}^{\infty} \rho^{(i-1)/2} I_{M+n+i+1}(2t\sqrt{\rho}) \}, \qquad (11.80)$$

where M is the initial queue size.

Another type of discrete distribution arising in queueing theory is the distribution of the number of customers served during a busy period for a queue in equilibrium. Consider, first, the distribution of X, the total number of customers served before a queue vanishes, given a single server queue with random arrival of customers at constant rate α and a constant length of time β occupied in serving each customer (i.e., the M/D/1 queue). The assumptions are that the probability of the arrival of a customer during the interval $(t, t+(\Delta t))$ is $\alpha(\Delta t) + o(\Delta t)$ and that the probability of the arrival of two (or more) customers in this period is $o(\Delta t)$. If there are initially M customers in the queue, then

$$\Pr[X = x] = \frac{M}{(x - M)!} x^{x-M-1} (\alpha\beta)^{x-M} e^{-\alpha\beta x}, \qquad (11.81)$$

where $x = M, M+1, \ldots$. This is the Borel-Tanner distribution (see Chapter 9, Section 11). For (11.81) to represent a proper distribution, it is necessary to have $\alpha\beta \leq 1$; if $\alpha\beta > 1$, then $\sum_{j=M}^{\infty} \Pr[X = j] < 1$. The distribution was obtained by Borel (1942) for the case $M = 1$, and for general values of M by Tanner (1953).

The corresponding busy-period discrete distribution for the M/M/1 queue described earlier in this section is the lost-games distribution; see Haight (1961a) and Section 10 of this chapter.

14 RECORD-VALUE DISTRIBUTIONS

Chandler (1952) studied the distribution of the serial numbers of record values in an infinite time-series comprising independent sample values y_1, y_2, $y_3, \ldots$ from a fixed continuous distribution. He was concerned with those members of the series that are smaller than all preceding members; he called these lower record values. For example, if the observed sequence of values is 2.1, 3.6, 1.9, 1.7, 1.4, 2.5, 4.3, etc., then the first lower record value is the first observation (2.1), the second record value is the third observation (1.9), the third record value is the fourth observation (1.7), and so on. Similarly, the first, second, third, etc., higher record values are the first, second, seventh, etc., observations. (By convention the first observation is regarded as both a low and a high record.)

Let X_r be the rth lower record value and let U_r be its serial number in the sequence of observations. Chandler obtained the distribution of U_r and also the distribution of $U_r - U_{r-1}$ (the number of observations between the $(r-1)$th and the rth lower record). Clearly the same distributions hold if high records are being considered. Chandler showed that

$$\Pr[U_r = j] = j^{-1} K_{r-2}(j-1), \qquad j = r, r+1, \ldots, \; r \geq 2, \tag{11.82}$$

where $K_0(y) = y^{-1}$ when $y \geq 1$, and $K_{r+1}(y) = y^{-1} \sum_{i=r+1}^{y-1} K_r(i)$ when $y \geq r+2$. He also showed that

$$\Pr[U_r - U_{r-1} \geq t] = M_{r-2}(t), \qquad t = 1, 2, 3, \ldots, \; r \geq 2, \tag{11.83}$$

where $M_0(y) = y^{-1}$ when $y \geq 1$, and $M_{r+1}(y) = y^{-1} \sum_{i=1}^{y} M_r(i)$ when $y \geq 1$.

Both distributions have infinite means; also under very general conditions they are independent of the distribution of the sample observations. Chandler (1952) provided tables from which the probabilities (11.82) and (11.83) can be obtained.

Foster and Stuart (1954) showed that for the *total* number of records in a sequence of length N the mean and variance are

$$\mu = 2 \sum_{j=2}^{N} j^{-1} \quad \text{and} \quad \mu_2 = 2 \sum_{j=2}^{N} j^{-1} - 4 \sum_{j=2}^{N} j^{-2}. \tag{11.84}$$

The pgf is $(N!)^{-1} \prod_{j=1}^{N-2}(j + 2z)$. The distribution is asymptotically normal as N increases.

A related distribution is that of the number of *local* records. These are values that are the largest in *some* sequence of k successive values. Thus x_i is a local record if it is the greatest value among $\{x_{i-k+1}, x_{i-k+2}, \ldots, x_i\}$, or among $\{x_{i-k+2}, \ldots, x_{i+1}\}$, or among $\{x_{i-k+3}, \ldots, x_{i+2}\}$, $\ldots$, or among $\{x_i, x_{i+1}, \ldots, x_{i+k-1}\}$. The distribution of the number R of such local maxima in a sequence of n values $x_1, x_2, \ldots, x_n$ is sometimes called a *Morse distribution* [Freimer et al. (1959)] (this name has also been applied to the distribution of $S = n - R$). The name arose from relevance to a method of machine-decoding hand-keyed Morse code, based upon the identification of the largest and the smallest of each successive sequence of six spaces.

Other relevant early references to distributions of this kind are Austin et al. (1957) and David and Barton (1962). Shorrock (1972a,b) and Resnick (1973) have studied record value times. A profound paper on the independence of record processes is that by Goldie and Rogers (1984). Important recent work on the position of records has concentrated on the random walks performed by the elements in an ordered sequence as new elements are inserted; see Blom and Holst (1986) and also Blom, Thorburn, and Vessey (1990).

A good elementary review paper concerning record-value distributions is Glick (1978). Two more recent surveys are Nevzorov (1987) and Nagaraja (1988).

15 SICHEL DISTRIBUTION: POISSON-INVERSE GAUSSIAN DISTRIBUTION

The *Poisson-inverse Gaussian distribution* is a two-parameter mixture of Poisson distributions, obtained by allowing the Poisson parameter λ to have an inverse Gaussian distribution with density function

$$f(\lambda) = \frac{(1-\theta)^{-1/4}\{(2/(\alpha\theta)\}^{-1/2}\lambda^{-3/2}}{2\,K_{1/2}(\alpha\sqrt{1-\theta})} \exp\left[\left(1 - \frac{1}{\theta}\right)\lambda - \frac{\alpha^2\theta}{4\lambda}\right], \quad \lambda > 0,$$

$$(11.85)$$

where $0 < \alpha$, $0 < \theta < 1$, and $K_\nu(\cdot)$ is a modified Bessel function (Chapter 1, Section A 12). Remembering that $K_{-1/2}(y) = K_{1/2}(y) = \sqrt{\pi/(2y)}e^{-y}$, the pmf is

$$\Pr[X = x] = \sqrt{\frac{2\alpha}{\pi}}\,\frac{\exp\{\alpha\sqrt{1-\theta}\}(\alpha\theta/2)^x}{x!}\,K_{x-1/2}(\alpha), \quad x = 0, 1, \ldots.$$

$$(11.86)$$

The distribution was introduced by Holla (1966) as a useful model in studies of repeated accidents and recurrent disease symptoms. It has also been researched by Sankaran (1968) and Sichel (1971). Shaban (1981) looked at limiting cases and approximations. Sichel (1982b) investigated the following parameter estimation methods: sample moments, using the sample mean and variance; use of the mean and zero frequency; and maximum likelihood. For $1.5 < \alpha < 20.0$ and $0.90 \leq \theta$, he found the first two of these methods to be very inefficient, and therefore recommended the use of ML estimation. Willmot (1987b) has discussed the Poisson-inverse Gaussian distribution as an alternative to the negative binomial.

The two-parameter Poisson-inverse Gaussian distribution is a special case of the more general *Sichel distribution*; this is a three-parameter mixture of

Poisson distributions, obtained by allowing the Poisson parameter λ to have the following generalization of an inverse Gaussian distribution [see Sichel [(1974, 1975)]:

$$f(\lambda) = \frac{(1 - \theta)^{\gamma/2}\{(2/(\alpha\theta)\}^{\gamma}\lambda^{\gamma-1}}{2\,K_{\gamma}(\alpha\sqrt{1 - \theta})}\exp\left[\left(1 - \frac{1}{\theta}\right)\lambda - \frac{\alpha^2\theta}{4\lambda}\right], \quad \lambda > 0 \tag{11.87}$$

where $0<\alpha$, $0<\theta<1$, $-\infty < \gamma < \infty$, and $K_\nu(\cdot)$ is a modified Bessel function. An alternative parameterization with $\beta = \alpha\theta/2$ [Atkinson and Yeh (1982)] leads to a simpler expression for the pmf of the Sichel distribution; we have

$$\Pr[X = x] = \frac{(\alpha^2 - 2\alpha\beta)^{\gamma/2}}{\alpha^\gamma\,K_\gamma(\sqrt{\alpha^2 - 2\alpha\beta})}\frac{\beta^x\,K_{\gamma+x}(\alpha)}{x!}, \quad x = 0, 1, \ldots, \tag{11.88}$$

with $0 < \beta < \alpha/2$.

Stein, Zucchini and Juritz (1987) have recommended a further reparameterization with $\xi = \beta(1 - 2\beta/\alpha)^{-1/2}$ and $\omega = (\xi^2 + \alpha^2)^{1/2} - \xi$. This gives

$$\Pr[X = 0] = \frac{(\omega/\alpha)^\gamma K_\gamma(\alpha)}{K_\gamma(\omega)},$$

$$\Pr[X = 1] = \frac{(\xi\omega/\alpha)K_{\gamma+1}(\alpha)}{K_\gamma(\alpha)}\Pr[X = 0],$$

and in general

$$\Pr[X = x] = \frac{(\omega/\alpha)^\gamma}{K_\gamma(\omega)}\frac{(\xi\omega/\alpha)^x\,K_{\gamma+x}(\alpha)}{x!}, \quad x = 0, 1, \ldots. \tag{11.89}$$

Use of the recurrence relationship $K_{n+1}(y) = (2n/y)K_n(y) + K_{n-1}(y)$, together with $K_{-n}(y) = K_n(y)$, yields the useful recurrence formula

$$\Pr[X=x] = \frac{2\beta}{\alpha}\left(\frac{\gamma+x-1}{x}\right)\Pr[X=x-1] + \frac{\beta^2}{x(x-1)}\Pr[X=x-2]$$

$$= \frac{2\xi\omega}{\alpha^2}\left(\frac{\gamma+x-1}{x}\right)\Pr[X=x-1] + \frac{(\xi\omega/\alpha)^2}{x(x-1)}\Pr[X=x-2] \tag{11.90}$$

for $x = 2, 3, \ldots$.

The pgf is
$$G(z) = \sum_{j\geq 0}\frac{\xi^j\,K_{\gamma+j}(\omega)(z - 1)^j}{K_\gamma(\omega)j!} \tag{11.91}$$

$$= \frac{K_\gamma\{\omega\sqrt{1 - 2\beta(z - 1)}\}}{K_\gamma(\omega)\{1 - 2\beta(z - 1)\}^{\gamma/2}}. \tag{11.92}$$

see Willmot (1986). The factorial moments are $\mu'_{[r]} = \xi^r \, K_{\gamma+r}(\omega)/K_\gamma(\omega)$; the mean and variance are therefore

$$
\mu = \frac{\xi \, K_{\gamma+1}(\omega)}{K_\gamma(\omega)};
$$

$$
\mu_2 = \frac{\xi \, K_{\gamma+1}(\omega)}{K_\gamma(\omega)} \left[1 + \xi \left\{ \frac{K_{\gamma+2}(\omega)}{K_{\gamma+1}(\omega)} - \frac{K_{\gamma+1}(\omega)}{K_\gamma(\omega)} \right\} \right]. \tag{11.93}
$$

The three-parameter Sichel distribution becomes the two-parameter Poisson-inverse Gaussian distribution when $\gamma = -1/2$. The above formulas simplify considerably in this special case; for instance the mean and variance simplify to $\mu = \xi$ and $\mu_2 = \xi(1 + \xi/\omega)$. For this special case the parameters ξ and ω can be regarded as mean and shape parameters; we refer the reader to Stein et al. (1987).

An important feature of the three-parameter distribution is its flexibility, and its very long positive tail combined with finite moments of all orders. This makes the distribution particularly suitable for modelling highly skewed data.

The three-parameter form was introduced by Sichel (1971) and studied further by him in Sichel (1973 a, b) in the context of diamondiferous deposits. In Sichel (1974, 1975) he applied the distribution to sentence length and to word frequencies, and in Sichel (1982a) he examined its use for modeling repeat-buying. Ord and Whitmore (1986) have used the two-parameter form as a model for species abundance data. Its application to the number of insurance claims per policy has been studied by Willmot (1987b).

Atkinson and Yeh (1982) developed an approximate ML method for the three-parameter distribution, using a grid of half-integer values of γ. This removed a major numerical handicap regarding the use of the more flexible three-parameter form of the distribution. Ord and Whitmore (1986) investigated minimum chi-square estimation as well as ML estimation. Stein et al. (1987) have re-examined estimation for both the two-parameter and three-parameter forms. Their reparameterization has enabled them to give an algorithm for an exact ML method in the general case. They pointed out that as $\alpha \to \infty$ the distribution tends to a modified logarithmic distribution.

16 SKELLAM'S GENE FREQUENCY DISTRIBUTION

Skellam (1949) has investigated the following branching process: Suppose that for each individual there are a large number of potential offspring whose chances of survival are independent of one another, and that generations do not overlap. Assume also that if a particular gene occurs once in a parental generation, then the number of times that it occurs in the next generation has a Poisson distribution with pgf $e^{c(z-1)}$, $c < 1$. If the gene frequencies are in equilibrium over generations, the probability distribution in successive generations is determined by

$$G(z) = G(e^{c(z-1)}).\tag{11.94}$$

Differentiating with respect to z and setting $z = 1$ gives $\mu = G'(1) = cG'(1)$; this is only satisfied for $c < 1$ by taking $\mu = 0$. In the equilibrium state therefore the gene has died out.

Skellam assumed in addition that the gene arises independently by mutation, according to a Poisson process, at an average rate of λ times per generation; the number of representations of the gene in the next generation now has the pgf

$$G(z) = e^{\lambda(z-1)}G(e^{c(z-1)}).\tag{11.95}$$

Skellam gave the formal solution

$$G(z) = \exp\{\lambda[(z-1) + (e^{c(z-1)} - 1) + (e^{c(e^{c(z-1)}-1)} - 1) + \cdots]\},\tag{11.96}$$

with cumulant generating function

$$\ln\{G(e^t)\} = \lambda[(e^t - 1) + (e^{c(e^t-1)} - 1) + \cdots].\tag{11.97}$$

This enabled him to show that the first three cumulants are

$$\kappa_1 = \frac{\lambda}{1-c},$$

$$\kappa_2 = \frac{\lambda}{(1-c)(1-c^2)},$$

$$\kappa_3 = \frac{\lambda(1+c^2)}{(1-c)(1-c^2)(1-c^3)}.\tag{11.98}$$

Note that the cumulants exist only for $c < 1$.

Skellam found that a negative binomial with pgf

$$G(z) = \left(\frac{1-k}{1-kz}\right)^{2\lambda/c},\tag{11.99}$$

where $k = c/(2-c)$, is a first approximation to (11.96). He remarked that this result is a discrete analogy to Kendall's (1948) exact derivation of the negative binomial as the outcome of a stochastic process in continuous time. Truncating the zero class and letting $\lambda \to 0$ gives a logarithmic distribution [cf. Fisher et al. (1943)].

As a second approximation Skellam adopted a negative binomial distribution with $\mu = \lambda/(1-c)$, $\mu_2 = \lambda/[(1-c)(1-c^2)]$, and pgf

$$G(z) = \left(\frac{1-c^2}{1-c^2 z}\right)^{\lambda(1+c)/c^2}. \tag{11.100}$$

He demonstrated the closeness of these approximations for the case $\lambda = 1.2$, $c = 0.6$.

17 STEYN'S TWO-PARAMETER POWER SERIES DISTRIBUTIONS

Multiparameter power series distributions were put forward by Steyn, Jr. (1980) as extensions of one-parameter power series distributions and modified power series distributions (see Chapter 2, Sections 2.1 and 2.2 concerning the latter). Steyn's distributions have pgf's of the form

$$G(z) = f(\sum_{i=0}^{m} \theta_i z^i)/f(\sum_{i=0}^{m} \theta_i), \tag{11.101}$$

where $f(\lambda z)/f(\lambda)$ is the pgf of a classical one-parameter power series distribution. Reparameterization so that (11.101) has the form

$$G(z) = \omega + (1-\omega)F(\sum_{i=1}^{m} \theta_i^\dagger z^i)/F(\sum_{i=1}^{m} \theta_i^\dagger) \tag{11.102}$$

appears to have certain advantages for estimation purposes; the estimate of ω depends only on the observed proportion of zeros in the data, and then there remain only m, not $m+1$, parameters to estimate.

Steyn obtained the moments of these distributions via their *exponential moments* — these he defined to be the coefficients of $u^i/i!$ in $\exp\{G(1+u)-1\}$. He found that

$$\mu = \left(\frac{\sum_{i=0}^{m} i\theta_i}{\sum_{i=0}^{m} \theta_i}\right) \mu_{[1]}^*(\sum_{i=0}^{m} \theta_i),$$

$$\mu_2 = \left(\frac{\sum_{i=0}^{m} i(i-1)\theta_i}{\sum_{i=0}^{m} \theta_i}\right) \mu_{[2]}^*(\sum_{i=0}^{m} \theta_i) + \left(\frac{\sum_{i=0}^{m} i\theta_i}{\sum_{i=0}^{m} \theta_i}\right)^2 \mu_{[1]}^*(\sum_{i=0}^{m} \theta_i), \tag{11.103}$$

where $\mu_{[r]}^*(\lambda)$ is the r'th factorial moment of the distribution with pgf $f(\lambda z)/f(\lambda)$.

Steyn investigated maximum likelihood, moment, and other estimation procedures. He was particularly interested in modes of genesis; he suggested cluster models, genesis as a univariate multinomial distribution compounded with a power series distribution, and also a derivation from a multivariate power series distribution. He commented on certain special forms of the distributions. Their relationship to distributions of order k awaits detailed exploration.

Steyn's (1984) paper concentrated on two-parameter versions of these distributions, in particular on

1. the two-parameter Poisson distribution (this is the Hermite distribution of Chapter 9, Section 4),
2. the two-parameter logarithmic distribution with pgf

$$G(z) = \frac{\ln(1 - \theta z - c\theta^2 z^2)}{\ln(1 - \theta - c\theta^2)} \qquad (11.104)$$

and pmf $\Pr[X = x] = -\theta^x \sum_{i=0}^{[x/2]} \frac{(x - i - 1)! c^i}{(x - 2i)! i!} \Big/ \ln(1 - \theta - c\theta^2)$

$$= \frac{-\theta^x {}_2F_1[-x/2, (1 - x)/2; -x; -4c]}{x \ln(1 - \theta - c\theta^2)}, \qquad (11.105)$$

(note that $[x/2]$ denotes the integer part of $x/2$),
3. the two-parameter geometric distribution (this is the Gegenbauer distribution; see Section 7 of this chapter).

Steyn gave recurrence relations for the probabilities and expressions for the factorial moments of these distributions.

In his 1984 paper he showed that such distributions are the outcome of a certain type of equilibrium birth-and-death process, and he interpreted this as a model for free-forming groups. He used ML estimation to fit a two-parameter logarithmic distribution and a truncated two-parameter Poisson (truncated Hermite) distribution to a group-size data set.

18 UNIVARIATE MULTINOMIAL-TYPE DISTRIBUTIONS

Univariate multinomial-type distributions have been developed by Steyn (1956); see also Patil et al. (1984).

Suppose, first, that there is a sequence of n independent trials, where each trial has $s + 1$ possible outcomes, $A_0, A_1, A_2, \ldots, A_s$, that are mutually exclusive and have probabilities $p_0, p_1, p_2, \ldots, p_s$, respectively. Let the occurrence of A_i, $i = 1, 2, \ldots, s$, be deemed to be equivalent to i successes, and the occurrence of A_0 be deemed to be a failure. The number of successes X

achieved in the n trials has then the *univariate multinomial distribution* with parameters $n, s, p_1, \ldots, p_s$, and pmf

$$\Pr[X = x] = \sum \frac{n!}{r_0! r_1! \cdots r_s!} p_0^{r_0} p_1^{r_1} \cdots p_s^{r_s}, \qquad x = 0, 1, 2, \ldots, ns, \quad (11.106)$$

where summation is over all nonnegative integers $r_0, r_1, \ldots, r_s$, such that $\sum_{i=0}^{s} r_i = n$ and $\sum_{i=0}^{s} i r_i = x$. The parameter constraints are as follows: $n = 1, 2, \ldots$; $s = 1, 2, \ldots$; $0 < p_i < 1$ for $i = 0, 1, 2, \ldots$; and $\sum_{i=0}^{s} p_i = 1$. The pgf is

$$G(z) = (p_0 + p_1 z + p_2 z^2 + \cdots + p_s z^s)^n, \qquad (11.107)$$

and the mean and variance are

$$\mu = n \sum_{i=1}^{s} i p_i \quad \text{and} \quad \mu_2 = n \sum_{i=1}^{s} i^2 p_i - \frac{\mu^2}{n}. \qquad (11.108)$$

The special case $s = 1$ is the binomial distribution.

Consider, second, a sequence of trials of the above type that has been continued until k failures have occurred. Then the number of successes X that has accumulated has the *univariate negative multinomial distribution* with pmf

$$\Pr[X = x] = \sum \frac{(k + r - 1)!}{(k - 1)! r_1! r_2! \cdots r_s!} p_0^k p_1^{r_1} \cdots p_s^{r_s}, \qquad x = 0, 1, 2, \ldots,$$

$$(11.109)$$

where summation is over all nonnegative integers $r_1, r_2 \ldots, r_s$, such that $\sum_{i=1}^{s} r_i = r$ and $\sum_{i=1}^{s} i r_i = x$. The constraints on the parameters are now as follows: $k = 1, 2, \ldots$; $s = 1, 2, \ldots$; $0 < p_i < 1$ for $i = 0, 1, 2, \ldots$; and $\sum_{i=0}^{s} p_i = 1$. The pgf is

$$G(z) = p_0^k (1 - p_1 z - p_2 z^2 - \cdots - p_s z^s)^{-k}, \qquad (11.110)$$

and the mean and variance are

$$\mu = k \sum_{i=1}^{s} \frac{i p_i}{p_0} \quad \text{and} \quad \mu_2 = k \sum_{i=1}^{s} \frac{i^2 p_i}{p_0} + \frac{\mu^2}{k}. \qquad (11.111)$$

The special case $s = 1$ gives the negative binomial distribution.

Third, let a finite population of N elements consist of $s + 1$ classes, A_0, $A_1, \ldots, A_s$, containing $N_0, N_1, \ldots, N_s$ elements, respectively, with $N_0 + N_1 + \cdots + N_s = N$. Let the selection of an element from A_i, $i = 1, 2, \ldots, s$ be treated as i successes and selection of an element from A_0 be treated as a failure. Let n elements be selected at random without replacement from the population. Then the number of successes X that have accrued has the *univariate factorial multinomial distribution* with pmf

$$\Pr[X = x] = \sum \binom{N_0}{r_0}\binom{N_1}{r_1}\cdots\binom{N_s}{r_s}\Big/\binom{N}{n}, \qquad x = 0, 1, 2, \ldots,$$

(11.112)

where summation extends over all nonnegative integers $r_0, r_1, \ldots, r_s$, such that $n = \sum_{i=0}^{s} r_i$ and $x = \sum_{i=0}^{s} i r_i$. The parameter constraints are as follows: $s = 1, 2, \ldots$; $n = 1, 2, \ldots$; $N_i = 1, 2, \ldots$, for $i = 0, 1, 2, \ldots, s$; and $\sum_{i=0}^{s} N_i = N$. The pgf is

$$G(z) = \frac{N_0!(N - n)!}{(N_0 - n)!N!} \, {}_{s+1}F_1[-n, -N_1, -N_2, \ldots, -N_s; N_0 - n + 1; z, z^2, \ldots, z^s];$$

(11.113)

for an explanation of the notation see Steyn (1956). The mean and variance are

$$\mu = n\sum_{i=1}^{s} \frac{iN_i}{N} \quad \text{and} \quad \mu_2 = \frac{N - n}{N - 1}\left[n\sum_{i=1}^{s}\left(\frac{i^2 N_i}{N}\right) - \frac{\mu^2}{n}\right]. \qquad (11.114)$$

The classical hypergeometric distribution with parameters n, N, N_1 is the special case with $s = 1$. The univariate factorial multinomial distribution tends to a univariate multinomial distribution as $N \to \infty$, $N_i \to \infty$, such that $N_i/N \to p_i$, $0 < p_i < 1$, $i = 0, 1, \ldots, s$, and $\sum_{i=0}^{s} p_i = 1$.

Fourth, consider the previous setup, but suppose now that sampling continues without replacement until exactly k failures have been obtained (i.e., inverse sampling). Then the number of successes that have accrued has the *univariate negative factorial multinomial distribution* with pmf

$$\Pr[X = x] = \frac{N_0 - k + 1}{N - k - r + 1}\sum\binom{N_0}{k-1}\binom{N_1}{r_1}\cdots\binom{N_s}{r_s}\Big/\binom{N}{k+r-1},$$

(11.115)

where $x = 0, 1, 2, \ldots, \sum_{i=1}^{s} iN_i$, and summation is over all nonnegative integers $r_1, r_2, \ldots, r_s$, such that $r = \sum_{i=1}^{s} r_i$ and $x = \sum_{i=1}^{s} i r_i$. The constraints on the parameters are as follows: $s = 1, 2, \ldots$; $k = 1, 2, \ldots$; $N_i = 1, 2, \ldots$, for $i = 0, 1, 2, \ldots, s$; and $\sum_{i=0}^{s} N_i = N$. The pgf is

$$G(z) = \frac{N_0!(N - k)!}{(N_0 - k)!N!} \, {}_{s+1}F_1[k, -N_1, -N_2, \ldots, -N_s; k - N; z, z^2, \ldots, z^s],$$

(11.116)

and the mean and variance are

$$\mu = k \sum_{i=1}^{s} \frac{iN_i}{N_0 + 1} \quad \text{and} \quad \mu_2 = \frac{N_0 - k + 1}{N_0 + 2} \left[k \sum_{i=1}^{s} \left(\frac{i^2 N_i}{N_0 + 1} \right) + \frac{\mu^2}{k} \right].$$

$$(11.117)$$

The inverse hypergeometric distribution with parameters k, N, N_1 is the special case with $s = 1$. The univariate negative factorial multinomial distribution tends to a univariate negative multinomial distribution as $N \to \infty$, $N_i \to \infty$, such that $N_i/N \to p_i$, $0 < p_i < 1$, $i = 0, 1, \ldots, s$, and $\sum_{i=0}^{s} p_i = 1$.

19 URN MODELS WITH STOCHASTIC REPLACEMENTS

Urn models with stochastic replacements can be constructed in many ways. Usually we are interested in either (1) the distribution of the number of balls of various kinds in the urn after a fixed number of trials, (2) the outcome of a fixed number of draws, and/or (3) the (discrete) waiting time until a specified set of conditions are fulfilled.

The concept of an urn model has a very long history dating back to biblical times [Rabinovitch (1973)]. Sambursky (1956) has discussed ideas concerning random selection in ancient Greece. The theory underlying urn models dates back at least to Bernoulli's (1713) *Ars Conjectandi*; see, e.g., David (1962). Of interest is not only the wide variety of discrete distributions that can arise from urn models but especially the understanding that is generated when two apparently quite different situations lead to closely related urn models. Here we do no more than indicate very briefly several different types of urn models, and various ways of handling them. Interested readers are referred to Berg's (1988b) review article and to the comprehensive monograph devoted exclusively to urn models by Johnson and Kotz (1977).

For the Pólya urn model (Chapter 6, Section 2.4) initially containing w white balls and b black ones, the replacement strategy is to return each drawn ball together with c extra balls of similar color, where c may be negative as well as positive. Let $P(n, x)$ denote the probability of x successes in n trials. Then

$$P(n+1, x+1) = \frac{w + cx}{w + b + cn} P(n, x) + \frac{b + c(n - x)}{w + b + cn} P(n, x+1). \quad (11.118)$$

Friedman's (1949) urn model is a generalization of Pólya's. Again initially the urn is assumed to contain w white balls and b black ones. Also each drawn ball is returned together with c of the same color, but *in addition* d balls of the opposite color are also added at the time of replacement. Friedman noted the following special cases: (1) $c = 0$, $d = 0$ gives a binomial model; (2) $c = -1$, $d = 0$ gives sampling without replacement; (3) $d = 0$ gives the Pólya urn model; (4) $c = -1$, $d = 1$ gives the Ehrenfest model of heat exchange; and (5) $c = 0$ gives a safety campaign model in which each draw of a white ball is penalized. Let W_n be the number of white balls in the urn after n draws, and let $X_n = 0$ or 1 according to whether the nth drawn ball was white or black. Friedman was able to obtain a formula for the moment generating function by the use of a difference-differential equation. He showed that there is an explicit solution when $\gamma = (c + d)/(c - d)$ takes the values $0, \pm 1$, and hence in cases (1) to (5) above. (For $\gamma = 1$ we have $d = 0$, $(c \neq 0)$, and for $\gamma = -1$ we have $c = 0$, $(d \neq 0)$.) Freedman (1965) obtained asymptotic results for the behaviour of W_n as n becomes large by treating the sampling scheme as a Markov chain.

A quite different model has been studied by Woodbury (1949) and Rutherford (1954). Here the probability of success depends on the number of previous successes but not on the number of previous trials. The following boundary conditions hold: $P(n, x) = 0$ for $x < 0$ or $x > n$, and $P(0, 0) = 1$. Let p_x denote the probability of a success given that x successes have already occurred. Then

$$P(n + 1, x + 1) = p_x P(n, x) + q_{x+1} P(n, x + 1) \qquad (11.119)$$

where $q_x = 1 - p_x$.

Woodbury was able to investigate this model by setting up the following generalization of (11.119),

$$P(n + 1, x + 1) = (q - q_x) P(n, x) + q_{x+1} P(n, x + 1), \qquad (11.120)$$

and putting

$$P(n, x) = (q - q_0)(q - q_1) \cdots (q - q_x) F(n, x). \qquad (11.121)$$

The use of partial fractions then gives

$$P(n, x) = (q - q_0)(q - q_1) \cdots (q - q_x)$$
$$\times \sum_{i=0}^{x} q_i^x / [(q - q_0) \cdots (q - q_{i-1})(q - q_{i+1}) \cdots (q - q_x)], \quad (11.122)$$

as the solution to (11.120); taking $q = 1$ gives the solution to (11.119).

Rutherford (1954) devoted particular attention to the case $p_x = p + cx$, where $0 < p < 1$. If $c > 0$, the restriction $p + cn < 1$ is needed, whereas, if $c < 0$, we require $p + cn > 0$. Under this replacement scheme the composition of the urn is altered only when a white ball is drawn. Using Woodbury's result, Rutherford showed that

$$P(n,x) = \frac{(p/c + x - 1)!}{x!(p/c - 1)!} \sum_{r=0}^{x} (-1)^r \binom{x}{r} (1 - p + cr)^n. \qquad (11.123)$$

This is the coefficient of $\omega^n/n!$ in $e^{\omega(1-p)}(1 - e^{-c\omega})^x$. The pgf for $P(n,x)$ is the coefficient of $\omega^n/n!$ in

$$e^{\omega(1-p)}[1 - (1 - e^{-c\omega})c]^{-p/c} = e^{\omega}[(1 - c)e^{c\omega} + c]^{-p/c}. \qquad (11.124)$$

Rutherford derived expressions for the first three factorial moments of the distribution. By equating their theoretical and observed values, he obtained good fits to two data sets. For Greenwood and Yule's (1920) data on women working on H.E. shells, the urn model has a clear physical meaning. Rutherford also obtained negative binomial and Gram-Charlier approximations to the distribution. He regarded this urn model as a possible explanation for the good descriptive fits that these distributions often provide.

Two other special cases of the Woodbury urn model, with (1) $p_x = (p + cx)/(1 + cx)$, and (2) $p_x = p/(1 + cx)$, have been investigated by Chaddha (1965), who related this choice of functions for p_x to the construction of a model representing number of attendances at a sequence of committee meetings.

Naor's (1957) urn model (see Section 11 above) was studied in the context of a machine-minding problem.

Wei (1979) has considered a realistic variant of the play-the-winner rule for use in the assignment of patients who present sequentially in a medical trial concerning k types of treatment. Here, if the response to treatment i is a success, then α (> 0) balls of color i are added to the urn; if the response to treatment i is a failure, then β (> 0) balls of every other color except colour i are added. Wei's analysis of the model assumes that the response of a patient is virtually instantaneous.

Another use of urn models (which may involve the simultaneous use of several urns) is to explain modes of genesis for certain Lagrangian distributions; see, e.g., Consul (1974), Consul and Mittal (1975), and Janardan and Schaeffer (1977). Applications of urn models in fields as diverse as genetics, capture-recapture sampling of animal populations, learning processes, and filing systems are discussed in Johnson and Kotz (1977).

20   ZIPF AND ZETA DISTRIBUTIONS

These are long-tailed distributions that have been found to be useful for size-frequency data and are related in different ways to the Riemann zeta function (Chapter 1, Section A12),

$$\zeta(s) = \sum_{j=1}^{\infty} j^{-s}, \qquad s > 1.$$

From observation and analysis of the frequency of words in long sequences of text, the approximate formula

$$n_x \propto x^{-(\rho+1)}, \qquad x = 1, 2, \ldots, \quad \rho > 0, \tag{11.125}$$

for n_x the number of words appearing x times, was found to give a usefully accurate fit. A probability distribution appropriate to such situations can be constructed by taking

$$\Pr[X = x] = c x^{-(\rho+1)}, \qquad x = 1, 2, \ldots, \tag{11.126}$$

with

$$c = \left[\sum_{x=1}^{\infty} x^{-(\rho+1)} \right]^{-1} = \frac{1}{\zeta(\rho+1)},$$

where $\zeta(\cdot)$ denotes the Riemann zeta function and $\rho > 0$. When used in linguistics, this is often called the *Zipf-Estoup law* [Estoup (1916), Zipf (1949)]; the distribution is then referred to as the *Zipf distribution*. The use of the distribution in this connection has also been studied by Good (1957). Seal (1947) has applied the distribution to the number of insurance policies held by individuals. Some interesting comments on Zipf and Yule distributions have been made by Kendall (1961); see Chapter 6, Section 10.3 for properties of the Yule distribution.

The distribution with pmf (11.126) is analogous to the continuous Pareto distribution (see Chapter 19), in the same way that the discrete rectangular is analogous to the continuous rectangular (uniform) distribution. It is therefore sometimes called the *discrete Pareto distribution*. Very often it is called the *Riemann zeta distribution*; some authors abbreviate this to "zeta distribution."

The rth moment about zero of the distribution is

$$\mu_r' = \frac{\zeta(\rho - r + 1)}{\zeta(\rho+1)} \tag{11.127}$$

for $r < \rho$. If $r \geq \rho$, the moment is infinite.

Very often ρ is found to have a value slightly in excess of unity. For low values of ρ the distribution has a very long positive tail. Table 11.1 compares Riemann zeta, logarithmic, and shifted geometric distributions with the same expected value, $\mu = 2$, and shows that the Riemann zeta distribution decays more rapidly at first, but later more slowly.

Table 11.1 Comparison of Riemann Zeta, Logarithmic, and Shifted Geometric Distributions, Each with Mean Equal to 2

x	Riemann zeta[a]	Pr[X = k] Logarithmic[b]	Geometric[c]
1	0.7409	0.5688	0.5000
2	0.1333	0.2035	0.2500
3	0.0488	0.0971	0.1250
4	0.0240	0.0521	0.0625
5	0.0138	0.0298	0.0312
6	0.0088	0.0178	0.0156
7	0.0060	0.0109	0.0078
8	0.0043	0.0068	0.0039
9	0.0032	0.0044	0.0020
10	0.0025	0.0028	0.0010
>10	0.0144	0.0060	0.0010

[a] $\Pr[X = x] = 0.74088x^{-2.4749}$.
[b] $\Pr[X = x] = 0.79475(0.71563)^x/x$.
[c] $\Pr[X = x] = 2^{-x}$.

Schreider (1967) has attempted to give a theoretical justification for the Zipf distribution, based on thermodynamical analogies. His arguments would lead one to expect the Zipf distribution to be approached asymptotically for "sufficiently long" extracts of "stable" text. They also indicate that greater deviations from the Zipf form may be expected at lower, rather than higher, frequencies; this agrees with empirical findings in a considerable number of cases. Nanapoulos (1977) has established a weak law of large numbers for the distribution. Devroye (1986) has proposed an interesting method for the computer generation of rv's from the Zipf (Riemann zeta) distribution.

Given values of N independent random variables, each having the Riemann zeta distribution (11.126), the maximum likelihood estimator, $\hat{\rho}$, of ρ satisfies the equation [Seal (1952)]

$$N^{-1}\sum_{i=1}^{N}\ln(x_i) = \frac{-\zeta'(\hat{\rho}+1)}{\zeta(\hat{\rho}+1)}. \tag{11.128}$$

Using appropriate tables, such as Table 11.2 below, it is not difficult to obtain an acceptable solution of this equation. For $\rho > 4$, $\zeta'(\rho+1)/\zeta(\rho+1) \doteq (1+2^{\rho+1})^{-1}\ln 2$. The variance of $\hat{\rho}$ is approximately $\{N g(\rho)\}^{-1}$, where

$$g(\rho) = \frac{d}{d\rho}\left\{\frac{\zeta'(\rho+1)}{\zeta(\rho+1)}\right\}. \tag{11.129}$$

Table 11.2   Values of $-\zeta'(\rho+1)/\zeta(\rho+1)$

ρ	$-\zeta'(\rho+1)/\zeta(\rho+1)$	ρ	$-\zeta'(\rho+1)/\zeta(\rho+1)$
0.1	9.441	1.6	0.256
0.2	4.458	1.7	0.228
0.3	2.808	1.8	0.204
0.4	1.990	1.9	0.183
0.5	1.505	2.0	0.164
0.6	1.186	2.2	0.134
0.7	0.961	2.4	0.110
0.8	0.796	2.6	0.0914
0.9	0.669	2.8	0.0761
1.0	0.570	3.0	0.0637
1.1	0.490	3.2	0.0535
1.2	0.425	3.4	0.0451
1.3	0.372	3.6	0.0382
1.4	0.327	3.8	0.0324
1.5	0.289	4.0	0.0276

Adapted from Walther (1926).

Values of $g(\rho)$ have been obtained by numerical differentiation and are given in Table 11.3 below.

An alternative, though usually considerably poorer, estimation method is based on the ratio f_1/f_2 of the observed frequencies of 1's and 2's among the $x_1, x_2, \ldots, x_N$. Equating f_j/N to $\Pr[X = j]$ for $j = 1, 2$, we obtain

$$\rho^* = \frac{\ln(f_1/f_2)}{\ln 2} - 1. \tag{11.130}$$

The variance of ρ^* is approximately

$$\frac{(1 + 2^{\rho+1})}{(\ln 2)^2 \zeta(\rho+1)N}. \tag{11.131}$$

(Note that (11.130) ignores the possibility that $f_1 = 0$ or $f_2 = 0$.) For $\rho = 0.5, 1.0, 1.5, 2.0, 2.5, 3.0$, the approximate values of $\mathrm{Var}(\rho^*)/\mathrm{Var}(\hat{\rho})$ are 11.8, 5.7, 3.7, 2.7, 2.0, 1.8, respectively. It will be observed that $\mathrm{Var}(\rho^*)$ is considerably larger than $\mathrm{Var}(\hat{\rho})$ when ρ is small. As ρ becomes large the ratio tends to unity.

A better estimator than ρ^* appears to be obtained by equating the population and sample means. This gives $\tilde{\rho}$, where $\tilde{\rho}$ satisfies the equation

Table 11.3 Values of $\frac{d}{d\rho}\{\zeta'(\rho+1)/\zeta(\rho+1)\} = g(\rho)$

ρ	$g(\rho)$	ρ	$g(\rho)$
0.5	3.860	1.8	0.225
0.6	2.638	1.9	0.196
0.7	1.909	2.0	0.172
0.8	1.436	2.1	0.152
0.9	1.114	2.2	0.134
1.0	0.904	2.3	0.119
1.1	0.716	2.4	0.106
1.2	0.588	2.5	0.095
1.3	0.490	2.6	0.085
1.4	0.412	2.7	0.076
1.5	0.354	2.8	0.069
1.6	0.300	2.9	0.062
1.7	0.258	3.0	0.056

Adapted from Moore (1956).

$$\bar{x} = \frac{\zeta(\tilde{\rho})}{\zeta(\tilde{\rho}+1)}. \tag{11.132}$$

Moore (1956) has provided tables to facilitate the solution of this equation. When $\rho = 3$, $\text{Var}(\tilde{\rho}) \doteq 23.1/N$, whereas $\text{Var}(\hat{\rho}) \doteq 17.9/N$, a ratio of about 1.3.

Haight's (1966) investigation was concerned with models for word association data (the number of response words elicited by a stimulus word). He made a comparison between the fits provided by the Yule, Borel-Tanner, and logarithmic distributions (Chapter 6, Section 10.3, Chapter 9, Section 11, and Chapter 7), and by two distributions related in a rather different way to Riemann's zeta function. The Yule distribution gave a markedly superior fit, and the logarithmic an inferior fit, compared with the Borel-Tanner distribution. Haight's new "zeta" distribution did even better than the Yule for two of the four data sets. He remarked that there are other phenomena that have, in common with word lists, both frequency listing (the Yule approach) and rank listing (the Zipf approach). Such phenomena include publication lists of scientists, states with numbers of constituent counties, chemical elements with their abundances, lists of species per genera, and surnames with their frequencies.

Haight (1966, 1969) returned to Zipf's conjecture concerning city sizes [cf. Simon (1955)], by considering a theoretical tabulation of categories $x = 1, 2, 3, \ldots$, with category x containing N_x entities such that

$$x - 0.5 \leq ZN_x^{-\beta} < x + 0.5, \tag{11.133}$$

where $Z \geq 1$, $\beta > 0$ are constants. Rearranging the inequalities gives

$$N_x = \left[\left(\frac{2Z}{2x-1}\right)^{1/\beta}\right] - \left[\left(\frac{2Z}{2x+1}\right)^{1/\beta}\right];$$ (11.134)

note that throughout this section $[\cdot]$ means the integer part. Haight argued that a discrete probability distribution can be constructed by taking

$$\Pr[X = x] = \frac{N_x}{(2Z)^{1/\beta}}, \qquad x = 1, 2, \ldots,$$ (11.135)

provided that $(2Z)^{1/\beta}$ is an integer. We note that the restriction on Z is necessary in order that

$$\sum_{x\geq1} \Pr[X = x] = \left[\left(\frac{2Z}{1}\right)^{1/\beta}\right] (2Z)^{-1/\beta} = 1.$$ (11.136)

Moreover, for the Zipf model to be meaningful, Z must be an integer.

  Haight's zeta distribution is obtained by letting $Z \to \infty$ and writing $\sigma = 1/\beta$. This gives

$$\Pr[X = x] = (2x - 1)^{-\sigma} - (2x + 1)^{-\sigma}, \qquad x = 1, 2, \ldots; \quad \sigma > 0.$$ (11.137)

The name "zeta" was adopted because the first two moments of the distribution can be expressed in terms of the Riemann zeta function:

$$\mu = \sum_{x\geq1} x \Pr[X = x] = (1 - 2^{-\sigma})\zeta(\sigma),$$ (11.138)

$$\mu_2 = \sum_{x\geq1} x^2 \Pr[X = x] - \mu^2 = (1 - 2^{1-\sigma})\zeta(\sigma - 1) - \mu^2.$$ (11.139)

The relationship to the zeta function is here via the mean and variance, not the probabilities. For $\sigma \leq 1$ the mean is infinite, and for $\sigma \leq 2$ the variance is infinite.

  When $\beta = \sigma = 1$ and Z is finite, the outcome is *Haight's harmonic distribution*, with

$$\Pr[X = x] = \frac{1}{2Z}\left\{\left[\frac{2Z}{2x-1}\right] - \left[\frac{2Z}{2x+1}\right]\right\},$$ (11.140)

where $2Z$ is a positive integer and the support of the distribution is $x = 1, 2, \ldots [Z+1/2]$. Patil et al. (1984) have given the following formulas for the mean and variance:

$$\mu = \sum_{x=1}^{[Z+1/2]} \left[\frac{2Z}{2x-1}\right] (2Z)^{-1},$$ (11.141)

$$\mu_2 = \sum_{x=1}^{[Z+1/2]} \left(\frac{2x-1}{2Z}\right) \left[\frac{2Z}{2x-1}\right] - \mu^2. \qquad (11.142)$$

Haight gave tables of the distribution for $Z = 1, 5, 10, 20, 30, 100$ and 1000. It has the unusual feature that $\Pr[X = x]$ is equal to zero for considerable ranges of values of x, interspersed with isolated nonzero probabilities. The distribution terminates. As Z tends to infinity, $\Pr[X = x]$ tends to $2(2x - 1)^{-1}(2x+1)^{-1}$, $x = 1, 2, \ldots$. (Similarity to the Yule distribution with $\rho = 1$ is evident.) The mean is infinite.

Note that Haight's zeta distribution and the harmonic distribution each have a single parameter.

The four data sets on word associations in Haight's (1966) paper were fitted very successfully with Haight's zeta distribution by equating the observed and expected first frequencies. Haight did not try to fit the harmonic distribution to these data sets; with hindsight we can see that equating the observed and expected first frequencies would give the estimator Z^* satisfying

$$\frac{f_1}{N} = 1 - \frac{1}{2Z^*}\left[\frac{2Z^*}{3}\right], \qquad (11.143)$$

which does not have a solution if $f_1/N \leq 2/3$ (as in all four data sets). In Haight (1969) the author attempted to fit both his zeta distribution and the harmonic distribution to two data sets on population sizes by equating observed and expected means. Because moment estimation of the parameter σ for this zeta distribution requires evaluation of the Riemann zeta function, he recommended the use of the approximation

$$(1-2^{-\sigma})\zeta(\sigma) = 0.5(\sigma-1)^{-1} + 0.635518142 + 0.11634237(\sigma-1)$$
$$-0.01876574(\sigma-1)^2 \qquad (11.144)$$

for σ close to unity.

Bibliography

Abate, J., and Whitt, W. (1988). Transient behaviour of the $M/M/1$ queue via Laplace transforms, *Advances in Applied Probability*, **20**, 145–178. [11.13]

Abdel-Aty, S. H. (1954). Ordered variables in discontinuous distributions, *Statistica Neerlandica*, **8**, 61–82. [4.4]

Abdul-Razak, R. S., and Patil, G. P. (1986). Power series distributions and their conjugates in statistical modeling and Bayesian inference, *Communications in Statistics-Theory and Methods*, **15**, 623–641. [2.2.1]

Abouammoh, A. M. (1987). On discrete α-unimodality, *Statistica Neerlandica*, **41**, 239–244. [1B.5]

Abouammoh, A. M., and Mashour, A. F. (1981). A note on the unimodality of discrete distributions, *Communications in Statistics-Theory and Methods*, **A10**, 1345–1354. [1B.5]

Abramowitz, M., and Stegum, I. A. (1965). *Handbook of Mathematical Functions*, New York: Dover. [1A.2, 1A.3, 1A.5, 1A.7, 1A.9, 1A.11, 1A.12]

Aczél, J. (1972). On a characterization of the Poisson distribution, *Journal of Applied Probability*, **9**, 852–856. [4.8]

Aczél, J. (1975). On two characterizations of Poisson distributions, *Abhandlungen aus dem Mathematischen Seminar der Universität Hamburg*, **44**, 91–100. [4.8]

Adatia, A. (1991). Computation of variances and covariances of order statistics from the geometric distribution, *Journal of Statistical Computation and Simulation*, **39**, 91–94. [5.2]

Ahmed, A. N. (1991). Characterization of beta, binomial and Poisson distributions, *IEEE Transactions on Reliability*, **R-40**, 290–295. [3.9]

Ahmed, M. S. (1961). On a locally most powerful boundary randomized similar test for the independence of two Poisson variables, *Annals of Mathematical Statistics*, **32**, 809–827. [9.4]

Ahmed, S.E. (1991). Combining Poisson means, *Communications in Statistics— Theory and Methods*, **20**, 771–789. [4.7.2]

Ahrens, J. H. (1987). A comparison of hypergeometric distributions with corresponding binomial distributions, *Ökonomie und Mathematik*, O. Opitz and B. Rauhut (editors), 253–265. Berlin: Springer-Verlag. [6.5]

Ahrens, J. H., and Dieter, U. (1974). Computer methods for sampling from gamma, beta, Poisson and binomial distributions, *Computing*, **12**, 223–246. [1C.3]

Ahrens, J. H., and Dieter, U. (1982). Computer generation of Poisson deviates from modified normal distributions, *ACM Transactions on Mathematical Software*, **8**, 163–179. [1C.4]

Ahsanullah, M., and Holland, B. (1984). Record values and the geometric distribution, *Statistische Hefte*, **25**, 319–327. [5.9.1]

Ahuja, J. C. (1971a). Certain properties of the Stirling distribution of the second kind, *Australian Journal of Statistics*, **13**, 133–136. [4.12.3]

Ahuja, J. C. (1971b). Distribution of the sum of independent decapitated negative binomial variables, *Annals of Mathematical Statistics*, **42**, 383–384. [5.11]

Ahuja, J. C., and Enneking, E. (1972). Certain properties of the distribution of the sum of independent zero-one-truncated Poisson variables, *Australian Journal of Statistics*, **14**, 50–53. [4.12.3]

Ahuja, J. C., and Enneking, E. A. (1974). Convolution of independent left-truncated negative binomial variates and limiting distributions, *Annals of the Institute of Statistical Mathematics, Tokyo*, **26**, 265–270. [5.11]

Aitken, A. C. (1939). *Statistical Mathematics* (First edition), Edinburgh: Oliver and Boyd. [3.12.2]

Aitken, A. C. (1945). *Statistical Mathematics* (Fourth edition), Edinburgh: Oliver and Boyd. [3.12.2]

Aki, S. (1985). Discrete distributions of order k on a binary sequence, *Annals of the Institute of Statistical Mathematics, Tokyo*, **37**, 205–224. [10.6.3]

Aki, S., and Hirano, K. (1988). Some characteristics of the binomial distribution of order k and related distributions, *Statistical Theory and Data Analysis II*, K. Matusita (editor), 211–222. Amsterdam: Elsevier [10.6.2]

Aki, S., and Hirano, K. (1989). Estimation of parameters in the discrete distributions of order k, *Annals of the Institute of Statistical Mathematics, Tokyo*, **41**, 47–61. [10.6.2]

Aki, S., Kuboki, H., and Hirano, K. (1984). On discrete distributions of order k, *Annals of the Institute of Statistical Mathematics, Tokyo*, **36**, 431–440. [10.6.2]

Al-Zaid, A. A. (1989). On the unimodality of mixtures, *Pakistan Journal of Statistics*, **5**, 205–209. [8.3.2, 8.3.3]

Alzaid, A. A., Rao, C. R., and Shanbhag, D. N. (1986). Characterization of discrete probability distributions by partial independence, *Communications in Statistics-Theory and Methods*, **15**, 643–656. [4.8]

Alt, F. B. (1985). The matching problem, *Encyclopedia of Statistical Sciences*, **5**, S. Kotz, N. L. Johnson and C. B. Read (editors), 293–296. New York: Wiley. [10.3]

Altham, P. M. E. (1978). Two generalizations of the binomial distribution, *Applied Statistics*, **27**, 162–167. [3.12.6]

Anderson, T. W. (1943). On card matching, *Annals of Mathematical Statistics*, **14**, 426–435. [2.4.2, 10.3]

Anderson, T. W., and Samuels, S. M. (1967). Some inequalities among binomial and Poisson probabilities, *Proceedings of the Fifth Berkeley Symposium on Mathematical Statistics and Probability*, **1**, 1–12. Berkeley: University of California Press. [3.6.1, 4.4]

Andrews, G. E. (1986). *q Series: Their Development and Application in Analysis, Number Theory, Combinatorics, Physics, and Computer Algebra*, Providence, RI: American Mathematical Society. [1A.12]

Angus, J. E., and Schafer, R. E. (1984). Improved confidence statements for the binomial parameter, *American Statistician*, **38**, 189–191. [3.8.3]

Anraku, K., and Yanagimoto, T. (1990). Estimation for the negative binomial distribution based on the conditional likelihood, *Communications in Statistics-Simulation and Computation*, **19**, 771–806. [5.8.3]

Anscombe F. J. (1948). The transformation of Poisson, binomial and negative binomial data, *Biometrika*, **35**, 246–254. [3.6.3, 4.5, 5.6, 5.8.3]

Anscombe, F. J. (1950). Sampling theory of the negative binomial and logarithmic series distributions, *Biometrika*, **37**, 358–382. [5.1, 5.8.3, 5.8.4, 7.2, 7.7.2, 7.12, 9.12]

Appell, P., and Kampé de Fériet, J. (1926). *Fonctions Hypergéométriques et Hypersphériques*, Paris: Gauthier-Villars. [8.2.2]

Arbous, A. G., and Kerrich, J. E. (1951). Accident statistics and the concept of accident proneness, *Biometrics*, **7**, 340–432. [4.12.5, 5.10]

Archibald, E. E. A. (1948). Plant populations I: A new application of Neyman's contagious distribution, *Annals of Botany, New Series*, **12**, 221–235. [9.6.5]

Arfwedson, G. (1951). A probability distribution connected with Stirling's second class numbers, *Skandinavisk Aktuarietidskrift*, **34**, 121–132. [10.4.1]

Arley, N., and Buch, K. R. (1950). *Introduction to the Theory of Probability and Statistics*, New York: Wiley. [5.3]

Arnold, B. C. (1980). Two characterizations of the geometric distribution, *Journal of Applied Probability*, **17**, 570–573. [5.9.1]

Arnold, B. C., Becker, A., Gather, U., and Zahedi, H. (1984). On the Markov property of order statistics, *Journal of Statistical Planning and Inference*, **9**, 147–154. [5.2]

Arnold, B. C., and Meeden, G. (1975). Characterization of distributions by sets of moments of order statistics, *Annals of Statistics*, **3**, 754–758. [1B.10]

Aroian, L. A. (1937). The Type B Gram-Charlier series, *Annals of Mathematical Statistics*, **8**, 183. [11.8]

Arratia, R., Goldstein, L., and Gordon, L. (1989). Two moments suffice for Poisson approximations: the Chen-Stein method, *Annals of Probability*, **17**, 9–25. [4.2]

Arratia, R., Goldstein, L., and Gordon, L. (1990). Poisson approximation and the Chen-Stein method, *Statistical Science*, **5**, 403–434. [4.2]

Atkinson, A. C., and Yeh, L. (1982). Inference for Sichel's compound Poisson distribution, *Journal of the American Statistical Association*, **77**, 153–158. [11.15]

Austin, T., Fagan, R., Lehrer, T., and Penney, W. (1957). The distribution of the number of locally maximal elements in a random sample from a continuous distribution, *Annals of Mathematical Statistics*, **28**, 786–790. [11.14]

Ayyangar, A. A. K. (1934). A note on the incomplete moments of the hypergeometrical series, *Biometrika*, **26**, 264–265. [6.3]

Bahadur, R. R. (1960). Some approximations to the binomial distribution function, *Annals of Mathematical Statistics*, **31**, 43–54. [3.6.2]

Bailey, N. T. J. (1951). On estimating the size of mobile populations from recapture data, *Biometrika*, **38**, 293–306. [6.9.2]

Bailey, N. T. J. (1954). A continuous time treatment of a simple queue, using generating functions, *Journal of the Royal Statistical Society, Series B*, **16**, 288–291. [11.13]

Bailey, N. T. J. (1964). *The Elements of Stochastic Processes with Applications to the Natural Sciences*, New York: Wiley. [3.12.5]

Bailey, W. N. (1935). *Generalized Hypergeometric Series*, London: Cambridge University Press. [1A.6, 1A.12]

Bain, L. J., Engelhardt, M., and Williams, D. H. (1990). Confidence bounds for the binomial N parameter, *Communications in Statistics-Simulation and Computation*, **19**, 335-348. [3.8.3]

Balakrishnan, N. (1986). Order statistics from discrete distributions, *Communications in Statistics-Theory and Methods*, **15**, 657–675. [1B.10, 3.5].

Balakrishnan, N., and Cohen, A. C. (1991). *Order Statistics and Inference: Estimation Methods*, San Diego, CA: Academic Press. [1B.10]

Baldessari, B., and Weber, J. (1987). Intrinsic hypergeometric random variable, *Statistica*, **47**, 275–286. [6.12]

Bardwell, G. E. (1960). On certain characteristics of some discrete distributions, *Biometrika*, **47**, 473–475. [2.3.1]

Bardwell, G. E., and Crow, E. L. (1964). A two-parameter family of hyper-Poisson distributions, *Journal of the American Statistical Association*, **59**, 133–141. [2.3.1, 4.12.4]

Barlow, R. E., and Proschan, F. (1965). *Mathematical Theory of Reliability*, New York: Wiley. [3.4]

Barlow, R. E., and Proschan, F. (1975). *Statistical Theory of Reliability and Life Testing*, New York: Holt, Rinehart & Winston. [1B.2]

Barnard, G. A. (1954). Sampling inspection and statistical decisions, *Journal of the Royal Statistical Society, Series B*, **16**, 151–174. [6.9.2]

Barndorff-Nielsen, O. E. (1965). Identifiability of mixtures of exponential families, *Journal of Mathematical Analysis and Applications*, **12**, 115–121. [8.2.1]

Barndorff-Neilsen, O. E. (1991). Likelihood theory, *Statistical Theory and Modelling: In Honour of Sir David Cox, FRS*, D. V. Hinkley, N. Reid and E. J. Snell (editors), 232–264. London: Chapman and Hall. [1B.15]

Barnett, V. D. (1973). *Comparative Statistical Inference*, New York: Wiley. [1B.15]

Barry, D. (1990). Empirical Bayes estimation of binomial probabilities in one-way and two-way layouts, *The Statistician*, **39**, 437–453. [3.8.2]

Bartko, J. J. (1961). The negative binomial distribution: a review of properties and applications, *Virginia Journal of Science*, **12**, 18–37. [5.1]

Bartko, J. J. (1966). Approximating the negative binomial, *Technometrics*, **8**, 345–350. [5.6]

Bartlett, M. S. (1947). The use of transformations, *Biometrics*, **3**, 39–52. [3.6.3]

Bartlett, M. S. (1960). *Stochastic Population Models*, London: Methuen. [7.10]

Barton, D. E. (1957). The modality of Neyman's contagious distribution of type A, *Trabajos de Estadística*, **8**, 13–22. [9.6.1]

Barton, D. E. (1958). The matching distributions: Poisson limiting forms and derived methods of approximation, *Journal of the Royal Statistical Society, Series B*, **20**, 73–92. [10.3]

Barton, D. E. (1961). Unbiased estimation of a set of probabilities, *Biometrika*, **48**, 227–229. [4.7.2]

Barton, D. E. (1966). Review of P. J. Staff: The displaced Poisson distribution, *Mathematical Reviews*, **33**, 6583. [4.12.4, 8.3.2, 8.3.4]

Barton, D. E., and David, F. N. (1959a). Contagious occupancy, *Journal of the Royal Statistical Society, Series B*, **21**, 120–133. [10.4.1]

Barton, D. E., and David, F. N. (1959b). The dispersion of a number of species, *Journal of the Royal Statistical Society, Series B*, **21**, 190–194 [10.4.1]

Barton, D. E., David, F. N. and Merrington, M. (1963). Tables for the solution of the exponential equation $\exp(b) - b/(1-p) = 1$, *Biometrika*, **50**, 169–172. [7.6, 7.7.2]

Basu, D. (1955). A note on the structure of a stochastic model considered by V. M. Dandekar, *Sankhyā*, **15**, 251–252. [8.2.2, 11.2]

Bateman, H. (1910). On the probability distribution of particles, *Philosophical Magazine, Series 6*, **20**, 704–707. [4.2]

Beall, G. (1940). The fit and significance of contagious distributions when applied to observations on larval insects, *Ecology*, **21**, 460–474. [9.6.1, 9.9]

Beall, G. (1942). The transformation of data from entomological field experiments so that the analysis of variance becomes applicable, *Biometrika*, **32**, 243–262. [5.6]

Beall, G., and Rescia, R. R. (1953). A generalization of Neyman's contagious distributions, *Biometrics*, **9**, 354–386. [5.12.6, 9.9]

Behboodian, J. (1975). Structural properties and statistics of finite mixtures, *Statistical Distributions in Scientific Work*, **1**: *Models and Structures*, G. P. Patil, S. Kotz, and J. K. Ord (editors), 103–112. Dordrecht: Reidel . [8.2.1]

Bellman, R. E. (1960). On the separation of exponentials, *Bolletino della Unione Matematica Italiana*, **15**, 38–39. [8.2.1]

Benkherouf, L., and Bather, J. A. (1988). Oil exploration: sequential decisions in the face of uncertainty, *Journal of Applied Probability*, **25**, 529–543. [4.12.6, 7.11]

Bennett, B. M., and Birch, B. (1964). Sampling inspection tables for comparison of two groups using the negative binomial model, *Trabajos de Estadística*, **15**, 1–12. [5.10]

Bennett, B. M., and Nakamura, E. (1970). Note on an approximation to the distribution of the range from Poisson samples, *Random Counts in Scientific Work*, **1**: *Random Counts in Models and Structures*, G. P. Patil (editor), 119–121. University Park: Pennsylvania State University Press. [4.5]

Bennett, W. S. (1965). *A New Binomial Approximation for Cumulative Hypergeometric Probabilities*, PhD. thesis. Washington, DC: American University. [6.5]

Berg, S. (1974). Factorial series distributions, with applications to capture-recapture problems, *Scandinavian Journal of Statistics*, **1**, 145–152. [2.6, 3.4]

Berg, S. (1975). A note on the connection between factorial series distributions and zero-truncated power series distributions, *Scandinavian Actuarial Journal*, 233–237. [2.6]

Berg, S. (1978). Characterization of a class of discrete distributions by properties of their moment distributions, *Communications in Statistics-Theory and Methods*, **A7**, 785–789. [2.6]

Berg, S. (1983a). Factorial series distributions, *Encyclopedia of Statistical Sciences*, **3**, S. Kotz, N. L. Johnson and C. B. Read (editors), 17-22. New York: Wiley. [2.6, 3.4]

Berg, S. (1983b). Random contact processes, snowball sampling and factorial series distributions, *Journal of Applied Probability*, **20**, 31–46. [2.6]

Berg, S. (1985). Generating discrete distributions from modified Charlier type B expansions, *Contributions to Probability and Statistics in Honour of Gunnar Blom*, J. Lanke and G. Lindgren (editors), 39–48. Lund: University of Lund. [11.8]

Berg, S. (1988a). Stirling distributions, *Encyclopedia of Statistical Sciences*, **8**, S. Kotz, N. L. Johnson and C. B. Read (editors), 773–776. New York: Wiley. [7.11]

Berg, S. (1988b). Urn models, *Encyclopedia of Statistical Sciences*, **9**, S. Kotz, N. L. Johnson and C. B. Read (editors), 424–436. New York: Wiley. [5.1, 11.19]

Berg, S., and Mutafchiev, L. (1990). Random mappings with an attracting centre: Lagrangian distributions and a regression function, *Journal of Applied Probability*, **27**, 622–636. [2.5.2]

Berg, S., and Nowicki, K. (1991). Statistical inference for a class of modified power series distributions with applications to random mapping theory, *Journal of Statistical Planning and Inference*, **28**, 247–261. [2.5.2]

Berger, J. O. (1985). *Statistical Decision Theory and Bayesian Analysis* (Second edition), New York: Springer-Verlag. [1B.15]

Berger, R. L. (1991). A remark on algorithm AS 152: cumulative hypergeometric probabilities, *Applied Statistics*, **40**, 374–375. [6.6]

Berljand, O. S., Nazarov, I. M., and Pressman, A. J. (1962). i^n erfc distribution — or composite Poisson distribution, *Soviet Mathematics*, **3**, 1744–1746. [9.4]

Bernardo, J. M. (1976). Algorithm AS103: Psi (digamma) function, *Applied Statistics*, **25**, 315–317. [11.3]

Bernado, J. M., DeGroot, M. H., Lindley, D. V., and Smith, A. F. M. (editors) (1988). *Bayesian Statistics*, **3**. *Proceedings of the Third Valencia International Meeting*. Amsterdam: North-Holland. [1B.15]

Bernoulli, J. (1713). *Ars Conjectandi*, Basilea: Thurnisius. [11.19]

Bertrand, J. L. F. (1889). *Calcul des Probabilités*, Paris: Gauthier-Villars. [3.3]

Best, D. J. (1974). The variance of the inverse binomial estimator, *Biometrika*, **61**, 385–386. [5.8.2]

Best, D. J., and Gipps, P. G. (1974). An improved gamma approximation to the negative binomial, *Technometrics*, **16**, 621–624. [5.6]

Bhalerao, N. R., and Gurland, J. (1977). A unified approach to estimating parameters in some generalized Poisson distributions, *Technical Report No. 494*, Madison, WI: University of Wisconsin. [9.9]

Bhalerao, N. R., Gurland, J., and Tripathi, R. C. (1980). A method of increasing power of a test for the negative binomial and Neyman type A distributions, *Journal of the American Statistical Association*, **75**, 934–938. [9.6.4]

Bhat, U. N. (1978). Theory of queues. *Handbook of Operations Research*, **1**, J. J. Moder and S. E. Elmaghraby (editors), 352–397. New York: Van Nostrand-Reinhold. [11.13]

Bhat, U. N. (1984). *Elements of Applied Stochastic Processes* (Second edition), New York: Wiley. [5.2, 5.3, 5.10, 11.13]

Bhattacharya, S. K. (1966). Confluent hypergeometric distributions of discrete and continuous type with applications to accident proneness, *Bulletin of the Calcutta Statistical Association*, **15**, 20–31. [3.12.5, 4.12.4, 5.3, 8.3.2, 8.3.4]

Bhattacharya, S. K., and Holla, M. S. (1965). On a discrete distribution with special reference to the theory of accident proneness, *Journal of the American Statistical Association*, **60**, 1060–1066. [8.3.2]

Bienaymé, I. J. (1874). Sur une question de probabilités, *Bulletin de la Société Mathématique de France*, **2**, 153-154. [10.5.2]

Binet, F. (1953). The fitting of the positive binomial distribution when both parameters are estimated from the sample, *Annals of Eugenics, London*, **18**, 117–119. [3.8.2]

Binns, M. R. (1974). Approximating the negative binomial via the positive binomial, *Technometrics*, **16**, 323–324. [5.6]

Binns, M. R. (1975). Sequential estimation of the mean of a negative binomial distribution, *Biometrika*, **62**, 433–440. [5.8.2]

Binns, M. R. (1986). Behavioral dynamics and the negative binomial distribution, *Oikos*, **47**, 315–318. [5.10]

Birch, M. W. (1963). An algorithm for the logarithmic series distribution, *Biometrics*, **19**, 651–652. [7.7.2]

Birnbaum, A. (1964). Median-unbiased estimators, *Bulletin of Mathematical Statistics*, **11**, 25–34. [3.8.2]

Bishop, Y. M. M., Fienberg, S. E., and Holland, P. W. (1975). *Discrete Multivariate Analysis: Theory and Practice*, Cambridge, MA: MIT Press. [4.9]

Bissell, A. F. (1972a). A negative binomial model with varying element sizes, *Biometrika*, **59**, 435–441. [5.12.4]

Bissell, A. F. (1972b). Another negative binomial model with varying element sizes, *Biometrika*, **59**, 691–693. [5.12.4]

Bissinger, B. H. (1965). A type-resisting distribution generated from considerations of an inventory decision problem, *Classical and Contagious Discrete Distributions*, G. P. Patil (editor), 15–17. Calcutta: Statistical Publishing Society; Oxford: Pergamon Press. [6.2.4, 11.12]

Bizley, M. T. L. (1951). Some notes on probability, *Journal of the Institute of Actuaries Students' Society*, **10**, 161–203. [3.6.1]

Blackman, G. E. (1935). A study by statistical methods of the distribution of species in a grassland association, *Annals of Botany (New Series)*, **49**, 749–778. [7.9]

Blischke, W. R. (1962). Moment estimation for the parameters of a mixture of two binomial distributions, *Annals of Mathematical Statistics*, **33**, 444–454. [8.2.4]

Blischke, W. R. (1964). Estimating the parameters of mixtures of binomial distributions, *Journal of the American Statistical Association*, **59**, 510–528. [8.2.3, 8.2.4]

Blischke, W. R. (1965). Mixtures of discrete distributions, *Classical and Contagious Discrete Distributions*, G. P. Patil (editor), 351–372. Calcutta: Statistical Publishing Society; Oxford: Pergamon Press. [8.2.1, 8.2.4, 8.3.1, 8.3.2]

Bliss, C. I. (1965). An analysis of some insect trap records, *Classical and Contagious Discrete Distributions*, G. P. Patil (editor), 385–397. Calcutta: Statistical Publishing Society; Oxford: Pergamon Press. [7.11]

Bliss, C. I., and Fisher, R. A. (1953). Fitting the negative binomial distribution to biological data and note on the efficient fitting of the negative binomial, *Biometrics*, **9**, 176–200. [5.8.3, 5.8.4, 5.10]

Bliss, C. I., and Owen, A. R. G. (1958). Negative binomial distributions with a common *k*, *Biometrika*, **45**, 37–58. [5.8.4]

Blom, G., and Holst, L. (1986). Random walks of ordered elements with applications, *American Statistician*, **40**, 271–274. [11.14]

Blom, G., Thorburn, D., and Vessey, T. A. (1990). The distribution of the record position and its applications, *American Statistician*, **44**, 151–153. [11.14]

Blomqvist, N. (1952). On an exhaustion process, *Skandinavisk Aktuarietidskrift*, **35**, 201–210. [11.1]

Blum, J. R., and Susarla, V. (1977). Estimation of a mixing distribution function, *Annals of Probability*, **5**, 200–209. [8.3.1]

Blumenthal, S., and Dahiya, R. C. (1981). Estimating the binomial parameter *n*, *Journal of the American Statistical Association*, **76**, 903–909. [3.8.2]

Blumenthal, S., Dahiya, R. C., and Gross, A. J. (1978). Estimating the complete sample size from an incomplete Poisson sample, *Journal of the American Statistical Association*, **73**, 182–187. [4.10.1]

Blyth, C. R. (1980). Expected absolute error of the usual estimator of the binomial parameter, *American Statistician*, **34**, 155–157. [3.8.2]

Blyth, C. R. (1986). Approximate binomial confidence limits, *Journal of the American Statistical Association*, **81**, 843–855. [3.8.3]

Blyth, C. R., and Hutchinson, D. W. (1960). Tables of Neyman-shortest confidence intervals for the binomial parameter, *Biometrika*, **47**, 381–391. [3.8.3]

Blyth, C. R., and Hutchinson, D. W. (1961). Tables of Neyman-shortest-unbiased confidence intervals for the Poisson parameter, *Biometrika*, **48**, 191–194. [4.7.3]

Blyth, C. R., and Still, H. A. (1983). Binomial confidence intervals, *Journal of the American Statistical Association*, **78**, 108–116. [3.8.3]

Bohman, H. (1963). Two inequalities for Poisson distributions, *Skandinavisk Aktuarietidskrift*, **46**, 47–52. [4.5]

Böhning, D. (1983a). Refined estimation of the logarithmic series distribution, *EDV in Medizin und Biologie*, **14**, 58–61. [7.7.2]

Böhning, D. (1983b). Maximum likelihood estimation of the logarithmic series distribution, *Statistische Hefte*, **24**, 121–140. [7.7.2]

Bollinger, R. C., and Salvia, A. A. (1982). Consecutive-*k*-out-of-*n*:F networks, *IEEE Transactions on Reliability*, **R-31**, 53–55. [10.6.2]

Bol'shev, L. N. (1964). Distributions related to the hypergeometric, *Teoriya Veroyatnostei i ee Primeneniya*, **9**, 687–692. [6.2.2, 6.5]

Bol'shev, L. N. (1965). On a characterization of the Poisson distribution, *Teoriya Veroyatnostei i ee Primeneniya*, **10**, 446–456. [4.8]

Bondesson, L. (1988). On the gain by spreading seeds: a statistical analysis of sowing experiments, *Scandinavian Journal of Forest Research*, 305–314. [8.2.4]

Boole, G. (1854). *An Investigation of the Laws of Thought on Which are Founded the Mathematical Theories of Logic and Probability* (Reprinted 1951), New York: Dover. [10.2]

Borah, M. (1984). The Gegenbauer distribution revisited: Some recurrence relations for moments, cumulants, etc, estimation of parameters and its goodness of fit. *Journal of the Indian Society for Agricultural Statistics*, **36**, 72–78. [11.7]

Borel, E. (1942). Sur l'emploi du théorème de Bernoulli pour faciliter le calcul d'un infinité de coefficients. Application au problème de l'attente à un guichet, *Comptes Rendus, Académie des Sciences, Paris, Series A*, **214**, 452–456. [9.11, 11.13]

Borenius, G. (1953). On the statistical distribution of mine explosions, *Skandinavisk Aktuarietidskrift*, **36**, 151–157. [11.1]

Borges, R. (1970). Eine Approximation der Binomialverteilung durch die Normalverteilung der Ordnung $1/n$, *Zeitschrift für Wahrscheinlichkeitstheorie und Verwandte Gebeite*, **14**, 189–199. [3.6.1]

Bortkiewicz, L. von (1898). *Das Gesetz der Kleinen Zahlen*, Leipzig: Teubner. [4.2]

Bortkiewicz, L. von (1915). Über die Zeitfolge zufälliger Ereignisse, *Bulletin de l'Institut International de Statistique*, **20**, 30–111. [4.2]

Bortkiewicz, L. von (1917). *Die Iterationen*, Berlin: Springer. [10.5.1]

Bosch, A. J. (1963). The Pólya distribution, *Statistica Neerlandica*, **17**, 201–213. [5.3, 6.2.4]

Boswell, M. T., Gore, S. D., Patil, G. P., and Taillie, C. (1993). The art of computer generation of random variables, *Handbook of Statistics*, **9**, C. R. Rao (editor). Amsterdam: North Holland. [1C.1]

Boswell, M. T., Ord, J. K., and Patil, G. P. (1979). Chance mechanisms underlying univariate distributions, *Statistical Ecology*, **4**: *Statistical Distributions in Ecological Work*, J. K. Ord, G. P. Patil, and C. Taillie (editors), 1–156. Fairland, MD: International Co-operative Publishing House. [3.10, 9.2]

Boswell, M. T., and Patil, G. P. (1970). Chance mechanisms generating the negative binomial distributions, *Random Counts in Scientific Work*, **1**: *Random Counts in Models and Structures*, G. P. Patil (editor), 3–22. University Park: Pennsylvania State University Press. [5.3, 5.10, 5.11]

Boswell, M. T., and Patil, G. P. (1971). Chance mechanisms generating the logarithmic series distribution used in the analysis of number of species and individuals, *Statistical Ecology*, **1**: *Spatial Patterns and Statistical Distributions*, G. P. Patil, E. C. Pielou, and W. E. Waters (editors), 99–130. University Park: Pennsylvania State University Press. [7.1, 7.2]

Boswell, M. T., and Patil, G. P. (1973). Characterization of certain discrete distributions by differential equations with respect to their parameters, *Australian Journal of Statistics*, **15**, 128–131. [4.8]

Bowerman, P. N., and Scheuer, E. M. (1990). Calculation of the binomial survivor function, *IEEE Transactions on Reliability*, **39**, 162–166. [3.7]

Bowers, N. L., Gerber, H. U., Hickman, J. C., Jones, D. A., and Nesbitt, C. J. (1986). *Actuarial Mathematics*, Chicago: Society of Actuaries. [9.12]

Bowman, K. O. (1984). Extended moment series and the parameters of the negative binomial distribution, *Biometrics*, **40**, 249–252. [5.8.3]

Bowman, K. O., Kastenbaum, M. A., and Shenton, L. R. (1992). The negative hypergeometric distribution and estimation by moments, *Communications in Statistics-Simulation and Computation*, **21**, 301–332 [6.7.2, 8.3.3]

Bowman, K. O., and Shenton, L. R. (1965). Asymptotic covariance for the maximum likelihood estimators of the parameters of a negative binomial distribution, *Union Carbide Corporation Report*, K-1643: Oak Ridge, TN. [5.8.3]

Bowman, K. O., and Shenton, L. R. (1966). Biases of estimators for the negative binomial distribution, *Union Carbide Corporation Report*, ORNL-4005: Oak Ridge, TN. [5.8.3]

Bowman, K. O., and Shenton, L. R. (1967). Remarks on estimation problems for the parameters of the Neyman type A distribution, *Oak Ridge National Laboratory Report*, ORNL-4102: Oak Ridge, TN. [9.6.3, 9.6.4]

Bowman, K. O., and Shenton, L. R. (1970). Properties of the maximum likelihood estimator for the parameter of the logarithmic series distribution, *Random Counts in Scientific Work*, **1**: *Random Counts in Models and Structures*, G. P. Patil (editor), 127–150. University Park: Pennsylvania State University Press. [7.7.2]

Bowman, K. O., Shenton, L. R., and Kastenbaum, M. A. (1991). Discrete Pearson distributions, *Oak Ridge National Laboratory Technical Report*, TM-11899: Oak Ridge, TN. [2.3.2]

Box, G. E. P., and Cox, D. R. (1964). An analysis of transformations, *Journal of the Royal Statistical Society, Series B*, **26**, 211–252. [5.10]

Box, G. E. P., and Tiao, G. C. (1973). *Bayesian Inference in Statistics*, Reading, MA: Addison-Wesley. [1B.3]

Boyer, C. B. (1950). Cardan and the Pascal triangle, *American Mathematical Monthly*, **57**, 387–390. [3.2]

Brainerd, B. (1972). On the relation between types and tokens in literary text, *Journal of Applied Probability*, **9**, 507–518. [3.12.2]

Brass, W. (1958). Simplified methods of fitting the truncated negative binomial distribution, *Biometrika*, **45**, 59–68. [5.11]

Bratley, P., Fox, B. L., and Schrage, L. E. (1987). *A Guide to Simulation* (Second edition), New York: Springer-Verlag. [1C.1]

Brenner, D. J., and Quan, H. (1990). Exact confidence limits for binomial proportions — Pearson and Hartley revisited, *The Statistician*, **39**, 391–397. [3.8.3]

Broderick, T. S. (1937). On some symbolic formulae in probability theory, *Proceedings of the Royal Irish Academy, Series A*, **44**, 19–28. [10.2]

Brooks, R. J., James, W. H., and Gray. E. (1991). Modelling sub-binomial variation in the frequency of sex combinations in litters of pigs, *Biometrics*, **47**, 403–417. [3.12.6]

Brown, B. (1965). Some tables of the negative binomial distribution and their use, *RAND Corporation Memorandum*, RM-4577-PR: Santa Monica, CA. [5.7]

Brown, M. B., and Bromberg, J. (1984). An efficient two-stage procedure for generating variates from the multinomial distribution, *American Statistician*, **38**, 216-219. [4.8]

Brunk, H. D., Holstein, J. E., and Williams, F. (1968). A comparison of binomial approximations to the hypergeometric distribution, *American Statistician*, **22**, 24–26. [6.5]

Bruns, H. (1906). *Wahrscheinlichkeitsrechnung und Kollektivmasslehre*, Leipzig and Berlin: Teubner. [10.5.1]

Bryc, W. (1987). A characterization of the Poisson process by conditional moments, *Stochastics*, **20**, 17–26. [4.8]

Burr, I. W. (1973). Some approximate relations between terms of the hypergeometric, binomial and Poisson distributions, *Communications in Statistics*, **1**, 297–301. [6.5]

Burrell, Q. L., and Cane, V. R. (1982). The analysis of library data, *Journal of the Royal Statistical Society, Series A*, **145**, 439–471. [5.10]

Bustoz, J., Feldstein, A., Goodman, R., and Linnainmaa, S. (1979). Improved trailing digits estimates applied to optimal computer arithmetic, *Journal of the ACM.*, **26**, 716–730. [7.1]

Buxton, P. A. (1940). Studies on populations of head lice, III: Material from South India, *Parasitology*, **32**, 296–302. [7.9]

Cacoullos, T. (1975). Multiparameter Stirling and C-type distributions, *Statistical Distributions in Scientific Work*, **1**: *Models and Structures*, G. P. Patil, S. Kotz, and J. K. Ord (editors), 19–30. Dordrecht: Reidel. [4.12.3, 7.11]

Cacoullos, T., and Charalambides, C. A. (1975). On MVUE for truncated binomial and negative binomial distributions, *Annals of the Institute of Statistical Mathematics, Tokyo*, **27**, 235–244. [5.11]

Caraco, T. (1979). Ecological response of animal group size frequencies, *Statistical Ecology*, **4**: *Statistical Distributions in Ecological Work*, J. K. Ord, G. P. Patil, and C.Taillie (editors), 371–386. Fairland, MD: International Co-operative Publishing House. [7.2]

Carroll, R. J., and Lombard, F. (1985). A note on *N* estimators for the binomial distribution, *Journal of the American Statistical Association*, **80**, 423–426. [3.8.2]

Carver, H. C. (1919). On the graduation of frequency distributions, *Proceedings of the Casualty Actuarial Society of America*, **6**, 52–72. [2.3.1]

Carver, H. C. (1923). Frequency curves, *Handbook of Mathematical Statistics*, H. L. Rietz (editor), 92–119. Cambridge MA: Riverside. [2.3.1, 2.3.2]

Casella, G. (1986). Refining binomial confidence intervals, *Canadian Journal of Statistics*, **14**, 113–129. [3.8.2]

Casella, G., and Robert, C. (1989). Refining Poisson confidence intervals, *Canadian Journal of Statistics*, **17**, 45–57. [4.7.3]

Cassie, R. M. (1962). Frequency distribution models in the ecology of plankton and other organisms, *Journal of Animal Ecology*, **31**, 65–92. [7.11]

Castoldi, L. (1963). Poisson processes with events in clusters, *Rendiconti del Seminaro della Facoltà di Scienze della Università di Cagliari*, **33**, 433–437. [9.3]

Castro, G. de (1952). Note on differences of Bernoulli and Poisson variables, *Portugaliae Mathematica*, **11**, 173–175. [4.12.3]

Catcheside, D. G. (1948). Genetic effects of radiation, *Advances in Genetics* **2**, M. Demerec (editor), 271–358. New York: Academic Press. [9.2]

Cernuschi, F., and Castagnetto, L. (1946). Chains of rare events, *Annals of Mathematical Statistics*, **17**, 53–61. [9.6.1]

Chaddha, R. L. (1965). A case of contagion in binomial distribution, *Classical and Contagious Discrete Distributions*, G. P. Patil (editor), 273–290. Calcutta: Statistical Publishing Society; Oxford: Pergamon Press. [11.19]

Chahine, J. (1965). Une généralization de la loi binomiale négative, *Revue Statistique Appliquée*, **13**, 33–43. [6.3]

Champernowne, D. G. (1956). An elementary method of solution of the queueing problem with a single server and constant parameters, *Journal of the Royal Statistical Society, Series B*, **18**, 125–128. [11.13]

Chan, B. (1982). Derivation of moment formulas by operator valued probability generating functions, *American Statistician*, **36**, 179–181. [1B.5],

Chandler, K. N. (1952). The distribution and frequency of record values, *Journal of the Royal Statistical Society, Series B*, **14**, 220–228. [11.14]

Chandra, S. (1977). On the mixtures of probability distributions, *Scandinavian Journal of Statistics*, **4**, 105–112. [8.2.1]

Chang, D. K. (1989). On infinitely divisible discrete distributions, *Utilitas Mathematica*, **36**, 215–217. [9.3]

Chao, M. T., and Strawderman, W. E. (1972). Negative moments of positive random variables, *Journal of the American Statistical Association*, **67**, 429–431. [3.3]

Chapman, D. G. (1948). A mathematical study of confidence limits of salmon populations calculated from sample tag ratios, *International Pacific Salmon Fisheries Commision Bulletin*, **2**, 69–85. [6.7.1]

Chapman, D. G. (1951). Some properties of the hypergeometric distribution with applications to zoological sample censuses, *University of California Publications in Statistics*, **1**, 131–159. [6.7.1]

Chapman, D. G. (1952a). Inverse, multiple and sequential sample censuses, *Biometrics*, **8**, 286–306. [6.9.1, 6.9.2]

Chapman, D. G. (1952b). On tests and estimates for the ratio of Poisson means, *Annals of the Institute of Statistical Mathematics, Tokyo*, **4**, 45–49. [4.7.2, 4.7.3]

Chapman, D. W. (1934). The statistics of the method of correct matchings, *American Journal of Psychology*, **46**, 287–298. [10.3]

Charalambides, C. A. (1977a). On the generalized discrete distributions and the Bell polynomials, *Sankhyā, Series B*, **39**, 36–44. [9.1]

Charalambides, C. A. (1977b). A new kind of numbers appearing in the *n*-fold convolution of truncated binomial and negative binomial distributions, *SIAM Journal of Applied Mathematics*, **33**, 279–288. [5.11]

Charalambides, C. A. (1986a). Gould series distributions with applications to fluctuations of sums of random variables, *Journal of Statistical Planning and Inference*, **14**, 15–28. [2.5.3, 5.12.3, 9.1, 11.10]

Charalambides, C. A. (1986b). On discrete distributions of order *k*, *Annals of the Institute of Statistical Mathematics, Tokyo*, **38**, 557–568. [10.6.2]

Charalambides, C. A. (1986c). Derivation of probabilities and moments of certain generalized discrete distributions via urn models, *Communications in Statistics-Theory and Methods*, **15**, 677–696. [9.1]

Charalambides, C. A. (1990). Abel series distributions with applications to fluctuations of sample functions of stochastic processes, *Communications in Statistics-Theory and Methods*, **19**, 317–335. [2.5.3, 3.12.3]

Charalambides, C. A. (1991). On a generalized Eulerian distribution, *Annals of the Institute of Statistical Mathematics, Tokyo*, **43**, 197–206. [2.5.3]

Charalambides, C. A., and Singh, J. (1988). A review of the Stirling numbers, their generalizations and statistical applications, *Communications in Statistics-Theory and Methods*, **17**, 2533–2595. [1A.3]

Charlier, C. V. L. (1905a). Die zweite Form des Fehlergesetzes, *Arkiv für Matematik, Astronomi och Fysik No. 15*, **2**, 1–8. [4.2]

Charlier, C. V. L. (1905b). Über die Darstellung willkürlicher Funktionen, *Arkiv für Matematik, Astronomi och Fysik No. 20*, **2**, 1–35. [4.2]

Charlier, C. V. L. (1920). *Vorlesungen über die Grundzüge der Mathematischen Statistik*, Hamburg. [3.12.2]

Chatfield, C. (1969). On estimating the parameters of the logarithmic series and negative binomial distributions, *Biometrika*, **56**, 411–414. [7.7.2]

Chatfield, C. (1970). Discrete distributions in market research, *Random Counts in Scientific Work*, **3**: *Random Counts in Physical Science, Geo Science, and Business*, G. P. Patil (editor), 163–181. University Park: Pennsylvania State University Press. [7.9]

Chatfield, C. (1975). A marketing application of a characterization theorem, *Statistical Distributions in Scientific Work*, **2**: *Model Building and Model Selection*, G. P. Patil, S. Kotz, and J. K. Ord (editors), 175–185. Dordrecht: Reidel. [5.10]

Chatfield, C. (1983). *Statistics for Technology* (Third edition), London: Chapman & Hall. [4.9]

Chatfield, C. (1986). Discrete distributions and purchasing models, *Communications in Statistics-Theory and Methods*, **15**, 697–708. [7.9]

Chatfield, C., Ehrenberg, A. S. C., and Goodhardt, G. J. (1966). Progress on a simplified model of stationary purchasing behaviour (with discussion), *Journal of the Royal Statistical Society, Series A*, **129**, 317–367. [5.10, 7.9, 7.10]

Chatfield, C., and Goodhardt, G. J. (1970). The beta-binomial model for consumer purchasing behaviour, *Applied Statistics*, **19**, 240–250. [6.7.2, 6.9.2]

Chatterji, S. D. (1963). Some elementary characterizations of the Poisson distribution, *American Mathematical Monthly*, **70**, 958–964. [4.8]

Chaundy, T. W. (1962). F. H. Jackson, (obituary), *Journal of the London Mathematical Society*, **37**, 126–128. [1A.12]

Chen, H.-C., and Asau, Y. (1974). On generating random variates from an empirical distribution, *AIIE Transactions*, **6**, 163–166. [1C.2]

Chen, L. H. Y. (1975). Poisson approximation for dependent trials, *Annals of Probability*, **3**, 534–545. [4.2]

Chen, R. (1978). A surveillance system for congenital malformations, *Journal of the American Statistical Association*, **73**, 323–327. [5.2]

Cheng, T.-T. (1949). The normal approximation to the Poisson distribution and a proof of a conjecture of Ramanujan, *Bulletin of the American Mathematical Society*, **55**, 396–401. [4.4]

Chew, V. (1964). Applications of the negative binomial distribution with probability of misclassification, *Virginia Journal of Science (New Series)*, **15**, 34–40. [5.10]

Chew, V. (1971). Point estimation of the parameter of the binomial distribution, *American Statistician*, **25**, 47–50 . [3.8.2]

Chiang, D., and Niu, S. C. (1981). Reliability of consecutive-k-out-of-n:F systems, *IEEE Transactions on Reliability*, **R-30**, 87–89. [10.6.2]

Chong, K.-M. (1977). On characterizations of the exponential and geometric distributions by expectations, *Journal of the American Statistical Association*, **72**, 160–161. [5.9.1]

Chukwu, W. I. E., and Gupta, D. (1989). On mixing generalized Poisson with generalized gamma distribution, *Metron*, **47**, 314–318. [8.3.2]

Chung, J. H., and DeLury, D. B. (1950). *Confidence Limits for the Hypergeometric Distribution*, Toronto: University of Toronto Press. [6.6, 6.7.1, 6.9.1]

Chung, K. L., and Feller, W. (1949). Fluctuations in coin tossing, *Proceedings of the National Academy of Sciences USA*, **35**, 605–608. [6.10.2]

Clark, C. R., and Koopmans, L. H. (1959). Graphs of the hypergeometric O.C., and A.O.Q. functions for lot sizes 10 to 225, *Sandia Corporation Monograph*, SCR-121. [6.6]

Clark, R. E. (1953). Percentage points of the incomplete beta function, *Journal of the American Statistical Association*, **48**, 831–843. [3.8.3]

Clark, S. J., and Perry, J. N. (1989). Estimation of the negative binomial parameter k by maximum quasi-likelihood, *Biometrics*, **45**, 309–316. [5.8.3, 5.10]

Clemans, K. G. (1959). Confidence limits in the case of the geometric distribution, *Biometrika*, **46**, 260–264. [5.2]

Clevenson, M. L. and Zidek, J. V. (1975). Simultaneous estimation of the means of independent Poisson laws, *Journal of the American Statistical Association*, **70**, 698–705. [4.7.2]

Cliff, A. D., and Ord, J. K. (1981). *Spatial Processes: Models and Applications*, London: Pion. [4.9]

Clopper, C. J., and Pearson, E. S. (1934). The use of confidence or fiducial limits illustrated in the case of the binomial, *Biometrika*, **26**, 404–413. [3.8.3]

Cochran, W. G. (1954). Some methods for strengthening the common χ^2 tests, *Biometrics*, **10**, 417–451. [6.9.1]

Cohen, A. C. (1954). Estimation of the Poisson parameter from truncated samples and from censored samples, *Journal of the American Statistical Association*, **49**, 158–168. [4.10.2]

Cohen, A. C. (1959). Estimation in the Poisson distribution when sample values of $c+1$ are sometimes erroneously reported as c, *Annals of the Institute of Statistical Mathematics, Tokyo*, **9**, 189–193. [4.10.3]

Cohen, A. C. (1960a). Estimating the parameter in a conditional Poisson distribution, *Biometrics*, **16**, 203–211. [4.10.1]

Cohen, A. C. (1960b). An extension of a truncated Poisson distribution, *Biometrics*, **16**, 446–450. [4.10.3]

Cohen, A. C. (1960c). Estimating the parameters of a modified Poisson distribution, *Journal of the American Statistical Association*, **55**, 139–143. [4.10.3, 8.2.2]

Cohen, A. C. (1960d). Estimation in a truncated Poisson distribution when zeros and some ones are missing, *Journal of the American Statistical Association*, **55**, 342–348. [4.10.3]

Cohen, A. C. (1961). Estimating the Poisson parameter from samples that are truncated on the right, *Technometrics*, **3**, 433–438. [4.10.2]

Cohen, A. C. (1965). Estimation in mixtures of discrete distributions, *Classical and Contagious Discrete Distributions*, G. P. Patil (editor), 373–378. Calcutta: Statistical Publishing Society; Oxford: Pergamon Press. [8.2.5]

Cohen, J. W. (1982). *The Single Server Queue*, Amsterdam: North Holland. [11.13]

Consael, R. (1952). Sur les processus de Poisson du type composé, *Académie Royale de Belgique Bulletin, Classe de Sciences 5e Série*, **38**, 442–461. [8.3.2]

Consul, P. C. (1974). A simple urn model dependent upon predetermined strategy, *Sankhyā, Series B*, **36**, 391–399. [2.5.3, 5.12.3, 11.19]

Consul, P. C. (1975). Some new characterizations of discrete Lagrangian distributions, *Statistical Distributions in Scientific Work*, **3**: *Characterizations and Applications*, G. P. Patil, S. Kotz, and J. K. Ord (editors), 279–290. Dordrecht: Reidel. [9.11]

Consul, P. C. (1981). Relation of modified power series distributions to Lagrangian probability distributions, *Communications in Statistics-Theory and Methods*, **A10**, 2039–2046. [2.2.2, 2.5.2]

Consul, P. C. (1983). Lagrange and related probability distributions, *Encyclopedia of Statistical Sciences*, **4**, S. Kotz, N. L. Johnson and C. B. Read (editors), 448–454. New York: Wiley. [2.5.2]

Consul, P. C. (1984). On the distributions of order statistics for a random sample size, *Statistica Neerlandica*, **38**, 249–256. [9.11]

Consul, P. C. (1989). *Generalized Poisson Distributions*, New York: Dekker. [2.5.2, 4.11, 5.12.3, 8.3.1, 9.11]

Consul, P. C. (1990a). Geeta distribution and its properties, *Communications in Statistics-Theory and Methods*, **19**, 3051–3068. [11.6]

Consul, P. C. (1990b). Two stochastic models for the Geeta distribution, *Communications in Statistics-Theory and Methods*, **19**, 3699–3706. [11.6]

Consul, P. C. (1990c). New class of location-parameter discrete probability distributions and their characterizations, *Communications in Statistics-Theory and Methods*, **19**, 4653–4666. [2.2.2, 11.6]

Consul, P. C., and Famoye, F. (1989). Minimum variance unbiased estimation for the Lagrange power series distributions, *Statistics*, **20**, 407–415. [2.5.2, 5.12.3]

Consul, P. C., and Gupta, H. C. (1980). The generalized negative binomial distribution and its characterization by zero regression, *SIAM Journal of Applied Mathematics*, **39**, 231–237. [5.12.3]

Consul, P. C., and Jain, G. C. (1973a). A generalization of the Poisson distribution, *Technometrics*, **15**, 791–799. [2.5.2, 2.5.3]

Consul, P. C., and Jain, G. C. (1973b). On some interesting properties of the generalized Poisson distribution, *Biometrische Zeitschrift*, **15**, 495–500. [2.5.3, 9.11]

Consul, P. C., and Mittal, S. P. (1975). A new urn model with predetermined strategy, *Biometrische Zeitschrift*, **17**, 67–75. [2.5.3, 3.12.3, 11.19]

Consul, P. C., and Shenton, L. R. (1972). Use of Lagrange expansion for generating generalized probability distributions, *SIAM Journal of Applied Mathematics*, **23**, 239–248. [2.5.2, 2.5.3, 3.12.3, 5.12.3]

Consul, P. C., and Shenton, L. R. (1973). Some interesting properties of Lagrangian distributions, *Communications in Statistics*, **2**, 263–272. [2.5.2, 3.12.3, 5.12.3]

Consul, P. C., and Shenton, L. R. (1975). On the probabilistic structure and properties of discrete Lagrangian distributions, *Statistical Distributions in Scientific Work*,

1: *Models and Structures*, G. P. Patil, S. Kotz, and J. K. Ord (editors), 41–48. Dordrecht: Reidel. [2.5.2, 9.11]

Consul, P. C., and Shoukri, M. M. (1984). Maximum likelihood estimation for the generalized Poisson distribution, *Communications in Statistics-Theory and Methods*, **13**, 1533–1547. [9.11]

Consul, P. C., and Shoukri, M. M. (1985). The generalized Poisson distribution when the sample mean is larger than the sample variance, *Communications in Statistics-Theory and Methods*, **14**, 667–681. [9.11]

Consul, P. C., and Shoukri, M. M. (1986). The negative integer moments of the generalized Poisson distribution, *Communications in Statistics-Theory and Methods*, **15**, 1053–1064. [9.11]

Cook, G. W., Kerridge, D. F., and Pryce, J. D. (1974). Estimation of functions of a binomial parameter, *Sankhyā, Series A*, **36**, 443–448. [3.8.2]

Coolidge, J. L. (1921). The dispersion of observations, *Bulletin of the American Mathematical Society*, **27**, 439–442. [3.12.2]

Cormack, R. M. (1979). Models for capture-recapture, *Statistical Ecology*, **5**, R. M. Cormack, G. P. Patil, and D. S. Robson (editors), 217–255. Burtonsville, MD: International Co-operative Publishing House. [6.9.1]

Cormack, R. M. (1989). Log-linear models for capture-recapture, *Biometrics*, **45**, 395–413. [4.9]

Cormack, R. M., and Jupp, P. E. (1991). Inference for Poisson and multinomial models for capture-recapture experiments, *Biometrika*, **78**, 911–916. [4.9]

Cournot, M. A. A. (1843). *Exposition de la Théorie des Chances et des Probabilités*, Paris: Librairie de L. Machette. [6.2.1]

Cox, D. R. (1975). Partial likelihood, *Biometrika*, **62**, 269–276. [1B.15]

Cox, D. R. (1983). Some remarks on overdispersion, *Biometrika*, **70**, 269–274. [8.3.2]

Cox, D. R., and Hinkley, D. V. (1974). *Theoretical Statistics*, London: Chapman and Hall. [1B.15]

Cox, D. R., and Isham, V. (1980). *Point Processes*, London: Chapman and Hall. [8.3.2]

Cox, D. R., and Miller, H. D. (1965). *The Theory of Stochastic Processes*, London: Methuen. [5.12.5]

Cox, D. R., and Reid, N. (1987). Parameter orthogonality and approximate conditional inference, *Journal of the Royal Statistical Society, Series B*, **49**, 1–39. [1B.15]

Cox, D. R., and Smith, W. L. (1961). *Queues*, London: Methuen. [11.13]

Cox, D. R., and Snell, E. J. (1981). *Applied Statistics: Principles and Examples*, London: Chapman and Hall. [3.10]

Cox, D. R., and Snell, J. E. (1989). *Analysis of Binary Data* (Second edition), London: Chapman and Hall. [4.9, 6.9.1]

Craig, A. T. (1934). Note on the moments of a Bernoulli distribution, *Bulletin of the American Mathematical Society*, **40**, 262–264. [2.2.1]

Craig, C. C. (1953). On the utilization of marked specimens in estimating populations of flying insects, *Biometrika*, **40**, 170–176. [10.1, 10.4.1].

Crawford, G. B. (1966). Characterization of geometric and exponential distributions, *Annals of Mathematical Statistics*, **37**, 1790–1795. [5.9.1]

Cressie, N. (1978). A finely tuned continuity correction, *Annals of the Institute of Statistical Mathematics, Tokyo*, **30**, 436–442. [3.6.1]

Cressie, N., Davis, A. S., Folks, J. L., and Policello, G. E. (1981). The moment generating function and negative integer moments, *American Statistician*, **35**, 148–150. [3.3]

Cresswell, W. L., and Froggatt, P. (1963). *The Causation of Bus Driver Accidents*, London: Oxford University Press. [9.6.5, 9.9]

Crick, F. H. C., and Lawrence, P. A. (1975). Compartments and polychones in insect development, *Science*, **189**, 340–347. [5.10]

Crow, E. L. (1956). Confidence limits for a proportion, *Biometrika*, **43**, 423–435 (corrigenda **45**, 291). [3.8.3]

Crow, E. L. (1958). The mean deviation of the Poisson distribution, *Biometrika*, **45**, 556–559. [4.3]

Crow, E. L. (1975). Confidence limits for small probabilities from Bernoulli trials, *Communications in Statistics*, **4**, 397–414. [3.8.2, 3.8.3]

Crow, E. L., and Bardwell, G. E. (1965). Estimation of the parameters of the hyper-Poisson distributions, *Classical and Contagious Discrete Distributions*, G. P. Patil (editor), 127–140. Calcutta: Statistical Publishing Society; Oxford: Pergamon Press. [2.3.1, 4.12.4]

Crow, E. L., and Gardner, R. S. (1959). Confidence intervals for the expectation of a Poisson variable, *Biometrika*, **46**, 444–453. [4.4, 4.7.3, 4.12.3]

Crowder, M. J. (1978). Beta-binomial anova for proportions, *Applied Statistics*, **27**, 34–37. [6.9.2]

D'Agostino, R. B., Chase, W., and Belanger, A. (1988). The appropriateness of some common procedures for testing the equality of two independent binomial populations, *American Statistician*, **42**, 198–202. [3.8.4]

D'Agostino, R. B., and Stephens, M. A. (editors) (1986). *Goodness-of-Fit Techniques*, New York: Dekker. [1B.15]

Daboni, L. (1959). A property of the Poisson distribution, *Bolletino della Unione Mathematica Italiana*, **14**, 318–320. [4.8]

Dacey, M. F. (1972). A family of discrete probability distributions defined by the generalized hypergeometric series, *Sankhyā, Series B*, **34**, 243–250. [2.4.1]

Dacey, M. F. (1975). Probability laws for topological properties of drainage basins, *Statistical Distributions in Scientific Work*, **2**: *Model Building and Model Selection*, G. P. Patil, S. Kotz, and J. K. Ord (editors), 327–341. Dordrecht: Reidel. [2.4.1]

Dagpunar, J. (1988). *Principles of Random Variate Generation*, Oxford: Oxford University Press. [1C.1]

Dahiya, R. C. (1981). An improved method of estimating an integer-parameter by maximum likelihood, *American Statistician*, **35**, 34–37. [3.8.2]

Dahiya, R. C. (1986). Integer-parameter estimation in discrete distributions, *Communications in Statistics-Theory and Methods*, **15**, 709–725. [3.8.2]

Dahiya, R. C., and Gross, A. J. (1973). Estimating the zero class from a truncated Poisson sample, *Journal of the American Statistical Association*, **68**, 731–733. [4.10.1]

Dallas, A. C. (1974). A characterization of the geometric distribution, *Journal of Applied Probability*, **11**, 609–611. [5.9.1]

Dandekar, V. M. (1955). Certain modified forms of binomial and Poisson distributions, *Sankhyā*, **15**, 237–250. [8.2.2, 11.2]

Daniels, H. E. (1961). Mixtures of geometric distributions, *Journal of the Royal Statistical Society, Series B*, **23**, 409–413. [5.2, 8.2.5]

Danish, M. B., and Hundley, J. T. (1979). A generalized method for computing the probability of killing X targets out of N targets, *Memorandum Report* , ARBL-MR-02890. Aberdeen Proving Ground, MD: Ballistic Research Laboratory. [3.12.2]

Darroch, J. N. (1974). Multiplicative and additive interaction in contingency tables, *Biometrika*, **61**, 207–214. [3.12.6]

Darwin, J. H. (1960). An ecological distribution akin to Fisher's logarithmic distribution, *Biometrics*, **16**, 51–60. [7.12]

David, F. N. (1950). Two combinatorial tests of whether a sample has come from a given population, *Biometrika*, **37**, 97–110. [10.4.1]

David, F. N. (1962). *Games, Gods, and Gambling*, London: Griffin. [11.19]

David, F. N., and Barton, D. E. (1962). *Combinatorial Chance*, London: Griffin. [1A.3, 10.2, 10.3, 10.4.1, 10.5.1, 11.14]

David, F. N., and Johnson, N. L. (1952). The truncated Poisson distribution, *Biometrics*, **8**, 275–285. [4.10.1, 5.11]

David, F. N., Kendall, M. G., and Barton, D. E. (1966). *Symmetric Function and Allied Tables*, London: Cambridge University Press. [1A.9, 10.5.1]

David, F. N., and Moore, P. G. (1954). Notes on contagious distributions in plant populations, *Annals of Botany, New Series*, **18**, 47–53. [9.6.1, 9.6.5]

David, H. A. (1981). *Order Statistics* (Second edition), New York: Wiley. [1B.10, 3.5]

Davies, O. L. (1933). On asymptotic formulae for the hypergeometric series I, *Biometrika*, **25**, 295–322. [6.2.4, 6.5]

Davies, O. L. (1934). On asymptotic formulae for the hypergeometric series II, *Biometrika*, **26**, 59–107. [6.2.4, 6.5]

Davis, H. T. (1933, 1935). *Tables of the Higher Mathematical Functions*, 2 volumes, Bloomington, IN: Principia Press. [1A.2]

Decker, R. D., and Fitzgibbon, D. J. (1991). The normal and Poisson approximations to the binomial: a closer look, *Department of Mathematics Technical Report No. 82.3*, Hartford, CT: University of Hartford. [3.6.1]

Deely, J. J., and Lindley, D. V. (1981). Bayes empirical Bayes. *Journal of the American Statistical Association*, **76**, 833–841. [3.8.2]

DeGroot, M. H. (1970). *Optimal Statistical Decisions*, New York: McGraw-Hill. [1B.15]

Delaporte, P. (1959). Quelques problémes de statistique mathématique posés par l'assurance automobile et le bonus pour non sinistre, *Bulletin Trimestriel de l'Institut des Actuaires Français*, **227**, 87–102. [5.12.5]

Denning, P. J., and Schwartz, S. C. (1972). Properties of the working set model, *Communications of the A.C.M.*, **15**, 191–198. [10.4.1]

DeRiggi, D. F. (1983). Unimodality of likelihood functions for the binomial distribution, *Journal of the American Statistical Association*, **78**, 181–183. [3.8.2]

DeRouen, T. A., and Mitchell, T. J. (1974). A G_1 minimax estimator for a linear combination of binomial probabilities, *Journal of the American Statistical Association*, **69**, 231–233. [3.8.2]

Desloge, E. A. (1966). *Statistical Physics*, New York: Holt, Reinhart and Winston. [10.1]

Devroye, L. (1986). *Non-Uniform Random Variate Generation*, New York: Springer-Verlag. [1C.1, 1C.3, 1C.4, 1C.5, 1C.6, 1C.7, 4.5, 8.1, 11.20]

Devroye, L. (1992). The branching process method in Lagrangian random variate generation, *Communications in Statistics-Simulation and Computation*, **21**, 1–14. [2.5.2]

Diaconis, P., and Zarbell, S. (1991). Closed form summation for classical distributions: Variations on a theme of De Moivre, *Statistical Science*, **6**, 284–302. [3.3, 3.4, 4.8]

Diggle, P. J. (1983). *Statistical Analysis of Spatial Point Patterns*, London: Academic Press. [5.10]

Domb, C. (1952). On the use of a random parameter in combinatorial problems, *Proceedings of the Physical Society, Series A*, **65**, 305–309. [10.4.1]

Doob, J. L. (1953). *Stochastic Processes*, New York: Wiley. [4.9]

Doss, S. A. D. C. (1963). On the efficiency of best asymptotically normal estimates of the Poisson parameter based on singly and doubly truncated or censored samples, *Biometrics*, **19**, 588–594. [4.10.2]

Douglas, J. B. (1955). Fitting the Neyman type A (two parameter) contagious distribution, *Biometrics*, **11**, 149–173. [9.6.1, 9.6.3, 9.6.4]

Douglas, J. B. (1965). Asymptotic expansions for some contagious distributions, *Classical and Contagious Discrete Distributions*, G. P. Patil (editor), 291–302. Calcutta: Statistical Publishing Society; Oxford: Pergamon Press. [9.6.3, 9.7]

Douglas, J. B. (1971). Stirling numbers in discrete distributions, *Statistical Ecology* 1: *Spatial Patterns and Statistical Distributions*, G. P. Patil, E. C. Pielou, and W. E. Waters (editors), 69–98. University Park: Pennsylvania State University Press. [7.11, 9.0]

Douglas, J. B. (1980). *Analysis with Standard Contagious Distributions*, Burtonsville, MD: International Co-operative Publishing House. [1B.6, 2.2.1, 3.8.1, 3.10, 4.9, 4.11, 8.3.2, 9.0, 9.1, 9.5, 9.6.4, 9.7, 9.8, 9.10]

Douglas, J. B. (1986). Pólya-Aeppli distribution, *Encyclopedia of Statistical Sciences*, **7**, S. Kotz, N. L. Johnson and C. B. Read (editors), 56–59. New York: Wiley. [9.7]

Draper, N., and Guttman, I. (1971). Bayesian estimation of the binomial parameter, *Technometrics*, **13**, 667–673. [3.8.2]

Dubey, S. D. (1966a). Compound Pascal distributions, *Annals of the Institute of Statistical Mathematics, Tokyo*, **18**, 357–365 [6.2.3, 6.7.3]

Dubey, S. D. (1966b). Graphical tests for discrete distributions, *American Statistician*, **20**, 23–24. [4.7.1]

Dunin-Barkovsky, I. V., and Smirnov, N. V. (1955). *Theory of Probability and Mathematical Statistics in Engineering*, Moscow: Nauka (in Russian). [3.6.1]

Dunkl, C. F. (1981). The absorption distribution and the q-binomial theorem, *Communications in Statistics-Theory and Methods*, **10**, 1915–1920. [11.1]

Edwards, A. W. F. (1960). The meaning of binomial distribution, *Nature, London*, **186**, 1074. [3.12.6]

Edwards, A. W. F. (1987). *Pascal's Arithmetical Triangle*, London: Griffin. [3.2]

Eggenberger, F., and Pólya, G. (1923). Über die Statistik verketteter Vorgänge, *Zeitschrift für angewandte Mathematik und Mechanik*, **3**, 279–289. [5.1, 5.3, 6.2.4]

Eggenberger, F., and Pólya, G. (1928). Calcul des probabilités - sur l'interprétation de certaines courbes de frequence, *Comptes Rendus, Académie des Sciences, Paris*, **187**, 870–872. [5.3, 6.2.4]

Eideh, A.-H. A.-H., and Ahmed, M. S. (1989). Some tests for the power series distributions in one parameter using the Kullback-Leibler information measure, *Communications in Statistics-Theory and Methods*, **18**, 3649–3663. [2.2.1]

Elliott, J. M. (1979). Some methods for the statistical analysis of samples of Benthic invertebrates, *Freshwater Biological Association Scientific Publication*, **25**, Ambleside, England: Freshwater Biological Association. [5.10]

Emerson, J. D., and Hoaglin, D. C. (1983). Stem-and-leaf displays, *Understanding Robust and Exploratory Data Analysis*, D. C. Hoaglin, F. W. Mosteller and J. W. Tukey (editors), 7–32. New York: Wiley. [1B.10]

Engen S. (1974). On species frequency models, *Biometrika*, **61**, 263–270. [5.12.2, 7.7.2]

Engen, S. (1978). *Stochastic Abundance Models*, London: Chapman and Hall. [5.12.2]

Erdélyi, A., Magnus, W., Oberhettinger, F., and Tricomi, F. G. (1953). *Higher Transcendental Functions*, **I, II**, New York: McGraw-Hill. [1A.6, 1A.7, 1A.8]

Erdélyi, A., Magnus, W., Oberhettinger, F., and Tricomi, F. G. (1954). *Tables of Integral Transforms*, **I, II**, New York: McGraw-Hill. [1A.8, 1A.10]

Erlang, A. K. (1918). Solutions of some problems in the theory of probabilities of significance in automatic telephone exchanges, *Post Office Electrical Engineer's Journal*, **10**, 189–197. [11.9]

Estoup, J. B. (1916). *Les Gammes Sténographiques*, Paris: Institut Sténographique. [11.20]

Evans, D. A. (1953). Experimental evidence concerning contagious distributions in ecology, *Biometrika*, **40**, 186–211. [5.1, 9.6.5, 9.7]

Everitt, B. S., and Hand, D. J. (1981). *Finite Mixture Distributions*, London: Chapman and Hall. [8.2.1, 8.2.3, 8.2.4]

Exton, H. (1978). *Handbook of Hypergeometric Integrals*, Chichester: Ellis Horwood. [1A.8]

Exton, H. (1983). *q-Hypergeometric Functions and Applications*, Chichester: Ellis Horwood. [1A.12]

Famoye, F., and Consul, P. C. (1989). Confidence interval estimation in the class of modified power series distributions, *Statistics*, **20**, 141–148. [2.2.2, 5.12.3]

Fang, K.-T. (1985). Occupancy problems, *Encyclopedia of Statistical Sciences*, **6**, S. Kotz, N. L. Johnson and C. B. Read (editors), 402–406. New York: Wiley. [10.4.1]

Fazal, S. S. (1976). A test for quasi-binomial distribution, *Biometrische Zeitschrift*, **18**, 619–622. [3.12.3]

Feix, M. (1955). Théorie d'enregistrement d'évenements aléatoires, *Journal Physique Radium*, **16**, 719–727. [11.9]

Feldman, D., and Fox, M. (1968). Estimation of the parameter n in the binomial distribution, *Journal of the American Statistical Association*, **63**. 150–158. [3.8.2]

Feller, W. (1943). On a general class of "contagious" distributions, *Annals of Mathematical Statistics*, **14**, 389–400. [4.11, 5.9.2, 8.2.3, 8.3.1, 8.3.2, 9.0, 9.3, 9.9]

Feller, W. (1945). On the normal approximation to the binomial distribution, *Annals of Mathematical Statistics*, **16**, 319–329. [3.6.2]

Feller, W. (1948). On probability problems in the theory of counters, *Studies presented to R. Courant*, New York: Interscience. 105–115. [11.9]

Feller, W. (1950). *An Introduction to Probability Theory and Its Applications*, **1**, New York: Wiley. [4.11]

Feller, W. (1957). *An Introduction to Probability Theory and Its Applications*, **1** (Second edition), New York: Wiley. [3.2, 4.2, 4.11, 5.3, 5.12.5, 6.4, 6.5, 6.10.2, 7.10, 8.1, 8.3.1, 9.3, 10.1, 10.2, 10.4.1, 10.6.1, 10.6.2]

Feller, W. (1968). *An Introduction to Probability Theory and Its Applications*, **1** (Third edition), New York: Wiley. [3.12.2, 4.2, 4.9, 4.11, 5.3]

Ferguson, T. S. (1964). A characterization of the exponential distribution, *Annals of Mathematical Statistics*, **35**, 1199–1207. [5.9.1]

Ferguson, T. S. (1965). A characterization of the geometric distribution, *American Mathematical Monthly*, **72**, 256–260. [5.9.1]

Ferguson, T. S. (1967). *Mathematical Statistics*, New York: Academic Press. [6.4]

Fienberg, S. E. (1982). Contingency tables, *Encyclopedia of Statistical Sciences*, S. Kotz, N. L. Johnson and C. B. Read (editors), **2**, 161–171. New York: Wiley. [4.9]

Finetti, B. de (1931). Le funzioni caratteristiche di legge instantanea dotate di valori eccezionali, *Atti della Reale Academia Nazionale dei Lincei, Serie VI*, **14**, 259–265. [9.3]

Finney, D. J. (1949). The truncated binomial distribution, *Annals of Eugenics, London*, **14**, 319–328. [3.11]

Finney, D. J., Latscha, R., Bennett, B. M., and Hsu, P. (1963). *Tables for Testing Significance in a 2×2 Contingency Table*, Cambridge: Cambridge University Press. [6.9.1]

Fisher, A. (1936). *The Mathematical Theory of Probabilitiesand Its Application to Frequency-curves and Statistical Methods*, **1**, New York: Macmillan. [3.12.2]

Fisher, R. A. (1929). Moments and product moments of sampling distributions, *Proceedings of the London Mathematical Society, Series 2*, **30**, 199–238. [1B.6]

Fisher, R. A. (1931). Properties of the functions, (Part of introduction to) *British Association Mathematical Tables*, **1**. London: British Association. [9.4]

Fisher, R. A. (1941). The negative binomial distribution, *Annals of Eugenics, London*, **11**, 182–187. [3.8.2, 5.1, 5.8.3]

Fisher, R. A. (1951). Properties of the functions (Part of introduction to) *British Association Mathematical Tables*, **1** (Third edition). London: British Association. [1A.7]

Fisher, R. A., Corbet, A. S., and Williams, C. B. (1943). The relation between the number of species and the number of individuals in a random sample of an animal population, *Journal of Animal Ecology*, **12**, 42–58. [5.12.2, 7.1, 7.2, 7.6, 7.7.1, 7.9, 11.16]

Fisher, R. A., and Mather, K. (1936). A linkage test with mice, *Annals of Eugenics, London,* **7**, 265–280. [3.10]

Fishman, G. S. (1973). *Concepts and Methods in Discrete Event Simulation,* New York: Wiley. [1C.6]

Fishman, G. S. (1978). *Principles of Discrete Event Simulation,* New York: Wiley. [1C.5, 1C.6]

Fishman, G. S. (1979). Sampling from the binomial distribution on a computer, *Journal of the American Statistical Association,* **74**, 418–423. [1C.3]

Fisz, M. (1953). The limiting distribution of the difference of two Poisson random variables, *Zastosowania Matematyki,* **1**, 41–45 (In Polish). [4.12.3]

Fisz, M., and Urbanik, K. (1956) The analytical characterisation of the composed non-homogeneous Poisson process, *Bulletin de L'Academie Polonaise des Sciences, Classe III,* **3**, 149–150. [4.2]

Foster, F. G. (1952). A Markov chain derivation of discrete distributions, *Annals of Mathematical Statistics,* **23**, 624–627. [5.3]

Foster, F. G., and Stuart, A. (1954). Distribution-free tests in time-series based on the breaking of records, *Journal of the Royal Statistical Society, Series B,* **16**, 1–13. [11.14]

Fox, B. L., and Glynn, P. W. (1988). Computing Poisson probabilities, *Communications of the ACM,* **31**, 440–445. [4.6]

Frame, J. S. (1945). Mean deviation of the binomial distribution, *American Mathematical Monthly,* **52**, 377–379. [3.3]

Francis, B. J. (1991). AS R88: A remark on Algorithm AS 121: The trigamma function, *Applied Statistics,* **40**, 514–515. [11.3]

Fréchet, M. (1940). *Les Probabilités Associées a un Système d'Événements Compatibles et Dépendants,* **1**: *Événements en nombre fini fixe.* Actualités scientifiques et industrielles, No. 859. Paris: Hermann. [1B.9, 2.4.2, 10.2, 10.3]

Fréchet, M. (1943). *Les Probabilités Associées a un Système d'Événements Compatibles et Dépendants,* **2**: *Cas particuliers et applications.* Actualités scientifiques et industrielles, No. 942. Paris: Hermann. [1B.9, 2.4.2, 10.2, 10.3]

Freedman, D. A. (1965). Bernard Friedman's urn, *Annals of Mathematical Statistics,* **36**, 956–970. [11.19]

Freeman, G. H. (1980). Fitting two-parameter discrete distributions to many data sets with one common parameter, *Applied Statistics,* **29**, 259–267. [5.8.4]

Freeman, M. F., and Tukey, J. W. (1950). Transformations related to the angular and the square root, *Annals of Mathematical Statistics,* **21**, 607–611. [3.6.3, 4.5]

Freeman, P. R. (1973). Algorithm AS 59: Hypergeometric probabilities, *Applied Statistics,* **22**, 130–133. [6.6]

Freimer, M., Gold, B., and Tritter, A. L. (1959). The Morse distribution, *Transactions of the IRE — Information Theory,* **5**, 25–31. [11.14]

Friedman, B. (1949). A simple urn model, *Communications in Pure and Applied Mathematics,* **2**, 59–70. [11.19]

Frisch, R. (1924). Solution d'un problém du calcul des probabilitités, *Skandinavisk Aktuarietidskrift,* **7**, 153–174. [3.3]

Froggatt, P. (1966). One day absence in industry, *Journal of the Statistical and Social Enquiry Society of Ireland,* **21**, 166–178. [9.9]

Furry, W. H. (1937). On fluctuation phenomena in the passage of high energy electrons through lead. *Physical Review*, **52**, 569–581. [5.2, 5.3, 5.10]

Gabriel, K. R., and Neumann, J. (1962). On a distribution of weather cycles by length, *Quarterly Journal of the Royal Meteorological Society*, **83**, 375–380, *ibid.*, **88**, 90–95. [5.2]

Galambos, J. (1973). A general Poisson limit theorem of probability theory, *Duke Mathematical Journal*, **40**, 581–586. [3.6.1]

Galambos, J. (1975). Methods for proving Bonferroni type inequalities, *Journal of the London Mathematical Society*, **9**, 561–564. [10.2]

Galambos, J. (1977). Bonferroni inequalities, *Annals of Probability*, **5**, 577–581. [10.2]

Galambos, J., and Kotz, S. (1978). *Characterizations of Probability Distributions*, Berlin: Springer-Verlag. [4.8]

Galliher, H. P., Morse, P. M., and Simond, M. (1959). Dynamics of two classes of continuous review inventory systems, *Operations Research*, **7**, 362–384. [4.11, 9.3, 9.7]

Galloway, T. (1839). *A Treatise on Probability*, Edinburgh: Adam and Charles Black. [5.3]

Gani, J. (1975). Stochastic models for type counts in a literary text, *Perspectives in Probability and Statistics*, J. Gani (editor), 313–324. London: Academic Press. [3.12.2]

Gart, J. J. (1963). A median test with sequential application, *Biometrika*, **50**, 55–62. [6.9.2]

Gart, J. J. (1968). A simple nearly efficient alternative to the simple sib method in the complete ascertainment case, *Annals of Human Genetics, London*, **31**, 283–291. [3.11]

Gart, J. J. (1970). Some simple graphically oriented statistical methods for discrete data, *Random Counts in Scientific Work*, **1**: *Random Counts in Models and Structures*, G. P. Patil (editor), 171–191. University Park: Pennsylvania State University Press. [3.8.1, 4.7.1, 7.7.1]

Gart, J. J. (1974). Exact moments of the variance test for left-truncated Poisson distributions, *Sankhyā, Series B*, **36**, 406–416. [4.7.4, 4.12.3]

Gart, J. J., and Pettigrew, H. M. (1970). On the conditional moments of the k-statistics for the Poisson distribution, *Biometrika*, **57**, 661–664. [4.4]

Gauss, C. F. (1900). Schönes Theorem der Wahrscheinlichkeitsrechnung, *Werke*, **8**, 88. Koniglichen Gesellschaft der Wissenschaften zu Göttingen. Leipzig: Teubner. [1B.8]

Gebhardt, F. (1969). Some numerical comparisons of several approximations to the binomial distribution, *Journal of the American Statistical Association*, **64**, 1638–1648. [3.6.1]

Gebhardt, F. (1971). Incomplete beta-integral $B(x; 2/3, 2/3)$ and $[p(1-p)]^{-1/6}$ for use with Borges' approximation to the binomial distribution, *Journal of the American Statistical Association*, **66**, 189–191. [3.6.1]

Geiringer, H. (1938). On the probability theory of arbitrarily linked events, *Annals of Mathematical Statistics*, **9**, 260–271. [10.2]

Geisser, S. (1984). On prior distributions for binary trials, *Annals of Statistics*, **38**, 244–247. [3.8.2]

Gelfand, A. E., and Soloman, H. (1975). Analysing the decision making process of the American jury, *Journal of the American Statistical Association*, **70**, 305–310. [8.2.4]

Gerrard, D. J., and Cook, R. D. (1972). Inverse binomial sampling as a basis for estimating negative binomial population densities, *Biometrics*, **28**, 971–980. [5.8.2]

Gerstenkorn, T. (1962). On the generalized Poisson distribution, *Prace Lódzkie Towarzystwa Naukowe, Wydzial*, **3**, No. 85. [4.12.5, 8.3.1, 11.12]

Ghosh, B. K. (1980). The normal approximation to the binomial distribution, *Communications in Statistics-Theory and Methods*, **A9**, 427–438. [3.6.1]

Ghosh, M., and Yang, M.-C. (1988). Simultaneous estimation of Poisson means under entropy loss, *Annals of Statistics*, **16**, 278–291. [4.7.2]

Gibbons, J. D. (1983). Fisher's exact test, *Encyclopedia of Statistical Sciences*, **3**, S. Kotz, N. L. Johnson and C. B. Read (editors), 130–133. New York: Wiley. [6.9.1]

Gibbons, J. D. (1986). Randomness, tests of, *Encyclopedia of Statistical Sciences*, **7**, S. Kotz, N. L. Johnson and C. B. Read (editors), 555–562. New York: Wiley. [10.5.1, 10.5.2]

Gilbert, E. J. (1956). The matching problem, *Psychometrika*, **21**, 253–266. [10.3]

Giltay, J. (1943). A counter arrangement with constant resolving time, *Physica*, **10**, 725–734. [11.9]

Glasser, G. J. (1962). Minimum variance unbiased estimators for Poisson probabilities, *Technometrics*, **4**, 409–418. [4.7.2]

Gleeson, A. C., and Douglas, J. B. (1975). Quadrat sampling and the estimation of Neyman type A and Thomas distributional parameters, *Australian Journal of Statistics*, **17**, 103–113. [9.10]

Glick, N. (1978). Breaking records and breaking boards, *American Mathematical Monthly*, **85**, 2–26. [11.14]

Gnedenko, B. V. (1961). *Probability Theory* (Third edition), Moscow: GITTL. [5.3]

Gnedenko, B. V., and Kovalenko, I. N. (1989). *Introduction to Queueing Theory* (Second edition), translated by S. Kotz. Boston: Birkhäuser Boston. [4.9]

Godambe, A. V. (1977). On representation of Poisson mixtures as Poisson sums and a characterization of the gamma distribution, *Mathematical Proceedings of the Cambridge Philosophical Society*, **82**, 297–300. [8.2.3]

Godambe, A. V., and Patil, G. P. (1975). Some characterisations involving additivity and infinite divisibility and their applications to Poisson mixtures and Poisson sums, *Statistical Distributions in Scientific Work*, **3**: *Characterizations and Applications*, G. P. Patil, S. Kotz, and J. K. Ord (editors), 339–351. Dordrecht: Reidel. [4.11, 8.3.1]

Godbole, A. P. (1990a). Specific formulae for some success run distributions, *Statistics and Probability Letters*, **10**, 119–124. [10.6.2]

Godbole, A. P. (1990b). On hypergeometric and related distributions of order k, *Communications in Statistics-Theory and Methods*, **19**, 1291–1301. [6.12, 10.6.3]

Godwin, H. J. (1967). Review of T. Gerstenkorn: On the generalized Poisson distribution, *Mathematical Reviews*, **34**, No. 3608. [4.12.5]

Gokhale, D. V. (1980). On power series distributions with mean-variance relationships, *Journal of the Indian Statistical Association*, **18**, 81–84. [2.2.2]

Gold, L. (1957). Generalized Poisson distributions, *Annals of the Institute of Statistical Mathematics, Tokyo*, **9**, 43–47. [4.12.5, 11.12]

Goldberg, K., Leighton, F. T., Newman, M., and Zuckerman, S. L. (1976). Tables of binomial coefficients and Stirling numbers, *Journal of Research of the National Bureau of Standards*, **80B**, 99–171. [1A.3]

Goldie, C. M., and Rogers, L. C. G. (1984). The k-record processes are i.i.d., *Zeitschrift für Wahrscheinlichkeitstheorie und Verwandte Gebeite*, **67**, 197–211. [11.14]

Goldstein, L. (1990). Poisson approximation and DNA sequence matching, *Communications in Statistics-Theory and Methods*, **19**, 4167–4179. [4.9, 10.1]

Good, I. J. (1953). The population frequencies of species and the estimation of population parameters, *Biometrika*, **40**, 237–264. [4.7.2]

Good, I. J. (1957). Distribution of word frequencies, *Nature*, **179**, 595. [11.20]

Goodhardt, G. J., Ehrenberg, A. S. C., and Chatfield, C. (1984). The Dirichlet: a comprehensive model of buying behaviour, *Journal of the Royal Statistical Society, Series A*, **147**, 621–655. [5.10]

Goralski, A. (1977). Distribution z-Poisson, *Publications of the Institute of Statistics, Paris*, **12**, 45–53. [8.2.2]

Gould, H. W. (1962). Congruences involving sums of binomial coefficients and a formula of Jensen, *Mathematical Notes, May 1962*, 400–402. [2.5.3]

Govindarajulu, Z. (1962a). The reciprocal of the decapitated negative binomial variable, *Journal of the American Statistical Association*, **57**, 906–913 (corrigendum **58**, 1162). [5.11]

Govindarajulu, Z. (1962b). First two moments of the reciprocal of a positive hypergeometric variable, *Case Institute of Technology Statistics Laboratory Report*, 1061. [6.12]

Gower, J. C. (1961). A note on some asymptotic properties of the logarithmic series distribution, *Biometrika*, **48**, 212–215. [7.4, 7.5]

Grab, E. L., and Savage, I. R. (1954). Tables of the expected value of $1/x$ for positive Bernoulli and Poisson variables, *Journal of the American Statistical Association*, **49**, 169–177. [3.11, 4.10.1]

Greenberg, R. A., and White, C. (1965). The detection of a correlation between the sexes of adjacent sibs in human families, *Journal of the American Statistical Association*, **60**, 1035–1045. [3.12.6]

Greenwood, M., and Yule, G. U. (1920). An inquiry into the nature of frequency distributions representative of multiple happenings with particular reference to the occurrence of multiple attacks of disease or of repeated accidents, *Journal of the Royal Statistical Society, Series A*, **83**, 255–279. [2.4.2, 4.12.5, 5.3, 5.10, 8.3.1, 8.3.2, 9.4, 11.19]

Greenwood, R. E., and Glasgow, M. O. (1950). Distribution of maximum and minimum frequencies in a sample drawn from a multinomial population, *Annals of Mathematical Statistics*, **21**, 416–424. [3.8.2]

Greig-Smith, P. (1964). *Quantitative Plant Ecology* (Second edition), London: Butterworth. [4.9]

Greville, T. N. E. (1941). The frequency distribution of a general matching problem, *Annals of Mathematical Statistics*, **12**, 350–354. [10.3]

Griffiths, D. A. (1973). Maximum likelihood estimation for the beta-binomial distribution and an application to the household distribution of the total number of cases of a disease, *Biometrics*, **29**, 637–648. [6.7.2, 6.9.2]

Grimm, H. (1962). Tafeln der negativen Binomialverteilung, *Biometrische Zeitschrift*, **4**, 239–262. [5.7]

Grimm, H. (1964). Tafeln der Neyman-Verteilung Typ A, *Biometrische Zeitschrift*, **6**, 10–23. [9.6.3]

Grimm, H. (1970). Graphical methods for the determination of type and parameters of some discrete distributions, *Random Counts in Scientific Work*, **1**: *Random Counts in Models and Structures*, G. P. Patil (editor), 193–206. University Park: Pennsylvania State University Press. [3.8.1, 4.7.1, 5.8.1, 9.6.4]

Guenther, W. C. (1969). Use of the binomial, hypergeometric and Poisson tables to obtain sampling plans, *Journal of Quality Technology*, **1**, 105–109. [6.9.2]

Guenther, W. C. (1971). The average sample number for truncated double sample attribute plans, *Technometrics*, **13**, 811–816. [6.9.2]

Guenther, W. C. (1972). A simple approximation to the negative binomial (and regular binomial), *Technometrics*, **14**, 385–389. [5.6]

Guenther, W. C. (1975). The inverse hypergeometric — a useful model, *Statistica Neerlandica*, **29**, 129–144. [6.2.2, 6.9.2]

Guenther, W. C. (1977). *Sampling Inspection in Statistical Quality Control*, London: Griffin. [6.7.1]

Guenther, W. C. (1983). Hypergeometric distributions, *Encyclopedia of Statistical Sciences*, **3**, S. Kotz, N. L. Johnson and C. B. Read (editors), 707–712. New York: Wiley. [6.2.4, 6.6, 6.7.1]

Guldberg, A. (1931). On discontinuous frequency functions and statistical series, *Skandinavisk Aktuarietidskrift*, **14**, 167–187. [2.3.1, 2.3.2, 2.4.1]

Gumbel, E. J. (1958). *Statistics of Extremes*, New York: Columbia University Press. [6.2.2]

Gumbel, E. J., and Schelling, H. von (1950). The distribution of the number of exceedances, *Annals of Mathematical Statistics*, **21**, 247–262. [6.2.2]

Gupta, D., and Gupta, R. C. (1984). On the distribution of order statistics for a random sample size, *Statistica Neerlandica*, **38**, 13–19. [9.11]

Gupta, M. K. (1967). Unbiased estimate for $1/p$, *Annals of the Institute of Statistical Mathematics, Tokyo*, **19**, 413–416. [3.8.2]

Gupta, P. L. (1982). Probability generating functions of a MPSD with applications, *Mathematische Operationforschung und Statistik, series Statistics*, **13**, 99–103. [2.2.2]

Gupta, P. L., and Gupta, R. C. (1981). Probability of ties and Markov property in discrete order statistics, *Journal of Statistical Planning and Inference*, **5**, 273–279. [5.2]

Gupta, P. L., and Singh, J. (1981). On the moments and factorial moments of a MPSD, *Statistical Distributions in Scientific Work*, **4**: *Models, Structures, and Characterizations*, C. Taillie, G. P. Patil, and B. A. Baldessari (editors), 189–195. Dordrecht: Reidel. [2.2.2, 9.11]

Gupta, R. C. (1970). *Some Characterizations of Probability Distributions*, Ph.D thesis. Detroit, MI: Wayne State University. [5.9.1]

Gupta, R. C. (1974). Modified power series distributions and some of its applications, *Sankhyā, Series B*, **35**, 288–298. [2.2.2, 5.12.3, 9.11]

Gupta, R. C. (1975a). Maximum-likelihood estimation of a modified power series distribution and some of its applications, *Communications in Statistics*, **4**, 689–697. [2.2.2]

Gupta, R. C. (1975b). Some characterizations of discrete distributions by properties of their moment distributions, *Communications in Statistics*, **4**, 761–765. [2.2.2]

Gupta, R. C. (1976). Distribution of the sum of independent generalized logarithmic series variables, *Communications in Statistics-Theory and Methods*, **5**, 45–48. [7.11]

Gupta, R. C. (1977). Minimum variance unbiased estimation in modified power series distribution and some of its applications, *Communications in Statistics-Theory and Methods*, **A6**, 977–991. [2.2.2, 7.11]

Gupta, R. C. (1981). On the Rao-Rubin characterization of the Poisson distribution, *Statistical Distributions in Scientific Work*, **4**: *Models, Structures, and Characterizations*, C. Taillie, G. P. Patil, and B. A. Baldessari (editors), 341–347. Dordrecht: Reidel. [4.8]

Gupta, R. C. (1984). Estimating the probability of winning (losing) in a gambler's ruin problem with applications, *Journal of Statistical Planning and Inference*, **9**, 55–62. [2.2.2, 11.10]

Gupta, R. C., and Singh, J. (1982). Estimation of probabilities in the class of modified power series distributions, *Mathematische Operationforschung und Statistik, Series Statistics*, **13**, 71–77. [2.2.2]

Gupta, R. C., and Tripathi, R. C. (1985). Modified power series distributions, *Encyclopedia of Statistical Sciences*, **5**, S. Kotz, N. L. Johnson and C. B. Read (editors). 593–599. New York: Wiley. [2.2.2]

Gupta, R. P., and Jain, G. C. (1974). A generalized Hermite distribution and its properties, *SIAM Journal of Applied Mathematics*, **27**, 359–363. [9.4]

Gupta, S. S. (1960a). Order statistics from the gamma distribution, *Technometrics*, **2**, 243–262. [5.5]

Gupta, S. S. (1960b). Binomial order statistics, *Bell Laboratory Report*: Allentown, PA. [3.5].

Gupta, S. S. (1965). Selection and ranking procedures and order statistics for the binomial distribution, *Classical and Contagious Discrete Distributions*, G. P. Patil (editor), 219–230. Calcutta: Statistical Publishing Society; Oxford: Pergamon Press. [3.5].

Gupta, S. S., and Panchapakesan, P. (1974). On moments of order statistics from independent binomial populations, *Annals of the Institute of Statistical Mathematics, Tokyo, Supplement*, **8**, 95–113. [3.5]

Gurland, J. (1957). Some interrelations among compound and generalized distributions, *Biometrika*, **44**, 265–268. [8.3.1, 9.0, 9.1]

Gurland, J. (1958). A generalized class of contagious distributions, *Biometrics*, **14**, 229–249. [2.4.2, 6.11, 8.3.2, 8.3.4, 9.9, 9.12]

Gurland, J. (1965). A method of estimation for some generalized Poisson distributions, *Classical and Contagious Discrete Distributions*, G. P. Patil (editor), 141–

158. Calcutta: Statistical Publishing Society; Oxford: Pergamon Press. [5.8.1, 5.8.3, 9.3, 9.6.4]

Gurland, J., and Tripathi, R. C. (1975). Estimation of parameters on some extensions of the Katz family of discrete distributions involving hypergeometric functions, *Statistical Distributions in Scientific Work*, **1**: *Models and Structures*, G. P. Patil, S. Kotz, and J. K. Ord (editors), 59–82. Dordrecht: Reidel. [2.3.1, 4.12.4, 5.8.3]

Guttman, I. (1958). A note on a series solution of a problem in estimation, *Biometrika*, **45**, 565–567. [5.8.2]

Guzman, J. (1985). Some comments on the Poisson approximation to the binomial distribution, *American Statistician*, **39**, 157–158. [3.6.2]

Haag, J. (1924). Sur un problème général de probabilités et ses applications, *Proceedings of the International Mathematical Congress, Toronto*, **1**, 659–676. [10.3]

Hadley, G., and Whitin, T. M. (1961). Useful properties of the Poisson distribution, *Operations Research*, **9**, 408–410. [4.4]

Haight, F. A. (1957). Queueing with balking, *Biometrika*, **44**, 360–369. [3.2, 5.3]

Haight, F. A. (1961a). A distribution analagous to the Borel-Tanner, *Biometrika*, **48**, 167–173. [2.5.2, 11.10, 11.13]

Haight, F. A. (1961b). Index to the distributions of mathematical statistics, *Journal of Research of the National Bureau of Standards*, **65**B, 23–60. [9.3, 11.12]

Haight, F. A. (1965). On the effect of removing persons with N or more accidents from an accident prone population, *Biometrika*, **52**, 298–300. [2.4.2, 8.3.2]

Haight, F. A. (1966). Some statistical problems in connection with word association data, *Journal of Mathematical Psychology*, **3**, 217–233. [6.10.3, 11.20]

Haight, F. A. (1967). *Handbook of the Poisson Distribution*, New York: Wiley. [4.2, 4.3, 4.4, 4.9, 8.3.1, 8.3.2]

Haight, F. A. (1969). Two probability distributions connected with Zipf's rank-size conjecture, *Zastosowania Matematyki*, **10**, 225–228. [11.20]

Haight, F. A. (1972). The characterizations of discrete distributions by conditional distributions, *Zastosowania Matematyki*, **13**, 207–213. [3.9, 4.8]

Haight, F. A., and Breuer, M. A. (1960). The Borel-Tanner distribution, *Biometrika* **47**, 145–150. [2.5.2, 9.11]

Hald, A. (1952). *Statistical Theory with Engineering Applications*, New York: Wiley. [3.6.1, 5.3]

Hald, A. (1960). The compound hypergeometric distribution and a system of single sampling inspection plans based on prior distributions and costs, *Technometrics*, **2**, 275–340. [6.8, 6.9.2, 8.3.4]

Hald, A. (1967). The determination of single sampling attribute plans with given producer's and consumer's risk, *Technometrics*, **9**, 401–415. [3.6.1]

Hald, A. (1968). The mixed binomial distribution and the posterior distribution of p for a continuous prior distribution, *Journal of the Royal Statistical Society, Series B*, **30**, 359–367. [8.3.3]

Hald, A. (1990). *A History of Probability and Statistics and Their Applications before 1750*, New York: Wiley. [3.2, 10.2, 10.4.1]

Hald, A., and Kousgaard, E. (1967). A table for solving the binomial equation $B(c,n,p) = P$, *Matematisk-fysiske Skrifter, Det Kongelige Danske Videnskabernes Selskab*, **3**, No. 4 (48 pp.). [3.8.3, 4.6]

Haldane, J. B. S. (1941). The fitting of binomial distributions, *Annals of Eugenics, London*, **11**, 179–181. [3.8.2]

Hall, W. J. (1966). Some hypergeometric series distributions occurring in birth-and-death processes at equilibrium (abstract), *Annals of Mathematical Statistics*, **27**, 221. [4.12.4]

Hamdan, M. A. (1975). Correlation between the numbers of two types of children when the family size distribution is zero-truncated negative binomial, *Biometrics*, **31**, 765–766. [5.11]

Hamedani, G. G., and Walter, G. G. (1990). Empirical Bayes estimation of the binomial parameter *n*. *Communications in Statistics-Theory and Methods*, **19**, 2065–2084. [3.8.2]

Hampton, J. M., Moore, P. G., and Thomas, H. (1973). Subjective probability and its assessment, *Journal of the Royal Statistical Society, Series A*, **136**, 21–42. [1B.3]

Hannan, J., and Harkness, W. (1963). Normal approximation to the distribution of two independent binomials, conditional on a fixed sum, *Annals of Mathematical Statistics*, **34**, 1593–1595. [3.4, 6.11]

Hansen, B. G., and Willekens, E. (1990). The generalized logarithmic series distribution, *Statistics and Probability Letters*, **9**, 311–316. [7.11]

Harkness, W. L. (1965). Properties of the extended hypergeometric distribution, *Annals of Mathematical Statistics*, **36**, 938–945. [6.11]

Harkness, W. L. (1970). The classical occupancy problem revisited, *Random Counts in Scientific Work*, **3**: *Random Counts in Physical Science, Geo Science, and Business*, G. P. Patil (editor), 107–126. University Park: Pennsylvania State University Press. [10.4.1]

Harris, C. M. (1983). On finite mixtures of geometric and negative binomial distributions, *Communications in Statistics-Theory and Methods*, **12**, 987–1007. [8.2.5]

Hart, J. F. (1968). *Computer Approximations*, New York: Wiley. [4.5]

Harter, H. L. (1964). *New Tables of Incomplete Gamma Function Ratio*, United States Air Force Aerospace Research Laboratories. [1A.5]

Harter, H. L. (1988). History and role of order statistics, *Communications in Statistics-Theory and Methods*, **17**, 2091–2107. [1B.10]

Hartley, H. O., and Fitch, E. R. (1951). A chart for the incomplete beta funtion and cumulative binomial distribution, *Biometrika*, **38**, 423–426. [3.7]

Haseman, J. K., and Kupper, L. L. (1979). Analysis of dichotomous response data from certain toxicological experiments, *Biometrics*, **35**, 281–293. [3.12.6, 6.9.2]

Hasselblad, V. (1969). Estimation of finite mixtures of distributions from the exponential family, *Journal of the American Statistical Association*, **64**, 1459-1471. [8.2.3, 8.2.4]

Hausman, J., Hall, B. H., and Griliches, Z. (1984). Econometric models for count data with an application to the patents R & D relationship, *Econometrica*, **52**, 909–938. [4.9]

Hawkins, D. M., and Kotz, S. (1976). A clocking property of the exponential distribution, *Australian Journal of Statistics*, **18**, 170–172. [5.9.1]

Hemelrijk, J. (1967). The hypergeometric, the normal and chi-squared, *Statistica Neerlandica*, **21**, 225–229. [6.5]

Herdan, G. (1961). A critical examination of Simon's model of certain distribution functions in linguistics, *Applied Statistics*, **10**, 65–76. [6.10.3]

Heyde, C. C., and Seneta, E. (1972). The simple branching process, a turning point test and a fundamental inequality: A historical note on I. J. Bienaymé, *Biometrika*, **59**, 680–683. [9.1]

Hinz, P., and Gurland, J. (1967). Simplified techniques for estimating parameters of some generalized Poisson distributions, *Biometrika*, **54**, 555–566. [3.8.1, 5.8.1, 5.8.3, 9.5, 9.6.4, 9.7, 9.8]

Hinz, P., and Gurland, J. (1968). A method of analysing untransformed data from the negative binomial and other contagious distributions, *Biometrika*, **55**, 163–170. [9.6.4]

Hinz, P., and Gurland, J. (1970). A test of fit for the negative binomial and other contagious distributions, *Journal of the American Statistical Association*, **65**, 887–903. [9.6.4]

Hirano, K. (1986). Some properties of the distributions of order k, *Fibonacci Numbers and their Applications*, A. N. Philippou, G. E. Bergum and A. F. Horadam (editors), 43–53. Dordrecht: Reidel. [10.6.2, 10.6.3]

Hirano, K., and Aki, S. (1987). Properties of the extended distributions of order k, *Statistics and Probability Letters*, **6**, 67–69. [4.11, 10.6.3]

Hirano, K., Aki, S., Kashiwagi, N., and Kuboki, H. (1991). On Ling's binomial and negative binomial distributions of order k, *Statistics and Probability Letters*, **11**, 503–509. [10.6.3]

Hirano, K., Kuboki, H., Aki, S., and Kuribayashi, A. (1984). Figures of probability density functions in statistics II — discrete univariate case, *Institute of Statistical Mathematics Computer Science Monograph*, **20**: Tokyo. [7.11]

Hitha, N., and Nair, N. U. (1989). Characterization of some discrete models by properties of residual life function, *Bulletin of the Calcutta Statistical Association*, **38**, 219–223. [11.12]

Hoaglin, D. C., Mosteller, F., and Tukey, J. W. (editors) (1985). *Exploring Data Tables, Trends, and Shapes*, New York: Wiley. [4.7.1, 5.8.1, 7.7.1]

Hoaglin, D. C., and Tukey, J. W. (1985). Checking the shape of discrete distributions, *Exploring Data Tables, Trends, and Shapes*, D. C. Hoaglin, F. Mosteller, and J. W. Tukey (editors), 345–416. New York: Wiley. [3.8.1]

Hoel, P. G. (1947). Discriminating between binomial distributions, *Annals of Mathematical Statistics*, **18**, 556–564. [3.8.2, 3.8.3]

Hogg, R. V., and Klugman, S. A. (1984). *Loss Distributions*, New York: Wiley. [7.11]

Holgate, P. (1964). A modified geometric distribution arising in trapping studies, *Acta Theriologica*, **9**, 353–356. [8.2.2]

Holgate, P. (1966). Contributions to the mathematics of animal trapping, *Biometrics*, **22**, 925–936. [8.2.2]

Holgate, P. (1970). The modality of some compound Poisson distributions, *Biometrika*, **57**, 666–667. [8.3.2]

Holgate, P. (1989). Approximate moments of the Adès distribution, *Biometrical Journal*, **31**, 875–883. [11.4]

Holla, M. S. (1966). On a Poisson-inverse Gaussian distribution, *Metrika*, **11**, 115–121. [8.3.2, 11.15]

Holla, M. S., and Bhattacharya, S. K. (1965). On a discrete compound distribution, *Annals of the Institute of Statistical Mathematics, Tokyo*, **15**, 377–384. [8.3.4]

Holst, L. (1986). On birthday, collectors', occupancy and other classical urn problems, *International Statistical Review*, **54**, 15–27. [10.4.1]

Hopkins, J. W. (1955). An instance of negative hypergeometric sampling in practice, *Bulletin of the International Statistical Institute*, **34**(4), 298–306. [6.9.2]

Hoppe, F. M., and Seneta, E. (1990). A Bonferroni-type identity and permutation bounds, *International Statistical Review*, **58**, 253–261. [10.2]

Horsnell, G. (1957). Economic acceptance sampling schemes, *Journal of the Royal Statistical Society, Series A*, **120**, 148–191. [8.3.3, 8.3.4]

Huang, M.-L., and Fung, K. Y. (1989). Intervened truncated Poisson distribution, *Sankhyā, Series B*, **51**, 302–310. [4.12.3]

Hudson, H. M., and Tsui, K.-W. (1981). Simultaneous Poisson estimators for á priori hypotheses about means, *Journal of the American Statistical Association*, **76**, 182–187. [4.7.2]

Hunter, D. (1976). An upper bound for the probability of a union, *Journal of Applied Probability*, **13**, 597–603. [10.2]

Hurwitz, H., and Kac, M. (1944). Statistical analysis of certain types of random functions, *Annals of Mathematical Statistics*, **15**, 173–181. [4.2]

Hwang, J. S., and Lin, G. D. (1984). Characterizations of distributions by linear combinations of order statistics, *Bulletin of the Institute of Mathematics, Academia Sinica*, **12**, 179–202. [1B.10]

Hwang, J. T. (1982). Improving upon standard estimators in discrete exponential families with applications to Poisson and negative binomial cases, *Annals of Statistics*, **10**, 857–867. [4.7.2]

Irony, T. Z. (1992). Bayesian estimation for discrete distributions, *Journal of Applied Statistics*, **19**. [3.8.2, 4.7.2, 5.8.2]

Irwin, J. O. (1937). The frequency distribution of the difference between two independent variates following the same Poisson distribution, *Journal of the Royal Statistical Society, Series A*, **100**, 415–416. [4.12.3]

Irwin, J. O. (1941). Discussion on Chambers and Yule's paper, *Journal of the Royal Statistical Society Supplement*, **7**, 101–107. [5.3]

Irwin, J. O. (1953). On the "transition probabilities" corresponding to any accident distribution, *Journal of the Royal Statistical Society, Series B*, **15**, 87–89. [2.4.2, 4.12.5]

Irwin, J. O. (1954). A distribution arising in the study of infectious diseases, *Biometrika*, **41**, 266–268. [6.2.2, 6.9.2, 6.10.1]

Irwin, J. O. (1955). A unified derivation of some well-known frequency distributions of interest in biometry and statistics, *Journal of the Royal Statistical Society, Series A*, **118**, 389–404. [10.2, 10.3]

Irwin, J. O. (1959). On the estimation of the mean of a Poisson distribution from a sample with the zero class missing, *Biometrics*, **15**, 324–326. [4.10.1]

Irwin, J. O. (1963). The place of mathematics in medical and biological statistics, *Journal of the Royal Statistical Society, Series A*, **126**, 1–44. [3.12.5, 6.2.3, 6.10.4, 9.4, 11.7]

Irwin, J. O. (1964). The personal factor in accidents — A review article, *Journal of the Royal Statistical Society, Series A*, **127**, 438–451. [9.6.5]

Irwin, J. O. (1967). William Allen Whitworth and a hundred years of probability, *Journal of the Royal Statistical Society, Series A*, **130**, 147–176. [10.2]

Irwin, J. O. (1968). The generalized Waring distribution applied to accident theory, *Journal of the Royal Statistical Society, Series A*, **131**, 205–225. [6.2.3, 6.9.3]

Irwin, J. O. (1975a). The generalized Waring distribution, Part I, *Journal of the Royal Statistical Society, Series A*, **138**, 18–31. [6.7.3]

Irwin, J. O. (1975b). The generalized Waring distribution, Part II, *Journal of the Royal Statistical Society, Series A*, **138**, 204–227. [6.2.3, 6.7.3]

Irwin, J. O. (1975c). The generalized Waring distribution, Part III, *Journal of the Royal Statistical Society, Series A*, **138**, 374–384. [6.2.3]

Ishii, G., and Hayakawa, R. (1960). On the compound binomial distribution, *Annals of the Institute of Statistical Mathematics, Tokyo*, **12**, 69–80. [6.9.2, 8.3.3]

Ising, E. (1925). Beitrag zür Theorie des Ferromagnetismus, *Zeitschrift für Physik*, **31**, 253–258. [10.5.1]

Ivchenko, G. I. (1974). On comparison of binomial and Poisson distributions, *Theory of Probability and Its Applications*, **19**, 584–586. [3.6.1]

Iwase, K. (1986). UMVU estimator of the probability in the geometric distribution with unknown truncation parameter, *Communications in Statistics-Theory and Methods*, **15**, 2449–2453. [5.2]

Iyer, P. V. K. (1949). Calculation of factorial moments of certain probability distributions, *Nature, London*, **164**, 282. [10.2]

Iyer, P. V. K. (1958). A theorem on factorial moments and its applications, *Annals of Mathematical Statistics*, **29**, 254–261. [10.2]

Jagers, P. (1973). How many people pay their tram fares? *Journal of the American Statistical Association*, **68**, 801–804. [5.2]

Jain, G. C., and Consul, P. C. (1971). A generalized negative binomial distribution, *SIAM Journal of Applied Mathematics*, **21**, 501–513. [2.5.2, 2.5.3, 3.12.3, 5.12.3]

Jain, G. C., and Gupta, R. P. (1973). A logarithmic series type distribution, *Trabajos de Estadística*, **24**, 99–105. [2.2.2, 2.5.2, 7.11]

Jain, G. C., and Plunkett, I. G. (1977). A Borel-Hermite distribution and its applications, *Biometrical Journal*, **19**, 347–354. [9.4]

Janardan, K. G. (1973a). A characterization of multivariate hypergeometric and inverse hypergeometric models, *Sangamon State University Mathematical Systems Program Report*. [6.8]

Janardan, K. G. (1973b). On an alternative expression for the hypergeometric moment generating function, *American Statistician*, **27**, 242. [6.3]

Janardan, K. G. (1975). Markov-Polya urn model with predetermined strategies, *Gujurat Statistical Review*, **2**, 17–32. [2.5.3]

Janardan, K. G. (1978). On generalized Markov-Polya distribution, *Gujurat Statistical Review*, **5**, 16–32. [6.12]

Janardan, K. G. (1984). Moments of certain series distributions and their applications, *SIAM Journal of Applied Mathematics*, **44**, 854–868. [9.11]

Janardan, K. G. (1988). Relationship between Morisita's model for estimating the environmental density and the generalized Eulerian numbers, *Annals of the Institute of Statistical Mathematics, Tokyo*, **40**, 439–450. [2.5.3]

Janardan, K. G., and Patil, G. P. (1970). Location of modes for certain univariate and multivariate distributions, *Random Counts in Scientific Work*, **1**: *Random Counts in Models and Structures*, G. P. Patil (editor), 57–76. University Park: Pennsylvania State University Press. [2.3.2]

Janardan, K. G., and Patil, G. P. (1972). A unified approach for a class of multivariate hypergeometric models, *Sankhyā, Series A*, **34**, 363–376. [6.3]

Janardan, K. G., and Rao, B. R. (1982). Characterization of generalized Markov-Polya and generalized Polya-Eggenberger distributions, *Communications in Statistics*, **11**, 2113–2124. [4.8]

Janardan, K. G., and Schaeffer, D. J. (1977). A generalization of Markov-Polya distribution its extensions and applications, *Biometrical Journal*, **19**, 87–106. [11.19]

Janardan, K. G., and Schaeffer, D. J. (1981). Application of discrete distributions for estimating the number of organic compounds in water, *Statistical Distributions in Scientific Work*, **6**: *Applications in Physical, Social, and Life Sciences*, C. Taillie, G. P. Patil, and B. A. Baldessari (editors), 79–94. Dordrecht: Reidel. [5.10]

Jani, P. N. (1977). Minimum variance unbiased estimation for some left-truncated modified power series distributions, *Sankhyā, Series B*, **39**, 258–278. [2.2.2]

Jani, P. N. (1978a). New numbers appearing in minimum variance unbiased estimation for decapitated negative binomial and Poisson distributions, *Journal of the Indian Statistical Association*, **16**, 41–48. [2.2.2]

Jani, P. N. (1978b). On modified power series distributions, *Metron*, **36**, 173–186. [2.2.2]

Jani, P. N. (1986).The generalized logarithmic series distribution with zeroes, *Journal of the Indian Society for Agricultural Statistics*, **38**, 345–351. [7.11]

Jani, P. N., and Shah, S. M. (1979a). Integral expressions for the tail probabilities of the modified power series distributions, *Metron*, **37**, 75–79. [2.2.2]

Jani, P. N., and Shah, S. M. (1979b). Misclassifications in modified power series distribution in which the value one is sometimes reported as zero, and some of its applications. *Metron*, **37**, 121–136. [2.2.2]

Janko, J. (editor) (1958). *Statistical Tables* (Republished by Soviet Central Statistical Bureau, 1961): Prague. [4.6]

Jánossy, L., Rényi, A., and Aczél, J. (1950). On composed Poisson distributions, I, *Acta Mathematica Academiae Scientiarum Hungaricae*, **1**, 209–224. [9.3]

Jeffreys, H. (1941). Some applications of the method of minimum chi-squared, *Annals of Eugenics, London*, **11**, 108–114. [5.1]

John, S. (1970). On analysing mixed samples, *Journal of the American Statistical Association*, **65**, 755–762. [8.2.5]

Johnson, N. L. (1951). Estimators of the probability of the zero class in Poisson and certain related populations, *Annals of Mathematical Statistics*, **22**, 94–101. [4.7.2]

Johnson, N. L. (1957). A note on the mean deviation of the binomial distribution, *Biometrika*, **44**, 532–533. [3.3]

Johnson, N. L. (1959). On an extension of the connexion between Poisson and χ^2 distributions, *Biometrika*, **46**, 352–362. [4.12.3]

Johnson, N. L., and Kotz, S. (1969). *Discrete Distributions* (First edition), Boston: Houghton Mifflin. [10.4.1]

Johnson, N. L., and Kotz, S. (1977). *Urn Models and Their Application*, New York: Wiley. [3.10, 4.2, 5.1, 10.4.1, 10.4.2, 11.19]

Johnson, N. L., and Kotz, S. (1981). Moments of discrete probability distributions derived using finite difference operators, *American Statistician*, **35**, 268. [1B.5]

Johnson, N. L., and Kotz, S. (1982). Developments in discrete distributions, 1969–1980, *International Statistical Review*, **50**, 71–101. [9.2]

Johnson, N. L., and Kotz, S. (1985). Some distributions arising as a consequence of errors in inspection, *Naval Research Logistics Quarterly*, **32**, 35–43. [6.12]

Johnson, N. L., and Kotz, S. (1989). Characterization based on conditional distributions, *Annals of the Institute of Statistical Mathematics, Tokyo*, **41**, 13–17. [6.10.3]

Johnson, N. L., and Kotz, S. (1990a). Use of moments in deriving distributions and some characterizations, *Mathematical Scientist*, **15**, 42–52. [1B.5]

Johnson, N. L., and Kotz, S. (1990b). Randomly weighted averages: some aspects and extensions, *American Statistician*, **44**, 245–249. [1B.5]

Johnson, N. L., Kotz, S., and Rodriguez, R. N. (1985). Statistical effects of imperfect inspection sampling: I Some basic distributions, *Journal of Quality Technology*, **17**, 1–31. [6.12]

Johnson, N. L., Kotz, S., and Rodriguez, R. N. (1986). Statistical effects of imperfect inspection sampling II. Double sampling and link sampling, *Journal of Quality Technology*, **18**, 198–199. [6.12]

Johnson, N. L., Kotz, S., and Sorkin, H. L. (1980). "Faulty inspection" distributions, *Communications in Statistics-Theory and Methods*, **A9**, 917–922. [6.12]

Johnson, N. L., Kotz, S., and Wu, X. (1991). *Inspection Errors for Attributes in Quality Control*, London: Chapman & Hall. [6.12]

Johnson, N. L., and Young, D. H. (1960). Some applications of two approximations to the multinomial distribution, *Biometrika*, **47**, 463–469. [4.5, 4.7.1]

Jones, H. G. (1933). A note on the *n*-ages method, *Journal of the Institute of Actuaries*, **64**, 318–324. [8.2.3]

Jones, M. C. (1987). Inverse factorial moments, *Statistics and Probability Letters*, **6**, 37–42. [3.3]

Jordan, C. (1927). Sur un cas généralisé de la probabilité des épreuves répétées, *Comptes Rendus, Académie des Sciences, Paris*, **184**, 315–317. [6.2.4]

Jordan, C. (1950). *Calculus of Finite Differences* (Second edition), New York: Chelsea. [1A3, 9.1]

Jordan, K. (1972). *Chapters on the Classical Calculus of Probability*, Budapest: Akadémiai Kiadó. (Hungarian edition, 1956). [10.2]

Jordan, M. C. (1867). De quelques formules de probabilité, *Comptes Rendus, Académie des Sciences, Paris*, **65**, 993–994. [10.2]

Joseph, A. W., and Bizley, M. T. L. (1960). The two-pack matching problem, *Journal of the Royal Statistical Society, Series B*, **22**, 114–130. [10.3]

Joshi, S. W. (1974). Integral expressions for the tail probabilities of the power series distributions, *Sankhyā, Series B*, **36**, 462–465. [2.2.1, 3.4]

Joshi, S. W. (1975). Some recent advances with power series distributions, *Statistical Distributions in Scientific Work*, **1**: *Models and Structures*, G. P. Patil, S. Kotz, and J. K. Ord (editors), 9–17. Dordrecht: Reidel. [2.2.1, 3.4]

Jowett, G. H. (1963). The relationship between the binomial and *F* distributions, *The Statistician*, **13**, 55–57. [3.6.1]

Kabe, D. G. (1969). Some distribution problems of order statistics from discrete populations, *Annals of the Institute of Statistical Mathematics, Tokyo*, **21**, 551–556. [5.2]

Kachitvichyanukul, V., and Schmeiser, B. W. (1985). Computer generation of hypergeometric random variates, *Journal of Statistical Computation and Simulation*, **22**, 127–145. [1C.6]

Kachitvichyanukul, V., and Schmeiser, B. W. (1988). Binomial random variate generation, *Communications of the A.C.M.*, **31**, 216–222. [1C.3]

Kagan, A.M., Linnik, Y. V., and Rao, C. R. (1973). *Characterization Problems in Mathematical Statistics*, New York: Wiley. [3.9, 5.9.1, 5.9.2, 6.8]

Kahn, W. D. (1987). A cautionary note for Bayesian estimation of the binomial parameter *n*, *American Statistician*, **41**, 38–39. [3.8.2]

Kalbfleisch, J. D. (1986). Pseudo-likelihood, *Encyclopedia of Statistical Sciences*, **7**, S. Kotz, N. L. Johnson and C. B. Read (editors), 324–327. New York: Wiley. [1B.15]

Kamat, A. R. (1965). Incomplete and absolute moments of some discrete distributions, *Classical and Contagious Discrete Distributions*, G. P. Patil (editor), 45–64. Calcutta: Statistical Publishing Society; Oxford: Pergamon Press. [2.3.1, 2.3.2, 5.4, 6.3, 7.3]

Kamat, A. R. (1966). A generalization of Johnson's property of the mean deviation for a class of distributions, *Biometrika*, **53**, 285–287. [2.3.2]

Kambo, N. S., and Kotz, S. (1966). On exponential bounds for binomial probabilities, *Annals of the Institute of Statistical Mathematics, Tokyo*, **18**, 277–287. [3.6.2]

Kapadia, C. H., and Thomasson, R. L. (1975). On estimating the parameter of a truncated geometric distribution by the method of moments, *Annals of the Institute of Statistical Mathematics, Tokyo*, **27**, 269–272. [5.2]

Kaplan, N., and Risko, K. (1982). A method for estimating rates of nucleotide substitution using DNA sequence data, *Theoretical Population Biology*, **21**, 318–328. [3.10]

Kaplansky, I., and Riordan, J. (1945). Multiple matching and runs by the symbolic method, *Annals of Mathematical Statistics*, **16**, 272–277. [10.3]

Kapur, J. N. (1978a). On generalized birth and death processes and generalized hypergeometric functions, *Indian Journal of Mathematics*, **20**, 57–69. [2.4.1]

Kapur, J. N. (1978b). Application of generalized hypergeometric functions to generalized birth and death processes, *Indian Journal of Pure and Applied Mathematics*, **9**, 1059–1069. [2.4.1]

Katti, S. K. (1960). Moments of the absolute difference and absolute deviation of discrete distributions, *Annals of Mathematical Statistics*, **31**, 78–85. [2.3.2, 3.3, 4.12.3]

Katti, S. K. (1965). Some estimation procedures for discrete distributions, *Florida State University Statistical Report*. [9.6.4]

Katti, S. K. (1966). Interrelations among generalized distributions and their components, *Biometrics*, **22**, 44–52. [2.4.2, 3.10, 5.3, 6.11, 7.11, 8.3.2, 8.3.4, 9.1, 9.3, 9.12]

Katti, S. K. (1967). Infinite divisibility of integer valued random variables, *Annals of Mathematical Statistics*, **38**, 1306–1308. [7.4, 9.3]

Katti, S. K., and Gurland, J. (1961). The Poisson Pascal distribution, *Biometrics*, **17**, 527–538. [9.8]

Katti, S. K., and Gurland, J. (1962a). Efficiency of certain methods of estimation for the negative binomial and the Neyman type A distributions, *Biometrika*, **49**, 215–226. [5.8.3, 9.6.4]

Katti, S. K., and Gurland, J. (1962b). Some methods of estimation for the Poisson binomial distribution, *Biometrics*, **18**, 42–51. [9.5]

Katti, S. K., and Rao, A. V. (1970). The log-zero-Poisson distribution, *Biometrics*, **26**, 801–813. [8.2.2, 8.3.2]

Katz, L. (1945). *Characteristics of Frequency Functions Defined by First Order Difference Equations*, Dissertation, Ann Arbor, MI: University of Michigan. [2.3.1, 3.4]

Katz, L. (1946). On the class of functions defined by the difference equation $(x + 1)f(x + 1) = (a + bx)f(x)$ (abstract), *Annals of Mathematical Statistics*, **17**, 501. [2.3.1]

Katz, L. (1948). Frequency functions defined by the Pearson difference equation (abstract), *Annals of Mathematical Statistics*, **19**, 120. [2.3.1]

Katz, L. (1965). Unified treatment of a broad class of discrete probability distributions, *Classical and Contagious Discrete Distributions*, G. P. Patil (editor), 175–182. Calcutta: Statistical Publishing Society; Oxford: Pergamon Press. [2.3.1, 3.4]

Kay, R. (1985). Partial likelihood, *Encyclopedia of Statistical Sciences*, **6**, S. Kotz, N. L. Johnson and C. B. Read (editors), 591–593. New York: Wiley. [1B.15]

Kemp, A. W. (1968a). *Studies in Univariate Discrete Distribution Theory Based on the Generalized Hypergeometric Function and Associated Differential Equations*, Ph.D thesis, Belfast: The Queen's University of Belfast (191pp). [2.4.1, 2.4.2, 3.4, 3.10, 4.12.4, 5.3, 6.2.4, 8.3.4]

Kemp, A. W. (1968b). A wide class of discrete distributions and the associated differential equations, *Sankhyā, Series A*, **30**, 401–410. [2.4.1, 2.4.2, 3.4, 3.10, 4.12.4, 6.2.4, 6.10.2, 11.10, 11.11]

Kemp, A. W. (1968c). A limited risk cPp, *Skandinavisk Aktuarietidskrift*, **51**, 198–203. [2.4.2, 8.3.2]

Kemp, A. W. (1974). Towards a unification of the theory of counts, *University of Bradford School of Mathematics, Statistics Reports and Preprints*, **12**. [2.4.2]

Kemp, A. W. (1978a). Cluster size probabilities for generalized Poisson distributions, *Communications in Statistics-Theory and Methods*, **7**, 1433–1438. [7.2, 7.4, 9.3]

Kemp, A. W. (1978b). On probability generating functions for matching and occupancy distributions, *Zastosowania Matematyki*, **16**, 207–213. [2.4.2, 10.3, 10.4.1]

Kemp, A. W. (1979). Convolutions involving binomial pseudo-variables, *Sankhyā, Series A*, **41**, 232– 243. [3.12.5, 5.12.5, 11.7]

Kemp, A. W. (1981a). Conditionality properties for the bivariate logarithmic distribution with an application to goodness of fit, *Statistical Distributions in Scientific Work*, **5**: *Inferential Problems and Properties*, C. Taillie, G. P. Patil, and B. A. Baldessari (editors), 57–73. Dordrecht: Reidel. [7.2]

Kemp, A. W. (1981b). Efficient generation of logarithmically distributed pseudorandom variables, *Applied Statistics*, **30**, 249–253. [1C.7, 7.4]

Kemp, A. W. (1986). Weighted discrepancies and maximum likelihood estimation for discrete distributions, *Communications in Statistics-Theory and Methods*, **15**, 783–803. [1B.15, 5.8.3, 7.7.2, 8.2.2]

Kemp, A. W. (1987a). A Poissonian binomial model with constrained parameters, *Naval Research Logistics*, **34**, 853–858. [3.12.2]

Kemp, A. W. (1987b). Families of discrete distributions satisfying Taylor's power law, *Biometrics*, **43**, 693–699. [11.4]

Kemp, A. W. (1988a). Families of distributions for repeated samples of animal counts: Response, *Biometrics*, **44**, 888–890. [11.4]

Kemp, A. W. (1988b). Simple algorithms for the Poisson modal cumulative probability, *Communications in Statistics-Simulation and Computation*, **17**, 1495–1508. [1C.4, 4.4, 4.5, 4.6]

Kemp, A. W. (1989). A note on Stirling's expansion for factorial n, *Statistics and Probability Letters*, **7**, 139–143. [4.5, 4.6]

Kemp, A. W. (1992a). Heine-Euler extensions of the Poisson distribution, *Communications in Statistics-Theory and Methods*, **21** [4.12.6, 7.11]

Kemp, A. W. (1992b). Steady-state Markov chain models for the Heine and Euler distributions, *Journal of Applied Probability*, **29** [4.12.6, 7.11]

Kemp, A. W. (1992c). On counts of individuals able to signal the presence of an observer, *Biometrical Journal*, **34**, [11.7]

Kemp, A. W., and Kemp, C. D. (1966). An alternative derivation of the Hermite distribution, *Biometrika*, **53**, 627–628. [9.4]

Kemp, A. W., and Kemp, C. D. (1969a). Branching and clustering models associated with the 'lost-games' distribution, *Journal of Applied Probability*, **6**, 700–703. [2.5.1, 11.10]

Kemp, A. W., and Kemp, C. D. (1969b). Some distributions arising from an inventory decision problem, *Bulletin of the International Statistical Institute*, **43**(2), 336–338. [2.4.2, 6.2.4, 11.12]

Kemp, A. W., and Kemp, C. D. (1971). On mixing processes and the lost-games distribution, *Zastosowania Matematyki*, **12**, 167–173. [11.10]

Kemp, A. W., and Kemp, C. D. (1974). A family of distributions defined via their factorial moments, *Communications in Statistics*, **3**, 1187–1196. [2.4.1, 2.4.2, 3.4]

Kemp, A. W., and Kemp, C. D. (1975). Models for Gaussian hypergeometric distributions, *Statistical Distributions in Scientific Work*, **1**: *Models and Structures*, G. P. Patil, S. Kotz, and J. K. Ord (editors), 31–40. Dordrecht: Reidel. [6.2.3, 6.2.4]

Kemp, A. W., and Kemp, C. D. (1988). A rapid and efficient estimation procedure for the negative binomial distribution, *Biometrical Journal*, **29**, 865–873. [5.8.3]

Kemp, A. W., and Kemp, C. D. (1989). Even-point estimation, *Encyclopedia of Statistical Sciences Supplement*, S. Kotz, N. L. Johnson and C. B. Read (editors). New York: Wiley. [9.4]

Kemp, A. W., and Kemp, C. D. (1990). A composition-search algorithm for low-parameter Poisson generation, *Journal of Statistical Computation and Simulation*, **35**, 239–244. [1C.4, 4.4]

Kemp, A. W., and Kemp, C. D. (1991). Weldon's dice data revisited, *American Statistician*, **45**, 216–222. [3.12.2]

Kemp, A. W., and Kemp, C. D. (1992). A group-dynamic model and the lost-games distribution, *Communications in Statistics-Theory and Methods*, **21**, [11.10]

Kemp, A. W., and Newton, J. (1990). Certain state-dependent processes for dichotomised parasite populations, *Journal of Applied Probability*, **27**, 251–258. [2.4.1, 3.12.2]

Kemp, C. D. (1967a). 'Stuttering-Poisson' distributions, *Journal of the Statistical and Social Enquiry Society of Ireland*, **21**(5), 151–157. [4.11, 5.1, 9.3, 9.7]

Kemp, C. D. (1967b). On a contagious distribution suggested for accident data, *Biometrics*, **23**, 241–255. [9.6.5, 9.9]

Kemp, C. D. (1970). "Accident proneness" and discrete distribution theory, *Random Counts in Scientific Work*, **2**: *Random Counts in Biomedical and Social Sciences*, G. P. Patil (editor), 41–65. University Park: Pennsylvania State University Press. [5.10]

Kemp, C. D. (1986). A modal method for generating binomial variables, *Communications in Statistics-Theory and Methods*, **15**, 805–813. [1C.3, 3.6.1, 3.7]

Kemp, C. D., and Kemp, A. W. (1956a). Generalized hypergeometric distributions, *Journal of the Royal Statistical Society, Series B*, **18**, 202–211. [2.4.1, 6.2.2, 6.2.3, 6.2.4, 6.4, 8.3.4]

Kemp, C. D., and Kemp, A. W. (1956b). The analysis of point quadrat data, *Australian Journal of Botany*, **4**, 167–174. [6.4, 6.7.2, 6.9.2]

Kemp, C. D., and Kemp, A. W. (1965). Some properties of the 'Hermite' distribution, *Biometrika*, **52**, 381–394. [9.3, 9.4]

Kemp, C. D., and Kemp, A. W. (1967). A special case of Fisher's 'modified Poisson series', *Sankhyā, Series A*, **29**, 103–104. [9.4]

Kemp, C. D., and Kemp, A. W. (1968). On a distribution associated with certain stochastic processes, *Journal of the Royal Statistical Society, Series B*, **30**, 401–410. [11.10]

Kemp, C. D., and Kemp, A. W. (1987). Rapid generation of frequency tables, *Applied Statistics*, **36**, 277–282. [1C.2]

Kemp, C. D., and Kemp, A. W. (1988). Rapid estimation for discrete distributions, *The Statistician*, **37**, 243–255. [4.10.1, 7.7.2, 8.2.2]

Kemp, C. D., and Kemp, A. W. (1991). Poisson random variate generation, *Applied Statistics*, **40**, 143–158. [1C.4, 4.6]

Kemp, C. D., Kemp, A.W., and Loukas, S. (1979). Sampling from discrete distributions, *University of Bradford Statistics Reports and Preprints*, **42**. [1C.7]

Kempton, R. A. (1975). A generalized form of Fisher's logarithmic series, *Biometrika*, **62**, 29–38. [7.11]

Kendall, D. G. (1948). On some modes of population growth leading to R. A. Fisher's logarithmic series distribution, *Biometrika*, **35**, 6–15. [5.3, 7.2, 7.10, 9.2, 11.16]

Kendall, D. G. (1949). Stochastic processes and population growth, *Journal of the Royal Statistical Society, Series B*, **11**, 230–282. [5.3, 5.10]

Kendall, M. G. (1943). *The Advanced Theory of Statistics*, **1**, London: Griffin. [3.3, 4.3, 11.8]

Kendall, M. G. (1960). The bibliography of operational research, *Operational Research Quarterly*, **11**, 31–36. [6.10.3]

Kendall, M. G. (1961). Natural law in the social sciences, *Journal of the Royal Statistical Society, Series A*, **124**, 1–18. [6.10.3, 6.10.4, 11.20]

Kendall, M. G. (1968). Studies in the history of probability and statistics XVIII, Thomas Young on coincidences, *Biometrika*, **55**, 249–250. [10.3]

Kendall, M. G., and Stuart, A. (1961). *The Advanced Theory of Statistics*, **2**, London: Griffin. [3.8.3]

Khalil, Z., Dimitrov, B., and Dion, J. P. (1991). A characterization of the geometric distribution related to random sums. *Stochastic Models*, **7**, 321–326. [5.9.2]

Khamis, S. H. (1960). Incomplete gamma function expansions of statistical distribution functions, *Bulletin of the International Statistical Institute*, **37**, 385–396. [1A.5]

Khamis, S. H., and Rudert, W. (1965). *Tables of the Incomplete Gamma Function Ratio: Chi-Square Integral, Poisson Distribution*, Darmstadt: von Liebig. [1A.5, 4.6]

Khan, A. H., and Ali, M. M. (1987). Characterizations of probability distributions through higher order gap, *Communications in Statistics-Theory and Methods*, **16**, 1281–1287. [1B.10]

Kharshikar, A. V. (1970). On the expected value of $S^2/\bar{x}$, *Biometrics*, **26**, 343–346. [4.8]

Khatri, C. G. (1959). On certain properties of power-series distributions, *Biometrika*, **46**, 486–490. [2.2.1, 7.3]

Khatri, C. G. (1961). On the distributions obtained by varying the number of trials in a binomial distribution, *Annals of the Institute of Statistical Mathematics, Tokyo*, **13**, 47–51. [7.10, 8.2.2]

Khatri, C. G. (1962). Distributions of order statistics for the discrete case, *Annals of the Institute of Statistical Mathematics, Tokyo*, **14**, 167–171. [3.5, 9.12].

Khatri, C. G. (1983). Some remarks on the moments of discrete distributions using difference operators, *American Statistician*, **37**, 96–97. [1B.5]

Khatri, C. G., and Patel, I. R. (1961). Three classes of univariate discrete distributions, *Biometrics*, **17**, 567–575. [3.10, 8.2.2, 9.3, 9.12]

King, G. (1902). *Institute of Actuaries' Text Book of the Principles of Interest, Life Annuities, and Assurances, and their practical Application, II, Life Contingencies* (Second edition). London: Leyton. [10.2]

Kitagawa, T. (1952). *Tables of Poisson Distribution*, Tokyo: Baifukan Publishing Company. [4.6]

Kleinrock, L. (1975). *Queueing Systems*, **1**, New York: Wiley. [11.13]

Kleinrock, L. (1976). *Queueing Systems*, **2**, New York: Wiley. [11.13]

Klotz, J. (1970). The geometric density with unknown location parameter, *Annals of Mathematical Statistics*, **41**, 1078–1082. [5.2]

Koch, G. G., Atkinson, S. S., and Stokes, M. E. (1986). Poisson regression, *Encyclopedia of Statistical Sciences*, **7**, S. Kotz, N. L. Johnson and C. B. Read (editors), 32–41. New York: Wiley. [4.9]

Kocherlakota, S., and Kocherlakota, K. (1986). Goodness of fit tests for discrete distributions, *Communications in Statistics-Theory and Methods*, **15**, 815–829. [1B.15]

Kocherlakota, S., and Kocherlakota, K. (1990). Tests of hypothesis for the weighted binomial distribution, *Biometrics*, **46**, 645–656. [3.12.4]

Kolchin, V. F., Sevast'yanov, B. A., and Chistyakov, V. P. (1978). *Random Allocations*, New York: Wiley. [10.4.1]

Korwar, R. M. (1975). On characterizing some discrete distributions by linear regression, *Communications in Statistics*, **4**, 1133–1147. [1B.7, 4.8]

Kosambi, D. D. (1949). Characteristic properties of series distributions, *Proceedings of the National Institute for Science, India*, **15**, 109–113. [2.2.1, 2.2.2, 3.4]

Kotz, S. (1974). Characterizations of statistical distributions: a supplement to recent surveys, *International Statistical Review*, **42**, 39–65. [4.8]

Kotz, S., and Johnson, N. L. (1990). Regression relations in random sum theory, *Proceedings of the 1990 International Statistical Symposium, Taipei*. [1B.7]

Kotz, S., and Johnson, N. L. (1991). A note on renewal (partial sums) distributions for discrete variables, *Statistics and Probability Letters*, **12**, 229–231. [11.12]

Kourouklis, S. (1986). Characterizations of some discrete distributions based on a variant of the Rao-Rubin condition. *Communications in Statistics-Theory and Methods*, **15**, 839–851. [4.8]

Krafft, O. (1969). A note on exponential bounds for binomial probabilities, *Annals of the Institute of Statistical Mathematics, Tokyo*, **21**, 219–220. [3.6.2]

Kreweras, G. (1979). Some finite distributions tending towards Poisson distributions, *Bulletin of the International Statistical Institute*, **49**, 296–299. [4.2]

Krishnaji, N. (1974). Characterization of some discrete distributions based on a damage model, *Sankhyā, Series A*, **36**, 204–213. [4.8]

Kronmal, R. A., and Peterson, A. V. (1979). On the alias method for generating random variables from a discrete distribution, *American Statistician*, **33**, 214–218. [1C.2, 1C.6]

Kumar, A., and Consul, P. C. (1979). Negative moments of a modified power series distribution and bias of the maximum likelihood estimator, *Communications in Statistics-Theory and Methods*, **A 8**, 151–166. [2.2.2, 5.12.3]

Kumar, A., and Consul, P. C. (1980). Minimum variance unbiased estimation for modified power series distribution, *Communications in Statistics-Theory and Methods*, **A9**, 1261–1275. [2.2.2, 5.12.3, 9.11]

Kupper, L. L., and Haseman, J. K. (1978). The use of a correlated binomial model for the analysis of certain toxicological experiments, *Biometrics*, **34**, 69–76. [3.12.6]

Kyriakoussis, A., and Papageorgiou, H. (1991a). Characterization of hypergeometric type distributions by regression, *Statistics*, **22**, 467–477. [6.8]

Kyriakoussis, A., and Papageorgiou, H. (1991b). Characterizations of logarithmic series distributions, *Statistica Neerlandica*, **45**, 1-8. [7.8]

Laha, R. G. (1982). Characteristic functions, *Encyclopedia of Statistical Sciences*, **1**, S. Kotz, N. L. Johnson and C. B. Read (editors), 415–422. New York: Wiley. [1B.8]

Laplace, P. S. (1820). *Théorie Analytique des Probabilités* (Third edition). Paris: Courvier. [3.6.1]

Larson, H. R. (1966). A nomograph of the cumulative binomial distribution, *Industrial Quality Control*, **23**, 270–278. [3.7]

Laubscher, N. F. (1961). On stabilizing the binomial and negative binomial variances, *Journal of the American Statistical Association*, **56**, 143–150. [5.6]

Laurent, A. G. (1965). Probability distributions, factorial moments, empty cell test, *Classical and Contagious Discrete Distributions*, G. P. Patil (editor), 437–442. Calcutta: Statistical Publishing Society; Oxford: Pergamon Press. [1B.9]

Lee, L. F. (1986). Specification test for Poisson regression models, *International Economic Review*, **27**, 689–706. [4.9]

Lehmann, E. L. (1986). *Testing Statistical Hypotheses* (Second edition), New York: Wiley. [1B.15]

Lepage, Y. (1978). Negative factorial moments of positive random variables, *Industrial Mathematics*, **28**, 95–100. [3.3]

Lessing, R. (1973). An alternative expression for the hypergeometric moment generating function, *American Statistician*, **27**, 115. [6.3]

Letac, G. (1991). Contre exemple au théorème de P C Consul sur la factorisation des distributions de Poisson généralisées, *Canadian Journal of Statistics*, **19**, 229–231. [9.11]

Levene, H. (1949). On a matching problem arising in genetics, *Annals of Mathematical Statistics*, **20**, 91–94. [10.3]

Lévy, P. (1937a). Sur les exponentielles de polynomes et sur l'arithmétique des produits de lois de Poisson, *Annales Scientifique de l'Ecole Normale Supérieure, Series III*, **54**, 231–292. [9.3]

Lévy, P. (1937b). *Théorie de l'Addition des Variables Aléatoires*, Paris: Gautier-Villars. [4.2]

Lévy, P. (1954). *Théorie de l'Addition des Variables Aléatoires* (Second edition), Paris: Gauthier Villars. [3.6.2]

Lewis, P. A., and Orav, E. J. (1989). *Simulation Methodology for Statisticians, Operations Analysts and Engineers*, **1**, Pacific Grove, CA: Wadsworth & Brooks/Cole. [1C.1]

Lexis, W. (1877). *Zur Theorie der Massenerscheiungen in der Menschlichen Gesellschaft*, Freiburg: Wagner. [3.12.2]

Lidstone, G. J. (1942). Notes on the Poisson frequency distribution, *Journal of the Institute of Actuaries*, **71**, 284–291. [4.4]

Lieberman, G. J., and Owen, D. B. (1961). *Tables of the Hypergeometric Probability Distribution*, Stanford: Stanford University Press. [6.4, 6.5, 6.6]

Lin, G. D. (1987). Characterizations of distributions via relationships between two moments of order statistics, *Journal of Statistical Planning and Inference*, **19**, 73–80. [1B.10]

Lindley, D. V. (1958). Fiducial distributions and Bayes' theorem, *Journal of the Royal Statistical Society, Series B*, **20**, 102–107. [8.3.2]

Lindley, D. V. (1990). The present position in Bayesian statistics, *Statistical Science*, **5**, 44–65. [1B.3]

Ling, K. D. (1988). On binomial distributions of order k, *Statistics and Probability Letters*, **6**, 371–376. [10.6.3]

Ling, K. D. (1989). A new class of negative binomial distributions of order k, *Statistics and Probability Letters*, **7**, 247–250. [10.6.3]

Ling, K. D. (1990). On geometric distributions of order $(k_1, \ldots, k_m)$, *Statistics and Probability Letters*, **9**, 163–171. [10.6.3]

Ling, R. F., and Pratt, J. W. (1984). The accuracy of Peizer approximations to the hypergeometric distribution with comparisons to some other approximations. *Journal of the American Statistical Association*, **79**, 49–60. [6.5]

Lingappiah, G. S. (1987). Some variants of the binomial distribution, *Bulletin of the Malaysian Mathematical Society*, **10**, 82–94. [3.12.3, 5.12.7]

Littlewood, J. E. (1969). On the probability in the tail of a binomial distribution, *Advances in Applied Probability*, **1**, 43–72. [3.6.1]

Lloyd, E. L. (1980). *Handbook of Applicable Mathematics* **II**: *Probability*, Chichester: Wiley. [4.11]

Lloyd, E. L. (editor) (1984). *Handbook of Applicable Mathematics* **VI**A: *Statistics*. Chichester: Wiley. [3.8.4]

Lo, H.-P., and Wani, J. K. (1983). Maximum likelihood estimation of the parameters of the invariant abundance distributions, *Biometrics*, **39**, 977–986. [7.1, 7.7.2]

Loève, M. (1963). *Probability theory* (Third edition), New York: van Nostrand. [10.2, 10.4.2]

Louis, T. A. (1981). Confidence intervals for a binomial parameter after observing no successes, *American Statistician*, **35**, 154. [3.8.3]

Louv, W. C., and Littell, R. C. (1986). Combining one-sided binomial tests, *Journal of the American Statistical Association*, **81**, 550–554. [3.8.4]

Lüders, R. (1934). Die Statistik der seltenen Ereignisse, *Biometrika*, **26**, 108–128. [5.3, 5.12.5, 7.2, 9.3]

Lukacs, E. (1956). Characterization of populations by properties of suitable statistics, *Proceedings of the Third Berkeley Symposium on Mathematical Statistics and Probability*, **2**, 195–214. Berkeley: University of California Press. [4.8]

Lukacs, E. (1965). Characterization problems for discrete distributions, *Classical and Contagious Discrete Distributions*, G. P. Patil (editor), 65–74. Calcutta: Statistical Publishing Society; Oxford: Pergamon Press. [3.9, 4.8, 5.9.1]

Lukacs, E. (1970). *Characteristic Functions* (Second edition), London: Griffin. [1B.8, 9.3]

Lukacs, E. (1983). *Developments in Characteristic Function Theory*, London: Griffin. [1B.8]

Luke, Y. L. (1975). *Mathematical Functions and Their Approximations*, New York: Academic Press. [1A.6]

Lund, R. E. (1980). Algorithm AS 152: Cumulative hypergeometric probabilities, *Applied Statistics*, **29**, 221–223. [6.6]

Lundberg, O. (1940). *On Random Processes and their Application to Sickness and Accident Statistics*, Uppsala: Almqvist and Wicksells (Reprinted 1964). [5.3, 8.3.1, 8.3.2]

Ma, Y.-L. (1982). A simple binomial approximation for hypergeometric distribution. *Acta Mathematicae Applicatae Sinica*, **5**, 418–425 (In Chinese). [6.5]

Maceda, E. C. (1948). On the compound and generalized Poisson distributions, *Annals of Mathematical Statistics*, **19**, 414–416. [8.3.1]

Magistad, J. G. (1961). Some discrete distributions associated with life testing. *Proceedings of the 7th National Symposium on Reliability and Quality Control*, 1–11. [5.2]

Mainland, D. (1948). Statistical methods in medical research, *Canadian Journal of Research*, **26** (section E), 1–166. [3.8.3]

Makuch, R. W., Stephens, M. A., and Escobar, M. (1989). Generalized binomial models to examine the historical control assumption in active control equivalence studies, *The Statistician*, **38**, 61–70. [3.12.6]

Mandelbrot, B. (1959). A note on a class of skew distribution functions: analysis and critique of a paper by H. A. Simon, *Information and Control*, **2**, 90–99. [6.10.3]

Mann, N. M., Schafer, R. E., and Singpurwalla, N. D. (1974). *Methods for Statistical Analysis of Reliability and Life Data*, New York: Wiley. [5.2]

Mantel, N. (1951). Evaluation of a class of diagnostic tests, *Biometrics*, **3**, 240–246. [3.11]

Mantel, N. (1962). (Appendix to Haenzel, W., Loveland, D. B., and Sorken, M. B.) Lung cancer mortality as related to residence and smoking histories, I: White males, *Journal of the National Cancer Institution*, **28**, 947–997. [4.7.3]

Mantel, N., and Pasternack, B. S. (1968). A class of occupancy problems, *American Statistician*, **22**, 23–24. [10.4.1]

Margolin, B. H., Resnick, M. A., Rimpo, J. H., Archer, P., Galloway, S. M., Bloom, A. D., and Zeiger, E. (1986). Statistical analyses for *in vitro* cytogenetics assays using Chinese hamster ovary cells, *Environmental Mutagenesis*, **8**, 183–204. [4.9]

Margolin, B. H., and Winokur, H. S. (1967). Exact moments of the order statistics of the geometric distribution and their relation to inverse sampling and reliability of redundant systems, *Journal of the American Statistical Association*, **62**, 915–925. [5.2]

Maritz, J. S. (1952). Note on a certain family of discrete distributions, *Biometrika*, **39**, 196–198. [5.3, 9.3]

Maritz, J. S. (1969). Empirical Bayes estimation for the Poisson distribution, *Biometrika*, **56**, 349–359. [4.7.2, 4.8]

Maritz, J. S., and Lwin, T. (1989). *Empirical Bayes Methods* (Second edition), London: Chapman and Hall. [1B.15, 4.7.2]

Marlow, W. H. (1965). Factorial distributions, *Annals of Mathematical Statistics*, **36**, 1066–1068. [2.6, 6.10.4]

Martin, D. C., and Katti, S. K. (1962). Approximations to the Neyman type A distribution for practical problems, *Biometrics*, **18**, 354–364. [9.6.3]

Martin, D. C., and Katti, S. K. (1965). Fitting of some contagious distributions to some available data by the maximum likelihood method, *Biometrics*, **21**, 34–48 (correction, **21**, 514). [5.10, 8.2.2, 9.5, 9.6.5]

Maruyama, G. (1955). On the Poisson distribution derived from independent random walks, *Ochanomizu University Natural Science Reports*, **6**, 1–6. [4.2]

Mathai, A. M., and Saxena, R. K. (1973). *Generalized Hypergeometric Functions with Applications in Statistics and Physical Sciences*, Berlin: Springer-Verlag. [1A.8]

Mathai, A. M., and Saxena, R. K. (1978). *The H Function with Applications in Statistics and other Disciplines*, New Delhi: Wiley Eastern. [1A.8]

Matsunawa, T. (1986). Poisson distribution, *Encyclopedia of Statistical Sciences*, **7**, S. Kotz, N. L. Johnson and C. B. Read (editors), 20-25. New York: Wiley. [4.5]

Matuszewski, T. I. (1962). Some properties of Pascal distribution for finite population, *Journal of the American Statistical Association*, **57**, 172–174 (Correction: **57**, 919.) [6.3]

Maynard, J. M., and Chow, B. (1972). An approximate Pitman-type 'close' estimator for the negative binomial parameter, *P*, *Technometrics*, **14**, 77–88. [5.8.2]

McCullagh, P. (1991). Quasi-likelihood and estimating functions, *Statistical Theory and Modelling: In Honour of Sir David Cox, FRS*, D. V. Hinkley, N. Reid and E. J. Snell (editors), 265–286. London: Chapman and Hall. [1B.15]

McGill, W. (1967). Neural counting mechanisms and energy detection in audition, *Journal of Mathematical Psycholology*, **4**, 351–376. [5.12.5]

McGrath, E. J., and Irving, D. C. (1973). Techniques for efficient Monte Carlo simulation 2: random number generators for selected probability distributions, *Technical Report SAI-72-590-LJ*, La Jolla, CA: Science Applications, Inc. [1C.6]

McGuire, J. U., Brindley, T. A., and Bancroft, T. A. (1957). The distribution of European corn borer *Pyrausta Nubilalis* (Hbn.) in field corn, *Biometrics*, **13**, 65–78 (errata and extensions **14** (1958), 432–434.). [9.4, 9.5]

McKendrick, A. G. (1914). Studies on the theory of continuous probabilities, with special reference to its bearing on natural phenomena of a progressive nature, *Proceedings of the London Mathematical Society, 2*, **13**, 401–416. [3.2, 4.2, 5.3]

McKendrick, A. G. (1926). Applications of mathematics to medical problems, *Proceedings of the Edinburgh Mathematical Society*, **44**, 98–130. [2.4.2, 3.12.5, 4.10.1, 4.12.5, 9.4, 11.7, 11.10]

McKenzie, E. (1986). Autoregressive moving-average processes with negative-binomial and geometric marginal distributions, *Advances in Applied Probability*, **18**, 679–705. [5.10]

McKenzie, E. (1991). Linear characterizations of the Poisson distribution, *Statistics and Probability Letters*, **11**, 459–461. [4.8]

McMahon, P. A. (1894). A certain class of generating functions in the theory of numbers, *Philosophical Transactions of the Royal Society of London, Series A*, **185**, 111–160. [10.3]

McMahon, P. A. (1898). A new method in combinatory analysis with applications to Latin Squares and associated questions, *Transactions of the Cambridge Philosophical Society*, **16**, 262–290. [10.3]

McMahon, P. A. (1902). The problem of derangement in the theory of permutations, *Transactions of the Cambridge Philosophical Society*, **21**, 467–481. [10.3]

McMahon, P. A. (1915). *Combinatory Analysis*, **I**, London: Cambridge University Press (reprinted by Chelsea, New York, 1960). [10.2, 10.3]

McMahon, P. A. (1916). *Combinatory Analysis*, **II**, London: Cambridge University Press (reprinted by Chelsea, New York, 1960). [10.2]

Medgyessy, P. (1977). *Decomposition of Superpositions of Density Functions and Discrete Distributions*, Budapest: Akadémia Kiadó; Bristol: Adam Hilger. [8.1, 8.2.1, 8.2.3, 8.2.4, 8.2.5]

Medhi, J. (1975). On the convolutions of left-truncated generalized negative binomial and Poisson variables, *Sankhyā, Series B*, **37**, 293–299. [9.11]

Medhi, J., and Borah, M. (1984). On generalized Gegenbauer polynomials and associated probabilities, *Sankhyā, Series B*, **46**, 157–165. [11.7]

Meelis, E. (1974). Testing for homogeneity of k independent negative binomial distributed random variables, *Journal of the American Statistical Association*, **69**, 181–186. [5.8.4]

Meilijson, I., Newborn, M. R., Tenenbein, A., and Yechieli, U. (1982). Number of matches and matched people in the birthday problem, *Communications in Statistics-Simulation and Computation*, **11**, 361–370. [10.1, 10.4.1]

Mendenhall, W., and Lehman, E. H. (1960). An approximation to the negative moments of the positive binomial useful in life testing, *Technometrics*, **2**, 233–239. [3.11]

Meredith, W. (1971). Poisson distributions of error in mental test theory, *British Journal of Mathematical and Statistical Psychology*, **24**, 49–82. [5.9.2]

Mertz, D. B., and Davies, R. B. (1968). Cannibalism of the pupal stage by adult flour beetles: an experiment and a stochastic model, *Biometrics*, **24**, 247–275. [10.4.1]

Meyer, A. (translated by E Czuber) (1879). *Vorlesungen über Wahrscheinlichkeitsrechnung*, Leipzig: Teubner. [5.3]

Mikulski, P. W., and Smith, P. J. (1976). A variance bound for unbiased estimation in inverse sampling, *Biometrika*, **63**, 216–217. [5.8.2]

Miller, A. J. (1961). A queueing model for road traffic flow, *Journal of the Royal Statistical Society, Series B*, **23**, 64–90. [6.10.3, 6.10.4]

Milne-Thompson, L. M. (1933). *The Calculus of Finite Differences*, London: Macmillan. [1A.3, 1A.9]

Mises, R. von (1921). Das Problem der Iterationen, *Zeitschrift für angewandte Mathematik und Mechanik*, **1**, 297–307. [10.5.1]

Mises, R. von (1939). Über Aufteilungs und Besetzungswahrscheinlichkeiten, *Revue of the Faculty of Science, University of Istanbul, NS*, **4**, 145–163 (reprinted in *Selected Papers of R. von Mises*, **2**, 313–331. Providence, R. I.: American Mathematical Society). [10.4.1]

Mishra, A., and Sinha, J. K. (1981). A generalization of binomial distribution, *Journal of the Indian Statistical Association*, **19**, 93–98. [3.12.3]

Mohanty, S. G. (1966). On a generalized two-coin tossing problem, *Biometrische Zeitschrift*, **8**, 266–272. [2.5.2]

Moivre, A. de (1711). De Mensura Sortis, *Philosophical Transactions of the Royal Society, No. 329*, **27**, 213–264. [4.2, 6.2.1]

Moivre, A. de (1718). *The Doctrine of Chances: or, A Method of Calculating the Probability of Events in Play*, London: Pearson. [10.2]

Moivre, A. de (1725). *Annuities upon Lives: or, The Valuation of Annuities upon any Number of Lives: as also, of Reversions. To which is added, An Appendix Concerning the Expectations of Life, and Probabilities of Survivorship*, London: Fayram, Motte and Pearson. [10.2]

Moivre, A. de (1738). *The Doctrine of Chances: or, A Method of Calculating the Probability of Events in Play* (Second edition). London: Woodfall (Reprinted by Cass, London, 1967). [10.6.1]

Molenaar, W. (1965). Some remarks on mixtures of distributions, *Mathematical Center Tract*, **S343**: Amsterdam [8.3.1]

Molenaar, W. (1970a). Approximations to the Poisson, binomial and hypergeometric functions, *Mathematical Center Tract*, **31**: Amsterdam. [3.6.1, 4.5, 6.4, 6.5]

Molenaar, W. (1970b). Normal approximations to the Poisson distribution, *Random Counts in Scientific Work*, **2**: *Random Counts in Biomedical and Social Sciences*, G. P. Patil (editor), 237–254. University Park: Pennsylvania State University Press. [4.5, 4.7.3]

Molenaar, W. (1973). Simple approximations to the Poisson, binomial, and hypergeometric distributions, *Biometrics*, **29**, 403–408. [4.5, 4.7.3, 6.5]

Molina, E. C. (1942). *Poisson's Exponential Binomial Limit*, New York: Van Nostrand. [4.6]

Montmort, P. R. de (1713). *Essai d'Analyse sur les Jeux de Hasard* (Second edition), Paris: Quillau (Reprinted by Chelsea, New York, 1980). [5.3, 10.2]

Mood, A. M. (1940). The distribution theory of runs, *Annals of Mathematical Statistics*, **11**, 367–392. [10.5.1]

Mood, A. M. (1943). On the dependence of sampling inspection plans upon population distributions, *Annals of Mathematical Statistics*, **14**, 415–425. [6.8]

Mood, A. M. (1950). *Introduction to the Theory of Statistics*, New York: McGraw-Hill. [6.9.2]

Mood, A. M., Graybill, F. A., and Boes, D. C. (1974). *Introduction to the Theory of Statistics* (Third edition), New York: McGraw-Hill. [1A.1]

Moore, P. G. (1954). A note on truncated Poisson distributions, *Biometrics*, **10**, 402–406. [4.10.1, 4.10.2]

Moore, P. G. (1956). The geometric, logarithmic and discrete Pareto forms of series, *Journal of the Institute of Actuaries*, **82**, 130–136. [11.20]

Moran, P. A. P. (1952). A characteristic property of the Poisson distribution, *Proceedings of the Cambridge Philosophical Society*, **48**, 206–207. [4.8]

Moran, P. A. P. (1968). *An Introduction to Probability Theory*, Oxford: Oxford University Press (Reprinted, with corrections, 1984). [10.2, 10.4.1]

Moran, P. A. P. (1973). Asymptotic properties of homogeneity tests, *Biometrika*, **60**, 79–85. [4.7.4]

Morgan, B. J. T. (1984). *Elements of Simulation*, London: Chapman and Hall. [1C.1]

Morice, E., and Thionet, P. (1969). Loi binomiale et loi de Poisson, *Revue de Statistique Appliquée*, **17**, 75–89. [3.6.1]

Morlat, G. (1952). Sur une généralisation de la loi de Poisson, *Comptes Rendus, Académie des Sciences, Paris, Series A* **235**, 933–935. [4.12.5, 8.3.1]

Morris, C. N. (1982). Natural exponential families with quadratic variance functions, *Annals of Statistics*, **10**, 65–80. [3.4]

Morris, C. N. (1983). Natural exponential families with quadratic variance functions: statistical theory, *Annals of Statistics*, **11**, 515–529. [3.4]

Morris, K. W. (1963). A note on direct and inverse sampling, *Biometrika*, **50**, 544–545. [5.6]

Morton, R. (1991). Analysis of extra-multinomial data derived from extra-Poisson variables conditional on their total, *Biometrika*, **78**, 1–6. [6.11]

Mosteller, F., and Youtz, C. (1961). Tables of the Freeman-Tukey transformations for the binomial and Poisson distributions, *Biometrika*, **48**, 433–440. [3.6.3, 4.5]

Muench, H. (1936). The probability distribution of protection test results, *Journal of the American Statistical Association*, **31**, 677–689. [6.9.2]

Muench, H. (1938). Discrete frequency distributions arising from mixtures of several single probability values, *Journal of the American Statistical Association*, **33**, 390–398. [6.9.2]

Murakami, M. (1961). Censored sample from truncated Poisson distribution, *Journal of the College of Arts and Sciences, Chiba University*, **3**, 263–268. [4.10.1]

Muthu, S. K. (1982). *Problems and Errors for the Physical Sciences*, New Delhi: Orient Longman; London: Sangam Books. [10.4.2]

Nagaraja, H. N. (1982). On the non-Markovian structure of discrete order statistics, *Journal of Statistical Planning and Inference*, **7**, 29–33. [5.2]

Nagaraja, H. N. (1988). Record values and related statistics — a review, *Communications in Statistics-Theory and Methods*, **17**, 2223–2238. [5.9.1, 11.14]

Nagaraja, H. N. (1990). Order statistics from discrete distributions, *Ohio State University Department of Statistics Technical Report*, **45**. [1B.10, 5.2, 5.9.1]

Nagaraja, H. N., and Srivastava, R. C. (1987). Some characterizations of geometric type distributions based on order statistics, *Journal of Statistical Planning and Inference*, **17**, 181–191. [5.9.1]

Nair, N. U., and Hitha, N. (1989). Characterization of discrete models by distribution based on their partial sum, *Statistics and Probability Letters*, **8**, 335–337. [11.12]

Nakagawa, T., and Osaki, S. (1975). The discrete Weibull distribution, *IEEE Transactions on Reliability*, **R-24**, 300–301. [11.12]

Nanapoulos, P. (1977). Zeta laws and arithmetical functions, *Comptes Rendus, Académie des Sciences, Paris, Series A* **285**, 875–878. [11.20]

Nandi, S. B., and Dutta, D. K. (1988). Some developments in the generalized Bell distribution, *Sankhyā, Series B*, **50**, 362–375. [2.5.3]

Naor, P. (1956). On machine interference, *Journal of the Royal Statistical Society, Series B*, **18**, 280–287. [11.11]

Naor, P. (1957). Normal approximation to machine interference with many repairmen, *Journal of the Royal Statistical Society, Series B*, **1**, 334–341. [11.11, 11.19]

National Bureau of Standards (1950). *Tables of the Binomial Probability Distribution*, Washington, DC: U. S. Government Printing Office. [3.7]

Nayatani,Y., and Kurahara, B. (1964). A condition for using the approximation by the normal and the Poisson distribution to compute the confidence intervals for the binomial parameter, *Reports of Statistical Application Research, JUSE*, **11**, 99–105. [3.8.3]

Nedelman, J., and Wallenius, T. (1986). Bernoulli trials, Poisson trials, surprising variances, and Jensen's inequality, *American Statistician*, **40**, 286–289. [3.12.2]

Nelson, D. L. (1975). Some remarks on generalizations of the negative binomial and Poisson distributions, *Technometrics*, **17**, 135–136. [9.11]

Nelson, W. C., and David, H. A. (1967). The logarithmic distribution: a review, *Virginia Journal of Science*, **18**, 95–102. [7.2]

Neuman, P. (1966). Uber den Median der Binomial und Poissonverteilung, *Wissenschaftliche Zeitschrift der Technischen Universität Dresden*, **15**, 223–226. [3.6.2]

Neville, A. M., and Kemp, C. D. (1975). On characterizing the hypergeometric and multivariate hypergeometric distributions, *Statistical Distributions in Scientific Work*, **3**: *Characterizations and Applications*, G. P. Patil, S. Kotz, and J. K. Ord (editors), 353–357. Dordrecht: Reidel. [6.8]

Nevzorov, V. B. (1987). "Records", *Theory of Probability and its Applications*, **32**, 201–228. [11.14]

Newell, D. J. (1965). Unusual frequency distributions, *Biometrics*, **21**, 159–168. [3.11, 4.10.2]

Neyman, J. (1939). On a new class of "contagious" distributions applicable in entomology and bacteriology, *Annals of Mathematical Statistics*, **10**, 35–57. [9.1, 9.9]

Neyman, J. (1965). Certain chance mechanisms involving discrete distributions, *Classical and Contagious Discrete Distributions*, G. P. Patil (editor), 4–14. Calcutta: Statistical Publishing Society; Oxford: Pergamon Press. [2.5.1]

Neyman, J., and Scott, E. L. (1964). A stochastic model of epidemics, *Stochastic Models in Medicine and Biology*, J. Gurland (editor), 45–83. Madison, WI: University of Wisconsin Press. [2.5.1, 11.10]

Neyman, J., and Scott, E. L. (1966). On the use of $C(\alpha)$ optimal tests of composite hypotheses, *Bulletin of the International Statistical Institute*, **41**(1), 477–497. [4.7.4]

Ng, T.-H. (1989). A new class of modified binomial distributions with applications to certain toxicological experiments, *Communications in Statistics-Theory and Methods*, **18**, 3477–3492. [3.12.6]

Nicholson, W. L. (1956). On the normal approximation to the hypergeometric distribution, *Annals of Mathematical Statistics*, **27**, 471–483. [6.4, 6.5]

Nicholson, W. L. (1961). Occupancy probability distribution critical points, *Biometrika*, **48**, 175–180. [10.4.1]

Nisida, T. (1962). On the multiple exponential channel queueing system with hyper-Poisson arrivals, *Journal of the Operations Research Society, Japan*, **5**, 57–66. [4.12.4]

Noack, A. (1950). A class of random variables with discrete distributions, *Annals of Mathematical Statistics*, **21**, 127–132. [2.2.1, 3.4, 6.2.4]

Nörlund, N. E. (1923). *Vorlesungen über Differenzenrechnung*, New York: Chelsea (1954 reprint). [1A.9, 11.3]

O'Carroll, F. M. (1962). Fitting a negative binomial distribution to coarsely grouped data by maximum likelihood, *Applied Statistics*, **11**, 196–201. [5.8.3]

Okamoto, M. (1958). Some inequalities relating to the partial sum of binomial probabilities, *Annals of the Institute of Statistical Mathematics, Tokyo*, **10**, 29–35. [3.6.2]

Olds, E. G. (1938). A moment generating function which is useful in solving certain matching problems, *Bulletin of the American Mathematical Society*, **44**, 407–413. [10.3]

Oliver, R. M. (1961). A traffic counting distribution, *Operations Research*, **10**, 105–114. [11.9]

Olkin, I., Petkau, A. J., and Zidek, J. V. (1981). A comparison of n estimators for the binomial distribution, *Journal of the American Statistical Association*, **76**, 637–642. [3.8.2]

Olshen, R. A., and Savage, L. J. (1970). A generalized unimodality, *Journal of Applied Probability*, **7**, 21–34. [1B.5]

Ong, S. H., and Lee, P. A. (1979). The non-central negative binomial distribution, *Biometrical Journal*, **21**, 611–628. [5.12.5]

Ong, S. H. and Lee, P. A. (1986). On a generalized non-central negative binomial distribution, *Communications in Statistics-Theory and Methods*, **15**, 1065–1079. [3.12.5, 5.12.5]

Ord, J. K. (1967a). Graphical methods for a class of discrete distributions, *Journal of the Royal Statistical Society, Series A*, **130**, 232–238. [2.3.1, 2.3.2, 3.8.1, 4.7.1, 5.8.1, 6.1, 7.7.1]

Ord, J. K. (1967b). On a system of discrete distributions, *Biometrika*, **54**, 649–656. [2.3.1, 2.3.2, 3.4, 6.2.4]

Ord, J. K. (1967c). On families of discrete distributions, *Ph.D. thesis, University of London*. [2.3.2, 11.5]

Ord, J. K. (1968a). Approximations to distribution functions which are hypergeometric series, *Biometrika*, **55**, 243–248. [6.5]

Ord, J. K. (1968b). The discrete Student's *t* distribution, *Annals of Mathematical Statistics*, **39**, 1513–1516. [11.5]

Ord, J. K. (1970). The negative binomial model and quadrat sampling, *Random Counts in Scientific Work*, **2**: *Random Counts in Biomedical and Social Sciences*, G. P. Patil (editor), 151–163. University Park: Pennsylvania State University Press. [3.8.1]

Ord, J. K. (1972). *Families of Frequency Distributions*, London: Griffin. [2.3.1, 2.3.2, 4.7.1, 5.8.1, 7.7.1, 9.10]

Ord, J. K. (1985). Pearson systems of distributions, *Encyclopedia of Statistical Sciences*, **6**, S. Kotz, N. L. Johnson and C. B. Read (editors). 655–659. New York: Wiley. [2.3.2]

Ord, J. K., and Whitmore, G. (1986). The Poisson-Inverse Gaussian distribution as a model for species abundance, *Communications in Statistics-Theory and Methods*, **15**, 853–871. [11.15]

Otter, R. (1949). The multiplicative process, *Annals of Mathematical Statistics*, **20**, 206–224. [2.5.1, 11.10]

Ottestad, P. (1939). On the use of the factorial moments in the study of discontinuous frequency distributions, *Skandinavisk Aktuarietidskrift*, **22**, 22–31. [2.3.1, 5.8.1]

Ottestad, P. (1943). On Bernoullian, Lexis, Poisson and Poisson-Lexis series, *Skandinavisk Aktuarietidskrift*, **26**, 15–67. [3.12.2]

Ottestad, P. (1944). On certain compound frequency distributions, *Skandinavisk Aktuarietidskrift*, **27**, 32–42. [5.9.2, 8.3.2]

Owen, A. R. G. (1965). The summation of class frequencies (Appendix to Bliss, C. I.: An analysis of some insect trap records), *Classical and Contagious Discrete Distributions*, G. P. Patil (editor), 385–397. Calcutta: Statistical Publishing Society; Oxford: Pergamon Press. [7.5]

Owen, D. B. (1962). *Handbook of Statistical Tables*, Reading, MA: Addison-Wesley. [6.7.1, 6.9.1, 9.11, 10.3, 10.4.1]

Pachares, J. (1960). Tables of confidence limits for the binomial distribution, *Journal of the American Statistical Association*, **55**, 521–533. [3.8.3]

Paloheimo, J. E. (1963). On statistics of search, *Bulletin of the International Statistical Institute*, **40(2)**, 1060–1061. [7.10]

Panaretos, J. (1982). An extension of the damage model, *Metrika*, **29**, 189–194. [9.2]

Panaretos, J. (1983a). A generating model involving Pascal and logarithmic series distributions, *Communications in Statistics-Theory and Methods*, **12**, 841–848. [7.11]

Panaretos, J. (1983b). On Moran's property of the Poisson distribution, *Biometrical Journal*, **25**, 69–76 [4.8]

Panaretos, J. (1987a). Convolution and mixing properties of the damage model, *Statistica*, **47**, 1–8. [9.2]

Panaretos, J. (1987b). On a functional equation for the generating function of the logarithmic series distribution, *Revue Roumaine de Mathématiques Pures et Appliquées*, **32**, 365–367. [7.2]

Panaretos, J. (1989a). A probability model involving the use of the zero-truncated Yule distribution for analysing surname data, *IMA Journal of Mathematics Applied in Medicine and Biology*, **6**, 133–136. [6.10.3]

Panaretos, J. (1989b). On the evolution of surnames, *International Statistical Review*, **57**, 161–167. [6.10.3]

Panaretos, J., and Xekalaki, E. (1986a). On some distributions arising from certain generalized sampling schemes, *Communications in Statistics-Theory and Methods*, **15**, 873–891. [6.12, 7.11, 9.4, 10.6.2, 10.6.3]

Panaretos, J., and Xekalaki, E. (1986b). On generalized binomial and multinomial distributions and their relation to generalized Poisson distributions, *Annals of the Institute of Statistical Mathematics, Tokyo*, **38**, 223–231. [10.6.3]

Panaretos, J., and Xekalaki, E. (1986c). The stuttering generalized Waring distribution, *Statistics and Probability Letters*, **4**, 313–318. [10.6.3]

Panaretos, J., and Xekalaki, E. (1989). A probability distribution associated with events with multiple occurrences, *Statistics and Probability Letters*, **8**, 389–396. [10.6.3]

Pandey, K. N. (1965). On generalized inflated Poisson distribution, *Banaras Hindu University Journal of Scientific Research*, **15**, 157–162. [8.2.2]

Papageorgiou, H. (1985). On characterizing some discrete distributions by a conditional distribution and a regression function, *Biometrical Journal*, **27**, 473–479. [1B.7, 6.8]

Parthasarathy, P. R. (1987). A transient solution to an M/M/1 queue: A simple solution, *Advances in Applied Probability*, **19**, 997–998. [11.13]

Parzen, E. (1960). *Modern Probability Theory and Its Applications*, New York: Wiley. [10.2, 10.4.1]

Parzen, E. (1962). *Stochastic Processes with Applications to Science and Engineering*, San Francisco: Holden-Day. [4.2, 4.9]

Pascal, B. (1679). *Varia opera Mathematica D. Petri de Fermat*: Tolossae. [5.1, 5.3]

Patel, J. (1973). A catalogue of failure distributions, *Communications in Statistics*, **1**, 281–284. [6.10.1, 7.4]

Patel, J. K., Kapardia, C. H., and Owen, D. B. (1976). *Handbook of Statistical Distributions*, New York: Dekker. [1B.5, 3.8.2, 3.8.3]

Patel, S. R., and Jani, P. N. (1977). On minimum variance unbiased estimation of generalized Poisson distributions and decapitated generalized Poisson distributions, *Journal of the Indian Statistical Association*, **15**, 157–159. [2.2.2]

Patel, Y. C. (1971). *Some Problems in Estimation for the Parameters of the Hermite Distribution*, Ph.D. dissertation. Athens, GA: University of Georgia. [9.4]

Patel, Y. C. (1976a). Estimation of the parameters of the triple and quadruple stuttering-Poisson distributions, *Technometrics*, **18**, 67–73. [9.3]

Patel, Y. C. (1976b). Even point estimation and moment estimation in Hermite distribution, *Biometrics*, **32**, 865–873. [9.4]

Patel, Y. C. (1977). Higher moments of moment estimators and even point estimators for the parameters of the Hermite distribution, *Annals of the Institute of Statistical Mathematics, Tokyo*, **29A**, 119–130. [9.4]

Patel, Y. C. (1985). An asymptotic expression for cumulative sum of probabilities of the Hermite distribution, *Communications in Statistics-Theory and Methods*, **14**, 2233–2241. [9.4]

Patel, Y. C., Shenton, L. R., and Bowman, K. O. (1974). Maximum likelihood estimation for the parameters of the Hermite distribution, *Sankhyā, Series B*, **36**, 154–162. [9.4]

Patil, G. P. (1961). *Contributions to Estimation in a Class of Discrete Distributions*, Ph.D. thesis. Ann Arbor, MI: University of Michigan. [2.2.1]

Patil, G. P. (1962a). Certain properties of the generalized power series distribution, *Annals of the Institute of Statistical Mathematics, Tokyo*, **14**, 179–182. [2.2.1, 2.2.2]

Patil, G. P. (1962b). On homogeneity and combined estimation for the generalized power series distribution and certain applications, *Biometrics*, **18**, 365–374. [2.2.1]

Patil, G. P. (1962c). Maximum-likelihood estimation for generalized power series distributions and its application to a truncated binomial distribution, *Biometrika*, **49**, 227–237. [2.2.1]

Patil, G. P. (1962d). Some methods of estimation for the logarithmic series distribution, *Biometrics*, **18**, 68–75. [7.6, 7.7.2]

Patil, G. P. (1963a). On the equivalence of the binomial and inverse binomial acceptance sampling plans and an acknowledgement, *Technometrics*, **5**, 119–121. [5.6]

Patil, G. P. (1963b). Minimum variance unbiased estimation and certain problems of additive number theory, *Annals of Mathematical Statistics*, **34**, 1050–1056. [2.2.1]

Patil, G. P. (1964a). Estimation for the generalized power series distribution with two parameters and its application to binomial distribution, *Contributions to Statistics*, C. R. Rao (editor), 335–344. Calcutta: Statistical Publishing Society; Oxford: Pergamon Press. [2.2.1]

Patil, G. P. (1964b). On certain compound Poisson and compound binomial distributions, *Sankhyā, Series A*, **26**, 293–294. [7.10. 8.2.2, 9.4]

Patil, G. P. (1985). Logarithmic series distribution, *Encyclopedia of Statistical Sciences*, **5**, S. Kotz, N. L. Johnson and C. B. Read (editors), 111–114. New York: Wiley. [7.6]

Patil, G. P. (1986). Power series distributions, *Encyclopedia of Statistical Sciences*, **7**, S. Kotz, N. L. Johnson and C. B. Read (editors), 130–134. New York: Wiley. [3.4]

Patil, G. P., and Bildikar, S. (1966). Identifiability of countable mixtures of discrete probability distributions using methods of infinite matrices, *Proceedings of the Cambridge Philosophical Society*, **62**, 485–494. [8.3.1]

Patil, G. P., Boswell, M. T., Joshi, S. W., and Ratnaparkhi, M. V. (1984). *Dictionary and Bibliography of Statistical Distributions in Scientific Work*, **1**, *Discrete Models*. Fairland, MD: International Co-operative Publishing House. [1A.3, 2.4.1, 2.4.2, 5.1, 10.6.1, 11.2, 11.18, 11.20]

Patil, G. P., and Joshi, S. W. (1968). *A Dictionary and Bibliography of Discrete Distributions*, Edinburgh: Oliver and Boyd. [2.4.1, 2.4.2, 5.3, 11.12]

Patil, G. P., and Joshi, S. W. (1970). Further results on minimum variance unbiased estimation and additive number theory, *Annals of Mathematical Statistics*, **41**, 567–575. [2.2.1]

Patil, G. P., Kamat, A. R., and Wani, J. K. (1964). Certain studies on the structure and statistics of the logarithmic series distribution and related tables, *Aerospace Research Laboratories*, ARL 64–197, Wright-Patterson Air Force Base: Ohio. [7.6, 7.7.2]

Patil, G. P., and Rao, C. R. (1978). Weighted distributions and size biased sampling with application to wildlife populations and human families, *Biometrics*, **34**, 179–189. [3.12.4]

Patil, G. P., Rao, C. R., and Ratnaparkhi, M. V. (1986). On discrete weighted distributions and their use in model choice for observed data, *Communications in Statistics-Theory and Methods*, **15**, 907–918. [2.4.1, 3.12.4]

Patil, G. P., Rao, C. R., and Zelen, M. (1986). *A Computerized Bibliography of Weighted Distributions and Related Weighted Methods for Statistical Analysis and Interpretations of Encountered Data, Observational Studies, Representativeness Issues, and Resulting Inferences*. University Park, PA: Centre for Statistical Ecology and Environmental Statistics, Pennsylvania State University. [3.12.4]

Patil, G. P., Rao, C. R., and Zelen, M. (1988). Weighted distributions, *Encyclopedia of Statistical Sciences*, **9**, S. Kotz, N. L. Johnson and C. B. Read (editors), 565–571. New York: Wiley. [3.12.4]

Patil, G. P., and Ratnaparkhi, M. V. (1977). Characterizations of certain statistical distributions based on additive damage models involving Rao-Rubin condition and some of its variants, *Sankhyā, Series B*, **39**, 65–75. [4.8, 6.8]

Patil, G. P., and Seshadri, V. (1964). Characterization theorems for some univariate probability distributions, *Journal of the Royal Statistical Society, Series B*, **26**, 286–292. [3.9, 4.8, 5.9.1, 5.9.2, 6.8, 6.10.1]

Patil, G. P., and Wani, J. K. (1965a). Maximum likelihood estimation for the complete and truncated logarithmic series distributions, *Classical and Contagious Discrete Distributions*, G. P. Patil (editor), 398–409. Calcutta: Statistical Publishing Society; Oxford: Pergamon Press. (Also *Sankhyā, Series A*, **27**, 281–292). [7.6, 7.7.2, 7.10]

Patil, G. P., and Wani, J. K. (1965b). On certain structural properties of the logarithmic series distribution and the first type Stirling distribution, *Sankhyā, Series A*, **27**, 271–280. [7.7.2, 7.8, 7.11]

Patil, S. A., and Raghunandanan, K. (1990). Compound Hermite and stuttering Poisson distributions, *Bulletin of the Calcutta Statistical Association*, **39**, 97–103. [9.4, 11.7]

Paul, S. R. (1985). A three-parameter generalization of the binomial distribution, *Communications in Statistics-Theory and Methods*, **14**, 1497–1506. [3.12.6]

Paul, S. R. (1987). On the beta-correlated binomial (BCB) distribution: a three-parameter generalization of the binomial distribution, *Communications in Statistics-Theory and Methods*, **16**, 1473–1478. [3.12.6]

Paulson, A. S., and Uppuluri, V. R. R. (1972). A characterization of the geometric distribution and a bivariate geometric distribution, *Sankhyā, Series A*, **34**, 297–301. [5.9.1]

Pearson, E. S. (1925). Bayes' theorem, examined in the light of experimental sampling, *Biometrika*, **17**, 388–442. [6.9.2]

Pearson, E. S., and Hartley, H. O. (1976). *Biometrika Tables for Statisticians*, **1** (Third edition). London: Biometrika Trust. [1A.5, 3.7, 3.8.3, 4.6, 4.7.3, 6.9.1]

Pearson, K. (1895). Contributions to the mathematical theory of evolution I. Skew distribution in homogeneous material, *Philosophical Transactions of the Royal Society of London, Series A*, **186**, 343–414. [2.3.1, 6.2.1]

Pearson, K. (1899). On certain properties of the hypergeometrical series, and on the fitting of such series to observation polygons in the theory of chance, *Philosophical Magazine, 5th series*, **47**, 236–246. [6.2.1, 6.3]

Pearson, K. (1906). On the curves which are most suitable for describing the frequency of random samples of a population, *Biometrika*, **5**, 172–175. [6.5]

Pearson, K. (1907). On the influence of past experience on future expectation, *Philosophical Magazine, Series b*, **13**, 365–378. [6.2.2]

Pearson, K. (1915). On certain types of compound frequency distributions in which the components can be individually described by binomial series, *Biometrika*, **11**, 139–144. [8.1]

Pearson, K. (editor) (1922). *Tables of the Incomplete Γ-Function*, London: H. M. Stationery Office. [1A.5]

Pearson, K. (1924). On the moments of the hypergeometrical series, *Biometrika*, **16**, 157–162. [6.2.1, 6.3]

Pearson, K. (editor) (1934). *Tables of the Incomplete Beta-Function*, London: Cambridge University Press. [1A.5, 3.7, 3.8.3]

Peizer, D. B., and Pratt, J. W. (1968). A normal approximation for binomial, F, beta and other common related tail probabilities, I, *Journal of the American Statistical Association*, **63**, 1416–1456. [3.6.1, 4.5, 5.6]

Peng, J. C. M. (1975). Simultaneous estimation of the parameters of independent Poisson distributions, *Stanford University Department of Statistics Technical Report*, No. 78. [4.7.2]

Pérez-Abreu, V. (1991). Poisson approximation to power series distributions, *American Statistician*, **45**, 42–45. [4.2]

Perry, J. N. (1984). Negative binomial model for mosquitoes, *Biometrics*, **40**, 863. [5.10]

Perry, J. N., and Taylor, L. R. (1985). Adès: New ecological families of species-specific frequency distributions that describe repeated spatial samples with an intrinsic power-law variance-mean property, *Journal of Animal Ecology*, **54**, 931–953. [11.4]

Perry, J. N., and Taylor, L. R. (1988). Families of distributions for repeated samples of animal counts, *Biometrics*, **44**, 881–888. [11.4]

Pessin, V. (1961). Some asymptotic properties of the negative binomial distribution, *Annals of Mathematical Statistics*, **32**, 922–923 (abstract). [5.5]

Pessin, V. (1965). Some discrete distribution limit theorems using a new derivative, *Classical and Contagious Discrete Distributions*, G. P. Patil (editor), 109–122. Calcutta: Statistical Publishing Society; Oxford: Pergamon Press. [5.5]

Peterson, A. V., and Kronmal, R. A. (1980). A representation for discrete distributions by equiprobable mixtures, *Journal of Applied Probability*, **17**, 102–111. [8.1]

Peterson, A. V., and Kronmal, R. A. (1982). On mixture methods for the computer generation of random variables, *American Statistician*, **36**, 184–191. [8.1]

Peterson, C. G. J. (1896). The yearly immigration of young plaice into the Linfjord from the German Sea, *Danish Biological Station Report*, **6**, 5–48. [6.9.1]

Peto, S. (1953). A dose-response equation for the invasion of micro-organisms, *Biometrics*, **9**, 320–335. [10.4.1]

Pettigrew, H. M., and Mohler, W. C. (1967). A rapid test for the Poisson distribution using the range, *Biometrics*, **23**, 685–692. [4.5, 4.7.1]

Philippou, A. N. (1983). The Poisson and compound Poisson distributions of order k and some of their properties, *Zapisky Nauchnyka Seminarov Leningradskogo Otdelinya Matematscheskogo Instituta im V. A. Steklova AN SSSR*, **130**, 175–180 (In Russian). [4.11, 10.6.2, 10.6.3]

Philippou, A. N. (1984). The negative binomial distribution of order k and some of its properties, *Biometrical Journal*, **26**, 789–794. [10.6.2]

Philippou, A. N. (1986). Distributions and Fibonacci polynomials of order k, longest runs, and reliability of consecutive-k-out-of-n:F systems, *Fibonacci Numbers and Their Applications*, A. N. Philippou, G. E. Bergum, and A. F. Horadam (editors), 203–227. Dordrecht: Reidel. [10.6.2, 10.6.3]

Philippou, A. N. (1988). On multiparameter distributions of order k, *Annals of the Institute of Statistical Mathematics, Tokyo*, **40**, 467–475. [10.6.3]

Philippou, A. N. (1989). Mixtures of distributions by the Poisson distribution of order k, *Biometrical Journal*, **31**, 67–74. [10.6.3]

Philippou, A. N., and Antzoulakos, D. L. (1990). Multivariate distributions of order k on a generalized sequence, *Statistics and Probability Letters*, **9**, 453–463. [10.6.3]

Philippou, A. N., Georghiou, C., and Philippou, G. N. (1983). A generalized geometric distribution and some of its probabilities, *Statistics and Probability Letters*, **1**, 171–175 [10.6.1, 10.6.2]

Philippou, A. N., and Makri, F. S. (1986). Success runs and longest runs, *Statistics and Probability Letters*, **4**, 101–105; corrected version 211–215. [10.6.2]

Philippou, A. N., and Muwafi, A. A. (1982). Waiting for the k-th consecutive success and the Fibonacci sequence of order k, *Fibonacci Quarterly*, **20**, 28–32. [10.6.1, 10.6.2]

Philippou, A. N., Tripsiannis, G. A., and Antzoulakos, D. L. (1989). New Polya and inverse Polya distributions of order k, *Communications in Statistics-Theory and Methods*, **18**, 2125–2137. [6.12, 10.6.3]

Philipson, C. (1960a). The theory of confluent hypergeometric functions and its application to Poisson processes, *Skandinavisk Aktuarietidskrift*, **43**, 136–162. [9.7]

Philipson, C. (1960b). Note on the application of compound Poisson processes to sickness and accident statistics, *ASTIN Bulletin*, **1**, 224–237. [8.3.2]

Phillips, M. J. (1978). Sums of random variables having the modified geometric distribution with application to two-person games, *Advances in Applied Probability*, **10**, 647–665. [3.12.5]

Philpot, J. W. (1964). *Orthogonal Parameters for Two Parameter Distributions*, M.S. thesis. Blacksburg, VA: Virginia Polytechnic Institute. [9.6.3]

Pichon, G., Merlin, M., Fagneaux, G., Riviere, F., and Laigret, J. (1976). Etude de la distribution des densités microfilariennes dans des foyers de Filariose Lymphatique, *Institut de Recherches Medicale "Louis Malarde" Technical Report*: Papeete-Tahiti. [5.11]

Piegorsch, W. W. (1990). Maximum likelihood estimation for the negative binomial dispersion parameter, *Biometrics*, **46**, 863–867. [5.8.3]

Pielou, E. C. (1957). The effect of quadrat size on the estimation of the parameters of Neyman's and Thomas's distributions, *Journal of Ecology*, **45**, 31–47. [9.6.4, 9.6.5, 9.10]

Pielou, E. C. (1962). Runs of one species with respect to another in transects through plant populations, *Biometrics*, **18**, 579–593. [5.2, 6.10.3, 6.10.4]

Pielou, E. C. (1963). Runs of healthy and diseased trees in transects through an infected forest, *Biometrics*, **19**, 603–614. [5.2]

Pieters, E. P., Gates, C. E., Matis, J. H., and Sterling, W. L. (1977). Small sample comparisons of different estimators of negative binomial parameters, *Biometrics*, **33**, 718–723. [5.8.3]

Plackett, R. L. (1953). The truncated Poisson distribution, *Biometrics*, **9**, 485–488. [4.10.1]

Plunkett, I. G., and Jain, G. C. (1975). Three generalized negative binomial distributions, *Biometrische Zeitschrift*, **17**, 276–302. [3.12.5, 9.4, 11.7]

Poisson, S. D. (1837). *Recherches sur la Probabilité des Jugements en Matière Criminelle et en Matière Civile, Précédées des Regles Générales du Calcul des Probabilitiés*. Paris: Bachelier, Imprimeur-Libraire pour les Mathematiques, la Physique, etc. [3.12.2, 4.2]

Pólya, G. (1930). Sur quelques points de la théorie des probabilités, *Annales de l'Institut H. Poincaré*, **1**, 117–161. [6.2.4, 9.7]

Potthof, R. F., and Whittinghill, M. (1966). Testing for homogeneity II. The Poisson distribution, *Biometrika*, **53**, 183–190. [4.7.4]

Potts, R. B. (1953). Note on the factorial moments of standard distributions, *Australian Journal of Physics*, **6**, 498–499. [2.4.1]

Prasad, A. (1957). A new discrete distribution, *Sankhyā*, **17**, 353–354. [6.10.3]

Pratt, J. W. (1968). A normal approximation for binomial, F, beta and other common related tail probabilities, II, *Journal of the American Statistical Association*, **63**, 1457–1483. [3.6.1, 5.6]

Prekopa, A. (1952). On composed Poisson distributions, IV, *Acta Mathematica Academiae Scientiarum Hungaricae*, **3**, 317–325. [4.12.3]

Prentice, R. L. (1986). Binary regression using an extended beta-binomial distribution, with discussion of correlation induced by measurement errors, *Journal of the American Statistical Association*, **81**, 321–327. [3.12.6]

Preston, F. W. (1948). The commonness, and rarity, of species, *Ecology*, **29**, 254–283. [7.11]

Prohorov, Y. V. (1953). The asymptotic behaviour of the binomial distribution, *Uspekhi Matematicheskii Nauk* (New Series), **8**, 135–142. [3.6.1, 3.6.2]

Pulskamp, R. (1990). A note on the estimation of binomial probabilities, *American Statistician*, **44**, 293–295. [3.8.2]

Puri, P. S. (1966). Probability generating function of absolute difference of two random variables, *Proceedings of the National Academy of Science*, **56**, 1059–1061. [5.9.1]

Puri, P. S. (1973). On a property of exponential and geometric distributions and its relevance to multivariate failure rate, *Sankhyā, Series A*, **35**, 61–78. [5.9.1]

Puri, P. S., and Rubin, H. (1970). A characterization based on the absolute difference of two i.i.d. random variables, *Annals of Mathematical Statistics*, **41**, 2113–2122. [5.9.1]

Puri, P. S., and Rubin, H. (1972). On a characterization of the family of distributions with constant multivariate failure rates, *Annals of Probability*, **2**, 738–740. [5.9.1]

Qu, Y., Beck, G. J., and Williams, G. W. (1990). Polya-Eggenberger distribution: parameter estimation and hypothesis tests. *Biometrical Journal*, **32**, 229–242. [6.7.2, 6.8]

Quenouille, M. H. (1949). A relation between the logarithmic, Poisson, and negative binomial series, *Biometrics*, **5**, 162–164. [5.3, 7.2]

Quine, M. P., and Seneta, E. (1987). Bortkiewicz's data and the law of small numbers, *International Statistical Review*, **55**, 173–181. [4.2]

Qvale, P. (1932). Remarks on semi-invariants and incomplete moments, *Skandinavisk Aktuarietidskrift*, **15**, 196–210. [2.4.1]

Rabinovitch, N. L. (1973). Studies in the history of probability and statistics. XXII. Probability in the Talmud, *Biometrika*, **56**, 437–441. [11.19]

Raff, M. S. (1956). On approximating the point binomial, *Journal of the American Statistical Association*, **51**, 293–303. [3.6.1]

Raiffa, H., and Schlaifer, R. (1961). *Applied Statistical Decision Theory*, Cambridge, MA: MIT Press, [3.4, 6.2.3, 6.4]

Raikov, D. (1938). On the decomposition of Gauss' and Poisson's laws, *Izvestia Akademie Nauk SSSR, Series A*, 91–124. [4.8]

Rainville, E. R. (1960). *Special Functions*, New York: Macmillan. [1A.8, 11.7]

Ramasubban, T. A. (1958). The generalized mean differences of the binomial and Poisson distributions, *Biometrika*, **45**, 549–556. [4.3]

Rao, B. L. S. P., and Sreehari, M. (1987). On a characterization of Poisson distribution through inequalities of Chernoff-type, *Australian Journal of Statistics*, **29**, 38–41. [4.8]

Rao, B. R. (1981). Correlation between the numbers of two types of children in a family with the MPSD for the family size, *Communications in Statistics-Theory and Methods*, **10**, 249–254. [7.11]

Rao, B. R., and Janardan, K. G. (1984). Use of the generalized Markov-Polya distribution as a random damage model and its identifiability, *Sankhyā, Series A*, **46**, 458–462. [6.12]

Rao, B. R., and Janardan, K. G. (1985). An analog of the Rao-Rubin condition for distributions other than the Poisson, *Pakistan Journal of Statistics*, **1**, 1–15. [9.2]

Rao, B. R., Mazumdar, S., Waller, J. H., and Li, C. C. (1973). Correlation between the numbers of two types of children in a family, *Biometrics*, **29**, 271–279. [5.10]

Rao, C. R. (1965). On discrete distributions arising out of methods of ascertainment, *Classical and Contagious Discrete Distributions*, G. P Patil (editor), 320–332. Calcutta: Statistical Publishing Society; Oxford: Pergamon Press. (Republished *Sankhyā*, **A27** (1965), 311–324.) [2.4.1, 3.12.4, 4.8, 9.2]

Rao, C. R. (1971). Some comments on the logarithmic series distribution in the analysis of insect trap data, *Statistical Ecology*, **1**: Spatial Patterns and Statistical Distributions, G. P. Patil, E. C. Pielou, and W. E. Waters (editors), 131–142. University Park: Pennsylvania State University Press. [7.1, 7.7.2]

Rao, C. R. (1985). Weighted distributions arising out of methods of ascertainment: What populations does a sample represent?, *A Celebration of Statistics: ISI Centenary Volume*, A. C. Atkinson and S. E. Fienberg (editors), 543–569. New York: Springer- Verlag. [3.12.4]

Rao, C. R., and Rubin, H. (1964). On a characterization of the Poisson distribution, *Sankhyā, Series A*, **26**, 295–298. [3.9, 4.8, 4.10.1, 9.2]

Rao, C. R., and Srivastava, R. C. (1979). Some characterizations based on a multivariate splitting model, *Sankhyā, Series A*, **41**, 124–128. [4.8]

Rao, C. R., Srivastava, R. C., Talwalker, S., and Edgar, G. A. (1980). Characterization of probability distributions based on a generalized Rao-Rubin condition, *Sankhyā, Series A*, **42**, 161–169. [4.8, 9.2]

Rao, M. B., and Shanbhag, D. N. (1982). Damage models, *Encyclopedia of Statistical Sciences*, **2**, S. Kotz, N. L. Johnson and C. B. Read (editors), 262–265. New York: Wiley. [9.2]

Ratcliffe, J. F. (1964). The significance of the difference between two Poisson variables, *Applied Statistics*, **13**, 84–86. [4.12.3]

Ratnaparkhi, M. V. (1981). On splitting model and related characterizations of some statistical distributions, *Statistical Distributions in Scientific Work*, **4**: Models, Structures, and Characterizations, C. Taillie, G. P. Patil, and B. A. Baldessari (editors), 349–355. Dordrecht: Reidel. [4.8]

Ray, S. K., and Sahai A. (1978). On variance of the MVU inverse binomial estimator, *Bulletin of the Calcutta Statistical Association*, **27**, 105–108. [5.8.2]

Rayner, J. C. W., and Best, D. J. (1989). *Smooth Tests of Goodness of Fit*, New York: Oxford University Press. [1B.15, 3.8.4]

Reed, S. W., and Reed, S. C. (1965). *Mental Retardation: A Family Study*, Philadelphia: W. B. Saunders. [5.11]

Relles, D. (1972). Simple algorithm for generating binomial random variables when N is large, *Journal of the American Statistical Association*, **67**, 612–613. [1C.3]

Rényi, A. (1964). On an extremal property of the Poisson process, *Annals of the Institute of Statistical Mathematics, Tokyo*, **16**, 129–133. [4.2]

Resnick, S. I. (1973). Extremal processes and record value times, *Journal of Applied Probability*, **10**, 864–868. [11.14]

Richards, P. I. (1968). A generating function (Problem 67–18), *SIAM Review*, **10**, 455–456. [10.4.1]

Rider, P. R. (1953). Truncated Poisson distributions, *Journal of the American Statistical Association*, **48**, 826–830. [4.10.1]

Rider, P. R. (1955). Truncated binomial and negative binomial distributions, *Journal of the American Statistical Association*, **50**, 877–883 (corrigendum **50**, 1332). [5.11]

Rider, P. R. (1962a). Estimating the parameters of mixed Poisson, binomial and Weibull distributions, *Bulletin of the International Statistical Institute*, **39**(2), 225–232. [8.2.3, 8.2.4, 8.2.5]

Rider, P. R. (1962b). The negative binomial distribution and the incomplete beta function, *American Mathematical Monthly*, **69**, 302–304. [5.11]

Riordan J. (1937). Moment recurrence relations for binomial, Poisson, and hypergeometric frequency distributions, *Annals of Mathematical Statistics*, **8**, 103–111. [4.3]

Riordan, J. (1958). *An Introduction to Combinatorial Analysis*, New York: Wiley. [1A.3, 9.1, 10.2, 10.4.1]

Ripley, B. D. (1981). *Spatial Statistics*, New York: Wiley. [4.9]

Ripley, B. D. (1987). *Stochastic Simulation*, New York: Wiley. [1C.1]

Robbins, H. (1956). An empirical Bayes approach to statistics, *Proceedings of the Third Berkeley Symposium on Mathematical Statistics and Probability*, **1**, 157–163. Berkeley: University of California Press. [4.7.2]

Robinson, G. K. (1982). Confidence intervals and regions, *Encyclopedia of Statistical Sciences*, **2**, S. Kotz, N. L. Johnson and C. B. Read (editors), 120–127. New York: Wiley. [1B.15]

Robson, D. S., and Regier, H. A. (1964). Estimation of population number and mortality rates, *Methods for Assessment of Fish Production in Fresh Waters*, W. E. Ricker (editor), IBP Handbook No 3, 124–158. Oxford: Blackwell Scientific. [6.7.1]

Rogers, A. (1965). A stochastic analysis of the spatial clustering of retail establishments, *Journal of the American Statistical Association*, **60**, 1094–1103. [9.6.5]

Rogers, A. (1969). Quadrat analysis of urban dispersion: 2 Case studies of urban retail systems, *Environment and Planning*, **1**, 155–171. [9.6.5]

Roman, S. M., and Rota, G. C. (1978). The umbral calculus, *Advances in Mathematics*, **27**, 95–188. [2.5.3]

Romani, J. (1956). Distribución de la suma algebraíca de variables de Poisson, *Trabajos de Estadística*, **7**, 175–181. [4.12.3]

Romanovsky, V. (1923). Note on the moments of the binomial $(p+q)^N$ about its mean, *Biometrika*, **15**, 410–412. (See also *Les Principes de la Statistique Mathematique* (1933), 39–40 and 320–321.). [3.3]

Romanovsky, V. (1925). On the moments of the hypergeometrical series, *Biometrika*, **17**, 57–60. [6.2.1]

Romanowska, M. (1978). Poisson approximation of some probability distributions, *Bulletin de l'Academie Polonaise des Sciences, Series des Sciences Mathematiques, Astronomiques et Physiques*, **26**, 1023–1026. [3.6.1]

Romig, H. G. (1953). *Binomial Tables*. New York: Wiley. [3.7]

Ross, G. J. S. (1980). *MLP: Maximum Likelihood Program* (a Manual), Harpenden: Rothamsted Experimental Station Statistics Department. [5.8.3, 11.4]

Ross, G. J. S. (1990). *Nonlinear Estimation*, New York: Springer-Verlag. [1B.15]

Ross, G. J. S., and Preece, D. A. (1985). The negative binomial distribution, *The Statistician*, **34**, 323–335. [5.8.3]

Rowe, J. A. (1942). Mosquito light trap catches from ten American cities, 1940, *Iowa State College Journal of Science*, **16**, 487–518. [7.9]

Roy, J., and Mitra, S. K. (1957). Unbiased minimum variance estimation in a class of discrete distributions, *Sankhyā*, **18**, 371–378. [2.2.1, 5.8.2]

Rudolpher, S. M. (1990). A Markov chain model of extrabinomial variation, *Biometrika*, **77**, 255–264. [3.12.6]

Rushton, S., and Lang, E. D. (1954). Tables of the confluent hypergeometric function, *Sankhyā*, **13**, 369–411. [1A.7]

Russell, K. G. (1978). Estimation of the parameters of the Thomas distribution, *Biometrics*, **34**, 95–99. [9.10]

Rutemiller, H. C. (1967). Estimation of the probability of zero failures in m binomial trials, *Journal of the American Statistical Association*, **62**, 272–277. [3.8.2]

Rutherford, E., Chadwick, J., and Ellis, C. D. (1930). *Radiation from Radioactive Substances*, London: Cambridge University Press. [4.2]

Rutherford, E., and Geiger, H. (1910). The probability variations in the distribution of α particles, *Philosophical Magazine, 6th Series*, **20**, 698–704 [4.2]

Rutherford, R. S. C. (1954). On a contagious distribution, *Annals of Mathematical Statistics*, **25**, 703–713. [11.19]

Sadooghi-Alvandi, S. M. (1990). Estimation of the parameter of a Poisson distribution using a LINEX loss function, *Australian Journal of Statistics*, **32**, 393–398. [4.7.2]

Sahai, A., and Buhrman, J. M. (1979). Bounds for the variance of an inverse binomial estimator, *Statistica Neerlandica*, **33**, 213–216. [5.8.2]

Said, A. S. (1958). Some properties of the Poisson distribution, *Journal of the American Institute of Chemical Engineering*, **4**, 290–292. [4.4]

Saleh, A. K. M., and Rahim, M. A. (1972). Distribution of the sum of variates from truncated discrete populations, *Canadian Mathematical Bulletin*, **15**, 395–398. [5.11]

Saleh, A. K. M. E. (1981). Decomposition of finite mixture of distributions by minimum chi-square method, *Aligarh Journal of Statistics*, **1**, 86–97. [8.2.3]

Samaniego, F. J. (1976). A characterization of convoluted Poisson distributions with applications to estimation, *Journal of the American Statistical Association*, **71**, 475–479. [4.8, 4.12.3]

Sambursky, S. (1956). On the possible and probable in ancient Greece, *Osiris*, **12**, 35–48. [11.19]

Samiuddin, M., and Mallick, S. A. (1970). On the logit transformation of binomial variate, *Journal of Natural Science and Mathematics*, **10**, 59–63. [3.6.1]

Sampford, M. R. (1955). The truncated negative binomial distribution, *Biometrika*, **42**, 58–69. [5.11]

Samuels, S. M. (1965). On the number of successes in independent trials, *Annals of Mathematical Statistics*, **36**, 1272–1278. [4.5]

Sandelius, M. (1952). A confidence interval for the smallest proportion of a binomial population, *Journal of the Royal Statistical Society, Series B*, **14**, 115–117. [3.8.2]

Sandiford, P. J. (1960). A new binomial approximation for use in sampling from finite populations, *Journal of the American Statistical Association*, **55**, 718–722. [6.5]

Sandland, R. (1974). A note on some applications of the truncated geometric distribution, *Australian Journal of Statistics*, **16**, 57–58. [5.2]

Sandland, R. L., and Cormack, R. M. (1984). Statistical inference for Poisson and multinomial models for capture-recapture experiments, *Biometrika*, **71**, 27–33. [4.9]

Sankaran M. (1968). Mixtures by the inverse Gaussian distribution, *Sankhyā, Series B*, **30**, 455–458. [11.15]

Sankaran, M. (1970). The discrete Poisson-Lindley distribution, *Biometrics*, **26**, 145–149. [3.12.5, 8.3.2]

Sarkadi, K. (1957a). Generalized hypergeometric distributions, *Magyar Tudományos Akadémia Matematikai Kutató Intézetének Közlenényei*, **2**, 59–68. [6.2.2, 6.2.4]

Sarkadi, K. (1957b). On the distribution of the number of exceedances, *Annals of Mathematical Statistics*, **28**, 1021–1022. [6.2.2]

Satterthwaite, F. E. (1942). Generalized Poisson distributions, *Annals of Mathematical Statistics*, **13**, 410–417. [9.3]

Satterthwaite, F. E. (1957). Binomial and Poisson confidence limits, *Industrial Quality Control*, **13**, 56–59. [3.8.3]

Schader, M., and Schmid, F. (1989). Two rules of thumb for the approximation of the binomial distribution by the normal distribution, *American Statistician*, **43**, 23–24. [3.6.1]

Scheaffer, R. L. (1976). A note on approximate confidence limits for the negative binomial model, *Communications in Statistics-Theory and Methods*, **A5**, 149–158. [5.8.2]

Scheaffer, R. L., and Leavenworth, R. S. (1976). The negative binomial model for counts in units of varying size, *Journal of Quality Technology*, **8**, 158–163. [5.12.4]

Schenzle, D. (1979). Fitting the truncated negative binomial distribution without the second sample moment, *Biometrics*, **35**, 637–639. [5.11]

Schmeiser, B. W., and Kachitvichyanukul, V. (1981). Poisson random variate generation, *Research Memorandum*, 81-4, West Lafayette, IN: Purdue University School of Industrial Engineering. [1C.4]

Schneider, B. E. (1978). Algorithm AS121: Trigamma function, *Applied Statistics*, **27**, 97–99. [11.3]

Schreider, Y. A. (1967). On the possibility of a theoretical model of statistical laws for texts (towards a basis for Zipf's law), *Problemy Peredachi Informatsii*, **3**, 57–63 (In Russian). [11.20]

Seal, H. L. (1947). A probability distribution of deaths at age x when policies are counted instead of lives, *Skandinavisk Aktuarietidskrift*, **30**, 18–43. [11.20]

Seal, H. L. (1949a). Mortality data and the binomial probability law, *Skandinavisk Aktuarietidskrift*, **32**, 188–216. [3.10]

Seal, H. L. (1949b). The historical development of the use of generating functions in probability theory, *Bulletin de l'Association des Actuaires Suisses*, **49**, 209–228. [1B.9]

Seal, H. L. (1952). The maximum likelihood fitting of the discrete Pareto law, *Journal of the Institute of Actuaries*, **78**, 115–121. [11.20]

Seber, G. A. F. (1982a). Capture-recapture methods, *Encyclopedia of Statistical Sciences*, **1**, S. Kotz, N. L. Johnson and C. B. Read (editors), 367–374. New York: Wiley. [6.9.1]

Seber, G. A. F. (1982b). *The Estimation of Animal Abundance* (Second edition), London: Griffin. [3.10, 4.9, 5.2, 6.7.1]

Seneta, E. (1988). Degree, iteration and permutation in improving Bonferroni-type bounds, *Australian Journal of Statistics*, **30A**, 27–38. [10.2]

Serfling, R. J. (1977). The role of the Poisson distribution in approximating system reliability of k-out-of-n structures, *Theory and Applications of Reliability: with emphasis on Bayesian and nonparametric methods*, **1**, C. P. Tsokas and I. N. Shimi (editors), 243–258. New York: Academic Press. [4.2]

Shaban, S. A. (1981). Computation of the Poisson-inverse Gaussian distribution, *Communications in Statistics-Theory and Methods*, **A10**, 1389–1399. [11.15]

Shaban, S. A. (1988). Poisson-lognormal distributions, *Lognormal Distributions: Theory and Applications*, E. L. Crow and K. Shimizu (editors), 195–210. New York: Dekker. [7.11]

Shah, B. V., and Venkataraman, V. K. (1962). A note on modified Poisson distribution, *Metron*, **22**, No. 3/4, 27–35. [4.10.3]

Shah, S. M. (1961). The asymptotic variances of method of moment estimates of the parameters of the truncated binomial and negative binomial distributions, *Journal of the American Statistical Association*, **56**, 990–994. [5.11]

Shah, S. M. (1966). On estimating the parameter of a doubly truncated binomial distribution, *Journal of the American Statistical Association*, **61**, 259–263. [3.11]

Shah, S. M. (1971). The displaced negative binomial distribution *Bulletin of the Calcutta Statistical Association*, **20**, 143–152. [5.11]

Shah, S. M., and Kabe, D. G. (1981). Characterizations of exponential, Pareto, power function, Burr and logistic distributions by order statistics, *Biometrical Journal*, **23**, 109–112. [1B.10]

Shaked, M. (1974). On the distribution of the minimum and of the maximum of a random number of i.i.d. random variables, *Statistical Distributions in Scientific Work*, **1**: *Models and Structures*, G. P. Patil, S. Kotz, and J. K. Ord (editors), 363–380. Dordrecht: Reidel. [5.9.1]

Shanbhag, D. N. (1970a). Another characteristic property of the Poisson distribution, *Proceedings of the Cambridge Philosophical Society*, **68**, 167–169. [4.8]

Shanbhag, D. N. (1970b). Characterizations for exponential and geometric distributions, *Journal of the American Statistical Association*, **65**, 1256–1259. [5.9.1]

Shanbhag, D. N. (1972). Some characterizations based on the Bhattacharya matrix, *Journal of Applied Probability*, **9**, 580–587. [4.8]

Shanbhag, D. N. (1973). Comments on Wang's paper, *Proceedings of the Cambridge Philosophical Society*, **73**, 473–475. [4.8]

Shanbhag, D. N. (1974). An elementary proof for the Rao-Rubin characterization of the Poisson distribution, *Journal of Applied Probability*, **11**, 211–215. [4.8, 5.9.1]

Shanbhag, D. N., and Clark, R. M. (1972). Some characterizations for the Poisson distribution starting with a power series distribution, *Proceedings of the Cambridge Philosophical Society*, **71**, 517–522. [4.8]

Shanbhag, D. N., and Panaretos, J. (1979). Some results related to the Rao-Rubin characterization of the Poisson distribution, *Australian Journal of Statistics*, **21**, 78–83. [4.8]

Shanbhag, D. N., and Rajamannar, G. (1974). Some characterizations of the bivariate distribution of independent Poisson variables, *Australian Journal of Statistics*, **16**, 119–125. [4.8]

Shane, H. D. (1973). A Fibonacci probability function, *Fibonacci Quarterly*, **11**, 517–522. [10.6.1]

Shanmugam, R., and Singh, J. (1981). On the Stirling distribution of the first kind, *Statistical Distributions in Scientific Work*, **4**: *Models, Structures, and Characterizations*, C. Taillie, G. P. Patil, and B. A. Baldessari (editors), 181–187. Dordrecht: Reidel. [7.11]

Shanthikumar, J. G. (1982). Recursive algorithm to evaluate the reliability of a consecutive-k-out-of-n:F system, *IEEE Transactions on Reliability*, **R-31**, 442–443. [10.6.2]

Shanthikumar, J. G. (1985). Discrete random variate generation using uniformization, *European Journal of Operational Research*, **21**, 387–398. [1C.7]

Shea, B. L. (1989). Remark AS R77–A remark on algorithm AS 152: Cumulative hypergeometric probabilities, *Applied Statistics*, **38**, 199–204. [6.6]

Shenton, L. R. (1949). On the efficiency of the method of moments and Neyman's type A distribution, *Biometrika*, **36**, 450–454. [9.6.2, 9.6.4]

Shenton, L. R., and Bowman, K. O. (1967). Remarks on large sample estimators for some discrete distributions, *Technometrics*, **9**, 587–598. [9.6.1, 9.6.4]

Shenton, L. R., and Bowman, K. O. (1977). *Maximum Likelihood Estimation in Small Samples*, London: Griffin. [9.6.4, 9.7]

Shenton, L. R., and Myers, R. (1965). Comments on estimation for the negative binomial distribution, *Classical and Contagious Discrete Distributions*, G. P. Patil (editor), 241–262. Calcutta: Statistical Publishing Society; Oxford: Pergamon Press. [5.8.3]

Shenton, L. R., and Skees, P. (1970). Some statistical aspects of amounts and duration of rainfall, *Random Counts in Scientific Work*, **3**: *Random Counts in Physical Science, Geo Science, and Business*, G. P. Patil (editor), 73–94. University Park: Pennsylvania State University Press. [7.11]

Sherbrooke, C. C. (1968). Discrete compound Poisson processes and tables of the geometric Poisson distribution, *Naval Research Logistics Quarterly*, **15**, 189–203. [9.7]

Sheu, S. S. (1984). The Poisson approximation to the binomial distribution, *American Statistician*, **38**, 206–207. [3.6.2]

Shimizu, R. (1968). Generalized hypergeometric distributions, *Proceedings of the Institute of Statistical Mathematics, Tokyo*, **16**, 147–165 (In Japanese). [6.2.4]

Shorrock, R.W. (1972a). A limit theorem for inter-record times, *Journal of Applied Probability*, **9**, 219–223. [11.14]

Shorrock, R.W. (1972b). On record values and record times, *Journal of Applied Probability*, **9**, 316–326. [11.14]

Shoukri, M. M., and Consul, P. C. (1987). Some chance mechanisms generating the generalized Poisson probability models, *Advances in the Statistical Sciences V: Biostatistics*, I. B. MacNeill and G. J. Umphrey (editors), 259–268. Dordrecht: Reidel. [9.11]

Shumway, R., and Gurland, J. (1960a). A fitting procedure for some generalized Poisson distributions, *Skandinavisk Aktuarietidskrift*, **43**, 87–108. [9.5, 9.8]

Shumway, R., and Gurland, J. (1960b). Fitting the Poisson binomial distribution, *Biometrics*, **16**, 522–533. [9.5, 9.8]

Sibuya, M. (1979). Generalized hypergeometric, digamma and trigamma distributions, *Annals of the Institute of Statistical Mathematics, Tokyo*, **31**, 373–390. [11.3]

Sibuya, M. (1983). Generalized hypergeometric distributions, *Encyclopedia of Statistical Sciences*, **3**, S. Kotz, N. L. Johnson and C. B. Read (editors), 330–334. New York: Wiley. [6.2.4]

Sibuya, M. (1988). Log-concavity of Stirling numbers and unimodality of Stirling distributions, *Annals of the Institute of Statistical Mathematics, Tokyo*, **40**, 693–714. [7.11]

Sibuya, M., and Shimizu, R. (1981). Classification of the generalized hypergeometric family of distributions. *Keio Science and Technology Reports*, **34**, 1–38. [6.2.4]

Sichel, H. S. (1951). The estimation of the parameters of a negative binomial distribution with special reference to psychological data, *Psychometrika*, **16**, 107–127. [5.10]

Sichel, H. S. (1971). On a family of discrete distributions particularly suited to represent long-tailed frequency data, *Proceedings of the Third Symposium on Mathematical Statistics*, N. F. Laubscher (editor), 51–97. Pretoria : C.S.I.R. [11.15]

Sichel, H. S. (1973a). Statistical evaluation of diamondiferous deposits, *Journal of the South African Institute of Mining and Metallurgy*, **73**, 235–243. [11.15]

Sichel, H. S. (1973b). The density and size distribution of diamonds, *Bulletin of the International Statistical Institute*, **45**(2), 420–427. [11.15]

Sichel, H. S. (1974). On a distribution representing sentence-length in written prose, *Journal of the Royal Statistical Society, Series A*, **137**, 25–34. [11.15]

Sichel, H. S. (1975). On a distribution law for word frequencies, *Journal of the American Statistical Association*, **70**, 542–547. [11.15]

Sichel, H. S. (1982a). Repeat-buying and the generalized inverse-Gaussian distribution, *Applied Statistics*, **31**, 193–204. [11.15]

Sichel, H. S. (1982b). Asymptotic efficiencies of three methods of estimation for the inverse Gaussian-Poisson distribution, *Biometrika*, **69**, 467–472. [11.15]

Silva, G. (1941). Una generalizazione del problema delle concordanze, *Instituto Veneto de Scienze, Lettere ed Arti, Venezia Classe de Scienze Matematiche e Naturali*, **100**, 689–709. [10.3]

Silverman, B. W. (1985). Penalized maximum likelihood estimation, *Encyclopedia of Statistical Sciences*, **6**, S. Kotz, N. L. Johnson and C. B. Read (editors), 664–667. New York: Wiley. [1B.15]

Sim, C. H., and Lee, P. A. (1989). Simulation of negative binomial processes, *Journal of Statistical Computation and Simulation*, **34**, 29–42. [1C.5]

Simar, L. (1976). Maximum likelihood estimation of a compound Poisson process, *Annals of Statistics*, **4**, 1200–1209. [8.2.3]

Simon, G., and Johnson, N. L. (1971). On the convergence of binomial to Poisson distributions, *Annals of Mathematical Statistics*, **42**, 1735–1736. [3.6.1]

Simon, H. A. (1955). On a class of skew distribution functions, *Biometrika*, **42**, 425–440. [6.10.3, 11.20]

Simon, H. A. (1960). Some further notes on a class of skew distribution functions, *Information and Control*, **3**, 80–88. [6.10.3]

Singh, J. (1978). A characterization of positive Poisson distribution and its statistical application, *SIAM Journal of Applied Mathematics*, **34**, 545–548. [4.8]

Singh, S. N. (1963). A note on inflated Poisson distribution, *Journal of the Indian Statistical Association*, **1**, 140–144. [8.2.2]

Singh, S. N. (1964). On the time of first birth, *Sankhyā, Series B*, **26**, 95–102. [11.9]

Singh, S. N. (1968). A chance mechanism of the variation in the number of births per couple, *Journal of the American Statistical Association*, **63**, 209–213. [11.9]

Singh, S. N., Bhattacharya, B. N., and Yadava, R. C. (1974). A parity dependent model for number of births and its applications, *Sankhyā, Series B*, **36**, 93–102. [11.9]

Sinha, B. K., and Bose, A. (1985). Unbiased sequential estimation of $1/p$: settlement of a conjecture, *Annals of the Institute of Statistical Mathematics, Tokyo*, **37**, 455–460. [3.8.2]

Sinha, B. K., and Sinha, B. K. (1975). Some problems of unbiased sequential binomial estimation, *Annals of the Institute of Statistical Mathematics, Tokyo*, **27**, 245–258. [3.8.2]

Siotani, M. (1956). Order statistics for discrete case with a numerical application to the binomial distribution, *Annals of the Institute of Statistical Mathematics, Tokyo*, **8**, 95–104. [3.5].

Siotani, M., and Ozawa, M. (1948). Tables for testing the homogeneity of k independent binomial experiments on a certain event based on the range, *Annals of the Institute of Statistical Mathematics, Tokyo*, **10**, 47–63. [3.5].

Siromoney, G. (1962). Entropy of logarithmic series distributions, *Sankhyā, Series A*, **24**, 419–420. [7.4]

Siromoney, G. (1964). The general Dirichlet's distribution, *Journal of the Indian Statistical Association*, **2**, 69–74. [2.2.1]

Skellam, J. G. (1946). The frequency distribution of the difference between two Poisson variates belonging to different populations, *Journal of the Royal Statistical Society, Series A*, **109**, 296. [4.12.3]

Skellam, J. G. (1948). A probability distribution derived from the binomial distribution by regarding the probability of success as variable between the sets of trials, *Journal of the Royal Statistical Society, Series B*, **10**, 257–261. [5.3, 6.7.2, 6.9.2]

Skellam, J. G. (1949). The probability distribution of gene-frequencies in relation to selection, mutation, and random extinction, *Proceedings of the Cambridge Philosophical Society*, **45**, 364–367. [11.16]

Skellam, J. G. (1952). Studies in statistical ecology I: Spatial pattern, *Biometrika*, **39**, 346–362. [9.4, 9.5, 9.8]

Skellam, J. G. (1958). On the derivation and applicability of Neyman's type A distribution, *Biometrika*, **45**, 32–36. [9.6.5]

Skibinsky, M. (1970). A characterization of hypergeometric distributions, *Journal of the American Statistical Association*, **65**, 926–929. [6.8]

Slater, L. J. (1960). *Confluent Hypergeometric Functions*, London: Cambridge University Press. (errata *Mathematics of Computation*, **17** (1963), 486–487). [1A.7]

Slater, L. J. (1966). *Generalized Hypergeometric Functions*, London: Cambridge University Press. [1A.6, 1A.8, 1A.12]

Slud, E. V. (1977). Distribution inequalities for the binomial law, *Annals of Probability*, **5**, 404–412. [3.6.2]

Smith, A. F. M. (1984). Present position and potential developments: some personal views: Bayesian statistics, *Journal of the Royal Statistical Society, Series A*, **147**, 245–259. [1B.3]

Smith, A. F. M. (1985). An overview of some problems relating to finite mixtures, *Rassegna di Metodi Statistici ed Applicazione 5 Cagliari*, 138–149. [8.2.1]

Sobel, M. and Huyett, M. J. (1957). Selecting the best one of several binomial populations, *Bell System Technical Journal*, **36**, 537–576. [3.5].

Society of Actuaries. (1954). *Impairment Study*, 1951. [4.7.3]

Solow, R. M. (1960). On a family of lag distributions, *Econometrica*, **28**, 392–406. [5.10]

Somerville, P. N. (1957). Optimum sampling in binomial populations, *Journal of the American Statistical Association*, **52**, 494–502. [3.5].

Springer, M. D. (1979). *The Algebra of Random Variables*, New York: Wiley. [1A.10, 3.4]

Sprott, D. A. (1957). Probability distributions associated with distinct hits on targets, *Bulletin of Mathematical Biophysics*, **19**, 163–170. [10.4.1]

Sprott, D. A. (1958). The method of maximum likelihood applied to the Poisson binomial distribution, *Biometrics*, **14**, 97–106. [9.4, 9.5]

Sprott, D. A. (1965). A class of contagious distributions and maximum likelihood estimation, *Classical and Contagious Discrete Distributions*, G. P. Patil (editor), 337–350. Calcutta: Statistical Publishing Society; Oxford: Pergamon Press. [9.2]

Srivastava, H. M., and Kashyap, B. R. K. (1982). *Special Functions in Queueing Theory and Related Stochastic Processes*, New York: Academic Press. [2.4.1]

Srivastava, M. S. (1965). Characterization theorems for some distributions, *Annals of Mathematical Statistics*, **36**, 361 (abstract). [5.9.1]

Srivastava, R. C. (1971). On a characterization of the Poisson process, *Journal of Applied Probability*, **8**, 615–616. [4.8]

Srivastava, R. C. (1974). Two characterizations of the geometric distribution, *Journal of the American Statistical Association*, **69**, 267–269. [5.9.1]

Srivastava, R. C. (1979). Two characterizations of the geometric distribution by record values, *Sankhyā, Series B*, **40**, 276–279. [5.9.1]

Srivastava, R. C., and Singh, J. (1975). On some characterizations of the binomial and Poisson distributions based on a damage model, *Statistical Distributions in*

Scientific Work, **3**: *Characterizations and Applications*, G. P. Patil, S. Kotz, and J. K. Ord (editors), 271–277. Dordrecht: Reidel. [4.8]

Srivastava, R. C., and Srivastava, A. B. L. (1970). On a characterization of the Poisson distribution, *Journal of Applied Probability*, **7**, 497–501. [4.8]

Sródka, T. (1963). On approximation of hypergeometric distribution, *Zeszyty Naukowe Politechniki Łódzkiej*, **10**(53), 5–17. [6.5]

Stadlober, E. (1991). Binomial random variate generation: a method based on ratio of uniforms, *The Frontiers of Statistical Computation, Simulation, & Modeling*, P. R. Nelson, E. J. Dudewicz, A. Öztürk, and E. C. van der Meulen (editors), 93–112. Columbus, Ohio: American Sciences Press. [1C.3]

Staff, P. J. (1964). The displaced Poisson distribution, *Australian Journal of Statistics*, **6**, 12–20. [4.12.4]

Staff, P. J. (1967). The displaced Poisson distribution-region 3, *Journal of the American Statistical Association*, **62**, 643–654. [4.12.4]

Stam, A. (1973). Regular variation of the tail of a subordinated probability distribution, *Advances in Applied Probability*, **5**, 308–327. [8.3.2]

Stancu, D. D. (1968). On the moments of negative order of the positive Bernoulli and Poisson variables, *Studia Universitatis Babes-Bolyai, Series Math-Phys*, **13**, 27–31. [3.3]

Steck, G. P. (1973). Bounding Poisson approximations to the binomial, *Communications in Statistics*, **2**, 189–203 . [3.6.1]

Steck, G. P., and Zimmer, W. J. (1968). The relationship between Neyman and Bayes confidence intervals for the hypergeometric parameter. *Technometrics*, **10**, 199–203. [6.7.1, 6.9.2]

Stein, C. (1986). *Approximate Computation of Expectations*. Hayward, CA: Institute of Mathematical Statistics. [4.8]

Stein, G., Zucchini, W., and Juritz, J. M. (1987). Parameter estimation for the Sichel distribution and its multivariate extension, *Journal of the American Statistical Association*, **82**, 938–944. [11.15]

Stephan, F. F. (1945). The expected value and variance of the reciprocal and other negative powers of a positive Bernoullian variate, *Annals of Mathematical Statistics*, **16**, 50–61. [3.11]

Stephens, M. A. (1986). Tests based on EDF statistics, *Goodness-of-Fit Techniques*, R. B. D'Agostino and M. A. Stephens (editors), 97–193. New York: Dekker. [4.7.4]

Steutel, F. W. (1970). Preservation of infinite divisibility under mixing, *Mathematical Centre Tract*, **33**. Amsterdam. [7.4, 9.3]

Steutel, F. W. (1983). Infinite divisibility, *Encyclopedia of Statistical Sciences*, **4**, S. Kotz, N. L. Johnson and C. B. Read (editors), 114–116. New York: Wiley. [9.3]

Steutel, F. W. (1988). Note on discrete α–unimodality, *Statistica Neerlandica*, **42**, 137–140. [1B.5]

Steutel, F. W. (1990). The set of geometrically infinitely divisible distributions, *Technische Universiteit Eindhoven Memorandum COSOR*, **90–42**. [9.3]

Steutel, F. W., and Thiemann, J. G. F. (1989a). On the independence of integer and fractional parts, *Statistica Neerlandica*, **43**, 53–59. [5.2]

Steutel, F. W., and Thiemann, J. G. F. (1989b). The gamma process and the Poisson distribution, *Colloquia Mathematica Societatis János Bolyai* **57**. *Limit Theorems in Probability and Statistics*, 477–490. Pécs, Hungary. [4.4]

Stevens, W. L. (1937). Significance of grouping, *Annals of Eugenics, London*, **8**, 57–60. [10.1, 10.4.1].

Stevens, W. L. (1939). Distribution of groups in a sequence of alternatives, *Annals of Eugenics, London*, **9**, 10–17. [10.1, 10.3, 10.5.1]

Stevens, W. L. (1950). Fiducial limits for the parameter of a discontinuous distribution, *Biometrika*, **37**, 117–129. [3.8.3]

Stevens, W. L. (1951). Mean and variance of an entry in a contingency table, *Biometrika*, **38**, 468–470. [3.4]

Stevens, W. L. (1957). Shorter intervals for the parameter of the binomial and Poisson distributions, *Biometrika*, **44**, 436–440. [4.7.3]

Steyn, H. S. (1956). On the univariable series $F(t) = F(a; b_1, b_2, \ldots, b_k; t, t^2, \ldots, t^k)$ and its applications in probability theory, *Proceedings Koninklijke Nederlandse Akademie van Wetenschappen, Series A*, **59**, 190–197. [11.18]

Steyn, H. S., Jr. (1980). A class of multiparameter power series distributions, *South African Statistical Journal*, **14**, 1–15. [2.2.2, 11.17]

Steyn, H. S., Jr. (1984). Two parameter power series distributions with applications, *South African Statistical Journal*, **18**, 29–44. [2.2.2, 11.17]

Stigler, S. M. (1986). *The History of Statistics: The Measurement of Uncertainty before 1900*, Cambridge, MA: Harvard University Press. [3.2]

Stoyan, D., Kendall, W. S., and Mecke J. (1987). *Stochastic Geometry and Its Applications*, Chichester: Wiley. [4.9]

Strackee, J., and van der Gon, J. J. D. (1962). The frequency distribution of the difference between two Poisson variates, *Statistica Neerlandica*, **16**, 17–23. [4.12.3]

Stuart, A. (1963). Standard errors for percentages, *Applied Statistics*, **12**, 87–101. [3.7]

Stuart, A., and Ord, J. K. (1987). *Kendall's Advanced Theory of Statistics* (Fifth edition), **1**, London: Griffin. [1A.1, 1A.7, 1A.11, 1B.5, 1B.6, 1B.13, 1B.15, 3.12.2, 4.4, 7.2, 9.1]

Student (1907). On the error of counting with a haemocytometer, *Biometrika*, **5**, 351–360. [4.2, 5.3, 5.12.5]

Student (1919). An example of deviations from Poisson's law in practice, *Biometrika*, **12**, 211–213. [3.8.2]

Subrahmaniam, K. (1966). On a general class of contagious distributions: the Pascal-Poisson distribution, *Trabajos de Estadística*, **17**, 109–127. [5.12.6, 9.9, 9.12]

Subrahmaniam, K. (1978). The Pascal-Poisson distribution revisited: estimation and efficiency, *Communications in Statistics-Theory and Methods*, **A7**, 673–683. [5.12.6, 9.9, 9.12]

Sundt, B., and Jewell, W. S. (1981). Further results on recursive evaluation of compound distributions, *ASTIN Bulletin*, **18**, 27–39. [2.3.1]

Svensson, A. (1969). Some remarks on the Moran-Chatterji theorem, *University of Stockholm Department of Mathematical Statistics*. [3.9, 4.8]

Syski, R. (1988). Further comments on the solution of the M/M/1 queue, *Advances in Applied Probability*, **20**, 693. [11.13]

Szegö, G. (1939, 1959, 1967). *Orthogonal Polynomials*, Providence, RI: American Mathematical Society (Three editions). [1A.11]

Taguti, G. (1952). Tables of 5% and 1% points for the Polya-Eggenberger distribution function, *Reports of Statistical Application Research, JUSE*, **2**, [5.7]

Taillie, C., and Patil, G. P. (1986). The Fibonacci distribution revisited, *Communications in Statistics-Theory and Methods*, **15**, 951–959. [10.6.1]

Takács, L. (1955). Investigation of waiting time problems by reduction to Markov processes, *Acta Mathematica Academiae Scientiarum Hungaricae*, **6**, 101–128. [11.10]

Takács, L. (1962). A generalization of the ballot problem and its applications in the theory of queues, *Journal of the American Statistical Association*, **57**, 327–337. [2.5.2]

Takács, L (1967). On the method of inclusion and exclusion, *Journal of the American Statistical Association*, **62**, 102–113. [10.2]

Takács, L. (1980). The problem of coincidences, *Archive for the History of Exact Sciences*, **21**, 229–244. [10.3]

Tallis, G. M. (1969). The identifiability of mixtures of distributions, *Journal of Applied Probability*, **6**, 389–398. [8.3.1]

Tallis, G. M. (1983). Goodness of fit, *Encyclopedia of Statistical Sciences*, **3**, S. Kotz, N. L. Johnson and C. B. Read (editors), 451–460. New York: Wiley. [4.7.4]

Tallis, G. M., and Chesson, P. (1982). Identifiability of mixtures, *Journal of the Australian Mathematical Society*, **A 32**, 339–348. [8.3.1]

Talwalker, S. (1970). A characterization of the double Poisson distribution, *Sankhyā, Series A*, **32**, 265–270. [4.8]

Talwalker, S. (1975). Models in medicine and toxicology, *Statistical Distributions in Scientific Work*, **2**: *Model Building and Model Selection*, G. P. Patil, S. Kotz, and J. K. Ord (editors), 263–274. Dordrecht: Reidel. [4.8]

Talwalker, S. (1980). A note on the generalized Rao-Rubin condition and characterization of certain discrete distributions, *Journal of Applied Probability*, **17**, 563–569. [4.8]

Talwalker, S. (1986). Functional equations in characterizations of discrete distributions by Rao-Rubin condition and its variants, *Communications in Statistics-Theory and Methods*, **15**, 961–979. [9.2]

Tanner, J. C. (1953). A problem of interference between two queues, *Biometrika*, **40**, 58–69. [9.11, 11.13]

Tate, R. F., and Goen, R. L. (1958). Minimum variance unbiased estimation for the truncated Poisson distribution, *Annals of Mathematical Statistics*, **29**, 755–765. [2.2.1, 4.10.1, 4.12.3]

Taylor, H. M., and Karlin, S. (1984). *An Introduction to Stochastic Modeling*, Orlando, FL: Academic Press. [4.9, 5.2, 5.10]

Taylor, L. R. (1961). Aggregation, variance and the mean, *Nature*, **189**, 732–735. [5.8.4]

Teich, M. C., and McGill, W. J. (1976). Neural counting and photon counting in the presence of dead time, *Physical Review Letters*, **36**, 754–758. [5.12.5]

Teicher, H. (1955). An inequality on Poisson probabilities, *Annals of Mathematical Statistics*, **26**, 147–149. [4.4, 4.5]

Teicher, H. (1960). On the mixture of distributions, *Annals of Mathematical Statistics*, **31**, 55–73. [8.2.1, 8.2.3, 8.3.2]

Teicher, H. (1961). Identifiability of mixtures, *Annals of Mathematical Statistics*, **32**, 244–248. [8.2.1, 8.2.4, 8.3.1]

Teicher, H. (1963). Identifiability of finite mixtures, *Annals of Mathematical Statistics*, **34**, 1265–1269. [8.2.1, 8.2.4]

Thiele, T. N. (1889). *Theory of Observations*. Reprinted in English in *Annals of Mathematical Statistics*, **2**, 165–308 (1931). [4.2]

Thomas, D. G., and Gart, J. J. (1971). Small sample performance of some estimators of the truncated binomial distribution, *Journal of the American Statistical Association*, **66**, 169–177. [3.11]

Thomas, H. A. (1948). Frequency of minor floods, *Journal of the Boston Society of Civil Engineering*, **35**, 425. (Graduate School of Engineering Publication No. 466, Harvard University, Cambridge, Massachusetts.) [6.2.2]

Thomas, M. (1949). A generalization of Poisson's binomial limit for use in ecology, *Biometrika*, **36**, 18–25. [9.10]

Thomas, M. A., and Taub, A. E. (1975). Binomial trials with variable probabilities, *Naval Surface Weapons Center, Dahlgren, Virginia, Technical Note*, DK-25/75. [3.12.2]

Thomas, M. A., and Taub, A. E. (1982). Calculating binomial probabilities when the trial probabilities are unequal, *Journal of Statistical Computation and Simulation*, **14**, 125–131. [3.12.2]

Thompson, H. R. (1954). A note on contagious distributions, *Biometrika*, **41**, 268–271. [5.3, 9.3]

Thyrion, P. (1960). Note sur les distributions "par grappes", *Association Royale des Actuaires Belges Bulletin*, **60**, 49–66. [4.11, 5.3, 9.3]

Tiago de Oliviera, J. (1965). Some elementary tests for mixtures of discrete distributions, *Classical and Contagious Discrete Distributions*, G. P. Patil (editor), 379–384. Calcutta: Statistical Publishing Society; Oxford: Pergamon Press. [8.2.3]

Tiku, M. L. (1964). A note on the negative moments of a truncated Poisson variate, *Journal of the American Statistical Association*, **59**, 1220–1224. [4.10.1]

Tiku, M. L. (1989). Modified maximum likelihood estimation, *Encyclopedia of Statistical Sciences, Supplement*, S. Kotz, N. L. Johnson and C. B. Read (editors), 98–103. New York: Wiley. [1B.15]

Tippett, L. H. C. (1932). A modified method of counting particles, *Proceedings of the Royal Society of London, Series A*, **137**, 434–446. [4.10.2]

Titterington, D. M. (1990). Some recent research in the analysis of mixture distributions, *Statistics*, **21**, 619–641. [8.2.1]

Titterington, D. M., Smith, A. F. M., and Makov, U. E. (1985). *Statistical Analysis of Finite Mixture Distributions*, Chichester: Wiley. [8.2.1, 8.2.3]

Todhunter, I. (1865). *A History of the Mathematical Theory of Probability*, Cambridge: Macmillan (Reprinted 1965, New York: Chelsea). [5.3, 6.2.1, 6.2.2, 10.6.1]

Tripathi, R. C., Gupta, P. L., and Gupta, R. C. (1986). Incomplete moments of modified power series distributions with applications, *Communications in Statistics-Theory and Methods*, **15**, 999–1015. [2.2.2, 7.11]

Tripathi, R. C., and Gupta, R. C. (1985). A generalization of the log-series distribution, *Communications in Statistics-Theory and Methods*, **14**, 1779–1799. [7.11]

Tripathi, R. C., and Gupta, R. C. (1988). Another generalization of the logarithmic series and the geometric distributions, *Communications in Statistics-Theory and Methods*, **17**, 1541–1547. [7.11]

Tripathi, R. C., and Gurland, J. (1977). A general family of discrete distributions with hypergeometric probabilities, *Journal of the Royal Statistical Society, Series B*, **39**, 349–356. [2.3.1, 3.4, 4.12.4]

Tripathi, R. C., and Gurland, J. (1979). Some aspects of the Kemp families of distributions, *Communications in Statistics-Theory and Methods*, **8**, 855–869. [2.3.1, 2.4.1, 2.4.2, 3.4, 4.12.4, 5.8.1]

Tripathi, R. C., Gurland, J., and Bhalerao, N. R. (1986). A unified approach to estimating parameters in some generalized Poisson distributions, *Communications in Statistics-Theory and Methods*, **15**, 1017–1034. [9.7, 9.9]

Tsui, K.–W., and Press, S. J. (1982). Simultaneous estimation of several Poisson parameters under K–normalized squared error loss. *University of California, Riverside, Technical Report* No. 38. [4.7.2]

Tukey, J. W. (1949). Moments of random group size distributions, *Annals of Mathematical Statistics*, **20**, 523–539. [10.4.1]

Tukey, J. W. (1977). *Exploratory Data Analysis*, Reading, MA: Addison-Wesley. [1B.10]

Tweedie, M. C. K. (1947). Functions of a statistical variate with given means, with special reference to Laplacian distributions, *Proceedings of the Cambridge Philosophical Society*, **43**, 41–49. [2.2.1]

Tweedie, M. C. K. (1965). Further results concerning expectation-inversion technique, *Classical and Contagious Discrete Distributions*, G. P. Patil (editor), 195–218. Calcutta: Statistical Publishing Society; Oxford: Pergamon Press. [2.2.1]

Tweedie, M. C. K., and Veevers, A. (1968). The inversion of cumulant operators for power series distributions, and the approximate stabilization of variance, *Journal of the American Statistical Association*, **63**, 321–328. [2.2.1]

Uhlmann, W. (1966). Vergleich der Hypergeometrischen mit der Binomial-Verteilung, *Metrika*, **10**, 145–158. [3.4, 6.5]

Umbach, D. (1981). On inference for a mixture of a Poisson and a degenerate distribution, *Communications in Statistics-Theory and Methods*, **A10**, 299–306. [8.2.2]

Uppuluri, V. R. R., and Blot, W. J. (1970). A probability distribution arising in a riff-shuffle, *Random Counts in Scientific Work*, 1: *Random Counts in Models and Structures*, G. P. Patil (editor), 23–46. University Park: Pennsylvania State University Press. [5.12.7]

Uppuluri, V. R. R., Feder, P. I., and Shenton, L. R. (1967). Random difference equations occuring in one-compartment models, *Mathematical Biosciences*, **1**, 143–171. [5.9.1]

Uppuluri, V. R. R., and Patil, G. P. (1983). Waiting times and generalized Fibonacci sequences, *Fibonacci Quarterly*, **21**, 342–349. [10.6.1]

Uspensky, J. V. (1937). On Ch. Jordan's series for probability, *Annals of Mathematics*, **32**, 306–312. [3.6.2]

Vaart, H. R. van der (1972). A note on a functional equation for the generating function of a Poisson distribution, *Sankhyā, Series A*, **34**, 191–193. [4.8]

Varian, H. R. (1975). A Bayesian approach to real estate assessment, *Studies in Bayesian Econometrics and Statistics in Honor of Leonard J. Savage*, S. E. Fienberg and A. Zellner (editors), 195–208. Amsterdam: North-Holland. [4.7.2]

Vernon, P. E. (1936). The matching method applied to investigations of personality, *Psychological Bulletin*, **33**, 149–177. [10.1, 10.3]

Verrall, R. J. (1989). The individual risk model: a compound distribution, *Journal of the Institute of Actuaries*, **116**, 101–107. [9.12]

Vervaat, W. (1969). Upper bounds for the distance in total variation between the binomial and negative binomial and the Poisson distributions, *Statistica Neerlandica*, **23**, 79–86. [3.6.1]

Vit, I. (1974). Testing for homogeneity: the geometric distribution, *Biometrika*, **61**, 565–568. [5.2]

Vogler, L. E. (1964). Percentage points of the beta distribution, *National Bureau of Standards Technical Note*, 215: Washington, DC. [1A.5]

Volodin, I. N. (1965). On distinguishing between Poisson and Pólya distributions on the basis of a large number of samples, *Theory of Probability and Its Applications*, **10**, 335–338. [4.8]

Wadley, F. M. (1950). Notes on the form of distribution of insect and plant populations, *Annals of the Entomological Society of America*, **43**, 581–586. [9.6.5]

Waerden B. L. van der (1960). Sampling inspection as a minimum loss problem, *Annals of Mathematical Statistics*, **31**, 369–384. [4.9]

Wald, A., and Wolfowitz, J. (1940). On a test whether two samples are from the same population, *Annals of Mathematical Statistics*, **11**, 147–162. [10.5.1]

Walker, A. J. (1974). New fast method for generating discrete random numbers with arbitrary frequency distributions, *Electronics Letters*, **10**, 127–128. [1C.2]

Walker, A. J. (1977). An efficient method for generating discrete random variables, *ACM Transactions on Mathematical Software*, **3**, 253–256. [1C.2, 1C.6]

Wallenius, K. T. (1963). *Biased Sampling: The Noncentral Hypergeometric Probability Distribution*, Ph.D thesis, Stanford, CA: Stanford University. [6.12]

Wallis, W. A., and Moore, G. H. (1941). A significance test for time series, *National Bureau of Economic Research, New York, Technical Paper*, **1** [10.5.2]

Walsh, J. E. (1955). The Poisson distribution as a limit for dependent binomial events with unequal probabilities, *Operations Research*, **3**, 198–209. [4.9]

Walter, G. G., and Hamedani, G. G. (1987). Empirical Bayes estimation of binomial probability, *Communications in Statistics-Theory and Methods*, **16**, 559–577. [3.8.2]

Walther, A. (1926). Anschauliches zür Riemannschen Zetafunktion, *Acta Mathematica*, **48**, 393–400. [11.20]

Wang, P. C. C. (1970). A characterization of the Poisson distribution based on random splitting and random expanding, *Stanford University Department of Statistics Technical Report*, **1160**, [4.8]

Wang, Y. H. (1972). On characterization of certain probability distributions, *Proceedings of the Cambridge Philosophical Society*, **71**, 347–352. [4.8]

Wani, J. K. (1967). Moment relations for some discrete distributions, *Skandinavisk Aktuarietidskrift*, **50**, 50–55. [7.8]

Wani, J. K. (1978). Measuring diversity in biological populations with logarithmic abundance distributions, *Canadian Journal of Statistics*, **6**, 219–228. [7.1]

Wani, J. K., and Lo, H.–P. (1975a). Large sample interval estimation for the logarithmic series distribution, *Canadian Journal of Statistics*, **3**, 277–284. [7.7.2]

Wani, J. K., and Lo, H.–P. (1975b). Clopper-Pearson system of confidence intervals for the logarithmic series distribution, *Biometrics*, **31**, 771–775. [7.7.2]

Wani, J. K., and Lo, H.–P. (1977). Comparing confidence intervals for the logarithmic series distribution, *Canadian Journal of Statistics*, **5**, 153–158. [7.7.2]

Warde, W. D., and Katti, S. K. (1971). Infinite divisibility of discrete distributions II, *Annals of Mathematical Statistics*, **42**, 1088–1090. [9.3]

Watanabe, H. (1956). On the Poisson distribution, *Journal of the Mathematical Society of Japan*, **8**, 127–134. [4.2]

Watson, H. W., and Galton, F. (1874). On the probability of extinction of families, *Journal of the Anthropological Institute*, **4**, 138–144. [9.1]

Wei, L. J. (1979). The generalized Polya's urn design for sequential medical trials, *Annals of Statistics*, **5**, 291–296. [11.19]

Weiss, G. H. (1965). A model for the spread of epidemics by carriers, *Biometrics*, **21**, 481–490. [10.4.1]

Weiss, L. (1976). Two sample tests and tests of fit, *Communications in Statistics-Theory and Methods*, **5**, 1275–1285. [10.5.1]

Weiss, L. (1983). Generalized maximum likelihood estimation, *Encyclopedia of Statistical Sciences*, **3**, S. Kotz, N. L. Johnson and C. B. Read (editors), 348–352. New York: Wiley. [1B.15]

Weiss, L. (1988). Runs, *Encyclopedia of Statistical Sciences*, **8**, S. Kotz, N. L. Johnson and C. B. Read (editors), 222–226. New York: Wiley. [10.1, 10.5.1]

Wesolowski, J. (1988). A remark on a characterization of the Poisson process, *Demonstratio Mathematica*, **21**, 555–557. [4.8]

Wesolowski, J. (1989). Characterizations of some processes by properties of conditional moments, *Demonstrato Mathematica*, **22**, 537–556. [5.9.1]

Wetherill, G. B., and Köllerström, J. (1979). Sampling inspection simplified, *Journal of the Royal Statistical Society, Series A*, **142**, 1–32 (correction, 404). [3.6.1]

Whittaker, L. (1914). On the Poisson law of small numbers, *Biometrika*, **10**, 36–71. [4.2, 5.3]

Whitworth, W. A. (1878). *Choice and Chance*, Cambridge: Deighton Bell. [10.2]

Whitworth, W. A. (1948). *Choice and Chance: with one thousand exercises* (Reprint of the Fifth edition, 1901), New York: Hafner. [10.2]

Widdra, W. (1972). Eine Verallgemeinerung des "Gestzes seltener Ereignisse", *Metrika*, **19**, 68–71. [4.2]

Wilks, S. S. (1962). *Mathematical Statistics*, New York: Wiley. [11.19]

Williams C. B. (1944). The number of publications written by biologists, *Annals of Eugenics*, **12**, 143–146. [7.9]

Williams, C. B. (1947). The logarithmic series and its application to biological problems, *Journal of Ecology*, **34**, 253–272. [7.2, 7.10]

Williams, C. B. (1964). *Patterns in the Balance of Nature*, London: Academic Press. [7.1, 7.2, 7.4, 7.9]

Williams, D. A. (1975). The analysis of binary responses from toxicological experiments involving reproduction and teratogenicity, *Biometrics*, **31**, 949–952. [6.7.2, 6.9.2]

Williamson, E., and Bretherton, M. H. (1963). *Tables of the Negative Binomial Distribution*, New York: Wiley. [5.7]

Williamson, E., and Bretherton, M. H. (1964). Tables of the logarithmic series distribution, *Annals of Mathematical Statistics*, **35**, 284–297. [7.6, 7.7.2, 7.9]

Willmot, G. E. (1986). Mixed compound Poisson distributions, *ASTIN Bulletin*, **16**, S59–S79. [8.3.2, 11.15]

Willmot, G. E. (1987a). On the probabilities of the log-zero-Poisson distribution, *Canadian Journal of Statistics*, **15**, 293–297. [8.2.2]

Willmot, G. E. (1987b). The Poisson-inverse Gaussian distribution as an alternative to the negative binomial. *Scandinavian Actuarial Journal*, 113–127. [11.15]

Willmot, G. E. (1988a). Sundt and Jewell's family of discrete distributions, *ASTIN Bulletin*, **18**, 17–29. [2.3.1, 5.12.2]

Willmot, G. E. (1988b). Parameter orthogonality for a family of discrete distributions, *Journal of the American Statistical Association*, **83**, 517–521. [1B.15]

Willmot, G. E. (1989). Limiting tail behaviour of some discrete compound distributions, *Insurance: Mathematics and Economics*, **8**, 175–185. [5.12.5, 8.3.2]

Willmot, G. E. (1990). Asymptotic tail behaviour of Poisson mixtures with applications, *Advances in Applied Probability*, **22**, 147–159. [8.3.2]

Willmot, G. E., and Sundt, B. (1989). On evaluation of the Delaporte distribution and related distributions, *Scandinavian Actuarial Journal*, 101–113. [5.12.2]

Willson, L. J., Folks, J. L., and Young, J. H. (1984). Multistage estimation compared with fixed-sample-size estimation of the negative binomial parameter k, *Biometrics*, **40**, 109–117. [5.8.3]

Wilson, L. T., and Room, P. M. (1983). Clumping patterns of fruit and arthropods in cotton, with implications for binomial sampling, *Environmental Entomology*, **12**, 50–54. [5.10]

Wise, M. E. (1946). The use of the negative binomial distribution in an industrial sampling problem, *Journal of the Royal Statistical Society, Series B*, **8**, 202–211. [5.8.3]

Wise, M. E. (1954). A quickly convergent expansion for cumulative hypergeometric probabilities, direct and inverse, *Biometrika*, **41**, 317–329. [6.5]

Wishart, J., and Hirschfeld, H. O. (1936). A theorem concerning the distribution of joins between line segments, *Journal of the London Mathematical Society*, **11**, 227–235. [10.5.1]

Witherby, H. F., Jourdain, F. C. R., Tichihurst, N. F., and Tucker, B. W. (1941). *The Handbook of British Birds*, **5**. London: H. T., and G. Witherby. [7.9]

Wittes, J. T. (1972). On the bias and estimated variance of Chapman's two-sample capture-recapture population estimate, *Biometrics*, **28**, 592–597. [6.7.1, 6.9.1]

Wood, C. L., and Altavela, M. M. (1978). Large sample results for Kolmogorov-Smirnov statistics for discrete distributions, *Biometrika*, **65**, 235–239. [4.7.4]

Woodbury, M. A. (1949). On a probability distribution, *Annals of Mathematical Statistics*, **20**, 311–313. [11.19]

Worsley, K. J. (1982). An improved Bonferroni inequality and applications, *Biometrika*, **69**, 297–302. [10.2]

Wyshak, G. (1974). Algorithm AS68: A program for estimating the parameters of the truncated negative binomial distribution, *Applied Statistics*, **23**, 87–91. [5.11]

Xekalaki, E. (1981). Chance mechanisms for the generalized Waring distribution and related characterizations, *Statistical Distributions in Scientific Work*, **4**: *Models, Structures, and Characterizations*, C. Taillie, G. P. Patil, and B. A. Baldessari (editors), 157–172. Dordrecht: Reidel. [6.2.3, 6.8]

Xekalaki, E. (1983a). Infinite divisibility, completeness and regression properties of the generalized Waring distribution, *Annals of the Institute of Statistical Mathematics, Tokyo*, **35**, 279–289. [6.2.3]

Xekalaki, E. (1983b). The univariate generalized Waring distribution in relation to accident theory: proneness, spells or contagion? *Biometrics*, **39**, 887–895 [6.2.3, 6.9.3]

Xekalaki, E. (1983c). A property of the Yule distribution and its applications, *Communications in Statistics-Theory and Methods*, **12**, 1181–1189. [6.2.3, 6.10.3]

Xekalaki, E. (1983d). Hazard functions and life distributions in discrete time, *Communications in Statistics-Theory and Methods*, **12**, 2503–2509. [6.2.3, 6.10.4]

Xekalaki, E. (1984). Linear regression and the Yule distribution, *Journal of Econometrics*, **24**, 397–403. [6.10.3]

Xekalaki, E. (1985). Some identifiability problems involving generalized Waring distributions, *Publicationes Mathematicae*, **32**, 75–84. [6.2.3]

Xekalaki, E., and Panaretos, J. (1988). On the association of the Pareto and the Yule distribution, *Teoriya Veroyatnostei i ee Primeneniya*, **33**, 206–210. [6.10.3]

Xekalaki, E., and Panaretos, J. (1989). On some distributions arising in inverse cluster sampling, *Communications in Statistics- Theory and Methods*, **18**, 355–366. [10.6.3]

Xekalaki, E., Panaretos, J., and Philippou, A. N. (1987). On some mixtures of distributions of order k, *Fibonacci Quarterly*, **25**, 151–160. [6.12, 10.6.3]

Yakowicz, S. J., and Spragins, J. D. (1968). On the identifiability of finite mixtures, *Annals of Mathematical Statistics*, **39**, 209–214. [8.2.1]

Yanagimoto, T. (1988). The conditional MLE in the two-parameter geometric distribution and its competitors, *Communications in Statistics-Theory and Methods*, **17**, 2779–2787. [5.2]

Yanagimoto, T. (1989). The inverse binomial distribution as a statistical model, *Communications in Statistics-Theory and Methods*, **18**, 3625–3633. [11.10]

Yang, M.–C. (1990). Compromise between generalized Bayes and Bayes estimators of Poisson means under entropy loss, *Communications in Statistics-Theory and Methods*, **19**, 935–951. [4.7.2]

Yates, F. (1934). Contingency tables involving small numbers and the χ^2 test, *Supplement to the Journal of the Royal Statistical Society*, **1**, 217–235. [6.9.1]

Yip, P. (1988). Inference about the mean of a Poisson distribution in the presence of a nuisance parameter, *Australian Journal of Statistics*, **30**, 299–306. [4.7.2]

Yoneda, K. (1962). Estimations in some modified Poisson distributions, *Yokohama Mathematical Journal*, **10**, 73–96. [8.2.2]

Young, D. H. (1970). The order statistics of the negative binomial distribution, *Biometrika*, **57**, 180–181. [5.5]

Young, T. (1819). Remarks on the probabilities of error in physical observations, and on the density of the earth, considered, especially with regard to the reduction of experiments on the pendulum. In a letter to Capt HENRY KATER, F.R.S., *By* Thomas Young, M.D., *For. Sec.* R. S., *Philosophical Transactions of the Royal Society of London*, 70–95. [10.3]

Yule, G. U. (1925). A mathematical theory of evolution based on the conclusions of Dr. J. C. Willis, F.R.S., *Philosophical Transactions of the Royal Society of London, Series B*, **213**, 21–87. [5.3, 6.10.3]

Zabell, S. (1983). Lexis, Wilhelm, *Encyclopedia of Statistical Sciences*, **4**, S. Kotz, N. L. Johnson and C. B. Read (editors), 624–625. New York: Wiley. [3.12.2]

Zacks, S., and Goldfarb, D. (1966). Survival probabilities in crossing a field containing absorption points, *Naval Research Logistics Quarterly*, **13**, 35–48. [11.1]

Zellner, A. (1986). Bayesian estimation and prediction using asymmetric loss functions, *Journal of the American Statistical Association*, **81**, 446–451. [4.7.2]

Zipf, G. K. (1949). *Human Behaviour and the Principle of Least Effort*, Cambridge, MA: Addison-Wesley. [11.20]

Abbreviations

cdf	cumulative distribution function
cf	characteristic function
cgf	cumulant generating function
cmgf	central (corrected) moment generating function
DF	distribution function
DHR	decreasing hazard rate
fcgf	factorial cumulant generating function
fmgf	factorial moment generating function
FSD	factorial series distribution
GHFD	generalized hypergeometric factorial distribution
GHPD	generalized hypergeometric probability distribution
GHRD	generalized hypergeometric recast distribution
GPSD	generalized power series distribution
HF	hypergeometric factorial
HP	hypergeometric probability
IHR	increasing hazard rate
ML	maximum likelihood
MLE	maximum likelihood estimator
MLP	maximum likelihood program
MM	method of moments
MPSD	modified power series distribution
MVUE	minimum variance unbiased estimator
NBU	new better than used
pdf	probability density function
pgf	probability generating function
pmf	probability mass function
$\Pr(E)$	probability of event E
PSD	power series distribution
rv	random variable
STER	Sums of Truncated forms of the Expected value of the Reciprocal
umgf	uncorrected moment generating function

Index

551

*Now available in a lower priced paperback edition in the Wiley Classics Library.